NUMERICAL METHODS FOR THE
EULER EQUATIONS OF FLUID DYNAMICS

Proceedings of the INRIA Workshop on
Numerical Methods for the
Euler Equations of Fluid Dynamics
Rocquencourt, France
December 7–9, 1983

Numerical Methods for the Euler Equations of Fluid Dynamics

edited by

F. Angrand, A. Dervieux, J. A. Desideri and R. Glowinski
INRIA, Centres de Rocquencourt et Sophia Antipolis

Philadelphia/1985

Library of Congress Catalog Card Number: 85-62674
ISBN: 0-89871-200-9

To the memory of G. Vijayasundaram

CONTENTS

CONTENTS

PREFACE

Solving the *Euler equations* of fluid dynamics appears to be one of the most challenging problems from both the mathematical and the numerical points of view. Despite these difficulties, many scientists have been focussing their interest on these equations, in recent years. One of the main motivations for this growing interest in the Euler equations originates in aerodynamics. For years aeronautical engineers have been using linear and nonlinear potential equations in order to simulate flows around aircraft (or at least parts of aircraft); these potential methods have now reached a high degree of sophistication, particularly for transonic flows, subsonic at infinity, around civil aircraft. Simulations for supersonic military aircraft (fighters or bombers) are more demanding, particularly for flows at high incidence and/or flows in air intakes; indeed, the potential hypothesis is no longer true for such flows which may contain strong shocks and/or vortices. This fact implies that more realistic mathematical models like the Euler or Navier–Stokes equations have to be used for computations.

Concentrating on the Euler equations and their numerical solution, the editors and some of the authors in this volume agreed that the time had come to organize a meeting in which the leading experts of the western world could discuss and compare their various approaches and methods of solution; a realization of such meeting was the workshop organized in 1983 by the *Institut National de Recherches en Informatique et Automatique* (INRIA), at Rocquencourt, France, from December 7 to December 9. About 30 participants attended this workshop, which consisted of some 20 one-hour lectures and many discussions. We strongly believed that publishing in book form the texts associated with these lectures would be useful for a large class of scientists, such as mathematicians, engineers, etc. . . ., and we were consequently delighted that SIAM agreed to publish these proceedings.

For simplicity we have divided this volume into five parts, clearly with many overlaps between them. Let us comment briefly on the contents of the book.

In Part I, we have collected the papers concerned with the fundamental aspects and principles of the numerical methods for solving the Euler equations for compressible fluids; modern versions of the Godunov scheme are discussed here together with the analysis of various upwind methods.

Part II is more concerned with special methods for the fast calculation of the steady-state solutions of the Euler equations; such methods are of fundamental importance in aeronautical engineering. Part III contains two papers on finite element methods for conservation laws and the compressible Euler equations. Part IV is concerned with incompressible flows and subdomain decompositions. Finally, Part V contains a paper devoted to the simulation of the Kelvin–Helmholtz instability with applications in astrophysics.

We hope that this book will be an helpful tool in the field of computational fluid dynamics. We would like to thank all the people who have been involved in the project (Workshop + Proceedings): most particularly the INRIA Directors and Administration, SIAM, and of course the speakers (even the very few of them who did not send their written contributions).

F. Angrand, A. Dervieux,
J.A. Desideri, R. Glowinski.

LIST OF AUTHORS

MATANIA BEN-ARTZI, Department of Mathematics, Technion-Israel Institute of Technology, Haifa 32000, Israel

YANN BRENIER, INRIA, Centre de Roquencourt, 78153 Le Chesnay Cedex, France

LAURENT CAMBIER, Office National d'Etudes et de Recherches Aérospatiales, 92322 Chatillon Cedex, France

V. DARU, Office National d'Etudes et de Recherches Aérospatiales, 92322 Chatillon Cedex, France

A. DERVIEUX, INRIA, Centre de Sophia Antipolis, 06560 Valbonne, France

RONALD J. DIPERNA, Department of Mathematics, Duke University, Durham, North Carolina 27706

LARS-ERIK ERIKSSON, FFA, The Aeronautical Research Institute of Sweden, S-161 11 Bromma, Sweden

JOSEPH FALCOVITZ, Computation Division, Rafael Ballistic Center, Haifa 31021, Israel

ERWAN LE GRUYER, Laboratoire d'Analyse Numérique, INSA, 35043 Rennes Cedex, France

AMI HARTEN, School of Mathematical Sciences, Tel-Aviv University, Tel Aviv, Israel, and New York University, New York, New York 10012

THOMAS J. R. HUGHES, Division of Applied Mechanics, Stanford University, Stanford, California 94305

ANTONY JAMESON, Department of Mechanical and Aerospace Engineering, Princeton University, Princeton, New Jersey 08544

BRAM VAN LEER, University of Technology, Delft, The Netherlands

A. LERAT, ENSAM, 75013 Paris Cedex, France

ALAIN YVES LE ROUX, UER Mathématique et Informatique, Université de Bordeaux, 33405 Talence Cedex, France

RANDALL J. LEVEQUE, Mathematics Department, University of California, Los Angeles, California 90024

MICHEL MALLET, Division of Applied Mechanics, Stanford University, Stanford, California 94305

PHILIPPE MORICE, Office National d'Etudes et de Recherches Aérospatiales, 92322 Chatillon Cedex, France

WILLIAM A. MULDER, University Observatory, Leiden, The Netherlands

ARTHUR RIZZI, FFA, The Aeronautical Research Institute of Sweden, S-161 11 Bromma, Sweden

P. L. ROE, Royal Aircraft Establishment, Bedford, England, and College of Aeronautics, Cranfield Institute of Technology, Cranfield, England

BRUNO STOUFFLET, INRIA, Centre de Roquencourt, 78153 Le Chesnay Cedex, France

P. K. SWEBY, Institute for Computational Fluid Dynamics, Oxford University Computing Laboratory, Oxford, England

YOSHIHIRO TAKI, Meijo University, Nagoya, Japan

TAYFUN E. TEZDUYAR, Department of Mechanical Engineering, University of Houston, Houston, Texas 77004

ELI TURKEL, Institute for Computer Applications in Science and Engineering, NASA Langley Research Center, Hampton, Virginia 23665 and Tel-Aviv University, Tel Aviv, Israel

JEAN-PIERRE VEUILLOT, Office National d'Etudes et de Recherches Aérospatiales, 92322 Chatillon Cedex, France

G. VIJAYASUNDARAM, T.I.F.R. Centre, P.O. Box 1234, Bangalore 560 012, India

HENRI VIVIAND, Office National d'Etudes et de Recherches Aérospatiales, 92322 Chatillon Cedex, France

R. F. WARMING, Computational Fluid Dynamics Branch, NASA Ames Research Center, Moffett Field, California 94305

PAUL R. WOODWARD, Lawrence Livermore National Laboratory, Livermore, California 94550

H. C. YEE, Computational Fluid Dynamics Branch, NASA Ames Research Center, Moffett Field, California 94305

ROBERTO ZANUTTA, Division of Applied Mechanics, Stanford University, Stanford, California 94305

PART I: FOUNDATIONS OF NUMERICAL SCHEMES

CONVERGENCE OF NUMERICAL METHODS FOR CONSERVATION LAWS

RONALD J. DI PERNA*

Abstract. We describe some recent work dealing with the convergence of approximate solutions to hyperbolic conservation laws generated by the viscosity method and by conservative difference schemes. Convergence is established using the compensated compactness method of L. Tartar and F. Murat.

1. Introduction. We shall discuss some recent work concerning the convergence of approximate solutions to hyperbolic systems of conservation laws. The general setting is provided by a system of n conservation laws in one space dimension,

$$u_t + f(u)_x = 0, \quad -\infty < x < \infty . \tag{1.1}$$

It is assumed that the state variable u is constrained to an open region of R^n,

$$u = u(x,t) \in G \subset R^n ,$$

and that the nonlinear flux function $f: G \to R^n$ is a smooth strictly hyperbolic map, i.e. the Jacobian matrix $\nabla f = (\partial f^i/\partial u_j)$ has n real and distinct eigenvalues,

$$\lambda_1(u) < \lambda_2(u) < \dots < \lambda_n(u),$$

at each state u in G. With regard to approximation, we shall be concerned with sequences of approximate solutions generated by finite difference schemes

$$D_t u + D_x f(u) = 0, \quad u = u(x,t,\Delta x), \tag{1.2}$$

which are conservative in the sense of Lax and Wendroff [8] and by parabolic systems

$$u_t + f(u)_x = \varepsilon D(u,u_x)_x, \quad u = u(x,t,\varepsilon). \tag{1.3}$$

*Department of Mathematics, Duke University, Durham, N.C. 27706

Two classical examples of special interest arise from fluid dynamics and dynamic elasticity. The compressible Euler equations describe the conservation of mass and momentum in the form

$$\rho_t + m_x = 0$$
$$(1.4) \qquad m_t + (m^2/\rho + p)_x = 0.$$

The prototypical response of the pressure p to the density ρ is presented by an increasing convex function, for example, a polytropic gas $p = A\,\rho^\gamma$, $\gamma > 1$. Second, the standard reformulation of the quasilinear wave equation for the displacement w of an elastic body,

$$w_{tt} = \sigma(w_x)_x ,$$

produces a first order system of two conservation laws involving the velocity $u = w_t$ and the strain $v = w_x$:

$$u_t - \sigma(v)_x = 0$$
$$(1.5) \qquad v_t - u_x \quad = 0$$

Here, the stress σ typically responds to the strain v in an increasing non-convex manner and changes from convex in the expansive phase to concave in the compressive phase:

$$v\,\sigma''(v) \geq 0.$$

Several preliminary remarks are in order before discussing the convergence problems associated with (1.2) and (1.3). In the general setting of nonlinear partial differential equations, fundamental difficulties stem from the fact that nonlinear maps are not continuous in the weak topology, i.e. in the topology of convergence of local averages. This property expresses itself in a variety of ways in problems dealing with existence and approximation of solutions. With reference to existence questions, for example, one of the classical approaches is to introduce a family of approximate solutions using processes of discretizing, regularization, extremization, penalization, etc. and then try to extract a subsequence that converges in the strong topology, i.e. in the topology of pointwise convergence or in L^p_{loc}. One may regard the problem of proving convergence for the entire family as one of establishing strong compactness together with uniqueness of the limit. The central problem concerns compactness. It is well known that the classical compactness arguments require some form of uniform control on both the amplitude and the derivatives of the family in question, in order to select a subsequence that converges in the strong topology.

In numerous situations both linear and nonlinear, uniform a priori bounds on the amplitude of exact and approximate solutions are available through maximum principles and energy estimates. This turns out to be the case, for example, for the equations of compressible motion presented by systems (1.4) and (1.5). Hence it is usually possible to generate without difficulty a sequence of approximations having uniformly bounded amplitude and then extract a subsequence which converges in the weak topology. For nonlinear problems, the tools of classical functional analysis require uniform bounds on all partial derivatives in order to convert weak convergence to strong convergence and pass to the limit in a general nonlinear map. Indeed, for this very reason, the use of the weak topology has been restricted for a long time to linear problems and to midly nonlinear problems with monotonic structure. In contrast, the theory of compensated compactness introduced by Tartar and Murat [10,11,12] provides an opportunity to pass to the limit in highly nonlinear problems using just the derivative control presented by the special linear combinations (of nonlinear functions) that define the equations themselves. We shall discuss some results below concerning the convergence of approximate solutions which have been obtained with the aid of compensated compactness.

The first considerations which arise in the convergence problems associated with (1.2) and (1.3) involve stability. In the context of conservation laws we recall that L^∞ norm and the total variation norm provide a natural pair of metrics in which to investigate stability in the sense of uniform boundedness of families of exact and approximate solutions. The L^∞ norm serves as an appropriate metric to measure the solution amplitude while the total variation norm serves as an appropriate metric to measure the distributional solution gradient. The relevance of these metrics for systems of conservation laws (1) is established by the following theorem of Glimm [5] concerning the stability and convergence of the approximate solutions generated by the random choice method applied to the Cauchy problem.

Theorem 1. If the total variation of the initial data $u_0(x)$ is sufficiently small then a sequence of random choice approximations $u(x,t;\Delta x)$ converges pointwise almost everywhere in x and t to a globally defined distributional solution u and maintains a uniform bound on both the amplitude and spatial total variation:

$$|u(\cdot,t,\Delta x)|_\infty \le \text{const.}\,|u_0|_\infty$$

$$\text{TV}\ u(\cdot,t,\Delta x) \le \text{const. TV}\ u_0,$$

where the constants are independent the data u_0 and the mesh length Δx.

The proof of the theorem is based upon a general study of elementary wave interactions in the exact solution and in the random choice approximations. It remains a basic open problem in area of approximation for conservation laws to prove or disprove the corresponding estimates for the classical finite difference schemes which are conservative in the sense of Lax and Wendroff and for the standard parabolic regularizations based on diffusion and, more generally, to provide a complete analysis of discrete weak wave interactions. In the latter direction we refer the reader to [1] for a treatment of discrete interactions in conservative difference schemes and for a proof of stability and convergence for a class of hybridized methods which max the random choice algorithm with certain first order accurate conservative algorithms. In the paragraphs below we shall discuss several compactness theorems for sequences of approximate solutions generated by conservative difference schemes and parabolic regularizations. The analysis appeals to the weak topology and averaged quantities rather than the strong topology and fine scale features and uses uniform bounds only on the amplitude.

2. Finite Difference Schemes. We shall be concerned with finite difference algorithms which are conservative in the sense of Lax and Wendroff [8]. For simplicity we shall restrict our attention to one-step methods with a three-point domain of dependence:

$$(2.1) \qquad u(x,t+\Delta t)-u(x,t) + \lambda\, G\{u(x,t),u(x+\Delta x,t),\lambda\} - \lambda\, G\{u(x-\Delta x,t),u(x,t),\lambda\} = 0,$$

under the standard consistency condition that the numerical flux G reduces on the diagonal of state space to the exact flux f, independently of the ratio of mesh lengths $\lambda = \Delta t/\Delta x$:

$$G: R^n \times R^n \times R \to R \qquad G(a,a,\lambda) = f(a).$$

For simplicity in notation, we shall suppress the dependence of G on λ, if any, and write (2.1) in the standard form

$$u_{j+1}^n - u_j^n + \lambda\, G(u_j^n,u_{j+1}^n) - \lambda\, G(u_{j-1}^n,u_j^n) = 0.$$

For background we recall that the Lax-Friedrichs scheme,

$$u_j^{n+1} - \frac{1}{2}(u_{j+1}^n + u_{j-1}^n) + \frac{\lambda}{2}\{f(u_{j+1}^n) - f(u_{j-1}^n)\} = 0,$$

is based upon a numerical flux

$$G(a,b) = \frac{1}{2}\{f(a) + f(b)\} + \frac{1}{2\lambda}(a - b),$$

and Godunov's method is based upon a numerical flux equal to the value of the wave point of the solution of the Riemann problem with data (a,b):

$$G(a,b) = w(0,1;a,b),$$

where $w(x,t;a,b)$ denotes the solution of the initial value problem to (1.1) with data

$$u_0(x) = a \quad \text{if} \quad x < 0, \quad u_0(x) = b \quad \text{if} \quad x > 0.$$

Both algorithms are based upon integral (zone) averaging of the Riemann problem and may be regarded as formal approximations of the parabolic system

$$u_t + f(u)_x = \Delta t\, u_{xx} ,$$

which diffuses each of the primitive components at an equal rate. We shall not review the standard information concerning the interpretation, linear stability and trunction analysis of these schemes. They provide the prototypes for a general calss of diffusive first order accurate methods.

The main results concerning the Lax-Friedrichs scheme and Godunov's scheme are the following. Consider a system of two conservation laws which is strictly hyperbolic and genuinely nonlinear in the sense of Lax [6], i.e. the eigenvalues λ_j are monotone in the corresponding eigendirection:

$$r_j \cdot \nabla\lambda_j \neq 0 \quad \text{where} \quad \nabla f\, r_j = \lambda_j\, r_j .$$

By way of example we mention that the compressible Euler equations for a polytropic gas provide a system which is strictly hyperbolic and genuinely nonlinear, if the state variable u is restricted to a compact subset K of the state space $G = \{(\rho,m): \rho > 0\}$. Strict hyperbolicity is lost at the boundary of the state space given by the vacuum set $\rho = 0$. A second example is provided by the quasilinear wave equation (1.5) in the case where σ is strictly increasing and either strictly convex or strictly concave:

$$\sigma' > 0, \qquad \sigma'' \neq 0 .$$

The following is a representative theorem from the general theory.

Theorem 2. Suppose that $u(x,t;\Delta x)$ is a sequence of approximate solutions generated by either the Lax-Friedrichs scheme or Godunov's scheme and suppose that the values of $u(x,t;\Delta x)$ lie in a fixed compact subset of the state space G in R^2, i.e.

$$u(x,t;\Delta x) \in K$$

for all $x,t,\Delta x$. Then there exists a subsequence which converges pointwise almost everywhere to a solution of (1) satisfying the entropy condition.

In the case of systems for which the flux function f is defined on all of R^n the main hypothesis of the theorem above simply takes the form of an L^∞ bound which is uniform in the mesh length Δx:

$$|u(\cdot,t,\Delta x)|_\infty \leq M$$

for some fixed M. A similar compactness result holds at the continuum level for a broad class of parabolic systems

$$u_t + f(u)_x = \varepsilon\, D(u,u_x)_x$$

where the diffusion operator D respects the entropy condition. In particular, for the standard method of artificial viscosity where D equals the identity matrix, the following result is available.

Theorem 3. Suppose f is a strictly hyperbolic and genuinely nonlinear map on R^2 and suppose u^ε is a sequence of solutions to the system

$$u_t + f(u)_x = \varepsilon u_{xx} \tag{2.3}$$

which is uniformly bounded in L^∞, i.e. $|u_\varepsilon|_\infty \leq M$ for some constant M independent of ε. Then there exists a subsequence which converges in L^1_{loc} to a solution u of (1.1) that satisfies the entropy condition.

Several remarks are in order concerning the interpretation of the results. In the case of the diffusion equation, it is a well-known fact that, for each fixed positive value of the perturbation parameter ε, the corresponding solution operator forms a compact map from L^∞ to L^1_{loc}. The assertion here is that the family of solution operators $\{S_\varepsilon : \varepsilon > 0\}$ is a compact family uniformly in ε and that the compactness can be established without estimating the derivatives by using compensated compactness.

In general the source of the compactness lies at the hyperbolic level and is embodied in the entropy condition. It turns out that the loss of information occurring in systems (1.1) through the entropy condition is not destroyed by a broad class of diffusion operators and associated first order accurate discretizations. The role of the entropy condition in compactness at the discrete and continuum levels will be discussed briefly below.

As we remarked in the introduction, one of the main open problems in the approximation theory for conservation laws is to establish uniform amplitude estimates for general systems of equations. For a broad class of parabolic systems with constant diffusion matrices D,

$$u_t + f(u)_x = \varepsilon D u_{xx}$$

one anticipates uniform stability of small amplitude solutions near a constant state, i.e. an estimate of the

$$|u(\cdot,t,\varepsilon) - \bar{u}|_\infty \leq \text{const.}\, |u_0 - \bar{u}|_\infty \tag{2.3}$$

where the constant depends only on the fixed state $\bar{u} \in R^n$ provided that the quantity $|u_0 - \bar{u}|_\infty$ is sufficiently small. Appropriate analogues for solutions with large oscillation are expected to hold under mild growth conditions on the flux function at infinity. For a broad class of first order accurate one-step three-point domain of dependence conservative difference schemes, uniform stability of small amplitude approximations of the type (23) is also anticipated. In contrast, for second order methods the conjecture is that the amplitude response measured in L^∞ will be bounded provided that the initial data is not driven widely by oscillations, i.e. that an estimate of the form

$$|u(\cdot,t,\Delta x) - \bar{u}|_\infty \leq \text{const}|u_0 - \bar{u}|_\infty \quad \text{if } TVu_0 \leq \text{const}, \tag{2.4}$$

provided that the right hand side is sufficiently small. Thus, under modest constraints on the initial data one expects that a wide variety of methods will generate approximate solutions with uniformly bounded amplitude. The alternatives are to estimate the derivatives of the approximations uniformly in Δx and pass to the limit with classical functional analysis, or avoid the derivatives and pass to the limit with compensated compactness.

There are several noteworthy physical systems for which uniform L^∞ estimates can be established using maximum principles, thus providing convergence theorems with hypotheses only on the L^∞ norm of the data. In this regard, we recall that the quasilinear wave equation (1.5) in the case of a hard spring with one infection point, i.e.

$$v\sigma'' > 0 \quad \text{if} \quad v \neq 0, \tag{2.5}$$

admits a family of invariant regions from which one may deduce a large-amplitude uniform L^∞ estimate of the form

$$|u(\cdot,t,\varepsilon)|_\infty \leq M \quad \text{and} \quad |u(\cdot,t,\Delta x)| \leq M, \tag{2.6}$$

for the method of artifical viscosity (2.3) and for the Lax-Friedrichs scheme and Godunov's scheme, respectively. Here the constant M depends only on the L^∞ norm of the data. One observes the presence of a family of bounded invariant regions for the underlying hyperbolic system (1.5) and remarks that these regions are preserved by precisely those approximation methods which diffuse each of the primitive variables at an equal rate, an effect induced by either first order diffusion or by integral averaging. Technically, this is a consequence of Jensen's inequality. In conclusion, one obtains the following result.

Theorem. Suppose u_0 is arbitrary initial data in L^∞ for either the method of artificial viscosity (2.3), the Lax-Friedrichs scheme or Godunov's scheme, applied to the quasilinear wave equation (1.5) with a strictly increasing stress function σ satisfying (2.5). Then the resulting approximations satisfy (2.6) and one may extract a subsequence that converges in L^1_{loc} to a solution of (1.5).

As a second example, we mention the compressible Euler equations (1.5) for a polytropic gas. In this situation one also finds a family of invariant regions at the hyperbolic level which are preserved by equal diffusion, leading to estimates of the form

$$0 \leq \rho^{\varepsilon} \leq M, \quad |u^{\varepsilon}| \leq M \tag{2.7}$$

for the density ρ and velocity $u = m/\rho$ for the method of artifical viscosity applied to (1.5). Thus a simple L^{∞} bound is obtained. The essential difficulty stems from the fact that one can not, in general, avoid the singular vacuum state $\rho = 0$, representing a line along which strict hyperbolicity is lost due to the colapse of the eigenvalues. If one assumes a uniform lower bound on the density, say

$$\rho_{\varepsilon} \geq \delta > 0,$$

then the analysis [2] for general strictly hyperbolic and genuinely nonlinear systems can be applied to extract a convergent subsequence. In the general case where only the weaker bound (2.7) holds, additional analysis is required. We refer the reader to [3] which establishes convergence of the method of artifical viscosity applied to the compressible Euler equations (1.5) for a polytropic gas with adiatic exponent γ taking values in the physical sequence $\gamma = 1 + 2/n$, n denoting the number of molecular degrees of freedom. A similar result holds for the Lax-Friedrichs scheme and Godunov's scheme. In particular, one obtains as a corollary the first large data existence theorem for physically meaningful systems of conservation laws in gas dynamics.

3. The Entropy Condition. The source of the compactness in the strong topology lies in the nonlinear structure of the wave speeds and in the dissipation of generalized entropy along propagating shocks. We shall first recall the definition and some basic properties of generalized entropy in the sense of Lax [7]. Consider a system of n conservation laws (1.1). A pair of real-valued maps on the state space R^n

$$\eta: R^n \to R; \quad q: R^n \to R$$

is called an entropy pair for (1.1) if all smooth solutions of (1.1) satisfy an additional conservation law of the form

$$\eta(u)_t + q(u)_x = 0.$$

In brief, (η,q) is an entropy pair if all smooth flows conserve the corresponding entropy field. This notation is familiar in mechanics: in the smooth motion of a material conserving mass and momentum, i.e. satisfying (1.5), one observes the conservation of mechanic energy in the form

$$(\tfrac{1}{2} u^2 + \Sigma(v))_t - (u\sigma)_x = 0, \quad \Sigma' = \sigma.$$

Here the mechanical energy serves as the generalized entropy η while the power produced by the stress tensor, $-u\sigma$, serves as the generalized entropy flux.

We recall that for systems of two conservation laws (with quasi-convex Riemann invariants) one can construct a broad class of entropy pairs (η,q) in which η is strictly convex function [9]. For systems of more than two equations, the existence of an entropy pair is an infrequent event which fortunately occurs for the basic systems of mechanics in a fashion where η is strictly convex. In the following discussion we shall restrict our attention to genuinely nonlinear systems of conservation laws that are endowed with an entropy pair (η,q) for which η is strictly convex.

Within this class of systems the traditional admissibility criterion introduced for the purpose of selecting the physically relevant solution from the totality of all distributional solution is based upon entropy dissipation.

Definition. A solution u of system (1.1) is called admissible if

$$\eta(u)_t + q(u)_x \leq 0.$$

Hence, a weak solution is admissible if and only if its shock waves dissipate generalized entropy. We note that the concept of admissibility is well-defined in the sense that if u is admissible with respect all convex entropies.

The main theorem describing compactness at the hyperbolic level is the following.

Theorem 3. Suppose f is a strictly hyperbolic and genuinely nonlinear map on R^2. If u_k is a sequence of admissible solutions to the corresponding system of two conservation laws (1.1) satisfying the uniform bound

$$|u_k|_\infty \leq M,$$

then there exists a subsequence which converges pointwise almost everywhere to an admissible solution u.

Loosely, the solution operator of the hyperbolic system (1.1) restricted to admissible solutions forms a compact map from L^∞ to L^1_{loc}. The source of the compactness stems from the loss of information associated with the dissipation of entropy along propagating shocks. Technically, the source of the compactness stems from the structure of the entropy fields: it may be shown that for all entropy pairs (η,q), whether η be convex or not, the corresponding entropy fields

$$E_k \equiv \eta(u_k)_t + q(u_k)_x$$

present a family of distributions that lies in a compact subset of $W^{-1,2}$, provide that the sequence u_k in question consists of admissible solutions. The consequence is that the Young measure ν associated with the sequence u_k commutes with the basic antisymmetric bilinear form acting on entropy pairs:

$$\langle \nu, \eta_1 q_2 - \eta_2 q_1 \rangle = \langle \nu, \eta_1 \rangle \langle \nu, q_2 \rangle - \langle \nu, \eta_2 \rangle \langle \nu, q_1 \rangle \tag{3.1}$$

for all entropy pairs (η_j, q_j) $j = 1,2$. This forces the reduction of the Young measure to a point mass and conclusion that a subsequence is converging strongly. We shall not attempt to review here the manner in which the theory of compensated compactness leads to the commutativity constraint (3.1) on the Young measure nor the proof of Tartar's conjecture that any probability measure satisfying (3.1) for all entropy pairs is a Dirac mass. In this regard we refer the reader to [2,3].

We shall conclude this section with a statement of a general compactness theorem for functions and brief discussion of its application to convergence problems for both first and second order accurate conservative finite difference schemes.

Theorem 4. Suppose that f is a strictly hyperbolic and genuinely nonlinear map on R^2 and suppose that $u_k = u_k(x,t)$ is an arbitrary sequence of measurable maps from R^2 to R^2 which is uniformly bounded in L^∞. If, for all entropy pair (η, q), the entropy fields

$$\eta(u_k)_t + q(u_k)_x$$

lie in a compact subset of $W^{-1,2}$ then there exists a subsequence that converges pointwise almost everywhere.

The applications of this compactness result to conservative finite difference schemes involve the analysis of the discrete entropy fields associated with appropriate numerical entropy fluxes. In this matter we refer the reader to [2] and to a forthcoming paper [4] dealing with second order accurate methods. The main result of [4] may be stated roughly as follows. Consider a second order accurate conservative scheme of the form (2.1) applied to a scalar convex conservation law (1.1) for which f is strictly convex and has subquadratic growth at infinity. If the scheme is non-degenerate, L^2-stable and entropy satisfying, then it is convergent. Here one may assert convergence of the entire family because of the well-known uniqueness results for the scalar law. The qualifier non-degenerate simply means that the change in the discrete entropy field across a discrete shock is third order in the strength of the shock, as it is for the problems of mechanics, as opposed to higher order. As corollary one obtains a convergence result for the modification of the Lax-Wendroff scheme introduced by Majda and Osher [9]. The work of Majda and Osher [9] motivated our work on general result mentioned above and illuminated a variety of features concerning the analysis of the associated truncation error equation.

REFERENCES

[1] DiPerna, R. J., Finite differences schemes for conservation laws, Comm. Pure Appl. Math. 35 (1982), pp. 379-450.

[2] DiPerna, R. J., Convergence of approximate solutions to conservation laws, Arch. Rat. Mech. Anal. 82 (1983), pp. 27-70.

[3] DiPerna, R. J., Convergence of the viscosity method for isentropic gas dynamics, Comm. Math. Physics 91 (1983), pp. 1-30.

[4] DiPerna, R. J., Convergence of conservative finite difference schemes for conservation laws, to appear.

[5] Glimm, J., Solutions in the large for nonlinear hyperbolic systems of equations, Comm. Pure Appl. Math. 18 (1965), pp. 697-715.

[6] Lax, P. D., Hyperbolic systems of conservation laws, II, Comm. Pure Appl. Math. 10 (1957), pp. 537-566.

[7] Lax, P. D., Shock waves and entropy, in Contributions to nonlinear functional analysis, ed. E. A. Zarantonello, New York, Academic Press (1971), pp. 603-634.

[8] Lax, P. D. and B. Wendroff, Systems of conservation laws, Comm. Pure Appl. Math. 13 (1960), pp. 217-237.

[9] Majda, A. and S. Osher, Numerical viscosity and the entropy condition, Comm. Pure Appl. Math. 32 (1979), pp. 797-838.

[10] Murat, F., Compacite par compensation, Ann. Scuola Norm. Sup. Pisa Sci. Fis. Mat. 5 (1978), pp. 69-102.

[11] Tartar, L., The compensated compactness method applied to systems of conservation laws, in Systems of NOnlinear Partial Differential Equations, ed. J. M. Ball, D. Reidel Publishing Co., (1983).

[12] Tartar, L., Compensated compactness and applications to partial differential equations, in Research Notes in Mathematics, Nonlinear analysis and Mechanics: Heriot-Watt Symposium, Vol. 4, ed. R. Knops, Pitman Press (1979).

UPWIND SCHEMES USING VARIOUS FORMULATIONS OF THE EULER EQUATIONS

P. L. ROE*

Abstract. A discussion is given of various aspects of 'upwinded' numerical methods for one- and two-dimensional compressible flow problems. The unifying viewpoint is that 'fluctuation', or lack of equilibrium, in the data, is to be explained by using local solutions which are superpositions of simple wave solutions. The analysis is shown to be much simplified by using non-conservative forms of the equations. This may help to make the methods more widely understood and accepted, and may also be directly applicable to some problems of technical interest.

1. Preface. "Although the fields practitioners were scientists, the net result of their activity was something less than science. Being able to take no common body of belief for granted, each writer ... felt forced to build his field anew from its foundations. In doing so, his choice of supporting observation and experiment was relatively free, for there was no standard set of methods or of phenomena that every ... writer felt forced to employ and explain. Under these circumstances, the dialogue of the resulting books was often directed as much to the members of other schools as it was to nature. That pattern is not unfamiliar in a number of creative fields today, nor is it incompatible with significant discovery and invention".

This passage might well serve as a description of the state in which Computational Fluid Dynamics finds itself in 1983. In fact it is a description, by the scientific philosopher and historian T.S.Kuhn [1], of physical optics during the long interval between Aristotle and Newton. Kuhn argues that all sciences pass through such periods before acquiring commonly accepted vocabularies and techniques. During such a period, a scientist may choose to adopt a fixed viewpoint and try to prove its superiority over all alternatives, or he may choose to retain some flexibility of outlook. Both policies can be productive. A workshop like the present one provides both a platform for the enthusiast, and information to help the uncommitted make up their minds.

* Royal Aircraft Establishment, Bedford, U.K. Present address, College of Aeronautics,Cranfield Institute of Technology, Cranfield.

The present intense interest in the Euler equations has led to significant advances in many of the rival techniques, and this is not the place to attempt a survey. One of the areas in which rapid progress is being made is perhaps most generally recognised by the title 'upwind differencing', although this is not a wholly accurate description, and many other keywords would be appropriate to an effective literature search. Amongst these would be 'high-resolution schemes', 'Godunov-type schemes', 'physically motivated schemes', 'lambda formulation', 'flux vector splitting', and 'flux difference splitting'. The literature on such methods has grown explosively over the last two or three years, and does not always make easy reading. In particular, the attempt to generate schemes which automatically include as many features as possible leads to complicated analysis which sometimes obscures the underlying ideas. The author's belief is that upwind-differencing schemes will eventually become both simple and efficient. Indeed, if this does not happen, it may be better to concentrate on less ambitious algorithms and rely on computing power to obtain highly resolved details of the flow. The present paper is not so much concerned with presenting new analysis, as with stressing the underlying similarities between ideas, and with setting straight a few misunderstandings which have arisen.

Basically, all the workers in this 'upwind' area share the belief that the key to successful modelling of wave propagation problems must lie in a careful matching of numerical and physical processes. The differential equations are analysed ('split') to identify those groups of terms which describe a particular physical effect, and the numerical modelling of those terms is designed to avoid whatever sources of error or instability they seem most likely to contribute. This program has been very successfully carried out for one-dimensional unsteady flows, and the concept which is central to the entire class of methods is the local representation of the solution as the superposition of simple wave solutions. The differences between the methods have to do with such matters as whether or not conservation is built into the algorithm, whether we bother to distinguish between compression and rarefaction waves, and what measures are taken to eliminate oscillations from shockwaves. All these things are of course important, but towards all of them it is possible to take two distinct attitudes. The first is that a single numerical scheme, to be applied everywhere in the flowfield, can be made sufficiently elaborate to cope with every possibility. The second attitude is to make the basic scheme as simple as possible, but capable of carrying out tests and setting indicators to mark those areas where it might prove inadequate. These areas are then fixed by locally applying a more accurate method. Shock-fitting methods belong to this end of the spectrum.

I feel influenced in favour of the second approach, emphasising underlying simplicity, by the fact that we still seem to be some distance away from a complete and physically motivated treatment of multidimensional flows, even though partial successes have been reported [2,3,7,8]. For example the concept of total variation, which has proved powerful in one-dimensional calculations [4], turns out to be very restrictive in more than one dimension [5]. Moreover, vorticity is a phenomenon entirely absent from one-dimensional flow, but surely of importance in higher dimensions. We may speculate that it should be a vital ingredient of multidimensional modelling, but so far it has been given no role in these methods. Meanwhile, the analysis given here continues to concentrate on wavelike phenomena.

2. Simple Waves in One Dimension. We shall write the Euler equations for one-dimensional unsteady flow of an ideal gas in the form:

$$\underline{w}_t + A(\underline{w})\,\underline{w}_x = 0 \tag{1.1}$$

where $\underline{w}$ is any set of independent variables sufficient to describe the flow. In what follows we shall give special attention to the following choices.

$$\underline{w}_C = (\rho, \rho u, e)^T \tag{1.2}$$

$$\underline{w}_R = (a, u, S)^T \tag{1.3}$$

$$\underline{w}_S = (p, u, S)^T \tag{1.4}$$

The notation here is ρ = density, u = velocity, p = pressure, $a = (\gamma p/\rho)^{\frac{1}{2}}$ = sound speed, $e = p/(\gamma-1) + \frac{1}{2}\rho u^2$ = total energy, $S = p/\rho^\gamma$ = entropy. The set $\underline{w}_C$ comprises the conserved variables, whose use offers the most secure basis for shock capturing. The set $\underline{w}_R$ has been adopted by Moretti [6] and associated workers [2,7,8] on the grounds that it offers a simple analogy with the use of Riemann invariants in isentropic flow. The set $\underline{w}_S$ is theoretically significant because the Euler equations are made symmetric, at least in the limited sense that the matrix $A_S(\underline{w}_S)$ has symmetrically placed zeroes.

For these cases, the matrices are

$$A_C(\underline{w}_C) = \begin{vmatrix} 0 & 1 & 0 \\ (\kappa-1)u^2 & 2(1-\kappa)u & 2\kappa \\ -\gamma ue/\rho+2\kappa u^3 & \gamma e/\rho-3\kappa u^2 & \gamma u \end{vmatrix} \tag{1.5}$$

$$A_R(\underline{w}_R) = \begin{vmatrix} u & \kappa a & 0 \\ a/\kappa & u & -a^2/\gamma(\gamma-1)S \\ 0 & 0 & u \end{vmatrix} \tag{1.6}$$

$$A_S(\underline{w}_S) = \begin{vmatrix} u & \rho a^2 & 0 \\ 1/\rho & u & 0 \\ 0 & 0 & u \end{vmatrix} \qquad (1.7)$$

Here and subsequently, we use κ to denote the commonly recurring constant $\frac{1}{2}(\gamma-1)$.

The general theory of simple waves for any hyperbolic equation of the form (1.1) is given by Lax [9], and also by Jeffrey and Taniuti [10], and Jeffrey [11]. We seek flows in which the state vector $\underline{w}$ depends only on a scalar parameter, say $\underline{w} = \underline{w}(v)$. If such flows can be found, $\underline{w}_t$ must be a scalar multiple of $\underline{w}_x$, and therefore, by (1.1), each must be an eigenvector of $A(\underline{w})$. Let $\underline{r}^k(\underline{w})$ be the kth right eigenvector of $A(\underline{w})$, and let W be a path in $\underline{w}$-space satisfying the differential equation

$$d\underline{w}/dv = \underline{r}^{(k)} \qquad (1.8)$$

Then we may speak of W as a wave path and use it to construct our solution as follows.

To each point along some arc of W there corresponds a state $\underline{w}(v)$, an eigenvector $\underline{r}^{(k)}(v)$, and an eigenvalue $\lambda^{(k)}(v)$. Over some region of (x,t) space draw a pattern of non-intersecting straight lines with various slopes. To every point on the line whose reciprocal slope $dx/dt = \lambda^{(k)}(v)$ assign the state $\underline{w}(v)$. The equation which describes the constancy of $\underline{w}$ along such a line is

$$\underline{w}_t + \lambda^{(k)} \underline{w}_x = 0 \qquad (1.9)$$

and since $\lambda^{(k)}$ is an eigenvector of A(v), then (1.1) is also satisfied. The flow field thus constructed is known as a k-simple wave.

It is evident that the class of simple wave solutions to, say, the Euler equations, cannot depend on the choice of variables used to describe the flow. The description of simple waves will, however, be easier in some sets of variables than in others. Particular attention to this point leads us to the study of characteristic variables, or Riemann invariants. It is, therefore, unfortunate that the term Riemann invariant should be used by different writers to refer to two quite different concepts. In what follows we shall attempt to clarify matters.

If a smooth flow is not a simple wave, it can always be represented, at least locally, as a sum of simple waves. Project $\underline{w}_x$ onto the eigenvectors of A, so that

$$\underline{w}_x = \Sigma_k \alpha^{(k)} \underline{r}^{(k)} (\underline{w}) \qquad (1.10)$$

where $\alpha^{(k)}$ is the local strength of the kth wave. Multiplication by

A yields

$$\underline{w}_t = -\Sigma_k \alpha^{(k)} \lambda^{(k)} \underline{r}^{(k)}(\underline{w}) \tag{1.11}$$

and the decomposition into simple waves is achieved. Consider the differentiation operator $D^{(k)} = \partial/\partial t + \lambda^{(k)}\partial/\partial x$, which acts along a characteristic $dx/dt = \lambda^{(k)}$. Then adding (1.10) and (1.11) shows us that $D^{(k)}\underline{w}$ has no component along $r^{(k)}$, and this statement can be written

$$\underline{\ell}^{(k)} \cdot D^{(k)}\underline{w} = 0 \tag{1.12}$$

where $\underline{\ell}^{(k)}$ is the left eigenvector of A. (Recall that $\underline{\ell}^{(m)}.\underline{r}^{(n)} = 0$ if $m \neq n$.) Eqn.(1.12) is the kth characteristic equation. Now it can be shown [10, 11] that for any system comprising just two unknowns, say w_1, w_2, the two characteristic equations can both be integrated giving functions $R_1(w_1,w_2)$, $R_2(w_1,w_2)$ which remain constant along the characteristic lines. These are the quantities which are widely described as Riemann invariants in much of the engineering literature on the method of characteristics, a method which is at its best when dealing with two unknowns. For examples, see [12].

If there are more than two unknowns, there are not, in general, any functions which remain constant along characteristics, unless the differential equations satisfy some rather special "integrability conditions" [10, 11]. In fact, it becomes necessary to specialise again to simple wave flow before any useful functions can be found. These new functions are such that they remain constant along wave paths in $\underline{w}$-space, that is, as we move through, rather than along, the characteristics in x-t space, and they possess this property only if the flow is a simple wave. Jeffrey refers to this second kind of function as "generalised Riemann invariants", but Lax [9] and many other authors refer to them as "Riemann invariants". We therefore have the confusing situation with regard to systems of two equations, that all authorities would assert the existence of Riemann invariants, but each would ascribe them to the opposite family of characteristics. It must be emphasised that there is no disagreement about the mathematical facts, merely a conflict of nomenclature. This needs to be pointed out, since the term "Riemann invariant" is often used descriptively, without its meaning being immediately apparent from context. A third coinage is due to Moretti and his school, who use the term "Riemann variable" to denote quantities which would be constant along characteristics in simpler circumstances, such as isentropic flow.

When we really do have isentropic flow, S = const, and only two equations are needed to describe the flow, so that Riemann invariants of the first kind exist. The quantities constant along the characteristics with speeds $(u-a)$,$(u+a)$ are $(a-\kappa u)$, $(a+\kappa u)$.

For the purposes of constructing general numerical methods, it is obviously better to base discussion on something which always exists, namely the second or generalised kind of Riemann invariant. To avoid ambiguity, we might refer to these as <u>simple wave invariants</u>. For the Euler equations, the quantities <u>invariant across</u> simple waves whose speed is $(u-a)$ are S and $(a+\kappa u)$; if the wave speed is $(u+a)$ the invariants are S and $(a-\kappa u)$; if the wave speed is u the invariants are u and p.

All the numerical methods which incorporate "upwinding" principles are based upon various discrete analogues of the expression (1.11) which analyses the rate of temporal change into components due to various simple waves. On the RHS of (1.11) there appear three different sorts of information. These are the strengths $\alpha^{(k)}$ of each wave, their speeds $\lambda^{(k)}$, and the eigenvectors $\underline{r}^{(k)}$ which describe how the effects of each wave are distributed over the chosen set of state variables $\underline{w}$. It is not possible to device a physically correct scheme without access to all this information; the various schemes differ in providing it with greater or lesser precision and convenience. The more elaborate schemes will be justifiable only to the extent that they allow the incorporation of additional constraints (such as conservation) and even then only where it can be shown that those constraints are important for the technical applications which are intended. It is perhaps in the efficient matching of algorithms and applications that our ignorance is at present most profound.

We now exhibit the basic analysis on which all upwinded methods are founded by carrying out the simple wave decomposition for the three different sets of independent variables. On both physical and mathematical grounds it is clear that the wavespeeds, or eigenvalues, must be the same in every case, viz $u-a$, u, $u+a$. The eigenvectors which describe the distribution of the effects due to each wave are of course different. For the conservative variables $\underline{w}_C$ we have, with $\rho H = e + p$,

$$\underline{r}^{(1)} = \begin{vmatrix} 1 \\ u-a \\ H-ua \end{vmatrix} \quad \underline{r}^{(2)} = \begin{vmatrix} 1 \\ u \\ \frac{1}{2}u^2 \end{vmatrix} \quad \underline{r}^{(3)} = \begin{vmatrix} 1 \\ u+a \\ H+ua \end{vmatrix} \tag{1.13}$$

for the variables $\underline{w}_R$ we have

$$\underline{r}^{(1)} = \begin{vmatrix} 1 \\ -1/\kappa \\ 0 \end{vmatrix} \quad \underline{r}^{(2)} = \begin{vmatrix} 1 \\ 0 \\ \gamma S/2a \end{vmatrix} \quad \underline{r}^{(3)} = \begin{vmatrix} 1 \\ 1/\kappa \\ 0 \end{vmatrix} \tag{1.14}$$

and for the variables $\underline{w}_S$ we have

$$\underline{r}^{(1)} = \begin{vmatrix} \gamma p/a \\ -1 \\ 0 \end{vmatrix} \quad \underline{r}^{(2)} = \begin{vmatrix} 0 \\ 0 \\ 1 \end{vmatrix} \quad \underline{r}^{(3)} = \begin{vmatrix} \gamma p/a \\ 1 \\ 0 \end{vmatrix} \tag{1.15}$$

We must also, in each case, compute the wave strengths. In the first place, we will do this for the continum problem (1.11). (Later, we indicate the natural generalisation of the analysis to discrete data). The calculation is routine and we merely quote results for the different sets of variables. In terms of $\underline{w}_C$ we find

$$\begin{aligned} \alpha^{(2)} &= (2\kappa/a^2)[(H-u^2)w_{1,x} + uw_{2,x} - w_{3,x}] \\ \alpha^{(3)} + \alpha^{(1)} &= w_{1,x} - \alpha^{(2)} \\ a(\alpha^{(3)} - \alpha^{(1)}) &= uw_{1,x} - w_{2,x} \end{aligned} \tag{1.16}$$

where $w_{n,x} = \partial w_n/\partial x$. In terms of $\underline{w}_R$ we find the simpler expressions,

$$\begin{aligned} \alpha^{(2)} &= (2a/\gamma S)S_x \\ \alpha^{(3)} + \alpha^{(1)} &= a_x - \alpha^{(2)} \\ \alpha^{(3)} - \alpha^{(1)} &= \kappa u_x \end{aligned} \tag{1.17}$$

and in terms of $\underline{w}_S$,

$$\begin{aligned} \alpha^{(2)} &= S_x \\ \alpha^{(3)} + \alpha^{(1)} &= ap_x/\gamma p \\ \alpha^{(3)} - \alpha^{(1)} &= u_x. \end{aligned} \tag{1.18}$$

All the upwind differencing schemes are based on various discretisations of the above procedure. In the simplest versions, the derivatives which appear in (1.16) - (1.18) are approximated by $(\cdot)_x = [(\cdot)_{i+1} - (\cdot)_i]/\Delta x$, and other terms by arithmetic mean values so that $(\cdot) = \frac{1}{2}[(\cdot)_i + (\cdot)_{i+1}]$. To achieve conservation one technique is to use $\underline{w}_C$ and then to ensure that (2.4) is met by using special mean values with particular properties. These were introduced in [13], and exploited in [4] and elsewhere. Some of the analysis has since been streamlined [14], but remains distinctly more complicated than anything based on (1.17) or (1.18) would be. This is the premium which apparently must be paid for a conservative method.

There are other ways of providing the required information. One way is to carry out the analysis using one set of variables, and then to translate the results into another set of variables. This is one way to understand the scheme originated by Osher [15]. Let us observe that, in terms of the variables $\underline{w}_R$, the simple wave paths are rectilinear (1.14). We define a compound wave path from state $\underline{w}_i$ to $\underline{w}_{i+1}$ to be the union of three independent simple wave paths. If it is specified that the simple wave paths shall be taken in a particular order, then it can be shown that the compound path is unique. In particular, the break points which Osher denotes by $\underline{w}_{i+1/3}$, $\underline{w}_{i+2/3}$ are unique, and so is the point $\underline{w}_S$ at which the compound path crosses the sonic surface $|u| = a$. Note that Osher actually chooses the seemingly unnatural order 3,2,1, as though the flow were going into the past. This ordering is at present required to prove certain properties of the method, but may not be essential. The quantities $\underline{F}(\underline{w}_{i+1/3})$, $\underline{F}(\underline{w}_{i+2/3})$, $\underline{F}(\underline{w}_s)$, where $\underline{F}$ is the vector of 'flux variables' $\underline{F} = (\rho u,\ p+\rho u^2,\ u(e+p))$, then become the building blocks of Osher's scheme. Again there is a computational premium to be paid, related to the number of properties which can be incorporated into the method [15, 16].

As a common name to unite all these approaches, as well as others which there is not space to discuss, I would like to propose the term 'fluctuation splitting'. This follows a previous suggestion by the author [17] and is, I think, shorter and more descriptive than any of the alternative keywords listed in the introduction. It stresses that what is being analysed is the physical cause of the change, that is $A(w)\underline{w}_x$, plus, if need be, source terms [18]. Also, it appears to generalise naturally to higher dimensions, as discussed in Section 4.

3. Numerical Implementation. As soon as the basic information of wave strength, wave speed, and eigenvector distribution has been generated, a first order scheme can immediately be implemented. The signal due to the kth wave is defined in the interval $(i, i+1)$ to be

$$\Phi^{(k)}_{i+\frac{1}{2}} = -\Delta t\ \alpha^{(k)}\lambda^{(k)}\underline{r}^{(k)} \tag{3.1}$$

where $(i+\frac{1}{2})$ subscripts are to be understood on the RHS. This signal is added to the value of $\underline{w}_i$ if $\lambda^{(k)} < 0$, or to $\underline{w}_{i+1}$ if $\lambda^{(k)} > 0$. If we are working with variables $\underline{w}_R$ this algorithm is completely equivalent to the first-order version of the 'lambda scheme'. Precisely the same arithmetic operations and logical decisions are involved, and the code need take no longer to execute.

It is in the construction of higher-order schemes that the most interesting possibilities lie. Numerical methods which cleanly capture discontinuities, even those which are rapidly moving, have naturally been most developed by those workers who use conservation form, but analogous methods could be applied to other sets of variables. In the absence of conservation, the outcome would be a flow with cleanly captured discontinuities, moving with speeds which are not quite correct. For some applications this might not matter. There has been much recent interest in using the Euler equations to model flows where the important discontinuities are contact surfaces or vortex sheets [6, 19]. It is therefore interesting to note that the system $\underline{w}_S$ has the ability to 'numerically recognise' these features, for if Δp, Δu are both zero, eqn.(1.18) predicts that $\alpha^{(1)}$, $\alpha^{(3)}$ are both zero.

There is definitely a divergence of opinion concerning the correct numerical domain of dependence (= 'support') for high-order algorithms.

This is most easily illustrated by specialising to the linear scalar advection equation

$$u_t + au_x = 0 \tag{3.2}$$

assuming that a is a positive constant, and asking what data values $u_i^n = u\,(i\Delta x,\, n\Delta t)$ can be allowed to influence the change $u_i^{n+1} - u_i^n$. What we might call the strict upwinding doctrine teaches that the support of u_i should contain no points to the right of i. If this is required, it can easily be implemented [20] by altering the domain of influence for the signal (3.1).

Define the Courant number $\nu^{(k)} = \Delta t \lambda^{(k)}/\Delta x$. If $\nu^{(k)} < 0$ then send a signal $\frac{1}{2}(3 + \nu_{i+\frac{1}{2}}^{(k)})\, \Phi_{i+\frac{1}{2}}^{(k)}$ to $\underline{w}_i$ and a signal $-\frac{1}{2}(1 + \nu_{i+\frac{1}{2}}^{(k)})\, \Phi_{i+\frac{1}{2}}^{(k)}$ to w_{i-1}. If $\nu^{(k)} > 0$ then send a signal $\frac{1}{2}(3 - \nu_{i+\frac{1}{2}}^{(k)})\, \Phi_{i+\frac{1}{2}}^{(k)}$ to $\underline{w}_{i+1}$ and a signal $-\frac{1}{2}(1 - \nu_{i+\frac{1}{2}}^{(k)})\Phi_{i+\frac{1}{2}}^{(k)}$ to $\underline{w}_{i+2}$. This scheme is second-order accurate, but suffers from a leading phase error and transmits wiggles ahead of moving discontinuities. For the model problem (3.2) it is equivalent to the upwind scheme of Warming and Beam [21], and also to the second-order lambda scheme of Gabutti [22]. Unlike these schemes it is not expressed in predictor-corrector form, and it is therefore quicker to execute, because the fluctuation splitting need only be done once per time step. It is stable for Courant numbers less than 2.0.

However, the strictly upwind approach does have mathematical limitations, which prevent it from achieving better than second-order accuracy [23, 24], and it is not clear that it can be justified on physical grounds either. If the Courant number lies between 0 and 1.0 then the natural domain of dependence for (3.2) is clearly {i-1, i}. The use of information from any other source can only be justified if the data is smooth, so that {i-2, i-1} seems to be on exactly the same level of legitimacy as {i, i+1}. If we accept both as equally (un)reliable, then a wide range of algorithms can be created which impose only very slight computational overheads.

One of these reproduces the excellent algorithm of Fromm [25], in a version applicable to systems. Begin with the simple scheme (3.1). Find the correction signal $\frac{1}{4}(1 - |\nu|)\Phi$. (Indices $i+\frac{1}{2}$ and (k) are to be understood). If $\nu < 0$, add this quantity to $\underline{w}_{i+1}$ and subtract it from $\underline{w}_{i-1}$. If $\nu > 0$, add it to $\underline{w}_i$ and subtract if from $\underline{w}_{i+2}$. This scheme produces much smaller overshoots than Lax-Wendroff and has very low phase error. It is stable (in the switched version described here) for $-1 < \nu < 1$.

Another algorithm, not quite so simple but with twice the stability range is due to Pike (private communication). Begin again with (3.1). Find the correction signal $1/12(1 + |\nu|)\Phi$. Multiply this by further factors and distribute it as follows. If $\nu < 0$, a proportion $-(6+\nu)$ goes to $\underline{w}_{i+1}$, $(6 + 2\nu)$ goes to $\underline{w}_i$, and $-\nu$ to $\underline{w}_{i-1}$. If $\nu > 0$, a proportion $(\nu-6)$ goes to $\underline{w}_i$, $(6-2\nu)$ goes to $\underline{w}_{i+1}$, and ν to $\underline{w}_{i+2}$. Since these corrections sum to zero, the modified algorithms of both Fromm and Pike are conservative, if (3.1) is conservative.

These algorithms allow discontinuities to be captured, in some limited sense that may actually be quite extensive, even if conservative variables are not used. To eliminate the oscillations which would accompany moving discontinuities, some further extension is needed, and the 'flux-limiting' techniques which have been worked out for conservation laws are directly applicable. It may be verified that the following algorithm is also equivalent to Fromm's scheme in the linear case. Begin with (3.1). Find the correction signal $b = \frac{1}{2}(1-|\nu|)\Phi$. Find the arithmetic mean B of this quantity with the corresponding quantity in the upwind cell; that is, if $\nu_{i+\frac{1}{2}} < 0$, $B = \frac{1}{2}\,(b_{i+\frac{1}{2}} + b_{i+3/2})$ whereas if $\nu_{i+\frac{1}{2}} > 0$, $B = \frac{1}{2}(b_{i+\frac{1}{2}} + b_{i-\frac{1}{2}})$. Then if $\nu_{i+\frac{1}{2}} < 0$, add B to $\underline{w}_{i-\frac{1}{2}}$ and subtract it from $\underline{w}_{i+\frac{1}{2}}$, whereas if $\nu_{i+\frac{1}{2}} > 0$, transfer B in the opposite direction. (As a practical aside these schemes are actually quite easy to program, but practice with some simple scalar examples is needed to get all the plus and minus signs sorted out). Monotonicity and control of total variation can be assured by working with averaging functions B other than the arithmetic mean [17, 26-31]. The latest developments along these lines [28 - 31] yield captured contact discontinuities which remain for all time of small finite width. This observation seems very pertinent in conjunction with remarks made above, about the ability of the $\underline{w}_S$ system to recognise contacts.

Some philosophical doubts about the validity of these capturing procedures may be laid to rest if it is noted that the various averaging functions $B(b_1,b_2)$ all have the property that $B(b,0) = B(0,b) = 0$. This means that in regions of strong local variation the algorithms all tend to revert to (3.1) and to become wholly upwinded. Under these circumstances, the loss of formal accuracy is of no consequence.

4. Simple Waves in Two Dimensions. We consider hyperbolic systems of the form

$$\underline{w}_t + A(\underline{w})\underline{w}_x + B(\underline{w})\underline{w}_y = 0 \tag{4.1}$$

It is probably possible to generalise the simple wave concept in more than one way to obtain a two-dimensional extension. Only a rather simple generalisation will be attempted here. As in one dimension, we will assume that the flow field is such that the state variables $\underline{w}$ depend only on a scalar parameter v. In an (x,y) plane contours of any state variable are therefore also contours of any other state variable. Let $\underline{n} = (\cos\theta, \sin\theta)$ be a unit vector in (x,y) normal to such a contour, and let n be a coordinate in that normal direction.

Then (4.1) can be written

$$\underline{w}_t + [A(\underline{w}) \cos\theta + B(\underline{w}) \sin\theta]\, \underline{w}_n = 0 \tag{4.2}$$

But $\underline{w}_t$, $\underline{w}_n$ must each be proportional to $\underline{w}_v$, and so all three are eigenvectors of the matrix $A \cos\theta + B \sin\theta$. Denote such an eigenvector by $\underline{r}(\underline{w}, \theta)$, and let W be a wave path which satisfies the differential equation

$$d\underline{w}/dv = \underline{r}(\underline{w}, \theta) \tag{4.3}$$

Compare (4.3) with (1.8). The difference is that whereas the solutions to (1.8) depended on a finite set of parameters **k**, we now have an infinitely variable parameter θ. In principle θ might vary along the wave path in any arbitrary way. Before taking matters any further, however, we will exhibit these eigenvectors, using the $\underline{w}_S$ system $(p, u, v, S)^T$ which seems to show their physical significance most clearly. In the $\underline{w}_S$ system we have

$$A\cos\theta + B\sin\theta = \begin{vmatrix} & \rho a^2 \cos\theta & \rho a^2 \sin\theta & 0 \\ \rho^{-1}\cos\theta & q_n & 0 & 0 \\ \rho^{-1}\sin\theta & 0 & q_n & 0 \\ 0 & 0 & 0 & q_n \end{vmatrix} \tag{4.4}$$

where

$$q_n = u\cos\theta + v\sin\theta$$

The eigenvalues of this matrix are $\lambda = q_n, q_n, q_n \pm a$. Corresponding to an eigenvalue $\lambda = q_n + a$ we find the eigenvector

$$\underline{v}_a(\underline{w}, \theta) = \begin{vmatrix} \gamma p/a \\ \cos\theta \\ \sin\theta \\ 0 \end{vmatrix} \tag{4.5}$$

By taking $\theta = m\pi/2$ with m integer we recover the $\underline{r}^{(1)}$, $\underline{r}^{(3)}$ eigenvectors for one-dimensional acoustic waves travelling in the x or y directions. Clearly (4.5) represents an acoustic wave, propagating normal to itself with velocity $q_n + a$. (See Fig.1). The eigenvalue $\lambda = q_n - a$ has no independent significance, because the solution which it leads to can equally be generated by taking $\theta' = \theta + \pi$.

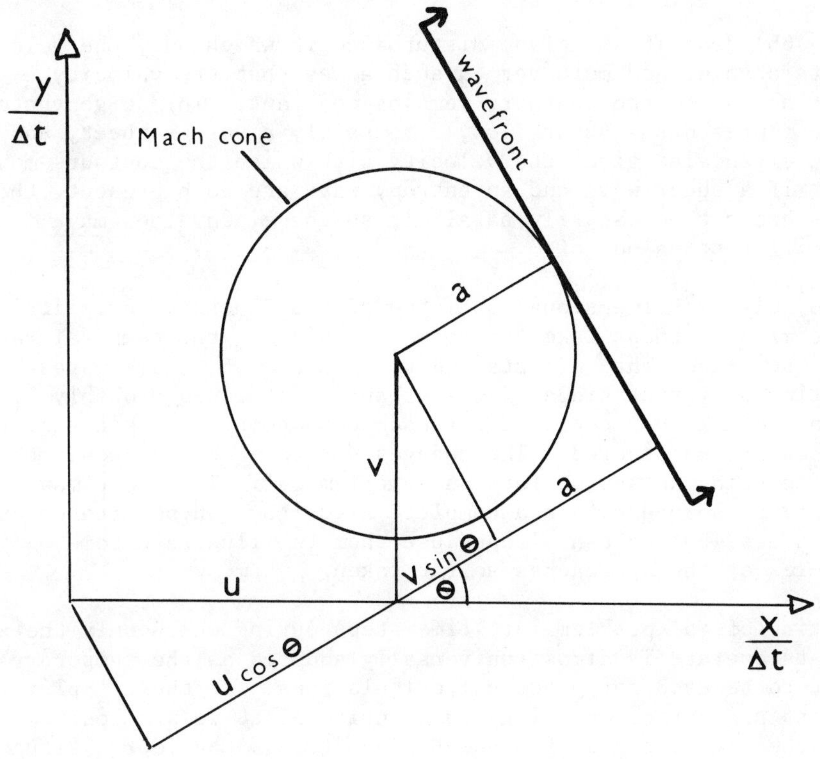

Figure 1.

The eigenvalue $\lambda = q_n$ is repeated twice and therefore its eigenvectors represent two different kinds of disturbance moving with the same velocity. Two eigenvectors which span the relevant nullspace are

$$\underline{r}_s(\underline{w}) = \begin{vmatrix} 0 \\ 0 \\ 0 \\ 1 \end{vmatrix}, \quad \underline{r}_v(\underline{w}, \theta) = \begin{vmatrix} 0 \\ -\sin\theta \\ \cos\theta \\ 0 \end{vmatrix} \tag{4.6a,b}$$

Eqn.(4.6a) is deceptively simple. It describes a plane disturbance in which everything is constant except entropy. But we know that entropy is carried passively with the fluid. Why then is the eigenvalue not equal to the fluid velocity $(u^2 + v^2)^{\frac{1}{2}}$? In fact, the eigenvalue we obtain is the velocity $q_n = \underline{u} \cdot \underline{n}$ with which the contours move.

Eqn.(4.6b) describes a plane disturbance in which only the velocity components change, and moreover in such a way that the velocity component normal to the contours remains constant. This eigenvector therefore represents a shear flow, or possibly a vortex sheet, and again the eigenvalue gives the velocity with which the contours move. Note that if a shear wave and an entropy wave are both present, their gradients are not necessarily parallel, so the eigenvalues may contain different values of θ.

A genuinely two-dimensional analogue of the fluctuation-splitting technique might perhaps take the form of analysing the temporal rate of change $Aw_x + Bw_y$ into effects due to a number of simple waves representing different kinds of disturbance. It would probably be sufficient to take θ = constant for each component wave, since only small areas are considered. The changes due to each wave would then be computed using an appropriate numerical model. There are many problems to be solved before a complete algorithm can be created out of these ideas, but we can already use them to illustrate some deficiencies of the approaches so far taken.

Splitting of the problem into time steps during which only the x or y gradients operate is almost universally adopted as the temporary expedient to be used for practical calculations. If these 'split time steps' are performed using upwind principles, we are led to consider the eigenvectors of A and B separately. Consider a situation in which the disturbance is actually an acoustic simple wave inclined at an angle θ, but it is being analysed only for the effect of the x-derivatives. The following calculation will ensue, in which the RHS gives the projection onto the eigenvectors of A.

$$\underline{r}_a(\underline{w},\theta) = \begin{bmatrix} \gamma p/a \\ \cos \\ \sin \\ 0 \end{bmatrix} = \tfrac{1}{2}(1+\cos\theta)\begin{bmatrix} \gamma p/a \\ 1 \\ 0 \\ 0 \end{bmatrix} + \sin\theta \begin{bmatrix} 0 \\ 0 \\ 1 \\ 0 \end{bmatrix} + \tfrac{1}{2}(1-\cos\theta)\begin{bmatrix} \gamma p/a \\ -1 \\ 0 \\ 0 \end{bmatrix} \qquad (4.7)$$

Unless $\theta = 0$ or $\theta = \pi$, the wave will be quite dreadfully misinterpreted as the sum of two acoustic waves and a shear wave. Similarly, if the true disturbance is a shear wave, the analysis will give

$$\underline{r}_v(\underline{w},\theta) = \begin{bmatrix} 0 \\ -\sin\theta \\ \cos\theta \\ 0 \end{bmatrix} = \tfrac{1}{2}\sin\theta \begin{bmatrix} \gamma p/a \\ -1 \\ 0 \\ 0 \end{bmatrix} + \cos\theta \begin{bmatrix} 0 \\ 0 \\ 1 \\ 0 \end{bmatrix} - \tfrac{1}{2}\sin\theta \begin{bmatrix} \gamma p/a \\ 1 \\ 0 \\ 0 \end{bmatrix} \tag{4.8}$$

and, again, the single oblique wave will be represented by a spurious model which invokes three one-dimensional waves. Only an entropy wave is correctly recognised. Similar results are obtained using any other formulation.

A partially successful attack on this aspect of the problem has been reported by Davis [3], and some pertinent observations have been made by Baines [31]. Davis attempts to identify those values of θ corresponding to waves which are actually present in the data. As reported, only a single value of θ is actually computed, and this is done on the assumption that the flow is dominated by a single acoustic wave. Nevertheless, preliminary numerical results seem quite promising.

It is also interesting to try and relate these concepts to the bicharacteristic form of (4.1) [32,33]. Let P be a typical point in (x,y,t) and let the circle C be its domain of dependence in the plane t-Δt. Let BP be a typical generator of the cone which has C as base and P as vertex. (See Figure 2).

Let T be the plane tangent to this cone along BP, and let ξ, η be local space coordinates as shown. Then changes $\Delta(\cdot) = (\cdot)_P - (\cdot)_B$ along the bicharacteristic BP can be shown to be related to non-uniformities in the data by

$$\Delta p + \rho a \Delta u_\eta = -\rho a^2 \Delta t \frac{\partial}{\partial \xi} u_\xi \tag{4.9}$$

where u_η, u_ξ are normal and tangential components of velocity.

This formula holds whether or not the flow is isentropic. If the flow is a simple wave and the point B has been chosen so that T contains the contours of the wave, then the RHS vanishes, and the LHS is consistent with (4.5). If the flow is a simple wave, but the point B has not been so well-chosen, various additional terms will arise in (4.9), which would seem detrimental to the clean modelling of the flow. Those papers which have attempted to extend the lambda formulation to multidimensional flow [2,7,8] have generally used combinations of bicharacteristic equations dictated by the geometry of the coordinates, rather than by the geometry of the flow.

I feel sure that if the ideas of upwind differencing can be used to build genuinely multi-dimensional algorithms, it will be necessary to create rather subtler pattern-recognition mechanisms

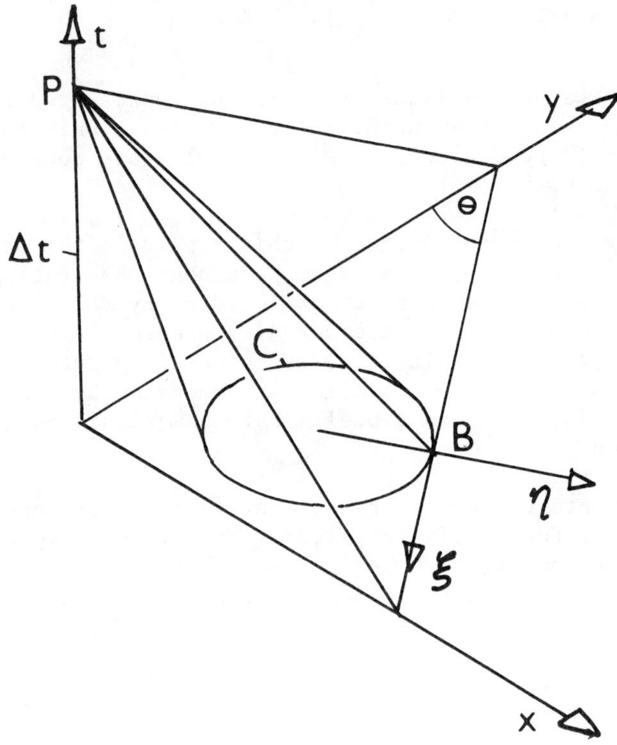

Figure 2.

than have yet been devised. It should be much easier to do this using simple non-conservative forms of the Euler equations, and then translating the results into conservation form.

5. Acknowledgements. Much of the stimulus for this paper arose out of a specialist seminar organised by Dr.A.di Carlo at Rome University on October 10th 1983, to debate the relationship between the lambda formulation and other upwind schemes. The hospitality and interest displayed on that occasion was much appreciated. Subsequent discussions with colleagues at the Royal Aircraft Establishment, particularly with Dr.D.E.Davies and Dr.C.M.Albone helped to clarify many of the issues.

REFERENCES

[1] T.S.KUHN, The Structure of Scientific Revolutions, Univ.of Chicago Press, 1970.

[2] G.MORETTI and L.ZANNETTI, A new, improved computational technique for two-dimensional unsteady compressible flows. AIAA paper 82-0168, 1982.

[3] S.F.DAVIS, A rotationally biased upwind difference scheme for the Euler equations. NASA CR 172179, ICASE, July 1983.

[4] A.HARTEN, High resolution schemes for hyperbolic conservation laws. J.Comp.Phys, 49, March 1983.

[5] R.J.LEVEQUE and J.B.GOODMAN, TVD schemes in one and two space dimensions. Proc.15th AMS-SIAM Summer Seminar in Applied Mathematics, to appear.

[6] G.MORETTI, Towards a closer cooperation between theoretical and numerical analysis in gas dynamics in Transonic, Shock and Multidimensional Flows, ed. R.E.Meyer, Academic Press, 1982.

[7] L.ZANNETTI and G.COLASURDO, Unsteady compressible flows; a computational method consistent with the physical phenomena. AIAA July, 19, 1981.

[8] A.DADONE and M.NAPOLITANO, An implicit lambda scheme, AIAA paper 82-0972, 1982.

[9] P.D.LAX, Hyperbolic systems of conservation laws II. Comm. Pure Appl.Math., X, p.537, 1957.

[10] A.JEFFREY and T.TANIUTI, Nonlinear Wave Propagation, Academic Press, 1964.

[11] A.JEFFREY, Quasilinear Hyperbolic Systems and Waves, Research Notes in Mathematics No.5, Pitman, 1976.

[12] M.B.ABBOTT, An Introduction to the Method of Characteristics, Thames and Hudson, 1966.

[13] P.L.ROE, Approximate Riemann solvers, parameter vectors and difference schemes, J.Comp.Phys., 43, 2, p.357, 1981.

[14] P.L.ROE and J.PIKE, Efficient construction and utilisation of approximate Riemann solutions. Proc.Sixth Int.Conf.Num.Meth. App.Sc. and Eng., North Holland, to appear, 1984.

[15] S.OSHER and F.SOLOMON, Upwind schemes for hyperbolic systems of conservation laws, Math.Comp., 38, 1982.

[16] S.OSHER and S.CHAKRAVARTHY, High resolution schemes and the entropy condition, NASA CR 172218, ICASE, September 1983.

[17] P.L.ROE, Fluctuations and signals - a framework for numerical problems, in Numerical Methods for Fluid Dynamics, eds.K.W.Morton M.J.Baines, Academic Press, 1982.

[18] M.PANDOLFI, A contribution to the numerical prediction of unsteady flows, AIAA paper 83-0121, 1983.

[19] W.SCHMIDT and A.JAMESON, Euler solutions as limit of infinite Reynolds number for separation flows and flows with vortices, in Proc.8th Int.Conf.Num.Meth.in Fluid Dyn.ed.E.Krause, Lecture Notes in Physics 170, Springer 1982.

[20] P.L.ROE, The use of the Riemann problem in finite difference schemes, in Proc.7th Int.Conf.Num.Meth.in Fluid Dyn.,eds W.C.Reynolds, R.W.MacCormack, Lecture Notes in Physics 141, Springer 1981.

[21] R.F.WARMING and R.W.BEAM, Upwind second-order difference schemes and applications in aerodynamic flows. AIAA Jul,14,p.1241,1976.

[22] B.GABUTTI, On two upwind finite-difference schemes for hyperbolic equations in non-conservative form. Computers and Fluids, 11, p.207, 1983.

[23] B.ENGQUIST and S.OSHER, One-sided difference approximations for non-linear conservation laws. Math.Comp., 36, p.321, 1981.

[24] A.ISERLES and G.STRANG. The optimal accuracy of difference schemes, Trans.Amer.Math.Soc. to appear.

[25] J.E.FROMM, A method for reducing dispersion in convective difference schemes. J.Comp.Phys., 3, p.176, 1968.

[26] B.van LEER, Towards the ultimate conservative differencing scheme, Part III, J.Comp.Phys., 23, 1976.

[27] G.D.van ALBADA, B.van LEER and W.W.ROBERTS, A comparative study of computational methods in cosmic gas dynamics. ICASE Report 81-24, August 1981.

[28] P.L.ROE and M.J.BAINES, Asymptotic behaviour of some non-linear schemes for linear advection. Proc.6th GAMM Conf.Num.Meth.Fl. Mech. eds.R.Piva, M.Pandolfi, Wieweg, to appear.

[29] P.L.ROE, Some contributions to the modelling of discontinuous flows. Proc.15th AMS-SIAM Summer Seminar in Applied Mathematics, to appear.

[30] P.K.SWEBY, High resolution TVD schemes using flux limiters. Proc.15th AMS-SIAM Summer Seminar in Applied Mathematics, to appear.

[31] M.J.BAINES, 2-Dimensional Shock Recognition and Roe's Scheme Report NA 10/83. Dept.of Mathematics, Univ.of Reading, 1983.

[32] R.D.RICHTMYER and K.W.MORTON, Difference Methods for Initial-Value Problems, Interscience, 1967.

[33] R.COURANT and K-O FRIEDRICHS, Supersonic Flow and Shock Waves, Interscience, 1948.

SOME PRELIMINARY RESULTS USING A LARGE TIME STEP GENERALIZATION OF GODUNOV'S METHOD

RANDALL J. LE VEQUE*

Abstract. It is possible to generalize Godunov's method to larger time steps (Courant number > 1) by "linearizing" shock interactions within each time step. This method is applied to a system of two conservation laws to observe the effects of linearization. We find that for moderate values of the Courant number it performs better than might be expected, giving better resolution of shocks and more accuracy in smooth regions than Godunov's method.

1. Introduction. In [2] a large timestep generalization of Godunov's method was proposed for solving scalar conservation laws in one space dimension. The use of larger timesteps results in considerably less smearing of discontinuities. In Godunov's method each jump discontinuity $u_{j+1}^n - u_j^n$ is propagated according to the Rankine-Hugoniot jump condition to its proper location at time t_{n+1}. The resulting solution is then averaged over intervals $[x_{j-1/2}, x_{j+1/2}]$ to give u_j^{n+1}. The usual Courant number restriction for Godunov's method ensures that the discontinuties do not interact with one another over the course of a single timestep.

In the large timestep generalization, discontinuities are allowed to propagate through several mesh points in each timestep. In general the discontinuities should then interact with one another. In practice we can either attempt to recognize and handle these interactions or we can ignore them. The former approach is clearly superior and for scalar problems it was found that this could be done easily and efficiently through a merging procedure. Unfortunately, that procedure does not generalize directly to systems and it is doubtful whether any such procedure exists that is sufficiently efficient to allow us to handle all interactions correctly. The question then arises as to whether most interactions in a typical problem can be ignored in each timestep, perhaps concentrating our effort on handling interactions between strong discontinuities.

*Mathematics Department, University of California, Los Angeles 90024

Ignoring interactions here means that we allow waves to pass through one another with no change in magnitude or speed just as they would in a linear problem. This method is described more precisely in [3] where it is also rewritten in conservation form (for arbitrarily large Courant number) in order to show that it enjoys the same convergence properties as the standard Godunov method. Any limit function obtained as the mesh is refined is a weak solution to the system of conservation laws. Here the method is described informally by example on a model problem.

Some numerical results are presented here which show that in smooth regions of the flow it may be quite reasonable to ignore interactions, at least with moderate timesteps. Moreover, shock propagation and interactions are handled remarkably well in some cases, often giving sharp and accurate results, although these results are highly variable and depend strongly on both the Courant number and the meshsize used.

An interesting phenomenon is observed in which gross inaccuracies in the solution caused by incorrectly handling interactions at one timestep will often by corrected in later timesteps. This gives further hope that a more sophisticated algorithm, in which the most important interactions are handled explicitly, will indeed be able to deal with realistic problems successfully and efficiently.

2. Numerical results. Computations have been performed only on a model system of two equations obtained by setting $\gamma = 1$ in the Euler equations (see Roe [5]). The equations are

$$\rho_t + m_x = 0 \tag{1}$$

$$m_t + (m^2/\rho + a^2\rho)_x = 0$$

where ρ is the density, m is the momentum, and a is the (constant) sound speed. The eigenvalues of the Jacobian are $m/\rho \pm a$. For simplicity of description we assume that the flow is subsonic everywhere, $|m/\rho| < a$, so that the general solution to a Riemann problem consists of one leftward moving wave and one rightward moving wave, each of which is either a shock or a centered rarefaction wave. The intermediate states and propagation speeds are easily computed for this example. The set of states (ρ,m) which can be connected to a given state (ρ_0,m_0) through a shock satisfy

$$m = \frac{\rho m_0}{\rho_0} \pm a\sqrt{\frac{\rho}{\rho_0}}\,(\rho - \rho_0),$$

the + and - signs corresponding to rightward and leftward moving shocks respectively. The velocity at which a shock propagates is $(m - m_0)/(\rho - \rho_0)$. The set of states which can be connected through a centered rarefaction wave satisfy

$$m = \frac{\rho m_0}{\rho_0} \pm a\rho \ln(\rho/\rho_0).$$

The large timestep algorithm is best described by considering a typical step. We begin by initializing $u_j^{n+1} = u_j^n$ for all j. Now consider a single Riemann problem between x_j and x_{j+1}. Suppose it has been determined that the states u_j^n and u_{j+1}^n are connected through a leftward moving rarefaction wave, an intermediate state u_m, and a rightward moving shock. The intermediate state u_m is obtained by solving for ρ_m and m_m from

$$m_m = \frac{\rho_m m_j}{\rho_j} - a\rho_m \ln(\rho_m/p_j)$$

and

$$m_{j+1} = \frac{\rho_{j+1} m_m}{\rho_m} + a\sqrt{\frac{\rho_{j+1}}{\rho_m}}\,(\rho_{j+1} - \rho_m)$$

The discontinuity $u_m - u_{j+1}^n$ propagates to the right with speed

$$c = \frac{m_{j+1} - m_m}{\rho_{j+1} - \rho_m}.$$

Setting $\mu = \lfloor ck/h \rfloor$, the integer part of the speed times the mesh ratio, we increment the values $u_{j+1}^{n+1}, u_{j+2}^{n+1}, \ldots, u_{j+\mu}^{n+1}$ by $u_m - u_{j+1}^n$ and increment $u_{j+\mu+1}^{n+1}$ by $(ck/h - \mu)(u_m - u_{j+1}^n)$. This is equivalent to propagating the discontinuity to the point $x_{j+1/2} + ck$ and projecting the resulting solution on to the grid by averaging, as in Godunov's method. If the Courant number is less than 1 then $\mu = 0$ and only the value u_{j+1}^{n+1} is affected.

Following [2], the rarefaction is split into several weaker discontinuities which are propagated as entropy-violating shocks. This allows the original discontinuity to spread out over several mesh points and gives a good approximation to the true rarefaction wave provided it is split into sufficiently many pieces. Specifically, we take r pieces where r is some integer roughly proportional to

$$\frac{k}{h}\left|\frac{m_m}{\rho_m} - \frac{m_j}{\rho_j}\right|,$$

the number of grid points the rarefaction wave is spread over at time t_{n+1}. There are several ways to choose the intermediate states $u_{j,i}$ for $i = 1,2,\ldots,r$. One possibility is to let the $\rho_{j,i}$, $j = 1,2,\ldots,r-1$ be equally spaced between ρ_j and ρ_m, setting

$$\rho_{j,i} = \rho_j + \frac{i}{r}(\rho_m - \rho_j),$$

$$m_{j,i} = \frac{\rho_{j,i} m_{j,i-1}}{\rho_{j,i-1}} - a\sqrt{\frac{\rho_{j,i-1}}{\rho_{j,i-1}}}\,(\rho_{j,i} - \rho_{j,i-1})$$

for $j = 1,2,\ldots,r-1$ (with $\rho_{j,0} = \rho_j$) and then determine $\rho_{j,r}$ and $m_{j,r}$ by solving

$$m_{j,r} = \frac{\rho_{j,r} m_{j,r-1}}{\rho_{j,r-1}} - a\sqrt{\frac{\rho_{j,r}}{\rho_{j,r-1}}}\,(\rho_{j,r} - \rho_{j,r-1})$$

and

$$m_{j+1} = \frac{\rho_{j+1} m_{j,r}}{\rho_{j,r}} + a\sqrt{\frac{\rho_{j+1}}{\rho_{j,r}}}\,(\rho_{j+1} - \rho_{j,r}).$$

Figure 1 shows these states in the phase plane (for $r = 3$), indicating how they relate to the true solution.

Note that in general $u_{j,r} \neq u_m$ since the shock and rarefaction curves do not coincide in the phase plane. In fact, it is the shock $u_{j,3} - u^n_{j+1}$ which should be propagated to the right rather than $u_m - u^n_{j+1}$. Assuming this is done, we are using an exact weak solution to the conservation laws and the resulting scheme is conservative. This weak solution does not satisfy the entropy condition but approximates the true rarefaction and, with the additional smearing due to averaging, should give a satisfactory solution.

As a numerical example, the following initial conditions have been used:

$$u(x,0) = \begin{cases} u_1, & 0 \leq x < 0.1, \\ u_2, & 0.1 \leq x < 0.5, \\ u_4, & 0.5 \leq x < 0.98, \\ u_5, & 0.98 \leq x < 1, \end{cases}$$

where

$$u = \begin{bmatrix} 0.5 \\ 0.111803 \end{bmatrix}, \quad u_2 = \begin{bmatrix} 0,4 \\ 0 \end{bmatrix}, \quad u_4 = \begin{bmatrix} 0.2 \\ 0 \end{bmatrix}, \quad u_5 = \begin{bmatrix} 0.35 \\ -0.19843 \end{bmatrix}.$$

We take $a = 1$. The states u_1 and u_2 are separated by a shock moving to the right, u_4 and u_5 by a shock moving to the left. The states u_2 and u_4 are separated by a rarefaction moving to the left, an intermediate state.

$$u_3 = \begin{bmatrix} 0.28260 \\ 0.098185 \end{bmatrix},$$

and a shock moving to the right. We compare computed and exact solutions at time 0.16, before any interaction has occured, and at time 0.32, after the shock separating u_1 and u_2 has interacted with the rarefaction separating u_2 and u_3 and the two shocks in the right half of the interval have also interacted (see Fig.2).

For this problem the Courant number is roughly $\nu \approx 1.5\lambda$, where $\lambda = k/h$. Figures 3 and 4 show computations with $h = 1/50$ and various values of λ. For $\lambda = 0.5$, $\nu < 1$ and we have Godunov's method. Note the excessive smearing of shocks. Taking $\lambda = 1$ gives a dramatic improvement. Another slight improvement is seen in going to $\lambda = 2$. For $\lambda = 4$ and 8 the results at $t = 0.16$ continue to improve but the interactions are handled poorly and the results at $t = 0.32$ are completely incorrect in some regions.

The same computations with $h = 1/100$ reveal an interesting phenomenon. With this smaller value of h the results with $\lambda = 4$ (Fig. 5a,b) are much better while the results for $\lambda = 8$ (Fig. 5c,d) have also improved and now look very similar to the previous results with $\lambda = 4$.

This will be explained in the next section, where we will see that errors caused by ignoring interactions tend to correct themselves in later timesteps, so that in this simple problem the accuracy is determined in part by the number of steps taken since the interaction. Since with fixed λ reducing the timestep increases the number of steps taken since the interaction, and thus increases the amount of "self-correction" which has taken place, a substantial improvement in the solution is seen.

This argument is valid in the right half of the interval, where only a single interaction takes place in the true solution, but comparing Figures 4b and 5b shows shows that the shock-rarefaction interaction is also handled more successfully, even though in this region new interactions occur (and are incorrectly handled) in every timestep.

Smooth solutions. As another test of the algorithm we have made a more quantitative comparison of the accuracy obtained with different Courant numbers on smooth solutions. The initial conditions used are

$$\rho(x,0) = 0.2 + 0.3\exp(-10(x-0.5)^2)$$

$$m(x,0) = -0.1\sin(2\pi x)$$

for which the Courant number is again roughly $\nu \approx 1.5\lambda$. The errors at time $t = 0.08$ are shown in Table 1. These show that for Courant numbers larger than 1 the method remains first order accurate. Moreover, in many cases the results obtained with large Courant numbers are in fact better than those obtained with smaller values. Some explanation of this will also be given in the next section.

These positive numerical results lead to the conjecture that the results obtained by the algorithm described here converge to the true solution as $k, h \to 0$ for any fixed value of the Courant number. This is demonstrated in [3] under the assumption that the total variation is bounded.

TABLE 1

Max norm of errors in smooth solution at $t = 0.08$. Errors in ρ and m are shown.

	h = 1/25	h = 1/50	h = 1/100
$\lambda = 0.5$	7.831(-3) 4.062(-3)	4.479(-3) 2.118(-3)	2.118(-3) 8.485(-4)
$\lambda = 1.0$	6.676(-3) 5.157(-3)	3.531(-3) 2.819(-3)	1.259(-3) 1.047(-3)
$\lambda = 2.0$	7.041(-3) 4.803(-3)	3.836(-3) 2.392(-3)	1.042(-3) 8.598(-4)
$\lambda = 4.0$		4.575(-3) 3.436(-3)	1.743(-3) 1.291(-3)

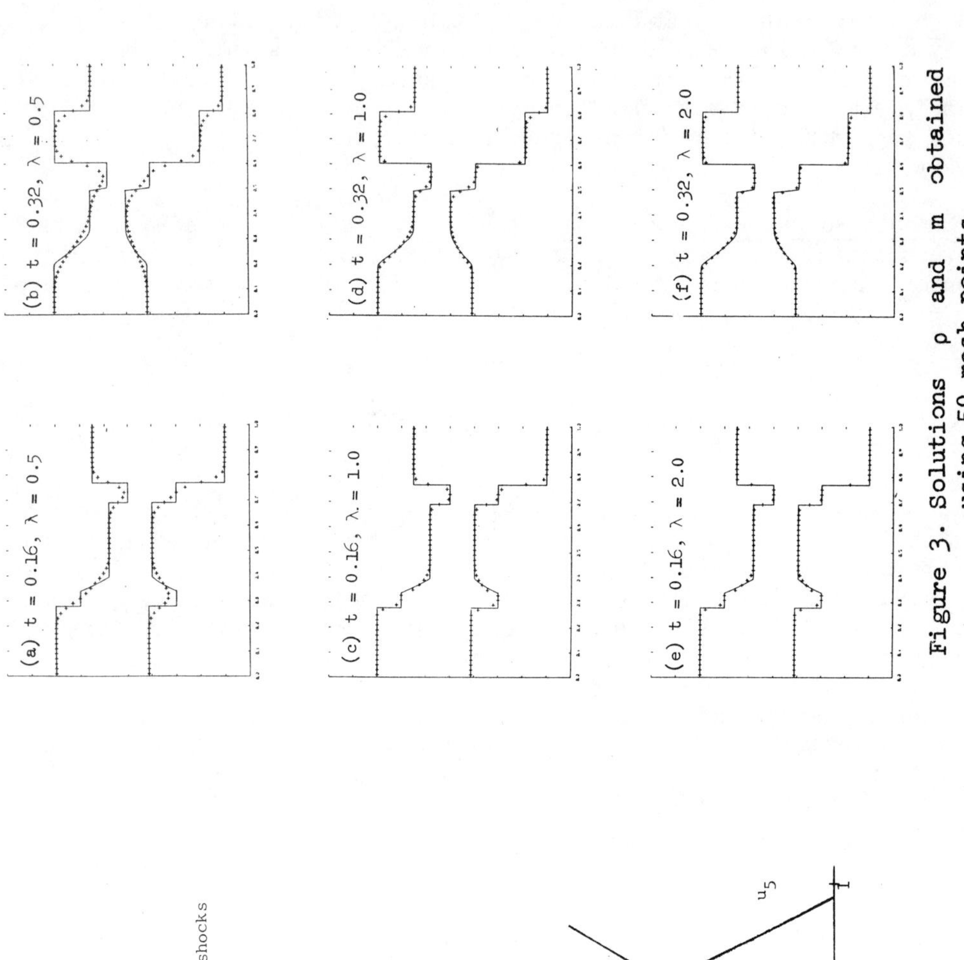

Figure 3. Solutions ρ and m obtained using 50 mesh points.

Figure 1

Figure 2

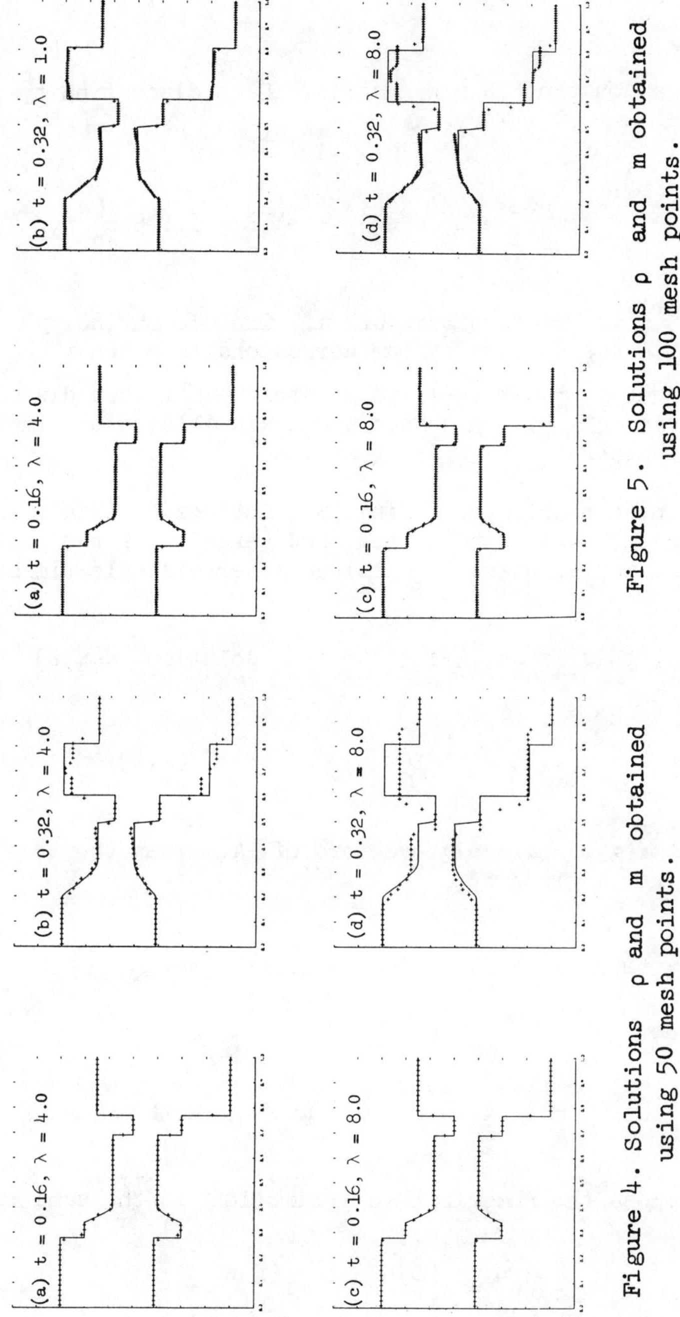

Figure 4. Solutions ρ and m obtained using 50 mesh points.

Figure 5. Solutions ρ and m obtained using 100 mesh points.

3. Analysis. First consider the effect of applying this algorithm to a linear system

$$u_t + Au_x = 0$$

where A is a constant $N \times N$ matrix. Each discontinuity $u_{j+1}^n - u_j^n$ is decomposed as

$$u_{j+1}^n - u_j^n = e_j^{(1)} + e_j^{(2)} + \dots + e_j^{(N)}$$

where the $e_j^{(i)}$ are eigenvectors of A. The discontinuity $e_j^{(i)}$ then propagates at a speed given by the corresponding eigenvalue μ_i. If the Courant number $\frac{k}{h} \max |\mu_i|$ is greater than 1 then discontinuities from neighboring Riemann problems are again allowed to simply pass through one another.

For the linear problem, ignoring interactions in this manner is in fact the correct way to handle them, and (except for the projection process) the exact solution is obtained after a single timestep of any length.

To see that this is so, write the true solution $u(x,t)$ has

$$u(x,t) = r^{(1)}(x,t) + \dots + r^{(N)}(x,t)$$

where the $r^{(i)}(x,t)$ are eigenvectors of A. Then the true solution at time $t+k$ is given by

$$u(x,t+k) = r^{(1)}(x-\mu_1 k,t) + \dots r^{(N)}(x-\mu_N k,t) .$$

In other words,

$$r^{(i)}(x,t+k) = r^{(i)}(x-\mu_i k,t).$$

We will decompose the numerical approximation in the same manner,

$$u_j^n = r_j^{(1)n} + \dots + r_j^{(N)n} .$$

Since our algorithm handles each eigenvector separately, it is sufficient to look at a single eigenvector, say $r^{(i)}$. For concreteness suppose $\mu_i > 0$ and set $\mu = \lfloor \mu_i k/h \rfloor$. Then from each grid point x_j the jump $e_j^{(i)}$ in the ith eigenvector propagates through μ mesh points, and part way through another. Turning this around and looking at what increments a fixed grid point receives when the algorithm is applied everywhere we find that

$$r_j^{(i)n+1} = r_j^{(i)n} - e_{j-1}^{(i)} - e_{j-2}^{(i)} - \ldots - e_{j-\mu}^{(i)} - (\mu_i k/h - \mu) e_{j-\mu-1}^{(i)}$$

$$= (1 - (\mu_i \frac{k}{h} - \mu)) r_{j-\mu}^{(i)n} + (\mu_i \frac{k}{h} - \mu) r_{j-\mu-1}^{(i)n}.$$

So $r_j^{(i)n+1}$ is obtained by linearly interpolating between $r_{j-\mu}^{(i)n}$ and $r_{j-\mu-1}^{(i)n}$ and hence is an $O(h^2)$ approximation to $r^{(i)}(x-\mu_i k, t)$. This is true for each of the eigenvectors and so the algorithm is simply the method of characteristics with linear interpolation.

For linear problems the only error is due to the projection (i.e. interpolation) process and so it is best to take very large timesteps, thus reducing the number of interpolations performed.

This behavior on the linear problem gives some indication of why the large timestep algorithm computes smooth solutions to nonlinear problems as well as it does. In a smooth solution the eigenvectors are locally nearly constant and the characteristics are nearly straight lines. There is a tradeoff between reducing the errors due to the nonlinearly by taking λ small, and reducing the errors due to interpolation by taking λ large. The optimal λ will depend on the deviation from linearity.

<u>Shock interactions</u>. In order to analyze the manner in which the algorithm handles shock interactions we return to the linear problem and view interactions there from a different standpoint. Consider a system with 2 variables which we again denote by ρ and m. Take initial conditions with two discontinuities:

$$u(x,0) = \begin{cases} u_1, & x < x_1, \\ u_3, & x_1 < x < x_2, \\ u_5, & x_2 < x. \end{cases}$$

The true solution is shown in Fig. 6.

If the u_j are decomposed into eigenvectors $r_j^{(1)}, r_j^{(2)}$ corresponding to the eigenvalues $\mu_1 < \mu_2$,

$$u_1 = r_1^{(1)} + r_1^{(2)},$$

$$u_3 = r_3^{(1)} + r_3^{(2)},$$

$$u_5 = r_5^{(1)} + r_5^{(2)},$$

then the intermediate states u_2, u_4 and $\tilde{u}_3$ are given by

$$u_2 = r_3^{(1)} + r_1^{(2)},$$

$$u_4 = r_5^{(1)} + r_3^{(2)},$$

$$\tilde{u}_3 = r_5^{(1)} + r_1^{(2)}.$$

The fact that the interaction is handled correctly is a conseqence of the fact that

$$\tilde{u}_3 = u_3 - (u_3 - u_2) + (u_4 - u_3),$$

the latter two quantities being the increments u_3 receives in the overlap region after propagating the discontinuities. Fig. 7 shows these states in the ρ-m plane. Adjacent states are connected by eigenvectors.

The fact that $\tilde{u}_3 = u_2 + u_4 - u_3$ will be expressed by saying that $\tilde{u}_3$ is the linear reflection of u_3 through the collision of u_2 and u_4.

For the nonlinear problem with similar initial conditions the solution is shown in Fig. 8 (for convenience we take $u_1 = u_2$ and $u_4 = u_5$). This corresponds to Fig. 9 in the ρ-m plane.

Because of the nonlinearity, the shock speeds change after the collision of u_2 and u_4 and the state $\tilde{u}_3$ is no longer the linear reflection of u_3. The large timestep algorithm, by propagating the discontinuities through one another, ignores these facts. After one timestep it produces the solution shown in Fig. 10, where

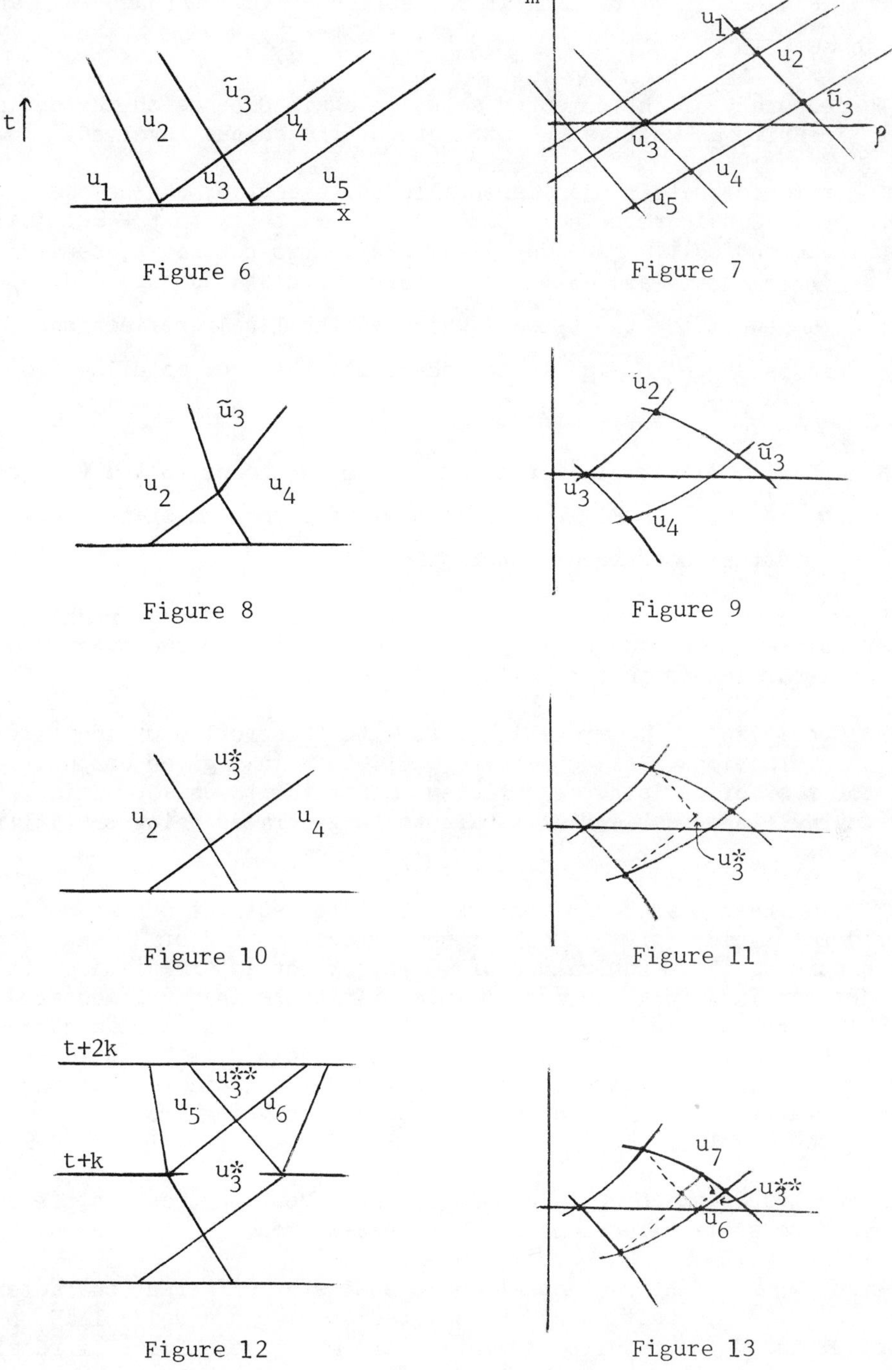

Figure 6

Figure 7

Figure 8

Figure 9

Figure 10

Figure 11

Figure 12

Figure 13

$u_3^* = u_2 + u_4 - u_3$ is the linear reflection of u_3 through u_2 and u_4 as seen in Fig. 11.

The accuracy of the computed solution again depends on the deviation from linearity as well as the strength of the shocks involved.

The self-correction phenomenon alluded to earlier can now be explained. Consider the next timestep, from $t + k$ to $t + 2k$. Each of the discontinuities present give rise to two new waves, separated by new intermediate states u_5 and u_6. When states u_5 and u_6 collide a new state u_3^{**} appears which is the linear reflection of u_3^*. This is shown in Fig. 12 together with the true solution from Fig. 8 as a dotted line.

From Fig. 13 it is clear that u_3^{**} is much closer to the true state $\tilde{u}_3$ than was u_3^*. In Fig. 4b one can clearly see the states u_5, u_3^{**} and u_6 in the shock-shock interaction.)

Further correction-waves are generated in later steps, pushing the intermediate state even closer to $\tilde{u}_3$ and restoring the outer shocks to their correct trajectories.

4. Conclusions. We are still left with the problem of identifying which interactions must be handled explicitly in a given computation and the task of devising an efficient algorithm to do so. This is necessary if dependable results are to be generated using truly large timesteps.

On the other hand, it may prove better to restrict our attention to moderate values of the Courant number, say $\nu = 2$ or 3, and ignore interactions. For such values of ν incorrect intermediate states will be confined to a few mesh points. If these inaccuracies are quickly corrected in subsdquent timesteps this may be a reasonable algorithm. Our goal of maintaining sharp shocks will be at least partly accomplished since the numerical results indicate that the most dramatic improvement in the solution occurs in going from $\nu < 1$ to $\nu \approx 1.5$.

Several further comments should be made about the results presented here and possible directions for future research.

Even ignoring interactions between shocks of different characteristic families, the propagation of an isolated shock can already lead to difficulties. The results presented here may give a false impression as to the ease with which this is handled. In general the shock is smeared over at least two intervals in order to represent it on the fixed grid. The intermediate state introduced in this manner is a convex combination of the left and right states, and in general will not lie on the Hugoniot curve between those states. In the next timestep each of the resulting discontinuities will generate waves of both characteristic families. The ones going backwards apparently cancel one another out. (The numerical evidence for this is supported by considering their positions in the phase plane.) Of the waves moving forward, the trailing shock travels faster than the leading shock and, for large Courant numbers, may end up several grid points in front, causing an unphysical smearing of the shock (see Fig. 4.1 in [2]). This has not happened in our examples because of the small relative difference in shock velocities (due to our choice of a subsonic example) and because only moderate values of the Courant number have been considered.

Further difficulties may appear when this method is applied to the full Euler equations, particularly in handling contact discontinuities. One expects that contact discontinuities will still tend to smear over time, lacking the restoring forces of shocks, but that at larger Courant number the smearing will proceed more slowly due to the reduction in the number of steps and hence the number of projections it undergoes.

Many of the difficulties stemming from averaging the solution at each timestep may be best avoided by eliminating the fixed grid altogether and representing the solution at each time by a list of discontinuities and their positions. This is almost certainly the best approach for scalar problems, but for a system of N equations each discontinuity may split into N new discontinuities at every timestep. It will be necessary to merge discontinuities in order to avoid an exponential growth in the amount of information retained.

We note that in general it is impossible to merge two discontinuities into a single discontinuity in a conservative manner but that it is always possible to merge an arbitrary number of discontinuities at points $x_1, x_2, \ldots, x_r$ into two discontinuities at the points x_1 and x_r. The correct intermediate state is obtained simply by averaging the original solution (i.e. integrating between x_1 and x_r). By ensuring that strong discontinuities always lie at the endpoint of some integration interval it may be possible to avoid a great deal of smearing and many of the associated problems. Harten and Hyman [1] have used a similar approach with good results.

For the model system considered here the exact solution to the Riemann problem was always used, except in rarefaction waves. For practical problems it may be desirable to use approximate Riemann solvers such as those advocated by Roe [4]. Each discontinuity is then split into eigenvectors of some locally defined matrix A which are propagated at velocities given by the corresponding eigenvalues. For the model system (1), Roe [5] recommends using the matrix

$$\tilde{A}(u_j, u_{j+1}) = \begin{bmatrix} 0 & 1 \\ a^2 - v^2 & 2v \end{bmatrix}$$

where v is the weighted average of velocities

$$v = \frac{m_j^{1/2}/\rho_j^{1/2} + m_{j+1}^{1/2}/\rho_{j+1}^{1/2}}{\rho_j^{1/2} + \rho_{j+1}^{1/2}}$$

In [3] it is shown that with large time steps the method can still be written in conservation form even when an approximate Riemann solver is used, although the risk of converging to a solution which does not satisfy the entropy condition is increased.

The use of an approximate Riemann solver leads to the calculation of incorrect intermediate states and propagation speeds for each discontinuity at every timestep. This shows up most clearly when larger timesteps are used but it is true in all calculations. It appears that these incorrect states are automatically corrected in later timesteps in much the same way as large timestep interactions are corrected. This suggests a close connection between approximate solutions to the Riemann problem and our approximate handling of interactions.

Acknowledgements. Much of this work was performed during a visit to the Mathematics Department of the University of Reading as a National Science Foundation Postdoctoral Fellow. I would like to express my appreciation to everybody there for their hospitality and particularly to M.J. Baines, K.W. Morton and P.L Roe of the RAE for many stimulating discussions.

REFERENCES

[1] A. HARTEN and J.M. HYMAN, A self-adjusting grid for the computa- of weak solutions of hyperbolic conservation laws: I. One-dimensional problems, Los Alamos report LA-9105, 1981.

[2] R.J. LEVEQUE, Large time-step shock capturing techniques for scalar conservation laws, SIAM J. Num. Anal. 19 (1982), pp. 1091-1109.

[3] R.J. LEVEQUE, Convergence of a large time step generalization of Godunov's method for conservation laws, to appear in Comm. Pure Appl. Math.

[4] P.L. ROE, Approximate Riemann solvers, parameter vectors, and difference schemes, J. Comp. Phys. 43 (1981), pp. 357-372.

[5] P.L. ROE, Numerical modelling of shock waves and other discontinuities, IMA Conference on Numerical Methods in Computational Aerodynamics, University of Reading, 1981.

FLUX LIMITERS

P. K. SWEBY*

Abstract. It is well known that first order accurate difference schemes for the numerical solution of conservation laws produce results which suffer from excessive numerical diffusion, whilst classical second order schemes, although giving better resolution, suffer from spurious oscillations. Recently much effort has been put into achieving high resolution without these oscillations, using a variety of techniques. Here we outline one class of such methods, that of flux limiting, together with the TVD constraint used to ensure oscillation free solutions. Brief numerical comparisons of different limiting functions are also presented.

1. Introduction. In recent years many authors have proposed a variety of techniques for obtaining high resolution schemes for hyperbolic conservation laws which do not suffer from the spurious oscillations, at discontinuities, which plague the more classical second or higher order accurate schemes such as Lax-Wendroff [6]. These techniques have ranged from the post-processing of first order results which suffer from excessive diffusion (e.g. Boris and Book's FCT [2], van Leer's averaging [9] and Roe's transfer functions [13], [15] to the pre-processing of data or data preparation, before applying a first order scheme (e.g. van Leer's MUSCL [10], LeVeque and Goodman's version of this [8] and Harten's application of a first order scheme to a modified flux function [5]).

In [20], [21] a framework was constructed which encompassed most (although not all) of the post-processing techniques in the form of flux limiters. It was shown how various of these post-processing techniques fitted into this framework and some simple numerical comparisons were given which demonstrated how a pertinent choice of limiter could greatly improve results.

* Institute for Computational Fluid Dynamics, Oxford University Computing Laboratory.
Part of this work was carried out by the author whilst at the University of California, Los Angeles, supported by NASA grant NAG1-273, the remainder of the work being carried out at Reading University while supported by the SERC.

In this paper we briefly recap, in Section 2, the flux limiter framework and the Total Variation Diminishing (TVD) criterion used to guarantee the lack of spurious oscillations. In Section 3 we give details of some specific limiters and present some more demanding numerical comparisons than those presented in [20], [21]. Finally in Section 4 a few concluding comments are made.

2. General Flux Limiting. In this section we briefly describe the framework of flux limiters constructed in [20], [21], the description of specific limiters being left to the next section. We begin by considering explicit approximations to the scalar conservation law

$$u_t + f(u)_x = 0 \quad . \tag{2.1}$$

In particular we confine our attention initially to 3-point first order accurate schemes, on a regular mesh, which we write in conservation form

$$u^k = u_k - \lambda(h_{k+\frac{1}{2}} - h_{k-\frac{1}{2}}) \quad , \tag{2.2}$$

where

$$\begin{aligned} h_{k+\frac{1}{2}} &= h(u_{k+1}, u_k) \\ \text{and} \quad & \\ h(u,u) &= f(u) \end{aligned} \tag{2.3}$$

is a consistent numerical flux function, λ is the usual mesh ratio

$$\lambda = \frac{\Delta t}{\Delta x} \tag{2.4}$$

and u_k are the piecewise constant values of the function $u_{\Delta x}(x,t)$ approximating $u(x,t)$ on the mesh. We use here the now common shorthand

$$u^k \equiv u_k^{n+1} \quad , \qquad u_k \equiv u_k^n \quad .$$

(Note that the assumption of a regular mesh is for ease of notation only; see below).

For each computational cell (x_k, x_{k+1}) we define the flux differences

$$\begin{aligned} (\Delta f_{k+\frac{1}{2}})^+ &= -(h_{k+\frac{1}{2}} - f(u_{k+1})) \\ (\Delta f_{k+\frac{1}{2}})^- &= (h_{k+\frac{1}{2}} - f(u_k)) \quad , \end{aligned} \tag{2.5}$$

noting that

$$(\Delta f_{k+\frac{1}{2}})^+ + (\Delta f_{k+\frac{1}{2}})^- = \Delta f_{k+\frac{1}{2}} \tag{2.6}$$

(the notation $\Delta g_{k+\frac{1}{2}} = g_{k+1} - g_k$ is used throughout), and local CFL numbers

$$\nu^{+}_{k+\frac{1}{2}} = \lambda\frac{(\Delta f_{k+\frac{1}{2}})^{+}}{\Delta u_{k+\frac{1}{2}}} \tag{2.7a}$$

$$\nu^{-}_{k+\frac{1}{2}} = \lambda\frac{(\Delta f_{k+\frac{1}{2}})^{-}}{\Delta u_{k+\frac{1}{2}}} \tag{2.7b}$$

with

$$\nu_{k+\frac{1}{2}} = \nu^{+}_{k+\frac{1}{2}} + \nu^{-}_{k+\frac{1}{2}} = \lambda\frac{\Delta f_{k+\frac{1}{2}}}{\Delta u_{k+\frac{1}{2}}} \quad . \tag{2.7c}$$

We pause here to note that, if the numerical flux $h_{k+\frac{1}{2}}$ is that of an E-scheme (Osher [11]), characterised by

$$\mathrm{sgn}(\Delta u_{k+\frac{1}{2}})[h_{k+\frac{1}{2}} - f(u)] \leqq 0 \tag{2.8}$$

for all u between u_k and u_{k+1}, then we have

$$\begin{aligned} \nu^{+}_{k+\frac{1}{2}} &\geqq 0 \\ \nu^{-}_{k+\frac{1}{2}} &\leqq 0 \quad , \end{aligned} \tag{2.9}$$

so justifying the superscripts.

In [11] Osher showed that semi-discrete E-schemes (i.e. where the time derivative of (2.1) is not discretised) satisfy an entropy inequality whilst Tadmor [23] has shown entropy satisfaction of fully-discrete E-schemes for the restrictive CFL condition of 1/2. Such entropy satisfaction ensures that the numerical scheme produces the correct physical solution, an important feature especially when used on problems involving expansion waves.

A crucial measure involved in convergence proofs of difference schemes approximating (2.1) is a uniform bound on the total variation of the solution

$$TV(u^{n+1}) = \sum_{k} |u^{n+1}_{k+1} - u^{n+1}_{k}| \tag{2.10}$$

(see e.g. [7], [17], [22]), and so an important class of difference schemes is those which are Total Variation Diminishing (TVD) (see Harten [5]), i.e.,

$$TV(u^{n+1}) \leqq TV(u^{n}) \quad , \tag{2.11}$$

for which the variation is then bounded by that of the initial data.

A more immediate practical consequence of TVD, however, is that it implies the absence of spurious oscillations in the solutions, such as those generated by the Lax-Wendroff and other classical second order accurate schemes near discontinuities of the solution.

If a difference scheme is written in the form

$$u^k = u_k - C_{k-\frac{1}{2}}\Delta u_{k-\frac{1}{2}} + D_{k+\frac{1}{2}}\Delta u_{k+\frac{1}{2}} \tag{2.12}$$

where $C_{k-\frac{1}{2}}$ and $D_{k+\frac{1}{2}}$ may be data dependent (i.e. functions of the set $\{u_k\}$), then it is easily shown ([5],[22]) that sufficient conditions for the scheme to be TVD are the inequalities

$$0 \leqq C_{k+\frac{1}{2}}, \quad 0 \leqq D_{k+\frac{1}{2}}, \quad C_{k+\frac{1}{2}} + D_{k+\frac{1}{2}} \leqq 1. \tag{2.13}$$

We notice from the definitions (2.5) that

$$h_{k+\frac{1}{2}} - h_{k-\frac{1}{2}} = (\Delta f_{k+\frac{1}{2}})^- + (\Delta f_{k-\frac{1}{2}})^+ \tag{2.14}$$

and therefore one way of writing (2.2) in the form (2.12) is by setting

$$C_{k-\frac{1}{2}} = \nu^+_{k-\frac{1}{2}} \quad \text{and} \quad D_{k+\frac{1}{2}} = -\nu^-_{k+\frac{1}{2}} \quad . \tag{2.15}$$

Observe that for an E-scheme, the TVD inequalities (2.13) are then satisfied if

$$\nu^+_{k+\frac{1}{2}} - \nu^-_{k+\frac{1}{2}} \leqq 1 \quad . \tag{2.16}$$

In general we shall assume that a given scheme is TVD under the CFL-like condition

$$\sup_{\xi}(\lambda|f'(\xi)|) \leqq \mu \leqq 1. \tag{2.17}$$

We now turn our attention to second order accurate schemes, using first (for simplicity) the linear advection equation

$$u_t + au_x = 0, \quad a = \text{constant} > 0. \tag{2.18}$$

In [20], [21] it was shown how the second order Lax-Wendroff [6] scheme could be constructed by adding an anti-diffusive flux to the simple first order upwind scheme, viz

$$u^k = \underbrace{u_k - \nu\Delta u_{k-\frac{1}{2}}}_{\text{first order upwind}} \; - \underbrace{\Delta_-(\tfrac{1}{2}(1-\nu)\nu\Delta u_{k+\frac{1}{2}})}_{\text{anti-diffusive flux}} \tag{2.19}$$

where $\nu = \lambda a$. It was then shown that, by limiting the amount of this anti-diffusive flux added, it is possible to obtain a class of second order accurate TVD schemes. The procedure of limiting this anti-diffusive flux was used by Boris and Book [2] and Zalesak [27] in their Flux Corrected Transport (FCT); however, this was a two-step procedure utilising u values at the intermediate step to determine the limitation of the anti-diffusive flux, whilst the flux limiters described in [20], [21] form part of an essentially one-step scheme, the limiters being functions only of data at time level n. We now describe this method of flux limitation, outlining the main points of [20], [21], the interested reader being referred to the original papers for full details.

The anti-diffusive flux (the second term on the right hand side of (2.19)) is limited via a limiting function ϕ_k,

$$-\Delta_-\{\phi_k \tfrac{1}{2}(1-\nu)\nu\Delta u_{k+\frac{1}{2}}\} \quad , \tag{2.20}$$

which is chosen to be a function of consecutive gradients of the data, i.e.

$$\phi_k = \phi(r_k) \quad \text{where} \quad r_k = \frac{\Delta u_{k-\frac{1}{2}}}{\Delta u_{k+\frac{1}{2}}} \quad . \tag{2.21}$$

This choice of argument for ϕ is in line with the approach used by van Leer [9], Roe [13] and others, whose limiters will be described in the next section.

On writing the scheme using (2.20) in the form of (2.12), with

$$\begin{aligned} C_{k-\frac{1}{2}} &= \nu^+_{k-\frac{1}{2}}(1 + \tfrac{1}{2}(1-\nu)[\phi(r_k)/r_k - \phi(r_{k-1})]) \\ D_{k+\frac{1}{2}} &\equiv 0 \quad , \end{aligned} \tag{2.22}$$

it is easily shown that such a scheme is TVD if

$$0 \leqq \left\{\frac{\phi(r)}{r} \quad \text{and} \quad \phi(r)\right\} \leqq 2, \quad \text{for all} \quad r. \tag{2.23}$$

The extra constraint of second order accuracy is imposed by requiring that the amount of limited flux added to the first order scheme be a convex average of that of the Lax-Wendroff scheme (obtained by $\phi \equiv 1$) and that of the second order upwind scheme of Warming and Beam [24] (obtained by $\phi \equiv r$). This restricts the choice of $\phi(r)$ to the region shown in Figure 1.

It is interesting to note here that numerical experiments with limiters which, although TVD, do not lie in the region of Figure 1 produce marked "squaring" of smooth data, a phenomenon encountered when using some forms of FCT.

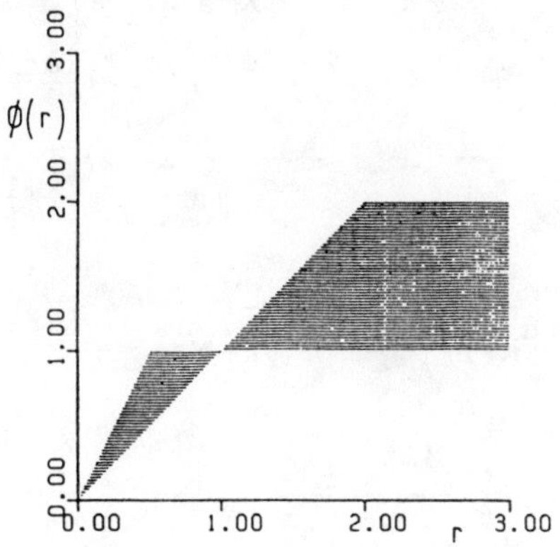

The second order TVD region

Figure 1.

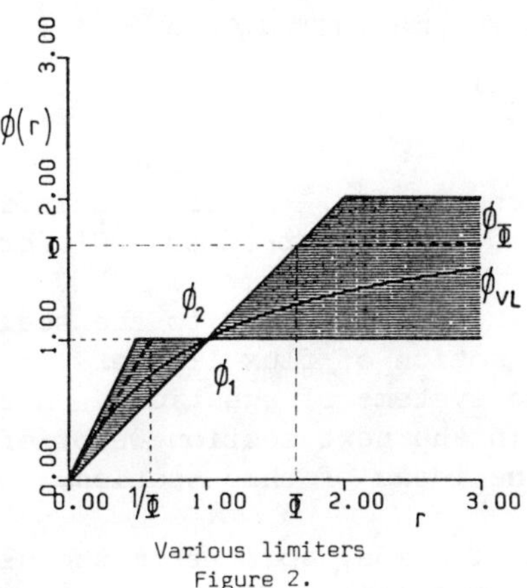

Various limiters

Figure 2.

This limiting of anti-diffusive flux is easily extended to the general non-linear equation (2.1) by replacing the constant ν by local CFL numbers and taking into account the direction of the flow (thus ensuring that the resulting scheme is always an average of a centered and an upwind scheme). In full:

$$u^k = u_k - \lambda\Delta_- h_{k+\frac{1}{2}} - \lambda\Delta_-\{\phi(r_k^+)\alpha^+_{k+\frac{1}{2}}(\Delta f_{k+\frac{1}{2}})^+ \tag{2.24a}$$

$$- \phi(r^-_{k+1})\alpha^-_{k+\frac{1}{2}}(\Delta f_{k+\frac{1}{2}})^-\} \tag{2.24b}$$

where

$$\alpha^\pm_{k+\frac{1}{2}} = \tfrac{1}{2}(1 \mp \nu^\pm_{k+\frac{1}{2}}) \tag{2.25}$$

and

$$r_k^\pm = \frac{\alpha^\pm_{k\mp\frac{1}{2}}(\Delta f_{k\mp\frac{1}{2}})^\pm}{\alpha^\pm_{k\pm\frac{1}{2}}(\Delta f_{k\pm\frac{1}{2}})^\pm} \tag{2.26}$$

Here $h_{k+\frac{1}{2}}$ is the numerical flux of any first order scheme, assumed to at least satisfy (2.9) (and therefore be TVD) although ideally it should be that of an E-scheme. The revised definition of the ratios $r_k^\pm$ ensures that the limiter scheme is an average of the Lax-Wendroff and Warming and Beam schemes.

Writing the scheme (2.24) in the form (2.12) in a similar manner to (2.22), now gives us the CFL restriction

$$\sup_\xi (\lambda|f'(\xi)|) \leq \left(\frac{2}{2+\Phi}\right) \mu \leq \frac{2}{3} \tag{2.27}$$

to guarantee that the scheme is TVD, where

$$0 \leq \frac{\phi(r)}{r}, \quad \phi(r) \leq \Phi \quad (\leq 2) \tag{2.28}$$

and the first order scheme used, (2.24a), satisfies the CFL condition (2.17). Again the reader is referred to [20] for full details.

So far attention has been directed to the scalar equation on a regular grid, but the notion of flux limiters is readily extended both to irregular grids and systems of equations. Before proceeding to look at specific limiters in the next section we briefly outline these and other extensions to the ideas of this section.

For irregular spatial grids, apart from the usual care needed to maintain conservation, the main difference is that the ratio $r_k^\pm$

should take into account the varying grid. This is achieved by defining

$$r_k^+ = \frac{\lambda_{k-\frac{1}{2}} \alpha^+_{k-\frac{1}{2}} (\Delta f_{k-\frac{1}{2}})^+}{\lambda_{k+\frac{1}{2}} \alpha^+_{k+\frac{1}{2}} (\Delta f_{k+\frac{1}{2}})^+} \tag{2.29}$$

and similarly r_k^- .

The simplest (but not the only) extension to systems of equations is to use the Roe Approximate Riemann Solver [14], which decomposes the system into m scalar equations corresponding to the characteristic fields of the system. The ratios $r_k^{\pm}$ are then defined in terms of the scalar flux differences associated with each field. Usually this is achieved by comparing values of the first components of the vectors for each field (e.g. the density component in the case of the Euler equations) but it is necessary to beware of fields whose eigenvector has zero first component. As we shall see in Section 3 (see also [20], [21]) different limiters may be used on different fields, sometimes to advantage.

Two (or more) dimensions can also be dealt with using either dimensional-splitting or truly two-dimensional techniques (see Baines [1]). It should be noted though that a result by Goodman and LeVeque [4] states that TVD is only attainable in two or more dimensions by first order schemes; however, we can still hope to gain advantages by using flux limited schemes.

Finally in this section we mention implicit methods. Because of the non-linear nature of flux limiter schemes (even for the linear equation (2.18)) they cannot be differentiated and used in conventional implicit solvers. However, for steady-state problems, limiters may be incorporated into non-conservative in time implicit methods (such as Harten's LNI [26]) to reach steady-state via artificial time. In such instances $\alpha^{\pm}$ in (2.25) is redefined as

$$\alpha^{\pm}_{k+\frac{1}{2}} \equiv 1/2$$

to remove dependence of the steady-state solution on the "time" step(s) used.

3. Limiters and Comparisons. We now give details of a few specific limiters which have been proposed and used by various authors and then present a few numerical comparisons.

In 1974 van Leer [9] proposed the combination of non-conservative, limited versions of the Lax-Wendroff and Warming and Beam schemes to yield a conservative, "limited" scheme which satisfied a local maximum principle. When reformulated in the notation of Section 2 (see [20] for details) the equivalent limiter is given by

$$\phi_{VL}(r) = \frac{|r|+r}{|r|+1}$$

or

(3.1)
$$\phi_{VL}(r) = \begin{cases} \dfrac{2r}{1+r} & r > 0 \\ 0 & r \leqq 0 \end{cases} .$$

which is a smooth curve in the second order TVD region of Figure 1 and is shown in Figure 2.

In the late 1970's Roe [13], [16], [19] proposed a second order accurate monotonicity preserving scheme using a new incremental approach. This involved the allocation of cell flux differences to update the values of u at the ends of the cells (giving a first order scheme) followed by the transfer of a fraction of this increment across the cell to achieve second order. This transfer was taken to be the lesser of the moduli of the anti-diffusive fluxes in adjacent cells. This recipe became known as the "minmod" transfer function or limiter, given by

(3.2)
$$\phi_1(r) = \mathrm{Max}(0,\mathrm{Min}(1,r)) .$$

(The reason for the notation ϕ_1 will become apparent soon.) Several other authors (e.g. Harten [5], LeVeque and Goodman [8] and Chakravarthy and Osher [3] have since used this limiter in one guise or another to obtain high resolution TVD schemes by a variety of methods. As can be seen in Figure 2, ϕ_1 corresponds to the bottom boundary of the second order TVD region and is the least compressive of possible limiters, in fact its effect is to switch directly between the Lax-Wendroff and Warming and Beam schemes depending on the value of r.

More recently Roe [15] has experimented with various transfer functions (or B-functions in his terminology), proposing his highly compressive "superbee" transfer function, which when written as a limiter is expressed as

(3.3)
$$\phi_2(r) = \mathrm{Max}(0,\mathrm{Min}(2r,1),\mathrm{Min}(2,r))$$

and corresponds to the upper boundary of the TVD region (see Figure 2). Roe has illustrated [15] that, for the linear advection equation (2.18), schemes using this transfer function will convect discontinuities of the data for many thousands of timestep without much loss of definition (also see below).

Finally it was shown in [20] that both the minmod limiter, ϕ_1, and the superbee limiter, ϕ_2, are members of the class of Φ limiters defined by

(3.4) $$\phi_{\Phi}(r) = \mathrm{Max}(0,\mathrm{Min}(\Phi r,1),\mathrm{Min}(\Phi,r))$$

for $1 \leq \Phi \leq 2$. These limiters span the whole second order TVD region in the manner shown in Figure 2, the minmod and superbee limiters being the extreme cases. Chakravarthy has suggested the use of this class of limiters varying Φ over the mesh to adjust the degree of compressibility of the limiter to local features of the flow (c.f. artificial compression/rarefaction used by Osher and Chakravarthy [12]).

We take the opportunity to note that all the limiters mentioned above possess the symmetry property

(3.5) $$\phi(1/r) = \phi(r)/r$$

which implies for example, that both the top and bottom of any discontinuity will be treated symmetrically. In [20], [21] it was demonstrated how a lack of such symmetry caused undesirable numerical results.

We now compare, numerically, van Leer's limiter ϕ_{VL}, with the two extreme Φ limiters, ϕ_1 and ϕ_2. The first and simplest test problem is that of linear advection, which clearly illustrates the differences between these limiters. Figures 3 and 4 show the results after 1000 timesteps of convecting discontinuous and smooth data respectively. In both cases the superiority of all the limiter schemes over the first order scheme is clearly evident as expected, with the most compressive limiter, ϕ_2, producing the best results. In particular, for the smooth data, Figure 4 shows that the superbee limiter exhibits a large improvement over the others despite a slight "squaring" of the wave form.

In [20], [21] Sod's shock-tube problem [18] was used to demonstrate various limiters on the Euler system of equations. Here we present results of the use of limiters on the more complicated blast-tube problem used by Woodward and Colella [25], which involves the interaction of shock waves and the reflection of rarefaction waves. Figure 5 shows the density and velocity profiles produced by the various limiters at time $t = 0.04$ (2000 timesteps) on a mesh of 500 points using Roe's approximate Riemann solver. As can be seen from the density profiles, the superbee limiter (ϕ_2) gives the best resolution of the discontinuities of the solution, but introduces a slight "kink" in the velocity profile ($\sim x = 0.75$) which is not apparent using the other limiters, nor in the results of Woodward and Colella. However, as can be seen in Figure 6, this dogleg can be decreased, without much loss of resolution elsewhere, by using the superbee limiter only on the linearly degenerate characteristic field and one of the other limiters on the remaining two fields. (Advantages of using different limiters on different fields were also shown in [20], [21]).

Finally we move briefly onto a two-dimensional problem, that of a

Figure 3 Discontinuous data

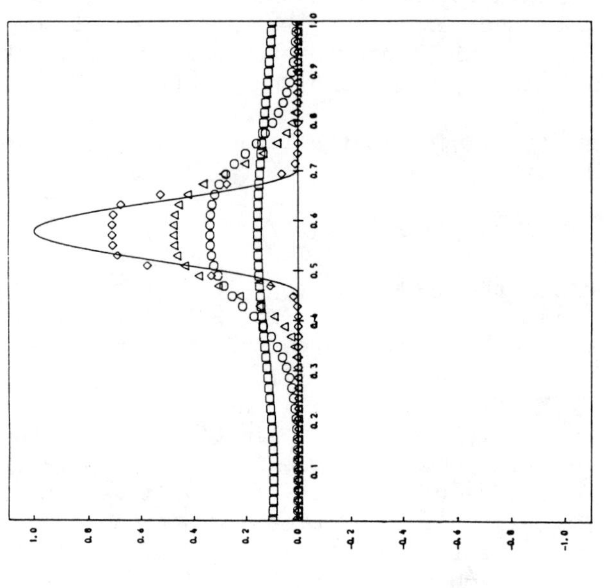

Figure 4 Smooth data

Linear advection equation

□:- first order
Θ:- minmod limiter
◇:- superbee limiter
Δ:- Van Leer's limiter

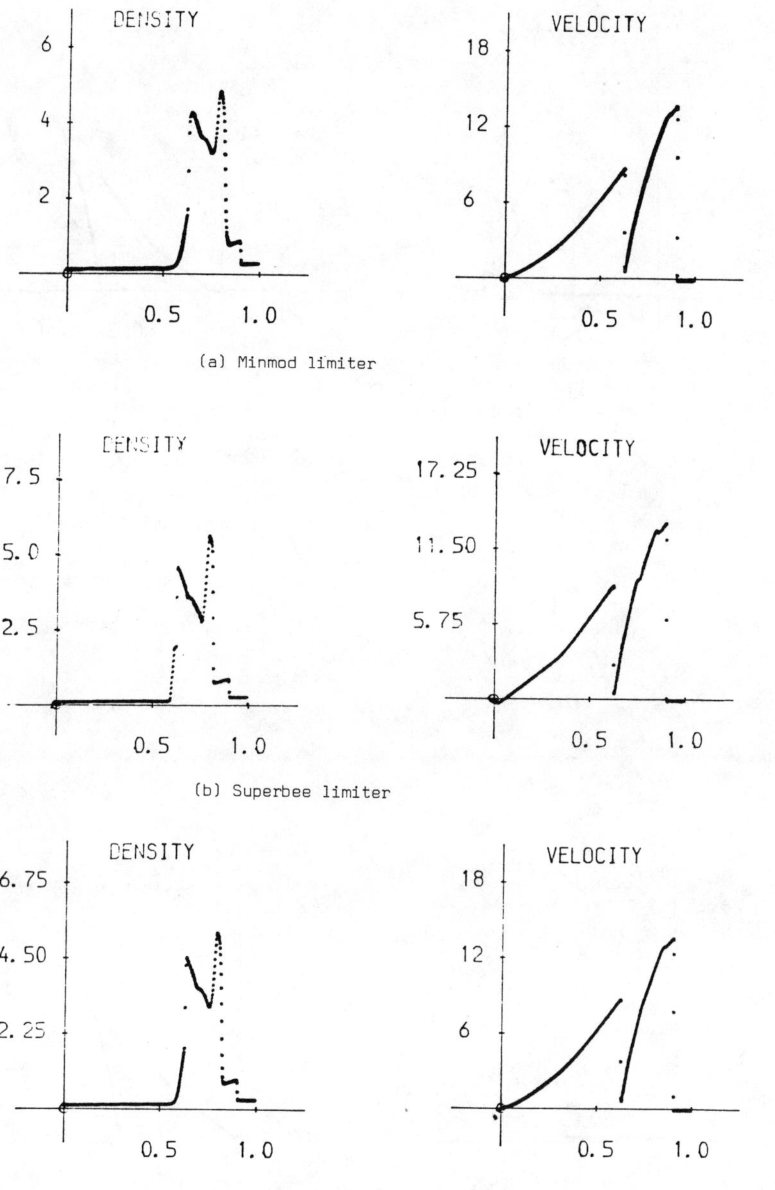

Figure 5. Blast-tube problem, t=0.04

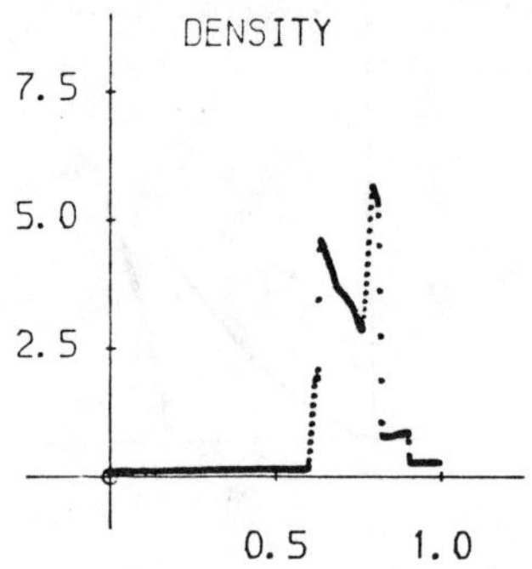

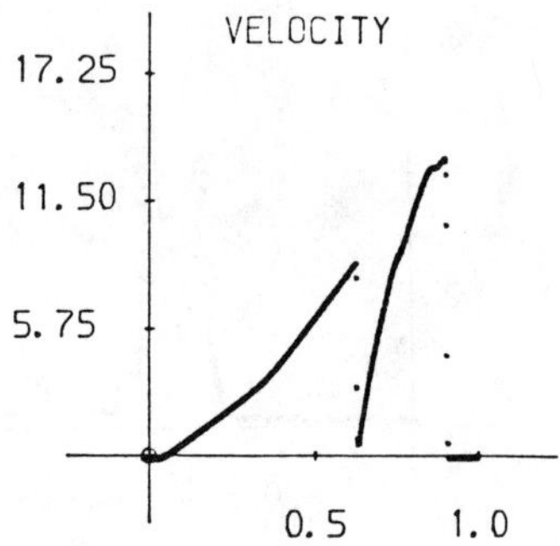

(a) superbee on linear field, minmod on others

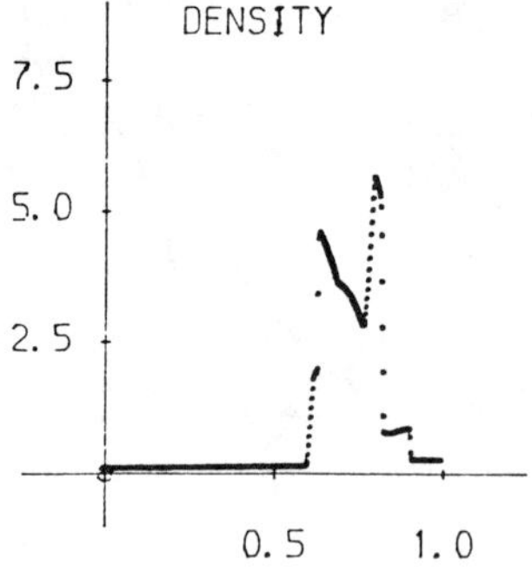

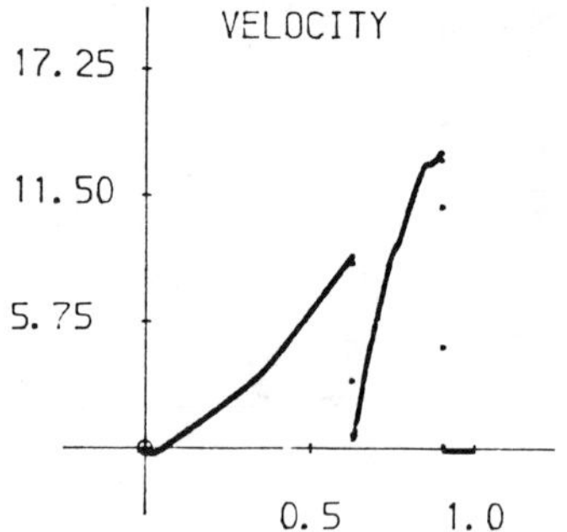

(b) superbee on linear field, van Leer on others

Figure 6. Blast-tube problem t=0.04

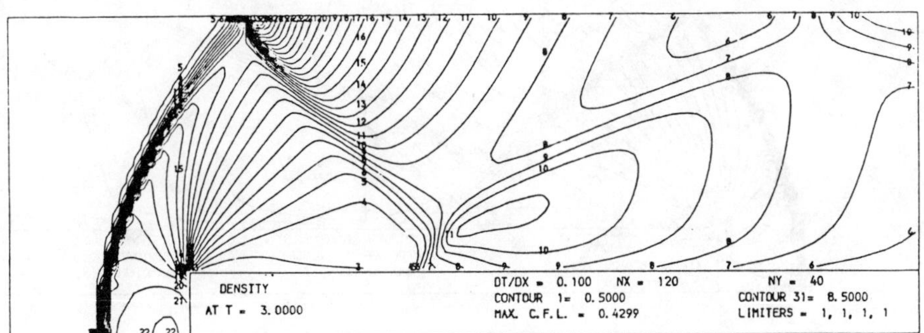

(a) No limiter (first order)

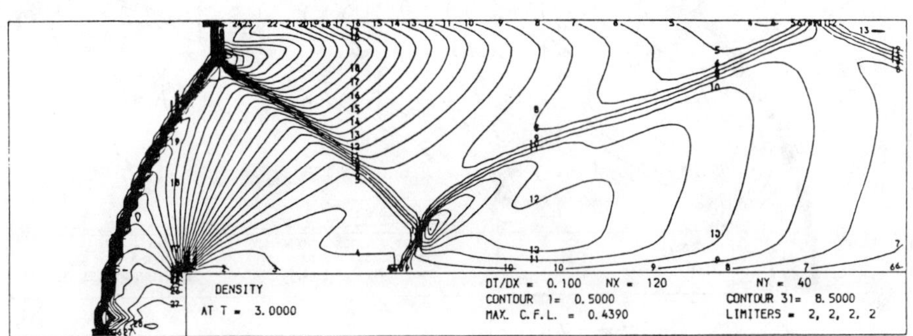

(b) Minmod limiter

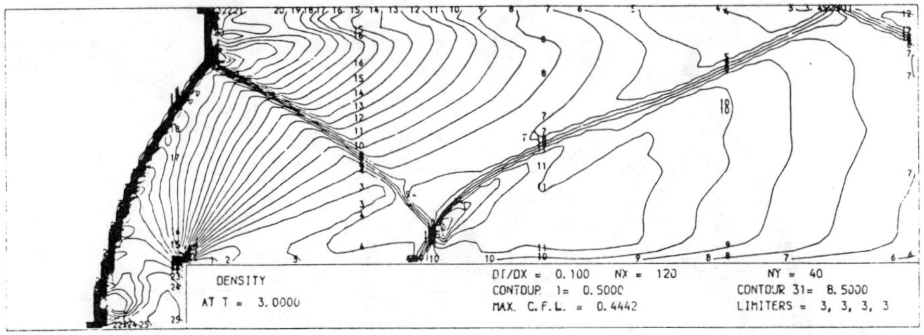

(c) Superbee limiter

Figure 7. Mach 3 windtunnel with step

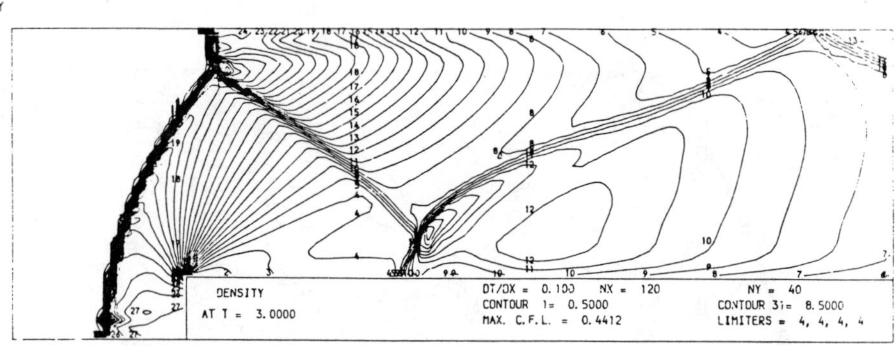

(d) Van Leer's limiter

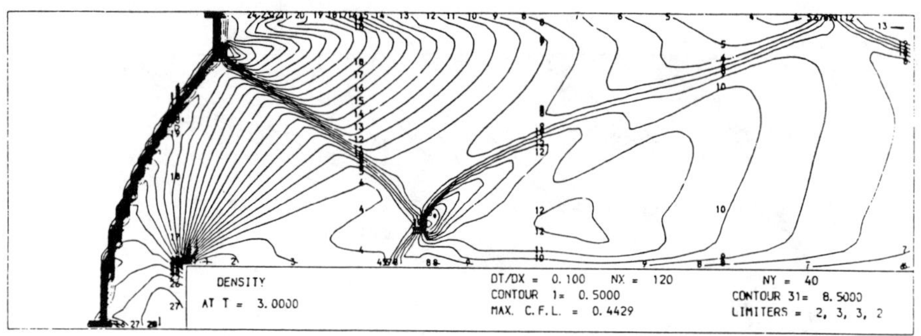

(e) Minmod / superbee combination

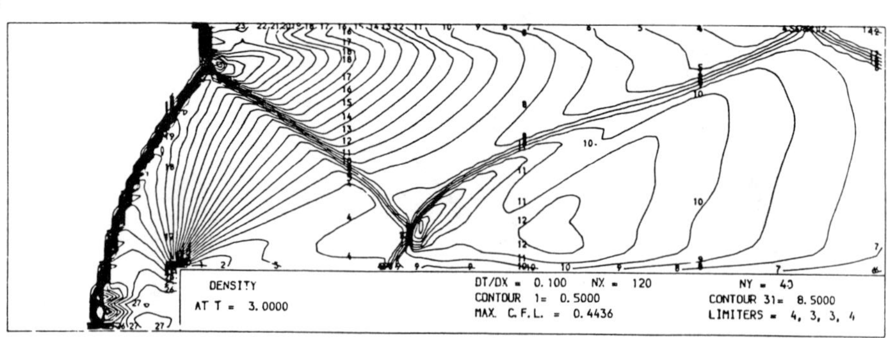

(f) Van Leer / superbee combination

Figure 7 continued

Mach 3 windtunnel containing a step, as also used by Woodward and Colella [25], the reader being referred to their paper for further details and references. Figure 7 displays preliminary results for this problem using the different limiters with dimensional splitting, and where no special treatment has been given to singular point at the corner of the step. At the moment readers are left to judge for themselves the merits of the individual limiters used, although it is hoped to present a more detailed and objective comparison in a future paper.

4. Comments. In this paper as well as in [20], [21], we have restricted numerical comparisons to those between different limiters. Although the test problems used have had results published elsewhere using other types of schemes, it would be desirable to compare the limiter schemes with schemes which do not fall into the flux limiter category. Such comparisons will be presented in a future paper.

We have not said anything here about the entropy satisfaction of limiter schemes. Although there is no general analytic result, numerical results suggest that if the underlying first order scheme is entropy satisfying, then so is the limiter scheme. In fact a remark by Tadmor [23] about schemes with non-vanishing numerical viscosity indicates that, except possibly for the extreme ϕ_2 limiter, limiter schemes are entropy satisfying for a sufficiently restrictive CFL condition, although it may be unnecessarily severe.

Acknowledgements

The author wishes to thank Professor S. Osher for the kind hospitality shown during his stay at UCLA.

The 2D results presented here were obtained using the author's modification of a program written by Paul Glaister at the University of Reading.

REFERENCES

[1] M.J. BAINES, Numerical algorithms for the non-linear scalar wave equation, Reading University Num. Anal. Rpt. 1/83, (1983).

[2] J.P. BORIS & D.L. BOOK, Flux corrected transport, I, SHASTA, a fluid transport algorithm that works, JCP 11 (1973), pp. 38-69.

[3] S. CHAKRAVARTHY & S. OSHER, High resolution applications of the Osher upwind scheme for the Euler equations, AIAA paper presented at 6th CFD Conference (1983).

[4] J.B. GOODMAN & R.J. LEVEQUE, On the accuracy of stable schemes for 2D scalar conservation laws, Preprint.

[5] A. HARTEN, High resolution schemes for conservation laws, JCP 49, (1983), pp. 357-393.

[6] P.D. LAX & R. WENDROFF, Systems of conservation laws, Comm. Pure Appl. Math. 13, (1960), pp. 217-237.

[7] A.Y. LeROUX, Convergence of an accurate scheme for first order quasi-linear equations, RAIRO 15, (1981), pp. 151-170.

[8] R.J. LEVEQUE & J.B. GOODMAN, TVD schemes in one and two space dimensions, Proc. 15th AMS-SIAM Summer Seminar in Appl. Math., to appear.

[9] B. VAN LEER, Towards the ultimate conservative difference scheme, II, monotonicity and conservation combined in a second order scheme, JCP 14, (1974), pp. 361-370.

[10] B. VAN LEER, Towards the ultimate conservative difference scheme, V, a second order sequel to Godunov's method, JCP 32, (1979), pp. 101-136.

[11] S. OSHER, Riemann solvers, the entropy condition, and difference approximations, submitted to SINUM.

[12] S. OSHER & S. CHAKRAVARTHY, High resolution schemes and the entropy condition, ICASE Report.

[13] P.L. ROE, Numerical algorithms for the linear wave equation, Royal Aircraft Establishment TR 81047, (1981).

[14] P.L. ROE, Approximate Rieman solvers, parameter vectors, and difference schemes, JCP 43, (1981), pp. 357-372.

[15] P.L. ROE, Some contributions to the modelling of discontinuous flows, Proc. 15th AMS-SIAM Summer Seminar in Appl. Math., to appear

[16] P.L. ROE & M.J. BAINES, Algorithms for advection and shock problems, Proc. 4th GAMM Conference on Numerical Methods in Fluid Mechanics, (1982).

[17] R. SANDERS, On convergence of monotone finite difference schemes with variable space differencing, Math. Comp. 40, (1983), pp. 91-106.

[18] G.A. SOD, A survey of several finite difference methods for systems of non-linear hyperbolic conservation laws, JCP 27, (1981), pp. 1-31.

[19] P.K. SWEBY, Shock capturing schemes, Reading University Ph.D. Thesis, (1982).

[20] P.K. SWEBY, High resolution schemes using flux limiters for hyperbolic conservation laws, SINUM, to appear.

[21] P.K. SWEBY, High resolution TVD schemes using flux limiters, Proc. 15th AMS-SIAM Summer Seminar in Appl. Math., to appear.

[22] P.K. SWEBY & M.J. BAINES, Convergence of Roe's scheme for the general non-linear scalar wave equation, Reading University Num. Anal. Rpt. 8/81, (1981).

[23] E. TADMOR, Numerical viscosity and the entropy condition for conservative difference schemes, ICASE Rpt. 172141, (1983).

[24] R.F. WARMING & R.M. BEAM, Upwind second order difference schemes and applications in aerodynamics, AIAA Journal 14, (1976), pp. 1241-1249.

[25] P.M. WOODWARD & P. COLELLA, The numerical simulation of two dimensional fluid flow with strong shocks, Report UCRL, Lawrence Livermore National Laboratory, (1982).

[26] H.C. YEE, R.F. WARMING & A. HARTEN, Implicit total variation diminishing (TVD) schemes for steady-state calculations, NASA TM 84342, (1983).

[27] S.T. ZALESAK, Fully multidimensional flux corrected transport algorithms for fluids, JCP 31, (1979), pp. 335-362.

A HIGH-RESOLUTION UPWIND SCHEME FOR QUASI 1-D FLOWS

MATANIA BEN-ARTZI* AND JOSEPH FALCOVITZ**

Abstract. A collection of upwind second order schemes are proposed for time dependent, inviscid, compressible fluid flow in one space dimension and variable cross-section. All these schemes provide high resolution of shocks and other discontinuities and singularities of the flow field, and can be applied both in the Lagrangian and Eulerian frameworks. The basic ingredient of the method is an exact solution of the corresponding GRP (Generalized Riemann Problem) leading to explicit expressions for the fluxes. Sonic lines are given a special treatment. The various schemes are derived on the basis of increasingly accurate approximations to the solution of the GRP. The simplest ones (denoted here by L_1, E_1) require practically no further computational effort beyond Godunov's (first order) method. The only accessory technique used is a simple monotonicity algorithm. Results of two Riemann problems (in the large) are shown, using slab symmetry for the first and cylindrical symmetry for the second.

1. Introduction. Consider the Euler equations that model the time-dependent flow of an inviscid, compressible fluid through a duct of smoothly varying cross-section. Denoting by A(r) the area of the cross-section at r, these equations are,

$$A \frac{\partial}{\partial t} U + \frac{\partial}{\partial r} [AF(U)] + A \frac{\partial}{\partial r} G(U) = 0, \tag{1.1}$$

$$U = \begin{pmatrix} \rho \\ \rho u \\ \rho E \end{pmatrix}, \quad F(U) = \begin{pmatrix} \rho u \\ \rho u^2 \\ (\rho E+p)u \end{pmatrix}, \quad G(U) = \begin{pmatrix} 0 \\ p \\ 0 \end{pmatrix}.$$

Here ρ, p, u, E are, respectively, tensity, pressure, velocity and total specific energy, where $E = e+\frac{1}{2}u^2$, e being the internal spcific energy. For simplicity we shall assume that,

* Department of Mathematics, Technion-Israel Institute of Technology, Haifa 32000, Israel

** Computation Division, Rafael Ballistic Center, P. O. Box 2250, Haifa 31021, Israel

(1.2) $$e = \frac{p}{(\gamma-1)\rho} \quad , \quad \gamma > 1,$$

even though our method applies to a general equation of state (see [3]).

In this paper we present a second-order generalization of Godunov's scheme [5] for the integration of (1.1). To give the basic idea, suppose that we use equally spaced grid-points $r_i = i\Delta r$ along the r-axis and equal time intervals of size Δt. By "cell i" we shall refer to the interval extending between the "cell-boundaries" $r_{i\pm\frac{1}{2}} = (i\pm\frac{1}{2})\Delta r$. We let Q_i^n denote the average value of the quantity Q over cell i at time $n\Delta t$. Similarly, we designate by $Q_{i+\frac{1}{2}}^{n+\frac{1}{2}}$ the value of Q at the cell boundary $r_{i+\frac{1}{2}}$, averaged over the time interval $(n\Delta t, (n+1)\Delta t)$. Generally speaking, a ("quasi-conservative") difference scheme for (1.1) is given by,

(1.3) $$\begin{aligned} U_i^{n+1} - U_i^n = &-\frac{\Delta t}{\Delta V_i}\left[A(r_{i+\frac{1}{2}})F(U)_{i+\frac{1}{2}}^{n+\frac{1}{2}} - A(r_{i-\frac{1}{2}})F(U)_{i-\frac{1}{2}}^{n+\frac{1}{2}}\right] \\ &-\frac{\Delta t}{\Delta r}\left[G(U)_{i+\frac{1}{2}}^{n+\frac{1}{2}} - G(U)_{i-\frac{1}{2}}^{n+\frac{1}{2}}\right], \end{aligned}$$

where $\Delta V_i = \int_{r_{i-\frac{1}{2}}}^{r_{i+\frac{1}{2}}} A(r)dr$. We must still give an appropriate interpretation to the "flux" values $F(U)_{i+\frac{1}{2}}^{n+\frac{1}{2}}$, $G(U)_{i+\frac{1}{2}}^{n+\frac{1}{2}}$. In Godunov's scheme, one solves a Riemann Problem (RP) at each cell boundary, namely, the initial value problem (1.1) with initial data,

(1.4) $$U(r,0) = \begin{cases} U_{i+1}^n \quad , & r > 0, \\ U_i^n \quad , & r < 0. \end{cases}$$

Let $U_{i+\frac{1}{2}}^n$ be the value of the solution of (1.1), (1.4) at $r = 0$, $t = 0+$ (i.e., $U_{i+\frac{1}{2}}^n = \lim_{t\to 0+} U(0,t)$). Godunov's (first-order) scheme is then obtained from (1.3) by setting,

(1.5) $$U_{i+\frac{1}{2}}^{n+\frac{1}{2}} = U_{i+\frac{1}{2}}^n \; , \; F(U)_{i+\frac{1}{2}}^{n+\frac{1}{2}} = F(U_{i+\frac{1}{2}}^{n+\frac{1}{2}}) \; , \; G(U)_{i+\frac{1}{2}}^{n+\frac{1}{2}} = G(U_{i+\frac{1}{2}}^{n+\frac{1}{2}}).$$

As is well known, jump discontinuities are poorly resolved by this scheme. In order to upgrade the order of accuracy and achieve better resolution of discontinuities we assume that the values of U in cell i at time $t_n = n\Delta t$ are linearly distributed and retain the notation U_i^n for the average values. Let $(\Delta U)_i^n$ be the variation of U over cell i. Hence, $U_+ = U_{i+1}^n - \frac{1}{2}(\Delta U)_{i+1}^n$ and $U_- = U_i^n + \frac{1}{2}(\Delta U)_i^n$ are the limiting values (from right and left respectively) of U at the discontinuity $r = r_{i+\frac{1}{2}}$. We are now led naturally to the following statement of the Generalized Riemann Problem (GRP).

Consider the initial value problem (1.1), with initial data,

$$U(r,0) = \begin{cases} U_+ + \frac{r}{\Delta r}(\Delta U)_{i+1}^n, & r > 0, \\ U_- + \frac{r}{\Delta r}(\Delta U)_i^n, & r < 0. \end{cases} \tag{1.6}$$

Let $U(r,t)$ be the solution for $t > 0$. Find,

$$\begin{aligned} &(a) \quad U(0,0) = \lim_{t\to 0+} U(0,t), \\ &(b) \quad \frac{\partial}{\partial t} U(0,0) = \lim_{t\to 0+} \frac{\partial}{\partial t} U(0,t). \end{aligned} \tag{1.7}$$

Once the GRP is resolved, its implementation in the scheme (1.3) is straightforward. In fact, we set, instead of (1.5),

$$\begin{aligned} U_{i+\frac{1}{2}}^n &= U(0,0), \quad \left[\frac{\partial}{\partial t} U\right]_{i+\frac{1}{2}}^n = \frac{\partial}{\partial t} U(0,0), \\ \left[\frac{\partial}{\partial t} F(U)\right]_{i+\frac{1}{2}}^n &= F'(U_{i+\frac{1}{2}}^n) \cdot \left[\frac{\partial}{\partial t} U\right]_{i+\frac{1}{2}}^n, \\ \left[\frac{\partial}{\partial t} G(U)\right]_{i+\frac{1}{2}}^n &= G'(U_{i+\frac{1}{2}}^n) \cdot \left[\frac{\partial}{\partial t} U\right]_{i+\frac{1}{2}}^n, \\ F(U)_{i+\frac{1}{2}}^{n+\frac{1}{2}} &= F(U_{i+\frac{1}{2}}^n) + \frac{\Delta t}{2}\left[\frac{\partial}{\partial t} F(U)\right]_{i+\frac{1}{2}}^n, \\ G(U)_{i+\frac{1}{2}}^{n+\frac{1}{2}} &= G(U_{i+\frac{1}{2}}^n) + \frac{\Delta t}{2}\left[\frac{\partial}{\partial t} G(U)\right]_{i+\frac{1}{2}}^n. \end{aligned} \tag{1.8}$$

In the above we have used $F'(U)$, $G'(U)$ to designate the Jacobians of F, G, respectively, with respect to U.

It follows by finite propagation that $U(0,0)$ in (1.7)(a) is obtained as the solution of the Riemann problem (1.1), (1.4), if we set $U_{i+1}^n = U_+$, $U_i^n = U_-$. Thus, the solution to the GRP is reduced to (1.7)(b). The resulting scheme (1.3), (1.8) is upwind and second-order accurate. Furthermore, let us point out the following features of the scheme, which make it particularly simple and easy to implement.

(a) A Riemann problem is solved once per cell per time step. It determines completely the initial wave structure which serves as a basis for all subsequent calculations.

(b) A "plug-in" procedure, using the explicit analytic solution, is used in order to compute time derivatives of flow quantities. Cell averages are then updated according to (1.3), (1.8), and the new slopes are obtained by a straightforward differencing of cell-boundary values at t_{n+1}.

(c) The only accessory technique used is a simple monotonicity algorithm, following van-Leer [11, Fig. 3]. In particular, no other dissipative mechanism is incorporated (see also the final paragraph of this Introduction).

In the rest of the paper we address solely the GRP (1.1), (1.6), aiming at an analytic derivation of (1.7). We shall keep the presentation on a rather expository level, indicating the basic ideas and omitting the more technical details, which may be found in [2],[3].

Observe that the scheme (1.3), (1.8), is already second-order accurate when $[\frac{\partial}{\partial t} U]^n_{i+\frac{1}{2}}$ is calculated with an $O(\Delta t)$ error (more precisely, we should add that the coefficient in $O(\Delta t)$ varies smoothly with r). This observation leads to what could be regarded as the "simplest" extension of Godunov's scheme. We label this scheme as L_1 or E_1, corresponding, respectively, to the Lagrangian or Eulerian frames. In contrast, L_∞,E_∞ denote, respectively, the Lagrangian and Eulerian schemes resulting from an exact determination of the time derivative in (1.7)(b). Following this terminology, van-Leer's MUSCL scheme [11] is actually L_2 (i.e., the derivative in (1.7)(b) is determined within $O(\Delta t^2)$. See [2, Appendix A] for the proof).

The plan of the paper is as follows.

Section 2. We switch to Lagrangian coordinates and discuss the obvious analogue of the GRP (1.1), (1.6). Note that here the time derivatives are taken along the contact discontinuity. We derive first the L_1 scheme and then proceed to present the L_∞ scheme. The exact solution of (1.7)(b) in this case is obtained via a "propagation of singularities" argument which is used in the resolution of a centered rarefaction wave.

Section 3. Taking advantage of the Lagrangian analysis of Section 2, we are able to complete the discussion of the GRP in the Eulerian case. Unlike the Lagrangian case, one encounters here the "sonic" case, namely, the case of vanishing slope in one of the characteristic fields ($u \pm c$) emerging from the discontinuity. This, of course, corresponds to the case where the grid line $r = r_{i+\frac{1}{2}}$ is contained in a centered rarefaction wave. Evidently, the solution in this case relies heavily upon the resolution of such a rarefaction wave. In analogy with the L_1, L_∞ schemes we have here also the E_1, E_∞ schemes. On this scale, it seems to us that E_2 is the equivalent of the PPM method of Colella-Woodward [4].

Section 4. The numerical results of two test problems are presented. Both problems are Riemann problems (or "shock tubes"). The first is Sod's example [9], where $A(r) \equiv 1$ (slab symmetry), and the second uses cylindrical symmetry ($A(r)=r$), giving rise to a converging shock which diverges after reflection from the axis of symmetry [1], [8], [10].

Finally, let us refer briefly to the problem of "entropy violating" (or "rarefaction shock") solutions that occur in some upstream and high-resolution methods, unless some additional dissipative mechanism is used [6], [7]. Such solutions are typically associated with sonic points, where the system (1.1) is replaced (locally) by a linear one. Being linear, it is obviously not capable of replacing the non-physical jump by a rarefaction fan and in this sense violates a basic idea underlying Godunov's method, namely, accounting "locally" for the non-linear interactions. On the other hand our procedure (1.3), (1.8) modifies Godunov's scheme and reduces to it when all slopes are set equal to zero. It therefore seems to us that the present method forces the entropy condition by using the dissipative mechanism inherent to the underlying Godunov scheme.

3. Lagrangian Analysis. Define a Lagrangian "mass" coordinate by,

$$d\xi = A\rho(r,0)\,dr \quad , \quad \xi(0) = 0, \tag{2.1}$$

In terms of ξ,t, the GRP (1.1), (1.6) takes the form,

$$\frac{\partial}{\partial t} V + \frac{\partial}{\partial \xi}[A\Phi(V)] + A\frac{\partial}{\partial \xi}\Psi(V) = 0,$$

$$V = \begin{pmatrix}\tau\\ u\\ E\end{pmatrix}, \quad \Phi(V) = \begin{pmatrix}-u\\ 0\\ pu\end{pmatrix}, \quad \Psi(V) = \begin{pmatrix}0\\ p\\ 0\end{pmatrix}, \quad \tau = \frac{1}{\rho}. \tag{2.2}$$

$$V(\xi,0) = \begin{cases} V_+ + \dfrac{\xi}{\Delta\xi}(\Delta V)^n_{i+1} & , \quad \xi > 0, \\[2ex] V_- + \dfrac{\xi}{\Delta\xi}(\Delta V)^n_i & , \quad \xi < 0. \end{cases} \tag{2.3}$$

Observe that in (2.3) we assume linearity in ξ. This will be justified by the fact that the solution to (1.7) depends only upon the limiting values of the derivatives of $\frac{\partial}{\partial r}U(r,0)$ as $r\to 0\pm$. Thus, in what follows we shall use the notation $(\frac{\partial Q}{\partial r})_\pm$ for the initial slopes of any variable Q on either side of the discontinuity and set,

$$(2.4) \qquad (\frac{\partial Q}{\partial \xi})_{\pm} = A(0)^{-1} \rho_{\pm}^{-1} (\frac{\partial Q}{\partial r})_{\pm}.$$

In what follows we shall assume that the situation is as displayed in Fig. 1, namely, a shock wave propagates to the right whereas a centered rarefaction wave travels to the left. These

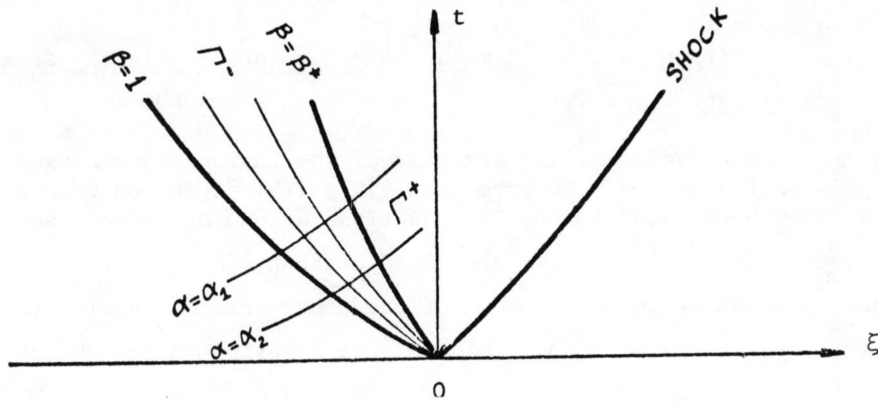

Figure 1. Structure of solution to the GRP in Lagrangian coordinates

waves are separated by a contact discontinuity ($\xi=0$), and bound regions of smooth flow (observe that the flow in each region is uniform only in the case that $A(r) \equiv$ const. and $(\frac{\partial V}{\partial \xi})_+ = (\frac{\partial V}{\partial \xi})_- = 0$). Let $V(\xi,t)$ be the solution to the GRP (2.2), (2.3), where,

$$(\frac{\partial V}{\partial \xi})_+ = \frac{(\Delta V)_{i+1}^n}{\Delta \xi}, \quad (\frac{\partial V}{\partial \xi})_- = \frac{(\Delta V)_i^n}{\Delta \xi}.$$

We set $V^* = \lim_{t\to 0+} V(0,t)$. Note that V^* is obtained as the solution along the contact discontinuity for the planar Riemann problem for (2.2) with initial data $V_{\pm}$. Hence, in complete analogy with (1.8), we need to compute

$$(2.5) \qquad (\frac{\partial V}{\partial t})^* = \lim_{t\to 0+} \frac{\partial}{\partial t} V(0,t).$$

Remark. Note that while u^*, p^* are continuous across $\xi = 0$, the density has there a jump discontinuity. We use there the obvious notation ρ_+^*, ρ_-^*.

In the following theorem we use an idea of van-Leer [11]. For a detailed proof see [3].

<u>Theorem 2.1. The derivatives</u> $(\frac{\partial p}{\partial t})^*$, $(\frac{\partial u}{\partial t})^*$ <u>are determined by a pair of linear equations,</u>

$$(2.6)\qquad \begin{aligned} a_+ (\frac{\partial u}{\partial t})^* + b_+ (\frac{\partial p}{\partial t})^* &= d_+ , \\ a_- (\frac{\partial u}{\partial t})^* + b_- (\frac{\partial p}{\partial t})^* &= d_- , \end{aligned}$$

<u>where</u> a_+, b_+, d_+ (<u>resp</u>. a_-, b_-, d_-) <u>can be determined explicitly from</u> V^*, V_+, $(\frac{\partial V}{\partial \xi})_+$ (<u>resp</u>. V^*, V_-, $(\frac{\partial V}{\partial \xi})_-$).

Clearly, once equations (2.6) are solved, the Lagrangian analogue of (1.8) is easily set up. Observe that $\Phi(V)$, $\Psi(V)$ depend on p,u only. As has already been noted in the Introduction, second order accuracy is achieved if $(\frac{\partial p}{\partial t})^*$, $(\frac{\partial u}{\partial t})^*$ are determined with an $O(\Delta t)$ error. In this case we speak of an L_1 scheme. At the other extreme, evaluating $(\frac{\partial u}{\partial t})^*$, $(\frac{\partial p}{\partial t})^*$ exactly yields an L_∞ scheme. We describe these schemes next.

A) <u>The L_1 Scheme</u>. We denote by c the speed of sound, $c^2 = (\frac{\partial p}{\partial \rho})_S$, and let $A'(0) = \frac{d}{dr} A(r)\big|_{r=0}$.

<u>Claim 2.2. Set in</u> equations (2.6),

$$(2.6)\qquad \begin{aligned} a_\pm &= \mp 1 , \quad b_\pm = \rho_\pm^{-1} c_\pm^{-1} , \\ d_\pm &= -A(0)\left(\rho_\pm c_\pm (\frac{\partial u}{\partial \xi})_\pm \mp (\frac{\partial p}{\partial \xi})_\pm\right) - \frac{A'(0)}{A(0)} u_\pm c_\pm . \end{aligned}$$

<u>Then, in regions of smooth flow,</u> $(\frac{\partial u}{\partial t})^*$, $(\frac{\partial p}{\partial t})^*$ <u>are obtained from</u> (2.6) <u>with an</u> $O(\Delta t)$ <u>error</u>.

<u>Proof</u>. We prove for the "+" sign. In a region of smooth flow we may assume that the right wave is, within an $O(\Delta t)$ error, an "acoustic" wave, namely, a single characteristic curve of slope $\frac{d\xi}{dt} = A\rho c$. All variables are continuous across this curve and so are their derivatives in that direction. We may therefore evaluate the directional derivative of p along the curve by using the values ahead and behind the wave, obtaining,

$$(2.8)\qquad A(0)\rho_+ c_+ (\frac{\partial p}{\partial \xi})^* + (\frac{\partial p}{\partial t})^* = A(0)\rho_+ c_+ (\frac{\partial p}{\partial \xi})_+ + (\frac{\partial p}{\partial t})_+ ,$$

where $(\frac{\partial p}{\partial \xi})^*$ is evaluated just behind the wave while $(\frac{\partial p}{\partial t})_+$ is evaluated ahead of it. From Equation (2.2) we get,

(2.9) $\left(\frac{\partial p}{\partial \xi}\right)^* = -A(0)^{-1}\left(\frac{\partial u}{\partial t}\right)^*$, $\left(\frac{\partial p}{\partial t}\right)_+ = c_+^2\left(\frac{\partial \rho}{\partial t}\right)_+ = -\rho_+^2 c_+^2\left[\frac{\partial}{\partial \xi} Au\right]_+$,

where we have used that the entropy is constant along a streamline ξ = const. Using (2.1) we have,

$$\left[\frac{\partial}{\partial \xi} Au\right]_+ = \frac{u_+}{\rho_+}\frac{A'(0)}{A(0)} + A(0)\left(\frac{\partial u}{\partial \xi}\right)_+ .$$

Inserting all these relations in (2.8) yields (2.7).

Q.E.D.

Note that there is practically no additional computational effort involved in upgrading Godunov's scheme by (2.6), (2.7). Furthermore, in this case $a_\pm$, $b_\pm$, $d_\pm$ depend only on the initial data $V_\pm$, $\left(\frac{\partial V}{\partial \xi}\right)_\pm$ (and not on V^*).

B) The L_∞ Scheme. Here we aim at an exact solution of (2.1) for the initial value problem (2.2), (2.3). We must distinguish now between a shock and a centered rarefaction wave and the latter must be carefully analyzed. To be specific, we assume the wave structure to be as in Figure 1 and map the rarefaction wave by characteristic coordinates as follows.

β = Normalized slope of Γ^- at the origin, $\beta = 1$ at the head characteristic.

α = Value of ξ at point of intersection of Γ^+ with the curve $\beta = 1$.

Every flow variable Q is now represented (throughout the rarefaction zone) by a smooth function $Q(\alpha,\beta)$, up to the singularity $\alpha = 0$. The directional derivative $\frac{\partial Q}{\partial \alpha}(0,\beta)$ (which, of course, vanishes for the planar Riemann problem) measures the local variation of Q at the singularity in the direction of the Γ^- characteristic of (normalized) slope β.

The following lemma serves as a basic tool in the derivation of the L_∞ scheme. We state it for the γ-law case. For a detailed proof (using a general equation of state) see [3].

Lemma 2.3. Let $a(\beta) = \frac{\partial u}{\partial \alpha}(0,\beta)$, $\beta^* \le \beta \le 1$, *where* $\beta = \beta^* = \frac{\rho^*_- c^*_-}{\rho_- c_-}$ *is the tail characteristic (Fig. 1). Then* $a(\beta)$ *is given by*

$$a(\beta) = a(1) + \frac{2}{\rho_- c_-(3\gamma-1)}\left[c_-\left(\frac{\partial(\rho c)}{\partial \xi}\right)_- - \frac{\gamma+1}{2}\left(\frac{\partial p}{\partial \xi}\right)_-\right]\cdot\left(\beta^{\frac{3\gamma-1}{2(\gamma+1)}} - 1\right)$$

(2.10)

$$+\frac{A'(0)}{A(0)^2}\left[\frac{4c_-}{(3\gamma-5)\rho_-}\left(\beta^{\frac{3\gamma-5}{2(\gamma+1)}}-1\right)-\frac{(\gamma-1)u_-+2c_-}{(\gamma-3)\rho_-}\left(\beta^{\frac{\gamma-3}{2(\gamma+1)}}-1\right)\right],$$

$$a(1) = \left(\frac{\partial u}{\partial \xi}\right)_- + \rho_-^{-1}c_-^{-1}\left(\frac{\partial p}{\partial \xi}\right)_- .$$

(*Equation* (2.10) *must be modified for* $\gamma = 5/3, 3$).

Remark. Observe that the expression (2.10) simplifies considerably at a "flat" point ($A'(0) = 0$). In particular, this applies to the planar case.

Consider once again the wave pattern of Figure 1. As is well-known the Lagrangian shock speed (i.e., $\frac{d\xi}{dt}$ along the shock trajectory) is given by,

$$W_+ = A(0)\,\frac{p^*-p_+}{u^*-u_+} .$$

The L_∞ scheme now builds upon the following theorem.

Theorem 2.4. Assume the configuration of Fig. 1 *and let* $a(\beta)$ *be determined by* (2.10) *and* $\mu^2 = \frac{\gamma-1}{\gamma+1}$. *Then the coefficients in* (2.6) *are given by,*

(2.11) $$a_- = 1\,,\ b_- = (\rho_-^*c_-^*)^{-1}\,,\ d_- = -A(0)(\rho_-\rho_-^*c_-c_-^*)^{\frac{1}{2}}\,a(\beta^*) - \frac{A'(0)}{A(0)}\,u^*c_-^*,$$

$$a_+ = -2 + \frac{1}{2}\,\frac{p^*-p_+}{p^*+\mu^2p_+},$$

(2.12) $$b_+ = -\frac{u^*-u_+}{2(p^*+\mu^2p_+)} + A(0)W_+^{-1} + A(0)^{-1}W_+(\rho_+^*c_+^*)^{-2},$$

$$d_+ = L_u\left(\frac{\partial u}{\partial \xi}\right)_+ + L_p\left(\frac{\partial p}{\partial \xi}\right)_+ + L_\rho\left(\frac{\partial \rho}{\partial \xi}\right)_+ + \frac{A'(0)}{A(0)}\,L_g ,$$

where,

$$L_u = \frac{A(0)}{2}\,(u_+-u^*)\left(\rho_+ + \mu^2\,\frac{\rho_+^2c_+^2}{p^*+\mu^2p_+}\right) - A(0)^2W_+^{-1}\rho_+^2c_+^2 - W_+,$$

$$L_p = 2A(0) + \frac{\mu^2}{2}\,A(0)\,\frac{p^*-p_+}{p^*+\mu^2p_+},$$

$$L_\rho = \frac{A(0)}{2}\,\frac{p^*-p_+}{\rho_+},$$

$$L_g = -W_+ A(0)^{-1} u^*(\rho_+^*)^{-1} - u_+ \rho_+^{-1} (u^* - u_+) \left[\rho_+^2 c_+^2 \left(\frac{1}{p^* - p_+} + \frac{\mu^2}{2} \frac{1}{p^* + \mu^2 p_+} \right) + \frac{\rho_+}{2} \right].$$

Sketch of proof (see [3] for details). To prove (2.11), we write the characteristic relation along Γ^- at $(0,\beta^*)$ in the form,

$$(2.13) \qquad \frac{\partial p}{\partial \alpha}(0,\beta^*) - \rho_-^* c_-^* \frac{\partial u}{\partial \alpha}(0,\beta^*) + \frac{A'(0)}{A(0)} \rho_-^* u^* (c_-^*)^2 \frac{\partial t}{\partial \alpha}(0,\beta^*) = 0.$$

It can be shown that,

$$(2.14) \qquad \frac{\partial t}{\partial \alpha}(0,\beta^*) = -(\rho_- c_- A(0))^{-1} (\beta^*)^{-\frac{1}{2}}, \quad \frac{\partial \xi}{\partial \alpha}(0,\beta^*) = (\beta^*)^{\frac{1}{2}}.$$

Since $a(\beta^*) = \frac{\partial u}{\partial \alpha}(0,\beta^*)$ is known, $\frac{\partial p}{\partial \alpha}(0,\beta^*)$ can be determined from (2.13). From the chain rule we get,

$$\frac{\partial p}{\partial \alpha}(0,\beta^*) = \left(\frac{\partial p}{\partial t}\right) \frac{\partial t}{\partial \alpha}(0,\beta^*) + \left(\frac{\partial p}{\partial \xi}\right)^* \frac{\partial \xi}{\partial \alpha}(0,\beta^*).$$

As in (2.9) we have $\left(\frac{\partial p}{\partial \xi}\right)^* = -A(0)^{-1} \left(\frac{\partial u}{\partial t}\right)^*$ and in conjunction with (2.13), (2.14), we get (2.11).

The derivation of (2.12) is even more straightforward. In fact, let u_r, p_r, ρ_r (u,p) be the pre- (post-) shock quantities at an arbitrary point. As is well-known, we have,

$$u - u_r = (p - p_r) \left[\frac{1 - \mu^2}{\rho_r (p + \mu^2 p_r)} \right]^{\frac{1}{2}}.$$

Differentiating this identity along the shock trajectory and letting the point under consideration tend to the singularity (i.e., $u \to u^*$, $p \to p^*$, $u_r \to u_+$ etc.) we obtain (2.12).

Q. E. D.

We may now sum up the Lagrangian procedure as follows.

(a) Set up the GRP (2.2), (2.3) and evaluate the time derivatives (2.5) by one of the above mentioned schemes.

(b) Compute average values at time t_{n+1} by the obvious analogue of (1.8).

(c) Compute new slopes in cells by

$$(\Delta V)_i^{n+1} = V_{i+\frac{1}{2}}^{n+1} - V_{i-\frac{1}{2}}^{n+1} ,$$

where the right-hand side is the difference of the new cell-boundary values.

(d) Apply a simple monotonicity algorithm to ensure that,

(i) $(\Delta V)_i^{n+1} = 0$ at an extremal point of $\{V_i^{n+1}\}$.

(ii) The sequence $V_{i-1}^{n+1}, V_i^{n+1} - \frac{1}{2}(\Delta V)_i^{n+1}, V_i^{n+1}, V_i^{n+1} + \frac{1}{2}(\Delta V)_i^{n+1}, V_{i+1}^{n+1}$ is monotone.

Specifically, we have used van-Leer's algorithm [11, Fig. 3].

3. Eulerian Analysis. In this section we take up the solution of the GRP (1.1), (1.6) in Eulerian coordinates, namely, the determination of the quantities in (1.7). As we have already noted, the limiting value $U(0,0)$ is obtained from the solution to the corresponding Riemann problem (set $(\Delta U)_{i+1}^n = (\Delta U)_i^n = 0$ in (1.6)). We shall henceforth denote $U_0 = U(0,0)$ and take it to be known. Thus we need to compute only $\frac{\partial}{\partial t} U(0,0)$, which we denote in the sequel by $(\frac{\partial U}{\partial t})_0$. The results of the previous section serve to determine $(\frac{\partial p}{\partial t})^*$, $(\frac{\partial u}{\partial t})^*$, the derivatives along the contact discontinuity. To fix the ideas, suppose that the configuration is as in Figure 2. In particular, we assume that the line

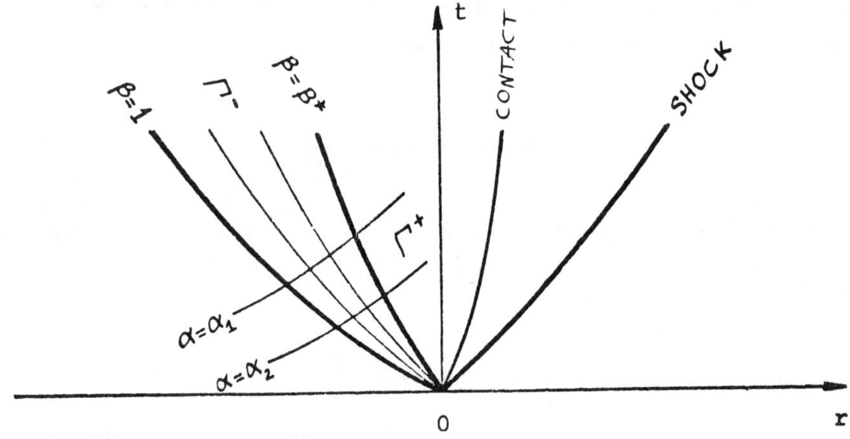

Figure 2. Structure of solution to the GRP in Eulerian coordinates, non-sonic case.

$r = 0$ is non-sonic, i.e., that it is not contained in a rarefaction fan. By means of the transformation (2.1) we switch locally to Lagrangian coordinates, adjusting the initial slopes as in (2.4). In this new system the line $r = 0$ is represented by the curve $\xi = \xi(t)$

satisfying,

$$\text{(3.1)} \qquad \frac{d\xi}{dt} = -A\left[r(\xi,t)\right]\rho(\xi,t)\,u(\xi,t)\;, \qquad \xi(0) = 0.$$

Indeed, this follows easily by differentiating the identity $r(\xi(t),t) = 0$. From the chain rule and (3.1) we obtain immediately the following formula for the time derivative $(\frac{\partial Q}{\partial t})_0$ of any flow variable Q ($Q = \rho, p, u$).

$$\text{(3.2)} \qquad \left(\frac{\partial Q}{\partial t}\right)_0 = \left(\frac{\partial Q}{\partial t}\right)^* - A(0)\,\rho_0 u_0 \left(\frac{\partial Q}{\partial \xi}\right)^*, \quad (U_0 = U_-^*).$$

In Equation (3.2) we must note the following.

(a) While $(\frac{\partial p}{\partial t})^*$, $(\frac{\partial u}{\partial t})^*$ are well defined, we have for the density $(\frac{\partial \rho}{\partial t})^*_\pm = (c^*_\pm)^{-2}(\frac{\partial p}{\partial t})^*$. Obviously, the sign is determined by the side on which $r = 0$ falls, relative to the contact discontinuity.

(b) The derivatives $(\frac{\partial p}{\partial \xi})^*$, $(\frac{\partial u}{\partial \xi})^*_\pm$, are determined from $(\frac{\partial u}{\partial t})^*$, $(\frac{\partial p}{\partial t})^*$ via Eq. (2.2). Note again that $(\frac{\partial u}{\partial \xi})^*$ is discontinuous.

(c) The derivatives $(\frac{\partial \rho}{\partial \xi})^*_\pm$ cannot be determined directly from (2.2). See [3] for the details of this derivation and (3.5) for a simplified version.

(d) In the case that the line $r = 0$ lies either to the left of the rarefaction wave or to the right of the shock wave, Eq. (3.2) is replaced simply by (1.1), i.e., the time derivatives are obtained directly from the initial data.

As was the case with the Lagrangian scheme, second order accuracy is ensured when $(\frac{\partial Q}{\partial t})_0$ is evaluated within an $O(\Delta t)$ error. This observation leads to the simplified E_1 scheme, which is discussed next.

A) <u>The E_1 Scheme</u>. As was pointed out in the proof of Claim 2.2, we may assume here (after solving the Riemann problem and obtaining U^*) that both waves are "acoustic", namely, characteristic curves of slopes $u_\pm \pm c_\pm$ along which the <u>derivatives</u> of flow quantities may experience jump discontinuities. In particular, there is no "sonic" case here and all possible cases fall into one of two categories.

(i) $u^* \geq 0$. The line $r = 0$ is to the left of the contact discontinuity.

(ii) $u^* \leq 0$. The line $r = 0$ is to the right of the contact discontinuity.

Figure 2 represents case (i) and we assume that this is the case in the discussion below. We summarize the E_1 scheme as follows.

Claim 3.1. Assume the configuration of Fig. 2. The derivatives $(\frac{\partial Q}{\partial t})_0$ are obtained with an $O(\Delta t)$ error if we use the following substitutions in the right hand side of (3.2).

$(\frac{\partial u}{\partial t})^*$, $(\frac{\partial p}{\partial t})^*$ are obtained from (2.6), (2.7) and $(\frac{\partial \rho}{\partial t})_-^* = (c_-^*)^{-2} (\frac{\partial p}{\partial t})^*$.

$$(3.3) \qquad (\frac{\partial p}{\partial \xi})^* = - A(0)^{-1} (\frac{\partial u}{\partial t})^*,$$

$$(3.4) \qquad (\frac{\partial u}{\partial \xi})_-^* = - A(0)^{-1}\left[(\rho_-^*)^{-2} (\frac{\partial \rho}{\partial t})_-^* + \frac{u^*}{\rho_-^*} \frac{A'(0)}{A(0)} \right],$$

$$(3.5) \qquad (\frac{\partial \rho}{\partial \xi})_-^* = (\frac{\partial \rho}{\partial \xi})_- - A(0)^{-1} (\rho_- c_-)^{-1} \left[(\frac{\partial \rho}{\partial t})_- - (\frac{\partial \rho}{\partial t})_-^* \right].$$

Proof. The time derivatives were treated in Claim 2.2. Equations (3.3) and (3.4) follow from (2.2) as in (2.9). To establish (3.5) we write the directional derivative of ρ along the left facing "acoustic" wave as follows (compare with (2.8)),

$$-A(0)\ \rho_- c_- (\frac{\partial \rho}{\partial \xi})_-^* + (\frac{\partial \rho}{\partial t})_-^* = - A(0)\ \rho_- c_- (\frac{\partial \rho}{\partial \xi})_- + (\frac{\partial \rho}{\partial t})_- .$$

Solving for $(\frac{\partial \rho}{\partial \xi})_-^*$ we obtain (3.5).

Q. E. D.

Remark. Note that the above equations could be further simplified without losing the $O(\Delta t)$ bound for the error. Thus, for example, we could replace u^*, ρ_-^* in (3.4) by u_-, ρ_-, respectively. However, since the Riemann problem solution V^* is available at this stage, we are trying to "deviate" as little as possible from the exact factors in (3.2). In fact, the only approximation here is in the "lumping" of the rarefaction fan into a single curve in the derivation of (3.5).

B) The E_∞ Scheme. Here we aim at an exact evaluation of $(\frac{\partial U}{\partial t})_0$ from Eq. (3.2), for the GRP (1,1), (1.6).

Suppose first that we are in the non-sonic case, as shown in Figure 2. Then we have the following

Claim 3.2. In the non-sonic case, the E_∞ scheme is implemented by taking the following steps.

(a) Compute $(\frac{\partial u}{\partial t})^*$, $(\frac{\partial p}{\partial t})^*$ by (2.6), (2.11), (2.12).

(b) Compute $(\frac{\partial \rho}{\partial t})_-^*$, $(\frac{\partial u}{\partial \xi})_-^*$ and $(\frac{\partial p}{\partial \xi})^*$ as in Claim 3.1.

(c) Compute $(\frac{\partial \rho}{\partial \xi})_-^*$ from the resolution of the centered rarefaction as is shown in [3, Lemma 5.1].

(d) Use Eq. (3.2).

Finally, we are left with the sonic case. Thus, we assume that the situation ia as in Figure 3.

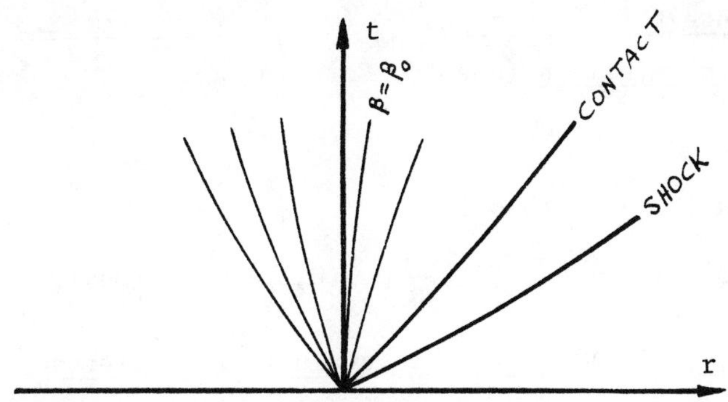

Figure 3. Structure of solution to the GRP in Eulerian coordinates, sonic case.

The representation (3.1) is of no use here since (3.2) is not valid any more. We are forced, therefore, to replace the (ξ,t) representation by the more refined characteristic parametrization which was introduced in the previous section (see the discussion preceding Lemma 2.3). Let $(\alpha(t),\beta(t))$ be the trajectory $r=0$ in the (α,β) plane. We have $(\alpha(0),\beta(0)) = (0,\beta_0)$, where β_0 is determined by $u(0,\beta_0) = c(0,\beta_0)$. Since the limiting values at the singularity $(\alpha=0)$ are those obtained for the planar Riemann problem, it follows immediately that,

$$\beta_0 = \left[\frac{\gamma-1}{\gamma+1}\left(\frac{u_-}{c_-} + \frac{2}{\gamma-1}\right)\right]^{\frac{\gamma+1}{\gamma-1}} . \tag{3.6}$$

Note that for the solution of the planar Riemann problem the characteristic curve $\beta = \beta_0$ coincides with $r = 0$.

In the domain covered by the rarefaction fan all flow variables are represented as functions of (α,β) and we replace Eq. (3.2) by,

$$(3.7)\qquad \left(\frac{\partial Q}{\partial t}\right)_0 = \frac{\partial Q}{\partial \alpha}(0,\beta_0)\cdot\alpha'(0) + \frac{\partial Q}{\partial \beta}(0,\beta_0)\cdot\beta'(0).$$

Parallel to Claim 3.2 we now have the following claim, which concludes the treatment of the E_∞ scheme.

Claim 3.3. *In the sonic case, the E_∞ scheme is implemented by taking the following steps.*

(a) *Compute the derivatives* $\frac{\partial Q}{\partial \beta}(0,\beta_0)$ *from the solution to the Riemann problem.*

(b) *Compute the derivatives* $\frac{\partial Q}{\partial \alpha}(0,\beta_0)$ *from* Lemma 2.3 *and the characteristic relations* (*see* [3, Corollary 3.3].

(c) *Set*

$$(3.8)\qquad \begin{aligned}\alpha'(0) &= -\,A(0)\rho_-c_-\beta_0^{\frac{1}{2}}\\ \beta'(0) &= \frac{1}{2}\,\beta_0^{\frac{1}{2}}\,A(0)\left[\frac{\partial}{\partial\alpha}(\rho c)(0,\beta_0) - \frac{\partial}{\partial\alpha}(\rho u)(0,\beta_0)\right].\end{aligned}$$

(d) *Use Eq.* (3.7) *in order to evaluate the Eulerian time derivatives.*

We shall not go further into details. The proof of (3.8) may be found in [3, Lemma 6.1]. Note that the expression for $\beta'(0)$ measures the "difference in curvature" between the characteristic curve (of slope $-A\rho c$) and the grid line (of slope $-A\rho u$) which coincide in the planar isentropic case.

4. *Numerical examples.* We show the results of two test cases, both of which are Riemann problems ("in the large"), using different types of spatial symmetries.

(a) Our first example is the shock tube problem used by Sod [9]. The tube extends from $x = 0$ to $x = 100$ and is divided into 100 equal cells. The gas is initially at rest ($u=0$) with $p = 1$, $\rho = 1$ for $0\le x<50$, $p = 0.1$, $\rho = 0.125$ for $50<x\le 100$. Numerical and exact solutions are shown (Fig. 4) at $t = 15$, when the shock wave moving to the right has approximately reached $x = 75$. The solid lines represent the exact solutions, whereas the dots represent cell averages in the numerical solution. The quantities shown are velocity, pressure, density and p/ρ^γ, which is a function of entropy (here $\gamma=1.4$). The numerical scheme used was the E_1 scheme (see Section 3,A) which requires very little computational effort (see Claim 3.1).

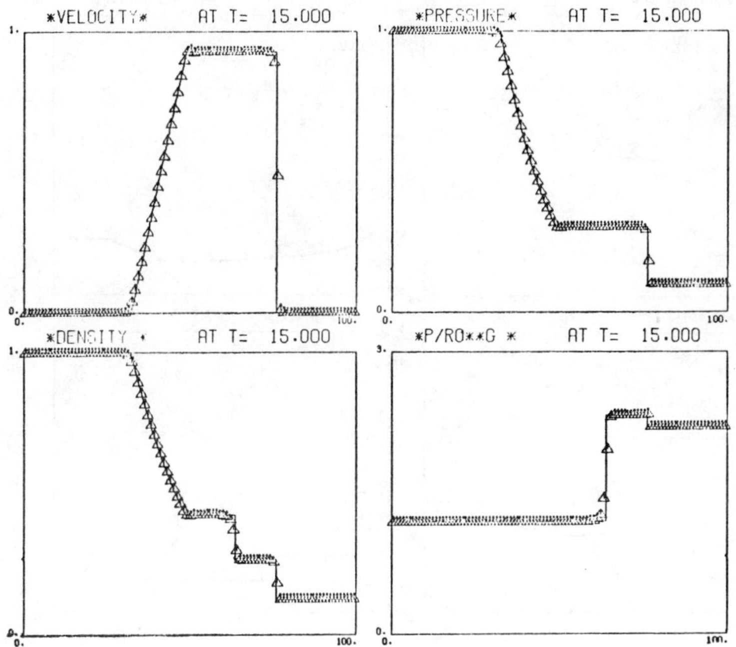

Figure 4. Results for Sod's example using the E_1 Scheme.

(b) Our second example is that of a converging cylindrical shock. Initially a cylindrical diaphragm of radius $r = 100$ separates two uniform regions of gas (γ=1.4) at rest. We take $p = \rho = 4$ in the outer region, $p = \rho = 1$ in the inner region. Removing the diaphragm at $t = 0$ leads to a shock wave followed by a contact discontinuity, both moving inward, and a diverging rarefaction wave. As is well-known, the shock accelerates as it approaches the axis of symmetry, is reflected from the axis, interacts with the (still converging) contact discontinuity, which results in a transmitted shock, a contact discontinuity (converging), and a weak reflected (converging) shock. This problem was treated by Lapidus [8], Abarbanel-Goldberg [1] and Sod [10]. We shall not discuss here the way we treat $r = 0$ in our scheme. Suffice it to say that it can be incorporated into the schemes described in the preceding section with only minor adjustments (see [3] for detailed results). In Figure 5 we give the flow profiles at two times. At $t = 56$ the shock is about to hit the axis. At $t = 110$, following the interaction between the reflected shock and the converging contact discontinuity, we see also a weak converging shock. In previous calculations this weak shock was detected only by Sod [10], using the sharp resolution of the random choice method. The scheme used here was E_∞, with 200 equally spaced grid-points.

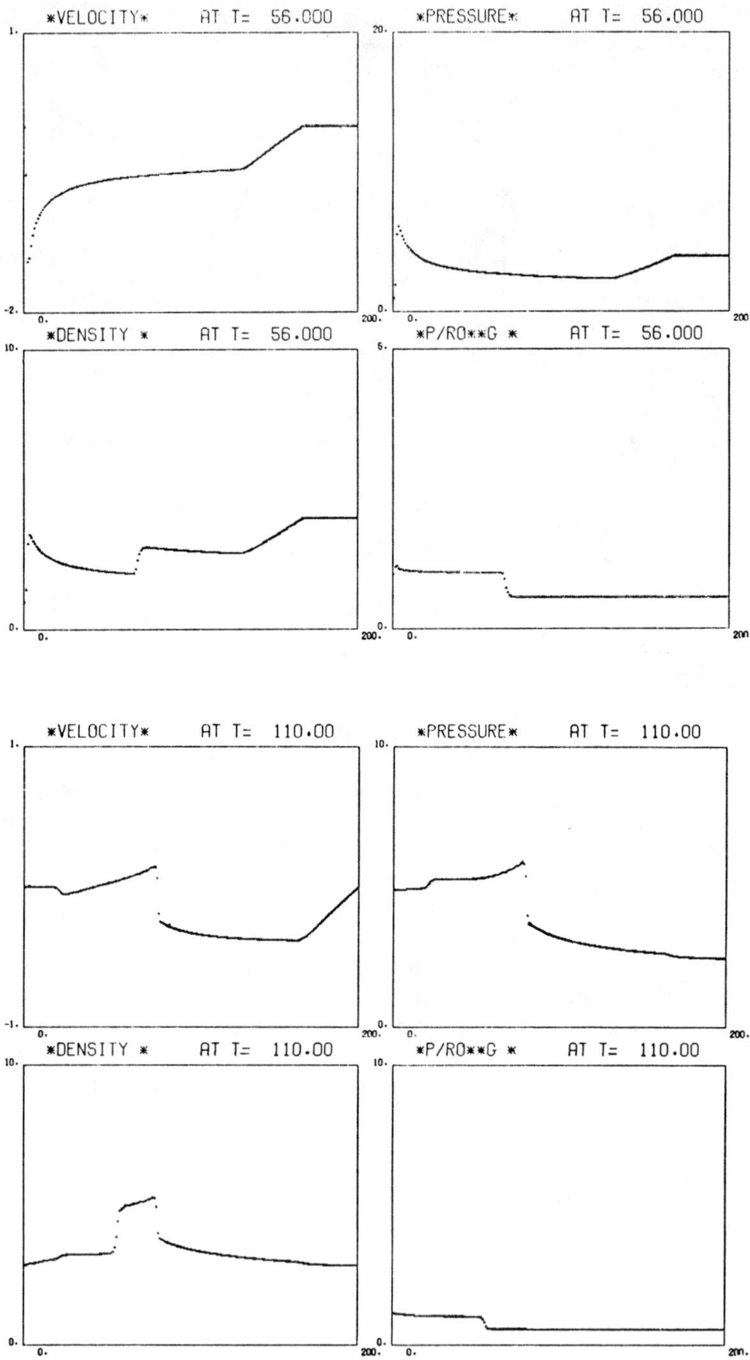

Figure 5. Results for the converging cylindrical shock.

REFERENCES

[1] S. ABARBANEL and M. GOLDBERG, Numerical solution of quasi-conservative hyperbolic systems - the cylindrical shock problem, J. Comp. Phys. 10(1972), pp. 1-21.

[2] M. BEN-ARTZI and J. FALCOVITZ, A second order Godunov-type scheme for compressible fluid dynamics, J. Comp. Phys. (1984, in press).

[3] M. BEN-ARTZI and J. FALCOVITZ, An upwind second order scheme for compressible duct flows, Technion preprint series MT-625.

[4] P. COLELLA and P. R. WOODWARD, The piecewise parabolic method (PPM) for gas dynamical simulations, Lawrence Berkeley Laboratory Report LBL-14661 (1982).

[5] S. K. GODUNOV, A finite difference method for the numerical computation and discontinuous solutions of the equations of fluid dynamics, Mat. Sbornik 47(1959), pp. 271-295.

[6] A. HARTEN, High resolution schemes for hyperbolic conservation laws, J. Comp. Phys. 49(1983), pp. 357-393.

[7] A. HARTEN, P.D. LAX and B. VAN-LEER, On upstream differencing and Godunov-type schemes for hyperbolic conservation laws, SIAM Review 25(1983), pp. 35-61.

[8] A. LAPIDUS, Computation of radially symmetric shocked flows, J. Comp. Phys. 8(1971), pp. 106-118.

[9] G. A. SOD, A survey of several finite difference methods for systems of non-linear hyperbolic conservation laws, J. Comp. Phys. 27(1978), pp. 1-31.

[10] G. A. SOD, A numerical study of a converging cylindrical shock, J. Fluid Mech. 83(1977), pp. 785-794.

[11] B. VAN-LEER, Towards the ultimate conservative difference scheme V, J. Comp. Phys. 32(1979), pp. 101-136.

ON A CLASS OF TVD SCHEMES FOR GAS DYNAMIC CALCULATIONS

H. C. YEE,† R. F. WARMING† AND AMI HARTEN‡

Abstract

The purpose of this paper is to review a class of explicit and implicit second-order accurate Total Variation Diminishing (TVD) schemes and to show by numerical experiments, the performance of these schemes to the Euler equations of gas dynamics. The method of constructing these second-order accurate TVD schemes is sometimes known as the modified flux approach.

1. Introduction

At the present stage of development, numerical computation of the Navier-Stokes equations of gas dynamics is concentrated on improving the method for the inviscid part of the equations; i.e., the Euler equations. The reason is that mathematical modelling of the viscous terms for most physical problems is not in a well developed stage. The viscous part of the various models for the Navier-Stokes equations currently being used can be easily handled numerically. Also,

†Research Scientist, Computational Fluid Dynamics Branch.
‡Associate Professor, School of Mathematical Sciences.

it seems that the slow convergence rate in solving both the Euler and the Navier-Stokes equations is associated with the inviscid terms when strong viscous effects are not present. Therefore, a good handle of numerically solving the Euler equations accurately and efficiently is essential for the development of numerical methods for the Navier-Stokes equations as well as for the truly inviscid fluid flows. The Euler equations, in conservative form, are a highly nonlinear system of hyperbolic conservation laws.

In one spatial dimension, the Euler equations of gas dynamics can be written in the conservative form as

$$\frac{\partial U}{\partial t} + \frac{\partial F(U)}{\partial x} = 0 \tag{1.1a}$$

where

$$U = \begin{bmatrix} \rho \\ m \\ e \end{bmatrix}, \quad F = \begin{bmatrix} m \\ m^2/\rho + p \\ (e + p)m/\rho \end{bmatrix} \tag{1.1b}$$

Here U is the vector of conservative variables, F is the flux vector, and $m = \rho u$. The primitive variables are the density ρ, the velocity u, and the pressure p. The total energy per unit volume e, is defined as

$$e = \rho\epsilon + \rho u^2/2 \tag{1.1c}$$

with ϵ as the internal energy per unit mass. The pressure p for a perfect gas is defined as

$$p = (\gamma - 1)[e - m^2/2\rho] \tag{1.1d}$$

where γ is the ratio of specific heats.

Let A denote the Jacobian matrix $\partial F(U)/\partial U$ whose eigenvalues are

$$(a_1, a_2, a_3) = (u - c, u, u + c) \tag{1.2}$$

where c is the local speed of sound. The quasi-linear equation associated with (1.1) is

$$\frac{\partial U}{\partial t} + A(U)\frac{\partial U}{\partial x} = 0 \tag{1.3}$$

If we assume that the matrix A is locally constant, equation (1.3) can be transformed into an uncoupled system of scalar hyperbolic equations

$$\frac{\partial u_l}{\partial t} + a_l \frac{\partial u_l}{\partial x} = 0, \qquad l = 1, 2, 3 \tag{1.4}$$

by defining a new vector $(u_1, u_2, u_3)^T = P^{-1}U$, where P is the transformation matrix that diagonalized A. When analyzing the stability properties of a linearized version of an algorithm for a partial differential equation associated with equation (1.3), we normally only apply the scheme to the scalar equation (1.4) provided the boundary conditions are applied to the characteristic variables u_l. Hereafter we refer to (1.4) as the hyperbolic model equations.

Stability analysis of difference schemes for linear hyperbolic partial differential equations (PDE's) (i.e., the model equations (1.4)) is very well established. However, stability analysis of difference schemes for nonlinear hyperbolic PDE's is still in its infancy. In general, the stability theory for linear difference equations is of use in checking the "local" stability of linearized equations (1.4) obtained from truly nonlinear equations. Practical experience shows that instabilities usually start as local phenomena, at least for mildly nonlinear problems. However, in many instances when strong discontinuities are present, local stability is neither necessary nor sufficient for the nonlinear problems. One traditional remedy for removing instabilities is to introduce "linear" numerical dissipation (or "artificial viscosity" or a "smoothing term") into the difference schemes. One can do so by designing the scheme to be dissipative [1].

The majority of practical applications in fluid dynamics is based on this traditional approach of adding an additional dissipation term [2,3] to the numerical scheme to improve nonlinear instability. However, this approach alone will not guarantee convergence to a physically correct solution in the nonlinear case.

Lax and Wendroff [4] showed that the limit solution of any finite difference scheme in a conservation form which is consistent with the conservation laws satisfies the jump conditions across a discontinuity *automatically*. This was a conceptual breakthrough which enabled the direct discretization of the conservation laws by introducing the notion of numerical dissipation. However, weak solutions (solutions with shocks and contact discontinuities) of hyperbolic conservation laws are not uniquely determined by their initial values; an entropy condition is needed to pick out the physically relevant solutions. The question arises whether finite-difference approximations converge to this particular solution. It is shown in reference [5,6] that in the case of a single conservation law, monotone schemes (to be defined later) always converge to the physically relevant solution. If the scheme is not monotone, then it must be consistent with an entropy inequality for the insurance of convergence to a physically relevant solution [5].

Thus monotone schemes possess many desirable properties for the calculation of discontinuous solutions. However, monotone schemes are only first-order accurate. Consequently, they are not accurate enough to compute complicated flow fields unless extremely small grid spacing is used.

To overcome the above difficulties, a new class of schemes that is more appropriate for the computation of weak solutions of nonlinear hyperbolic conservation laws is considered. These schemes are required (a) to be total variation diminishing in the nonlinear scalar case and the constant coefficient system case (see [7,8] or next section), (b) to be consistent with the conservation law and an entropy inequality [5], and (c) to be second-order accurate away from shocks. The first property guarantees that the scheme does not generate spurious oscillations. We refer to schemes with this property (a) as *Total Variation Diminishing* (TVD) schemes. The second property guarantees that the weak solutions are physical ones. This class of schemes is one step beyond the linear theory for solving the nonlinear scalar equation. We want to emphasize that a TVD scheme does not generate spurious oscillations for one-dimensional nonlinear scalar hyperbolic conservation laws and constant coefficient hyperbolic systems. Nothing is said about nonlinear systems or two-dimensional scalar hyperbolic conservation laws. Furthermore, Goodman and Leveque [9] recently have obtained the following result: for a specific norm, a two-dimensional scalar approximation cannot be TVD and still be more than first-order accurate. Maybe what we need is a different norm or a more relaxed definition than the TVD property for one dimension. Nevertheless, numerical experiments with these schemes which are based on a one-dimensional concept applied to two-dimensional problems via local one-dimensional splitting show that these types of schemes do perform well for fairly complex shock structures [10,11]. It is the purpose of this paper to show the application of a class of explicit and implicit TVD schemes to the one- and two-dimensional Euler equations of gas dynamics. Transient and steady-state numerical calculations are also included.

The following is an introduction to monotone, and explicit TVD schemes.

2. Monotone and First-Order Explicit TVD Schemes

Consider the scalar hyperbolic conservation law

$$\frac{\partial u}{\partial t} + \frac{\partial f(u)}{\partial x} = 0, \tag{2.1}$$

where $a(u) = \partial f / \partial u$ is the characteristic speed. Let u_j^n be the numerical solution of (2.1) at $x = j\Delta x$ and $t = n\Delta t$, with Δx the spatial mesh size and Δt the time step. A general three-point explicit difference scheme in conservation form can be written as

$$u_j^{n+1} = u_j^n - \lambda(h_{j+\frac{1}{2}}^n - h_{j-\frac{1}{2}}^n), \tag{2.2}$$

where $h^n_{j+\frac{1}{2}} = h(u^n_j, u^n_{j+1})$, $\lambda = \Delta t/\Delta x$. Here, h is commonly called a numerical flux function. The numerical flux function h is required to be consistent with the conservation law in the following sense

$$h(u_j, u_j) = f(u_j). \tag{2.3}$$

Rewrite equation (2.2) as

$$u^{n+1}_j = H(u^n_{j-1}, u^n_j, u^n_{j+1}). \tag{2.4}$$

The numerical scheme (2.4) is said to be monotone if H is a monotonic increasing function of each of its arguments. Examples of monotone schemes are: Lax-Friedrick scheme [6], Godunov scheme [1], and Engquist and Osher scheme [12].

Consider a numerical scheme with a numerical flux function of the following form

$$h_{j+\frac{1}{2}} = \frac{1}{2}[f_j + f_{j+1} - Q(a_{j+\frac{1}{2}})\Delta_{j+\frac{1}{2}}u], \tag{2.5a}$$

where $f_j = f(u_j)$, $\Delta_{j+\frac{1}{2}}u = u_{j+1} - u_j$ and

$$a_{j+\frac{1}{2}} = \begin{cases} (f_{j+1} - f_j)/\Delta_{j+\frac{1}{2}}u & \Delta_{j+\frac{1}{2}}u \neq 0 \\ a(u_j) & \Delta_{j+\frac{1}{2}}u = 0. \end{cases} \tag{2.5b}$$

Here Q is a function of $a_{j+\frac{1}{2}}$ and λ. The function Q is sometimes referred to as the coefficient of numerical viscosity.

Since the form of the numerical flux function (2.5) is not a common notation, we will give an example here which we refer to quite frequently for the rest of the paper. A scheme with a numerical flux of the form (2.5) is the first-order accurate upwind scheme [13,14].

$$u^{n+1}_j = u^n_j - \frac{\lambda}{2}[1 - \text{sgn}(a^n_{j+\frac{1}{2}})](f^n_{j+1} - f^n_j) - \frac{\lambda}{2}[1 + \text{sgn}(a^n_{j-\frac{1}{2}})](f^n_j - f^n_{j-1}). \tag{2.6}$$

If we define $a_{j+\frac{1}{2}}$ as (2.5b), then (2.6) can be written as

$$u^{n+1}_j = u^n_j - \frac{\lambda}{2}[f^n_{j+1} - f^n_{j-1} - |a^n_{j+\frac{1}{2}}|\Delta_{j+\frac{1}{2}}u^n + |a^n_{j-\frac{1}{2}}|\Delta_{j-\frac{1}{2}}u^n]. \tag{2.7a}$$

Here, the numerical flux function is

$$h_{j+\frac{1}{2}} = \frac{1}{2}[f_j + f_{j+1} - |a_{j+\frac{1}{2}}|\Delta_{j+\frac{1}{2}}u] \tag{2.7b}$$

with

$$Q(a_{j+\frac{1}{2}}) = |a_{j+\frac{1}{2}}|. \tag{2.7c}$$

This scheme is sometimes known as the Huang scheme [13], the Roe scheme [14] or the Murman scheme [15]. It is well known that (2.6) or (2.7) is not consistent with an entropy inequality, and the scheme might converge to a nonphysical solution. A slight modification of the numerical viscosity term

$$Q(z) = \begin{cases} |z| & |z| \geq \epsilon \\ (z^2 + \epsilon^2)/2\epsilon & |z| < \epsilon, \end{cases} \tag{2.8}$$

can remedy the entropy violating problem [7], where $\epsilon > 0$ is a parameter.

If we define

$$C^{\pm}(z) = \frac{1}{2}[Q(z) \pm z], \tag{2.9}$$

then, this upwind scheme can be written as

$$u_j^{n+1} = u_j^n + \lambda C^{-}(a_{j+\frac{1}{2}}^n)\Delta_{j+\frac{1}{2}}u^n - \lambda C^{+}(a_{j-\frac{1}{2}}^n)\Delta_{j-\frac{1}{2}}u^n. \tag{2.10}$$

In other words, this scheme can be viewed as a generalization of the Courant, Isaacson and Rees upwind scheme [16]. This scheme plays a major role in the development of a subclass of second- order accurate TVD schemes. The notation of the numerical flux function (2.5) will be used heavily for the rest of the paper.

The total variation of a mesh function u_j^n is defined to be

$$TV(u^n) = \sum_{j=-\infty}^{\infty} |u_{j+1}^n - u_j^n| = \sum_{j=-\infty}^{\infty} |\Delta_{j+\frac{1}{2}}u^n|. \tag{2.11}$$

The numerical scheme (2.2) for the initial-value problem of (2.1) is said to be TVD if

$$TV(u^{n+1}) \leq TV(u^n). \tag{2.12}$$

It can be shown that a sufficient condition [7] for (2.2), together with (2.5), to be a TVD scheme is

$$\frac{\lambda}{2}\left[-a_{j+\frac{1}{2}} + Q(a_{j+\frac{1}{2}})\right] \geq 0 \tag{2.13a}$$

$$\frac{\lambda}{2}\left[a_{j+\frac{1}{2}} + Q(a_{j+\frac{1}{2}})\right] \geq 0 \tag{2.13b}$$

$$\lambda Q(a_{j+\frac{1}{2}}) \leq 1. \tag{2.13c}$$

That is, for (2.10), $C^{\pm}$ have to be positive and their sum multiplied by λ must be less than or equal to 1. Applying the above conditions to equation (2.7), it can be easily shown that (2.7) is a TVD scheme. Therefore, for the scheme (2.2) with a general Q in (2.5) (other than (2.7c)) to be TVD, we have to pick Q such that (2.13) is satisfied.

It should be emphasized that condition (2.13) is only a sufficient condition; i.e., schemes that fail this test might still be TVD.

3. Second-Order Accurate (almost everywhere) Explicit TVD Scheme

Several techniques for the construction of nonlinear, explicit, second-order accurate almost everywhere, high-resolution schemes for hyperbolic conservation laws have been developed in recent years [7,8,17-20]. These schemes were constructed under a common theme; i.e., to achieved second-order accuracy without introducing spurious oscillations near discontinuities by employing some kind of feedback mechanism. Some of these schemes are based on physical and geometric arguments, and some are based on more formal mathematical approaches. Most of these schemes were independently derived, and thus they are very much different in form, methodology and design principle. However, from the standpoint of numerical analysis, these schemes are total variation diminishing (TVD) for nonlinear scalar hyperbolic conservation laws and for constant coefficient hyperbolic systems. Under certain asssumptions, some of these schemes turn out to be identical. The notion of TVD schemes was introduced by Harten in 1981 [7]. For convenience, from here on, second-order accurate means second-order accurate almost everywhere. The purpose of this section is to review Harten's method of constructing the second-order TVD schemes via the modified flux approach.

The modified flux approach is a technique to design a second-order accurate TVD scheme, without having to explicitly specify its switching mechanism. To do so we start with the first-order TVD scheme (2.2) together with (2.5) and (2.8) and apply it to a modified flux $\tilde{f}(u) = [f(u) + g(u)]$. The new numerical flux function $\tilde{h}_{j+\frac{1}{2}}$ depends on $(f+g)$ instead of f alone, the coefficient of the numerical viscosity term Q is a function of a modified characteristic speed $(a+\gamma)$, and $\tilde{h}_{j+\frac{1}{2}}$ can be written as

$$\tilde{h}_{j+\frac{1}{2}} = \frac{1}{2}[\tilde{f}_j + \tilde{f}_{j+1} - Q(a_{j+\frac{1}{2}} + \gamma_{j+\frac{1}{2}})\Delta_{j+\frac{1}{2}}u], \tag{3.1a}$$

where $\tilde{f}_j = f_j + g_j$ and $\gamma_{j+\frac{1}{2}}$ is defined almost the same way as $a_{j+\frac{1}{2}}$ except it is a function of the g_j's instead of the f_j's and it can be expressed as

$$\gamma_{j+\frac{1}{2}} = \begin{cases} (g_{j+1} - g_j)/\Delta_{j+\frac{1}{2}}u & \Delta_{j+\frac{1}{2}}u \neq 0 \\ 0 & \Delta_{j+\frac{1}{2}}u = 0. \end{cases} \tag{3.1b}$$

The requirements on g are: (i) The function g should have a bounded γ in (3.1b) so that the scheme (2.2) together with (3.1) is TVD with respect to the modified flux $(f + g)$. (ii) The modified scheme should be second-order accurate (except at points of extrema). In [7,8], Harten devised a recipe for g that satisfies the above two requirements. We will use this particular form of g for the discussion here. It can be written as

$$\begin{aligned} g_j &= S \cdot \max\left[0, \min(\sigma_{j+\frac{1}{2}}|\Delta_{j+\frac{1}{2}}u|, S \cdot \sigma_{j-\frac{1}{2}}\Delta_{j-\frac{1}{2}}u)\right] \\ S &= \operatorname{sgn}(\Delta_{j+\frac{1}{2}}u) \end{aligned} \tag{3.1c}$$

with $\sigma_{j+\frac{1}{2}} = \sigma(a_{j+\frac{1}{2}})$ and we choose

$$\sigma(z) = \frac{1}{2}[Q(z) - \lambda z^2] \geq 0 \tag{3.1d}$$

for time accurate calculations. With this choice of $\sigma(z)$, the scheme (2.2) together with (3.1) is second-order accurate in both time and space (except at points of extrema); see [7]. Also, with this choice of the g function, the second-order TVD scheme will automatically switch itself to first-order at points of extrema. This is one of the vehicles to prevent spurious oscillation near a shock. Other more general forms for g can be obtained following the line of argument of Sweby [21] and Roe [19].

The second-order TVD scheme can be written as

$$u_j^{n+1} = u_j^n + \lambda C^-(\tilde{a}_{j+\frac{1}{2}}^n)\Delta_{j+\frac{1}{2}}u^n - \lambda C^+(\tilde{a}_{j-\frac{1}{2}}^n)\Delta_{j-\frac{1}{2}}u^n, \tag{3.2}$$

with $\tilde{a}_{j+\frac{1}{2}}^n = a_{j+\frac{1}{2}}^n + \gamma_{j+\frac{1}{2}}^n$. This, again is an upwind scheme with the modified flux $\tilde{f} = (f + g)$. Another interesting observation is that equation (3.2) together with (2.7c) can be rewritten as

$$u_j^{n+1} = u_j^n - \frac{\lambda}{2}[1 - \text{sgn}(\tilde{a}_{j+\frac{1}{2}}^n)](\tilde{f}_{j+1}^n - \tilde{f}_j^n) - \frac{\lambda}{2}[1 + \text{sgn}(\tilde{a}_{j-\frac{1}{2}}^n)](\tilde{f}_j^n - \tilde{f}_{j-1}^n). \quad (3.3)$$

This is a straightforward extension of Huang or Roe's entropy violating first-order upwind scheme to second-order accuracy. The scheme looks identical to their first-order scheme (2.6) except the arguments a's and f's are different.

4. Extensions to Systems and Two-Dimensional Conservation Laws

Systems

At the present development, the concept of TVD schemes, like monotone schemes is only defined for nonlinear scalar conservation laws or constant coefficient hyperbolic systems. The main difficulty stems from the fact that, unlike the scalar case, the total variation in x of the solution to a system of nonlinear conservation laws is not necessarily a monotonic decreasing function of time. The total variation of the solution may actually increase at moments of interaction between waves. Not knowing a diminishing functional that bounds the total variation in x in the system case makes it impossible to fully extend the theory of the scalar case to the system case. What we can do is to extend the scalar TVD scheme to system cases so that the resulting scheme is TVD for the "*locally frozen*" constant coefficient system. To accomplish this, we define at each point a "*local*" system of characteristic fields (e.g.equation (1.4)), then apply the nonlinear scheme to each of the m scalar characteristic equations. Here m is the dimension of the hyperbolic system. This extension technique is a somewhat generalized version of the procedure suggested by Roe in [14].

We can further improve the resolution of the contact discontinuity by an artificial compression method (on the linearly degenerate field $a_2 = u$ of equation (1.2)). We can do so by increasing the value of g_j by

$$\tilde{g}_j = [1 + \omega\theta_j]g_j, \qquad \omega > 0 \quad (4.1)$$

with

$$\theta_j = \frac{|\Delta_{j+\frac{1}{2}}w - \Delta_{j-\frac{1}{2}}w|}{|\Delta_{j+\frac{1}{2}}w| + |\Delta_{j-\frac{1}{2}}w|} \quad (4.2)$$

where $\Delta_{j+\frac{1}{2}}w = w_{j+1} - w_j$, and w_j is the characteristic variable corresponding to the linearly degenerate field [22,23]. From numerical experiments for the Euler equations, $\omega = 2$ seems to be a good choice. For a detailed implementation of the scheme to one-dimensional Euler equations of gas dynamics, see reference [10].

Two Spatial Dimensions

Consider a two-dimensional scalar conservation law

$$\frac{\partial u}{\partial t} + \frac{\partial f(u)}{\partial x} + \frac{\partial g(u)}{\partial y} = 0. \tag{4.3}$$

Let $u^n_{j,k}$ be a numerical solution of (4.3) at $x = j\Delta x$, $y = k\Delta y$, and $t = n\Delta t$, with Δx the mesh size in the x-direction and Δy the mesh size in the y-direction. Also, let $\tilde{h}_{j+\frac{1}{2},k}$ and $\tilde{q}_{j,k+\frac{1}{2}}$ be the second-order accurate numerical fluxes in the x- and y-directions. The explicit TVD scheme can be implemented in two space dimensions by the method of fractional steps as follows:

$$\overset{*}{u}_{j,k} = \overset{n}{u}_{j,k} - \frac{\Delta t}{\Delta x}(\tilde{h}^n_{j+\frac{1}{2},k} - \tilde{h}^n_{j-\frac{1}{2},k}) = L_x u^n_{j,k} \tag{4.4a}$$

$$u^{n+1}_{j,k} = \overset{*}{u}_{j,k} - \frac{\Delta t}{\Delta y}(\overset{*}{\tilde{q}}_{j,k+\frac{1}{2}} - \overset{*}{\tilde{q}}_{j,k-\frac{1}{2}}) = L_y \overset{*}{u}_{j,k} \tag{4.4b}$$

that is

$$u^{n+1}_{j,k} = L_y L_x u^n_{j,k}. \tag{4.5}$$

In order to retain the original time accuracy of the method, we use a Strang type of fractional step operators, namely

$$u^{n+2}_{j,k} = \overset{*}{L}_x L_y L_x L_y \overset{*}{L}_x u^n_{j,k} \tag{4.6}$$

where $\overset{*}{L}_x$ denotes the operator with the time step equal to $\Delta t/2$.

The above method is a formal extension of the one-dimensional TVD scheme to two dimensions. In order to get a truly two-dimensional shock-capturing TVD scheme, we have to include a two-dimensional modified flux. Davis [24], and Roe and Baines [25] provide some interesting ideas on this subject.

On the other hand, We may enhance resolution for the two-dimensional scheme by an artificial compression method. Let g_j, g_k be the modified flux in the x- and y-directions, and let $\gamma_{j+\frac{1}{2}}$, $\gamma_{k+\frac{1}{2}}$ be the corresponding (3.1b) in the x- and y-directions. One way of ultilizing the artificial compression method is to increase the value of $\gamma_{j+\frac{1}{2}}$ and $\gamma_{k+\frac{1}{2}}$. To accomplish this, we increase the size of the corresponding g_j in (3.1c), for example, by multiplying g_j by $[1+\omega\theta_j]$; i.e.

$$\tilde{g}_j = [1+\omega\theta_j]g_j, \qquad \omega > 0 \tag{4.8a}$$

with

$$\theta_j = \frac{|\alpha_{j+\frac{1}{2}} - \alpha_{j-\frac{1}{2}}|}{|\alpha_{j+\frac{1}{2}}| + |\alpha_{j-\frac{1}{2}}|} \tag{4.8b}$$

where $\alpha_{j+\frac{1}{2}} = u_{j+1,k} - u_{j,k}$. Similarily, we can obtain $\tilde{g}_k = [1+\omega\theta_k]g_k$ for the y-direction. Again this is not an appropriate way of applying artificial compression to two-dimensional problems. To do it correctly, we have to devised a truly two-dimensional artificial compression term. However, from numerical experiments, it does enhance the shock resolution for the two-dimensional gas dynamic problems we are considering.

Detailed implementation of this scheme for the two-dimensional Euler equations can be found in reference [10].

5. Numerical Experiments with a Second-Order Explicit TVD Scheme

In this section, we show, by numerical experiments, the performance of the second-order accurate explicit TVD schemes (with the numerical flux defined in (3.1) together with (2.16)) when applied to gas dynamics problems. The problems consist of a one-dimensional shock tube problem, a quasi-one-dimensional divergent nozzle problem, a two-dimensional regular shock reflection problem, and a two-dimensional shock wave impinging on a cylinder problem. All these calculations have been reported separately in references [7,10,11,26]. Here we give a complete summary of these numerical experiments.

(a) The Shock Tube Problem (transient calculation)

The first experiment is a shock tube or Riemann problem with initial data

$$\begin{bmatrix}\rho\\ m\\ e\end{bmatrix} = \begin{bmatrix}0.445\\ 0.311\\ 8.928\end{bmatrix} \quad x<0, \qquad \begin{bmatrix}\rho\\ m\\ e\end{bmatrix} = \begin{bmatrix}0.5\\ 0.0\\ 1.4275\end{bmatrix} \quad x>0. \tag{5.1}$$

Figure (5.1) shows the result obtained by the second-order accurate TVD scheme. The numerical solution after 100 time steps is shown by circles; the exact solution is shown by the solid line. The calculation was performed with a CFL number $(=\max(a_l\Delta t/\Delta x))$ of 0.95, and 140 grid points. The CFL or Courant number is a measure of the maximum permissible time step for a stable solution. An artificial compression was added to the linearly degenerate field $a_2 = u$ of (1.2). Very accurate contact discontinuity as well as shock resolution was obtained. Comparison with other schemes can be found in [7].

(b) A Quasi-One-Dimensional Nozzle Problem (steady-state calculation)

The governing equation for the quasi-one-dimensional nozzle problem can be written as

$$\frac{\partial \overline{U}}{\partial t} + \frac{\partial F(\overline{U})}{\partial x} + H(\overline{U}) = 0, \tag{5.2a}$$

where

$$\overline{U} = \begin{bmatrix} \rho\kappa \\ m\kappa \\ e\kappa \end{bmatrix}, \quad F = \begin{bmatrix} m\kappa \\ (m^2/\rho + p)\kappa \\ (e+p)m\kappa/\rho \end{bmatrix}, \quad H(\overline{U}) = \begin{bmatrix} 0 \\ p\frac{\partial \kappa}{\partial x} \\ 0 \end{bmatrix} \tag{5.2b}$$

with κ, the area of the nozzle, a function of x. The nozzle we consider is a divergent nozzle [27] with

$$\kappa(x) = 1.398 + 0.347\tanh(0.8x - 4) \tag{5.2c}$$

The steady flow condition was supersonic inflow, subsonic outflow with a shock. The computational domain was $0 \leq x \leq 10$. We used a very coarse equal mesh spacing of $\Delta x = 0.5$ (i.e., 20 spatial intervals), to evaluate the resolution of the scheme. A linear interpolation between the exact steady-state boundary values is used as initial conditions.

Figure (5.2) shows the converged steady-state density distribution after 700 time steps. The solid line is the exact solution and the diamonds are the computed solution. Only 12 points are plotted. The 8 points not shown on both ends of the x-axis are equal to the exact solution. Very accurate shock resolution was obtained. Comparison with other schemes can be found in reference [26]. From numerical experiments, we found that artificial compression is not necessary for this one-dimensional problem.

(c) Shock Wave Cylinder Interaction (transient calculation)

Since the shock structures of this problem are less known in text book literature, we will devote more space to this problem. This work was done jointly with P. Kutler of NASA Ames Research Center.

A good test problem for assessing the capabilities of any shock-capturing scheme is the shock-diffraction problem; i.e., the computation of the unsteady flow field resulting from a planar moving shock wave striking an obstacle. The shock-diffraction problem contains most of the flow discontinuities of the Euler equations. In the present numerical experiment the diffraction process is determined over a cylinder. The shock patterns at two instances in time t_1 and t_2 after initial impingement are sketched in figure (5.3).

When the incident shock first collides with the cylinder, regular reflection occurs at the shock impingement point. As the impingement point of the incident shock propagates around the body, the reflection process makes a transition from regular to Mach reflection. It should be pointed out that during the transition process, complex and double Mach reflection shock structures are possible. Their occurrence is dependent on the initial strength of the incident shock wave. For single Mach reflection, a triple point forms and the incident shock no longer touches the body. Emanating from the triple point are three waves: 1) a Mach stem which strikes the body perpendicularly, 2) a slip surface or shear layer which strikes the body and results in a vortical singularity (nodal point of streamlines), and 3) the reflected shock which propagates away from the body. In addition to the above flow field characteristics, a stagnation point (saddle point of streamlines) exists at the plane of symmetry, both forward and aft on the body.

Both MacCormack's explicit method [28] and the explicit TVD scheme were applied to the shock wave-cylinder interaction problem. For a fair comparison, the TVD scheme was implemented in an existing computer code [29] which also contained MacCormack's method, so that same initial conditions, boundary conditions and coordinate transformation were used. A cylindrical grid consisting of 50 points around the half-cylindrical (ξ-direction) body and 51 points between the body and outer boundary (η-direction) was used. The body radius is one and the distance from the body to the outer boundary is between 2 to 4 (depending on the incident shock Mach number). Rays from the coordinate system origin are spaced at equal angles with points uniformly placed in the radial direction between the body and the outer boundary (see figure (5.4)).

Figure (5.4) shows a schematic representation of the grid with its boundaries and initial conditions. The nodal points to the right of the planar moving shock are initialized to free stream values while those to the left are set equal to the post moving shock conditions. In the outer boundary, it is necessary to track the moving planar shock as a function of time along this boundary surface.

At the planes of symmetry, the reflection principle is used; i.e., the pressure, density and u-velocity component are treated as even functions across the plane of symmetry while the v-velocity component is treated as an odd function. The boundary condition at the surface of the cylinder must satisfy the tangency condition which requires that the velocity in the radial-direction be equal to zero at the body. Furthermore, for convenience, an image line of nodal points is considered which falls one mesh interval inside the body, so that the reflection principle can be applied.

MacCormack's method with a fourth-order dissipation term was run at a Courant (CFL) number of 0.6 for stability while the TVD method (with one-dimensional artificial compression terms added to both the ξ- and the η-directions) was run at a Courant number of 0.9 for efficiency. The stability and accuracy of the TVD method is insensitive to Courant number between 0.5 and 1. Both methods were run to approximately the same total time (100 steps for MacCormack's method, 70 steps for the TVD method). The results in the form of pressure and density contour plots are shown in figure (5.5) at a time for which Mach reflection of the incident shock exists. The incident shock Mach number was 2. The results from MacCormack's method are shown in figures (5.5a) and (5.5b). Those for the TVD method are shown in figures (5.5c) and (5.5d). It can be seen that the TVD scheme results in a better defined flow field; i.e., "crisper" shocks and hardly any associated spurious oscillations. The slip surface which emanates from the triple point is not captured by either method. However, if we use more grid points, better control of artificial compression and more appropriate intermediate boundary condition procedures for the fractional step method, we think the TVD method will be able to capture the slip surface.

To test the shock-capturing capability of the TVD method at higher incident shock Mach numbers using the same grid size as before, cases were run at Mach numbers between 3 and 10. Figure (5.6) shows the pressure and density contour plots for the TVD scheme with an incident shock Mach number of 10 at a time when the incident shock was already passed the cylinder. For this case the Mach stem that extended from the triple point to the body has struck the plane of symmetry and reflected from it. This will eventually make a transition to a Mach reflection just as it did at the body. The result shows that the TVD scheme is very stable and produces high resolution shock waves. The MacCormack method, on the other hand, was unstable under the same flow condition and CFL ≥ 0.2.

The numerical result shown in figures (5.5c), (5.5d) and (5.6) used a simple form of the $Q(z)$ function [7,10]

$$Q(z) = \lambda z^2 + 1/4 \tag{5.3a}$$

$$\sigma(z) = 1/8 \tag{5.3b}$$

We found that with equation (5.3), a slightly better shock resolution was obtained than $Q(z)$ in (2.7c) or (2.8). Note that one can get (5.3b) by simply substituting (5.3a) into (3.1d).

The TVD scheme requires approximately twice the CPU time per time step as MacCormack's method but results in enhanced numerical stability and solution

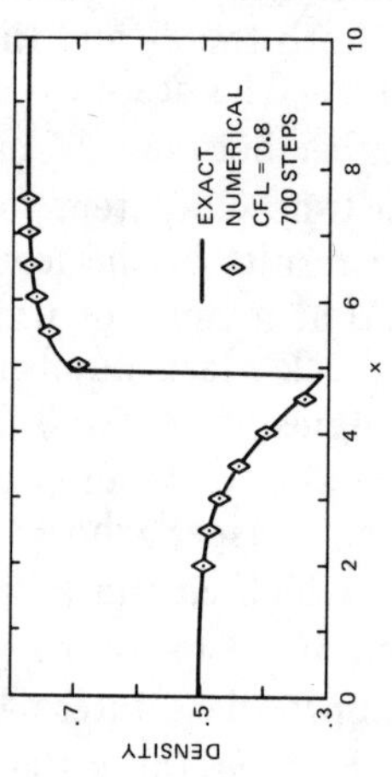

Fig. 5.2 Divergent nozzle: second-order explicit TVD scheme.

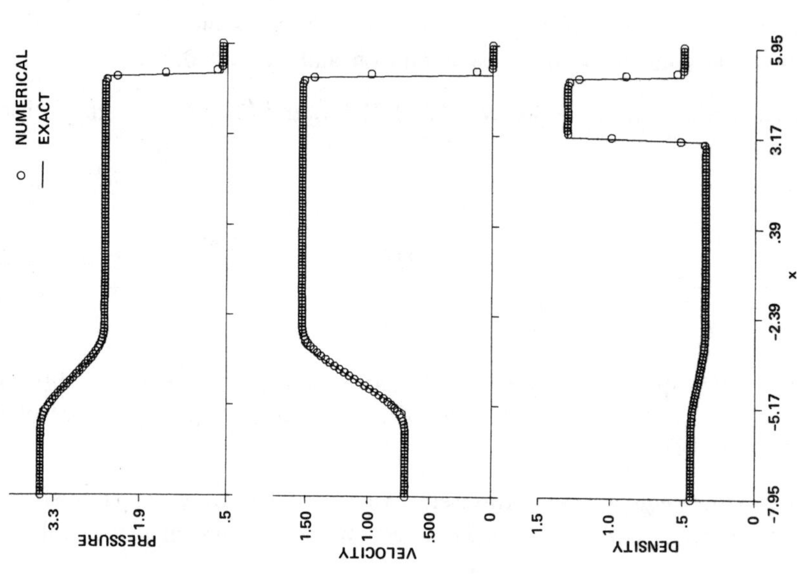

Fig. 5.1 Shock tube problem: pressure, velocity and density distribution of a second-order explicit TVD scheme.

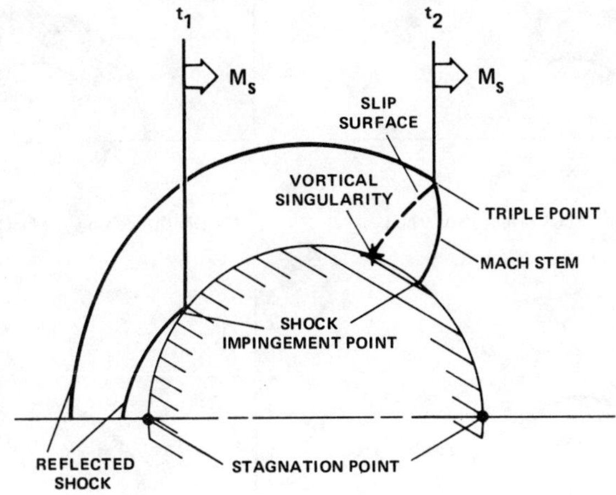

Fig. 5.3 Shock structure for shock diffraction over cylinder at two different times.

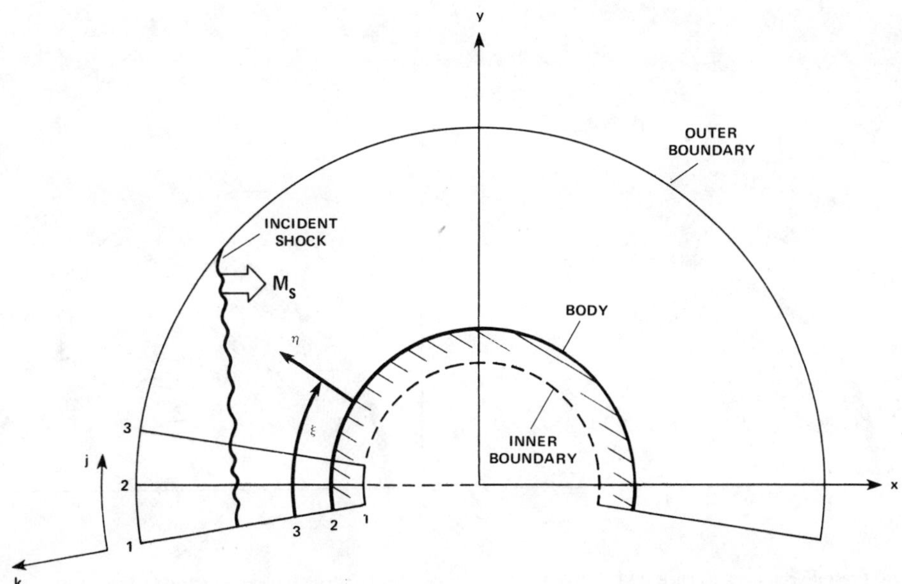

Fig. 5.4 Schematic of the computational grid.

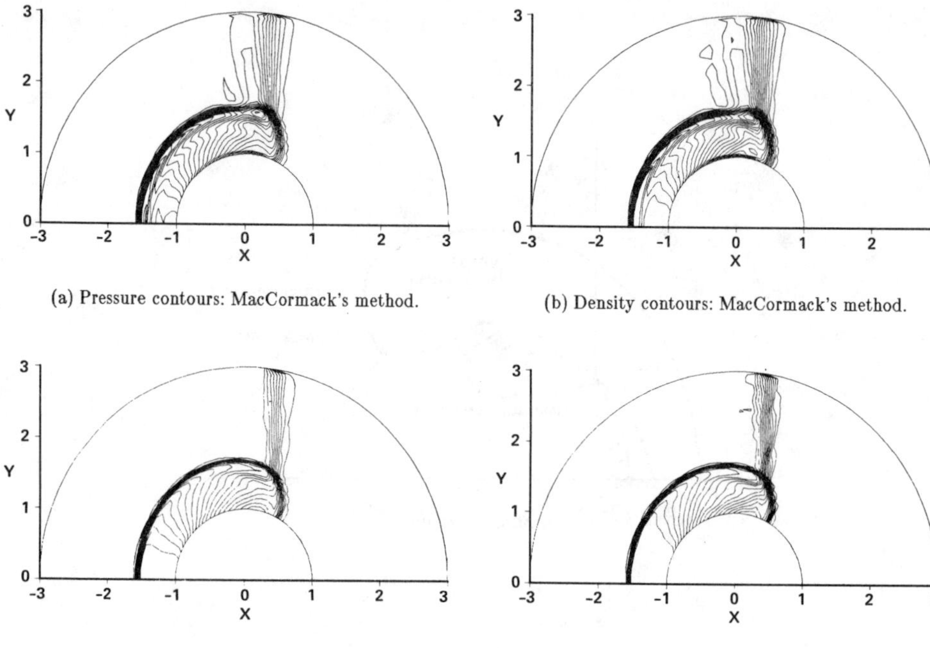

(a) Pressure contours: MacCormack's method.

(b) Density contours: MacCormack's method.

(c) Pressure contours: Explicit TVD method.

(d) Density contours: Explicit TVD method.

Fig. 5.5 Pressure and density contours for the shock-wave cylinder interaction. ($M_s = 2$)

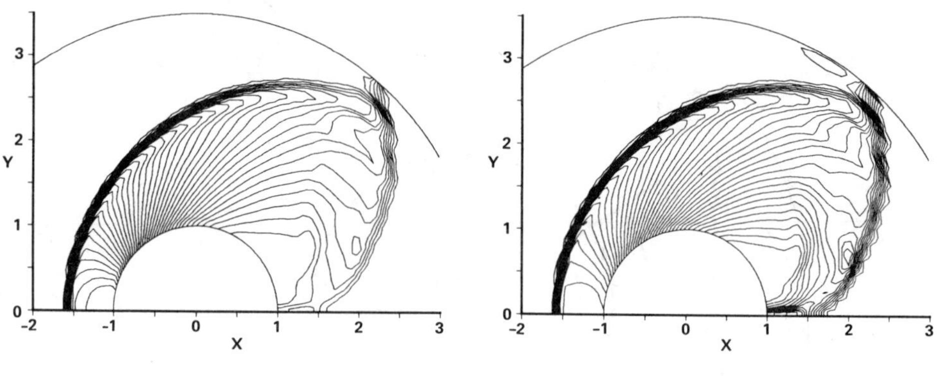

(a) Pressure contours: Explicit TVD method.

(b) Density contours: Explicit TVD method.

Fig. 5.6 Pressure and density contours for the shock-wave cylinder interaction. ($M_s = 10$)

accuracy. Implementation of the scheme in generalized coordinates can be found in [11].

(d) A Two-Dimensional Shock Reflection Problem (steady-state calculation)

In order to examine the applicability of the method for two-dimensional steady-state shock calculations, we consider a simple inviscid flow field developed by a shock wave reflecting from a rigid surface (figure (5.7)). The steady-state solution can be calculated exactly and thus can aid us in evaluating the quality of the numerical method. Figure (5.7) shows the indexing of the computational mesh. The incident shock angle ψ was 29° and the freestream Mach number M_{∞} was 2.9. The computational domain was $0 \leq x \leq 4.1$, and $0 \leq y \leq 1$, with a uniform grid size of 61x21. Initially, the entire flow field is set equal to the freestream supersonic inflow values. The appropriate analytical boundary conditions were applied along the boundaries of the domain. A more detailed description of the boundary condition procedures can be found in [30].

The exact minimum pressure corresponding to $\psi = 29^{\circ}$ and $M_{\infty} = 2.9$ is 0.714286 and the exact maximum pressure is 2.93398. The exact pressure solution and the computation domain are shown in figure (5.8). Forty-one pressure contour levels between the values of 0 and 4 with uniform increment 0.1 were used for the contour plots. The pressure coefficient was evaluated at $y = 0.5$ for $0 \leq x \leq 4.1$.

Figure (5.9) shows the numerical result of the explicit TVD scheme by a fractional step method with Q function defined by (5.3a) Again, a one-dimensional artificial compression term ($\omega = 2$ of equation (4.8)) was added to both the x- and the y-direction. It took approximately 350 steps to converge with a fixed $CFL = 0.8$. The average smearing of the shocks is two points. We also found that with equation (5.3), a slightly better shock resolution was obtain than $Q(z)$ in (2.7c) or (2.8).

6. Implicit TVD Schemes

Now we consider a one-parameter family of three-point conservative schemes of the form

$$u_j^{n+1} + \lambda\eta(h_{j+\frac{1}{2}}^{n+1} - h_{j-\frac{1}{2}}^{n+1}) = u_j^n - \lambda(1-\eta)(h_{j+\frac{1}{2}}^n - h_{j-\frac{1}{2}}^n) \tag{6.1}$$

where η is a parameter, $h_{j+\frac{1}{2}}^n = h(u_j^n, u_{j+1}^n)$, $h_{j+\frac{1}{2}}^{n+1} = h(u_j^{n+1}, u_{j+1}^{n+1})$, and $h(u_j, u_{j+1})$ is the numerical flux (2.5). This one-parameter family of schemes contains implicit as well as explicit schemes. When $\eta = 0$, (6.1) reduces to (2.2), the explicit method. When $\eta \neq 0$, (6.1) is an implicit scheme. For example: if $\eta = 1/2$, the time differencing is the trapezoidal formula, and if $\eta = 1$, the time differencing

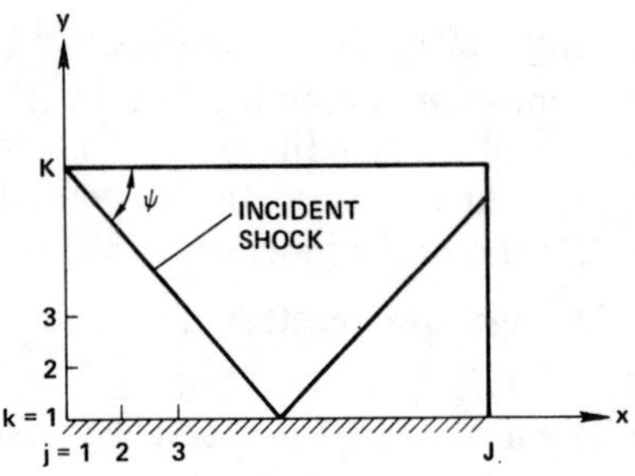

Fig. 5.7 Indexing of computational mesh for shock reflection problem.

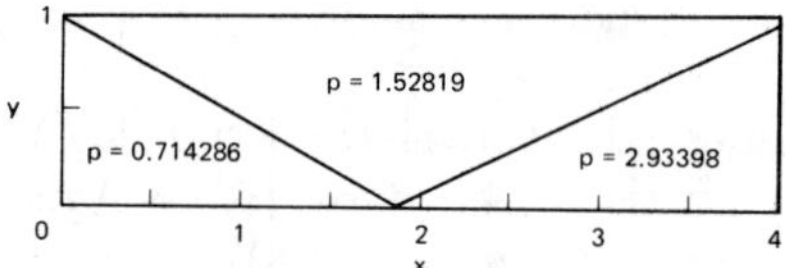

Fig. 5.8 Exact pressure solution.

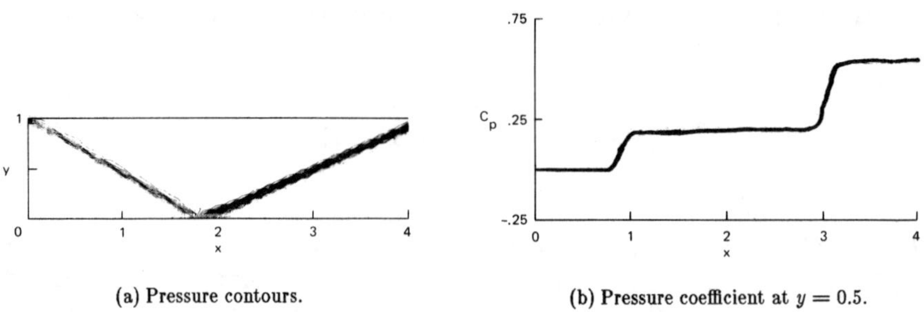

(a) Pressure contours.

(b) Pressure coefficient at $y = 0.5$.

Fig. 5.9 Pressure contours and pressure coefficient for the shock reflection problem.

is the backward Euler method. To simplify the notation, we will rewrite equation (6.1) as

$$L \cdot u^{n+1} = R \cdot u^n, \tag{6.2}$$

where L and R are the following finite-difference operators:

$$(L \cdot u)_j = u_j + \lambda\eta(h_{j+\frac{1}{2}} - h_{j-\frac{1}{2}}) \tag{6.3a}$$
$$(R \cdot u)_j = u_j - \lambda(1-\eta)(h_{j+\frac{1}{2}} - h_{j-\frac{1}{2}}). \tag{6.3b}$$

Sufficient conditions for (6.1) to be a TVD scheme are that

$$TV(R \cdot u^n) \leq TV(u^n) \tag{6.4a}$$

and

$$TV(L \cdot u^{n+1}) \geq TV(u^{n+1}). \tag{6.4b}$$

A sufficient condition for (6.4) is the CFL-like restriction

$$|\lambda a_{j+\frac{1}{2}}| \leq \lambda Q(a_{j+\frac{1}{2}}) \leq \frac{1}{1-\eta} \tag{6.5}$$

where $a_{j+\frac{1}{2}}$ is defined in equation (2.5b). Therefore, for the scheme to be TVD, we have to pick $Q(a_{j+\frac{1}{2}})$ such that (6.5) is satisfied. For a detailed proof of equations (6.4) and (6.5), see [8]. Observe that the backward Euler implicit scheme, $\eta = 1$ in (6.1) is unconditionally TVD, while the trapezoidal formula, $\eta = 1/2$ is TVD under the CFL-like restriction of 2. The forward Euler explicit scheme, $\eta = 0$ or equation (2.2), is TVD under the CFL restriction of 1.

7. Second-Order Implicit TVD Schemes

We can obtain a second-order accurate implicit TVD scheme by replacing the numerical flux function h of (6.1) with $\tilde{h}$ of equation (3.1). However, $\sigma_{j+\frac{1}{2}}$ is different from (3.1d). Instead, we choose

$$\sigma(z) = \begin{cases} \frac{1}{2}Q(z) + \lambda(\eta - \frac{1}{2})z^2 & \text{time dependent calculations} \\ \frac{1}{2}Q(z) & \text{steady-state calculations.} \end{cases} \tag{7.1}$$

The second choice in (7.1) makes the scheme second-order accurate in space, but first-order accurate in time. This choice of $\sigma(z)$ ensures that the steady-state solution does not depend on the time step Δt.

For example, the unconditionally TVD backward Euler scheme is of the form

$$u_j^{n+1} + \lambda(\tilde{h}_{j+1/2}^{n+1} - \tilde{h}_{j-1/2}^{n+1}) = u_j^n. \tag{7.2}$$

This is a highly nonlinear implicit scheme. An efficient procedure to solve this set of nonlinear equations is needed. Here we discuss a linearized form of the implicit scheme that is suitable for steady-state calculations.

For steady-state calculations, we can use the following unconditionally TVD linearized version of (7.2)

$$\begin{aligned} d_j - \lambda(C^-)^n(d_{j+1} - d_j) + \lambda(C^+)^n(d_j - d_{j-1}) \\ = -\lambda\left[\tilde{h}_{j+\frac{1}{2}} - \tilde{h}_{j-\frac{1}{2}}\right] \end{aligned} \tag{7.3a}$$

with $d_j = u_j^{n+1} - u_j^n$, $\tilde{h}_{j+\frac{1}{2}}$ from (3.1), and

$$(C^{\pm})^n = \frac{1}{2}[Q(a+\gamma) \pm (a+\gamma)]_{j+\frac{1}{2}}^n. \tag{7.3b}$$

One can obtain equation (7.3) by simply rewriting (7.2) into an upwind form so that the resulting equation is a function of $C^{\pm}(a+\gamma)_{j+\frac{1}{2}}^{n+1}$, $C^{\pm}(a+\gamma)_{j+\frac{1}{2}}^n$, $\Delta_{j+\frac{1}{2}}u^{n+1}$ and $\Delta_{j+\frac{1}{2}}u^n$, and then by dropping the time index of the $C^{\pm}$ from $(n+1)$ to n.

Although (7.3) is formally a five-point scheme, the coefficient matrix associated with it is tridiagonal with a dominant diagonal. See [10] for more details. We can also obtain another TVD linearized form

$$\begin{aligned} d_j - \lambda(B^-)^n(d_{j+1} - d_j) + \lambda(B^+)^n(d_j - d_{j-1}) \\ = -\lambda\left[\tilde{f}_{j+\frac{1}{2}}^n - \tilde{f}_{j-\frac{1}{2}}^n\right] \end{aligned} \tag{7.4a}$$

with

$$(B^{\pm})^n = \frac{1}{2}[Q(a) \pm a]_{j+\frac{1}{2}}^n \tag{7.4b}$$

Scheme (7.4) is spatially first-order accurate for the implicit operator and spatially second-order accurate for the explicit operator. It can be shown that (7.4) is still TVD. Detailed implementation for the one- and two-dimensional Euler equations can be found in reference [10].

To show the efficiency and accuracy of the implicit method for steady-state application, we apply this method to the same quasi-one-dimensional nozzle problem as in figure (5.2). Figure (7.1) shows the converged density distribution after 25 steps at a CFL number of 10^6. Figure (7.1) used 20 equally spaced grid points. Only 14 points are plotted. The 6 points not shown on both ends of the x-axis are equal to the exact solution. The solution looks very much like the explicit TVD scheme except it has a tremendous gain in efficiency.

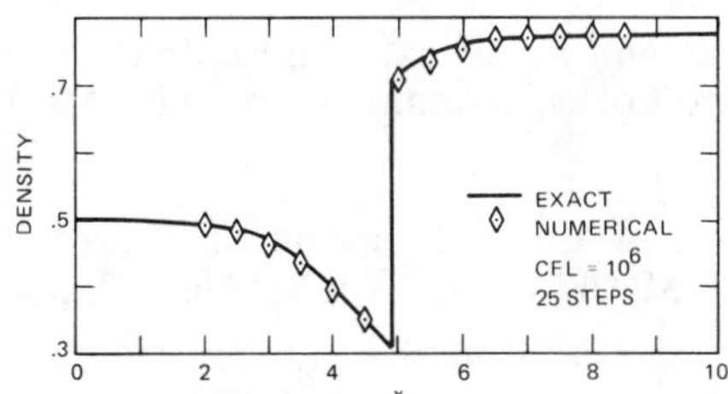

Fig. 7.1 Divergent nozzle: second-order implicit TVD scheme.

Formal extension of the implicit TVD scheme to two dimensions is straightforward. However, the method of solving the resulting nonlinear system of equations efficiently remains an open question. Research is underway to study this problem. Mulder and van Leer [31] suggested some useful ideas in this area.

8. Concluding Remarks

Numerical experiments for the Euler equations show that the application of the second-order explicit TVD schemes generate good shock resolution for both transient and steady-state one-dimensional and two-dimensional problems. Numerical experiments for a quasi-one-dimensional nozzle problem show that the second-order implicit TVD scheme produces a fairly rapid convergence rate and remains stable even when running with a CFL of a million. Research is underway to investigate an efficient extension of the implicit TVD method to two dimensions.

References

[1] R.D. Richtmyer and K.W. Morton, *Difference Methods for Initial-Value Problems*, Interscience-Wiley, New York, 1967.

[2] P.L. Roe, "An Introduction to Numerical Methods Suitable for the Euler Equations," *von Karman Institute for Fluid Dynamics Lecture Series: Introduction to Computational Fluid Dynamics*, Jan 24-28, 1983, Belgium.

[3] R.M. Beam and R.F. Warming, "Implicit Numerical Methods for the Compressible Navier-Stokes and Euler Equations," *von Karman Institute for Fluid Dynamics Lecture Series: Computational Fluid Dynamics*, March 29 - April 2, 1982, Belgium.

[4] P.D. Lax, "Hyperbolic Systems of Conservation Laws and the Mathematical Theory of Shock Waves," SIAM, Philadelphia, 1972.

[5] A. Harten, J.M. Hyman and P.D. Lax, "On Finite-Difference Approximations and Entropy Conditions for Shocks," Comm. Pure Appl. Math., Vol. 29, 1976, pp. 297-322.

[6] M.G. Crandall and A. Majda, "Monotone Difference Approximations for Scalar Conservation Laws," Math. Comp. Vol 34, No 149, Jan. 1980, pp. 1-21.

[7] A. Harten, "A High Resolution Scheme for the Computation of Weak Solutions of Hyperbolic Conservation Laws," NYU Report, Oct., 1981, and J. Comp. Phys., Vol. 49, 1983, pp. 357-393.

[8] A. Harten, "On a Class of High Resolution Total-Variation-Stable Finite-Difference Schemes," NYU Report, Oct., 1982, also to appear in SIAM J. Num. Anal.

[9] J.B. Goodman and R.J. LeVeque, "On the Accuracy of Stable Schemes for 2D Scalar Conservation Laws," NYU Report, May 1983, New York.

[10] H.C. Yee, R.F. Warming and A. Harten, "Implicit Total Variation Diminishing (TVD) Schemes for Steady-State Calculations," AIAA Paper No. 83-1902, Proc. of the AIAA 6th Computational Fluid Dynamics Conference, Danvers, Mass., July, 1983, also to appear in J. Comp. Phys., 1984.

[11] H.C. Yee, and P. Kutler, "Application of Second-Order-Accurate Total Variation Diminishing (TVD) Schemes to the Euler Equations in General Geometries," NASA TM-85845, August, 1983.

[12] B. Engquist and S. Osher, "Stable and Entropy Satisfying Approximations for Transonic Flow Calculations," Math. Comp. 34, 1980, pp. 45-75.

[13] L.C. Huang, "Pseudo-Unsteady Difference Schemes for Discontinuous Solutions of Steady-State, One-Dimensional Fluid Dynamics Problems," J. Comp. Phys., Vol. 42, 1981, pp. 195-211.

[14] P.L. Roe, "Approximate Riemann Solvers, Parameter Vectors, and Difference Schemes," J. Comp. Phys. Vol. 43, 1981, pp. 357-372.

[15] E.M. Murman, "Analysis of Embedded Shock Waves Calculated by Relaxation Methods," AIAA J., Vol. 12, No. 5, 1974, pp. 626-632.

[16] R. Courant, E. Isaacson, and M. Rees, "On the Solution of Nonlinear Hyperbolic Differential Equations by Finite Differences," Comm. Pure Appl. Math., Vol. 5, 1952, pp 243-255.

[17] B. van Leer, "Towards the Ultimate Conservative Difference Scheme. V. A Second-Order Sequel to Godunov's Method," J. Comp. Phys., Vol. 32, 1979, pp. 101-136.

[18] P. Colella and P.R. Woodward, "The Piecewise-Parabolic Method (PPM) for Gas-Dynamical Simulations," LBL report no. 14661, July 1982.

[19] P.L. Roe, "Some Contributions to the Modelling of Discontinuous Flows," to appear in Proc. AMS-SIAM Summer Seminar on Large Scale Computations in Fluid Mechanics, Univ. of Calif. at San Diego, June 27- July 8 1983.

[20] S. Osher, "Shock Modeling in Transonic and Supersonic Flow," to appear in Recent Advances in Numerical Methods in Fluids, Vol. 4, Advances in Computational Transonics, W.G. Habashi Ed., Pineridge Press, 1984.

[21] P.K. Sweby, "High Resolution Schemes Using Flux Limiters for Hyperbolic Conservation Laws," U.C.L.A. Report, June 1983, Los Angeles, Calif.

[22] A. Harten, "The Artificial Compression Method for Computation of Shocks and Contact Discontinuities. I. Single Conservation Laws," Comm. Pure Appl. Math., Vol. XXX, 1977, pp. 611-638.

[23] A. Harten, "The Artificial Compression Method for Computation of Shocks and Contact Discontinuities: III. Self-Adjusting Hybrid Schemes," Math. Comp., Vol. 32, No. 142, 1978, pp. 363-389.

[24] S.F. Davis, "A Rotationally Biased Upwind Difference Scheme for the Euler Equations," ICASE Contractor Report 172179, July, 1983, Hampton Virginia.

[25] P.L. Roe and M.J. Baines, "Alogorithms for Advection and Shock Problems," Proc. 4th GAMM Conference on Numerical Methods in Fluid Mechanics, H. Viviand Ed., Vieweg, 1982.

[26] H.C. Yee, R.F. Warming and A. Harten, "A High-Resolution Numerical Technique for Inviscid Gas-Dynamic Problems with Weak Solutions," Proc. Eighth International Conference on Numerical Methods in Fluid Dynamics, *Lecture Notes in Physics 170*, E. Krause ed., Springer-Verlag, 1982.

[27] G.R. Shubin, A.B. Stephens, H.M. Glaz, "Steady Shock Tracking and Newton's Method Applied to One-Dimensional Duct Flow," J. Comp. Phys., Vol. 39, 1981, pp 364-374.

[28] R.W. MacCormack, "The Effect of Viscosity in Hypervelocity Impact Cratering," AIAA Paper 69-354, Cincinnati, Ohio, 1969.

[29] P. Kutler, and A.R. Fernquist, "Computation of Blast Wave Encounter with Military Targets," Flow Simulations, Inc. Report No. 80-02, April 1980.

[30] H.C. Yee, R.F. Warming and A. Harten, "On the Application and Extension of Harten's High-Resolution Scheme," NASA TM-84256, June 1982.

[31] W.A. Mulder and B. van Leer, "Implicit Upwind Methods for the Euler Equations," AIAA Paper No. 83-1930, Proc. of the AIAA 6th Computational Fluid Dynamics Conference, Danvers, Mass., July, 1983.

A TIME DISCRETIZATION FOR CONSERVATION LAWS

YANN BRENIER*

Abstract. We introduce a time discretization for non linear hyperbolic systems of conservation laws unifying many recent numerical schemes, each of them being obtained by a different full discretization, Among them we recognize the Osher scheme, the Roe scheme and its generalization by Le Veque for large Courant numbers [14, 13, 16]. By using a non classical technique of full discretization suggested to us by the works of Chorin [5, 6, 7], we also obtain a particle scheme (which was previously introduced in [2] for the scalar case).

The time discretization described here is closely related to the Boltzmann type schemes considered in [10] and was previously studied in [1, 2, 3, 4, 8] for scalar case.

1. Construction of the time discretization operator T_t. We consider a hyperbolic system of conservation laws.

$$u_t + (B(u))_x = 0, \quad u = u(t,x) \in \mathbb{R}^m, \quad t > 0,\ x \in \mathbb{R} \tag{1.1}$$

where B is a given smooth application from $\mathbb{R}^m$ into itself, satisfying the hyperbolicity condition :

$$\forall\, w \in \mathbb{R}^m, \quad B'(w) \text{ has } m \text{ real distinct eigenvalues.} \tag{1.2}$$

We denote the eigenvalues by $\lambda_1(w), \ldots, \lambda_m(w)$ and the corresponding projectors by $P_1(w), \ldots, P_m(w)$ in order to have :

$$\forall\, p \in \mathbb{R}^m, \quad p = \sum_{k=1,m} P_k(w)p \ ; \ B'(w)\, p = \sum_{k=1,m} \lambda_k(w)\, P_k(w)p \tag{1.3}$$

It will be very useful to denote by $f(B'(w))$ the matrix defined by :

$$\forall\, p \in \mathbb{R}^m, \quad f(B'(w))\, p = \sum_{k=1,m} f(\lambda_k(w))\, P_k(w)p, \tag{1.4}$$

for some real function $f : \mathbb{R} \to \mathbb{R}$

Now we are going to define a time discretization for (1.1) by constructing an operator T_t which transforms any initial value $u(x)$ onto some approximation of the solution to (1.1) at time t : $T_t u(x)$.

We first define $T_t u$ when u is smooth and of compact support. The first derivative $u'(x)$ of the initial value may be seen as a continuous distribution of Dirac masses :

$$(1.5) \qquad u'(x) = \int_{\mathbb{R}} \delta(x-y)\, u'(y)\, dy$$

For each fixed $y \in \mathbb{R}$, we consider the constant coefficient linear hyperbolic system :

$$(1.6) \qquad z_t + B'(u(y))z_x = 0, \quad z = z(t,x) \in \mathbb{R}^m, \; t > 0, \; x \in \mathbb{R}$$

with

$$(1.7) \qquad z(0,x) = \delta(x-y)\, u'(y)$$

as the initial value.

The solution of (1.6,7) at time t is very well known (and obtained by putting the system in diagonal form) :

$$(1.8) \qquad z(t,x) = \sum_{k=1,m} \delta(x-y-t\lambda_k(u(y)))\, P_k(u(y))u'(y)$$

that is using the compact notation (1.4) :

$$(1.9) \qquad z(t,x) = \delta(x-y-t\, B'(u(y)))\, u'(y)$$

Now we define the first derivative of $T_t u(x)$ with respect to x, by summing these solutions with respect to y that is :

$$(1.10) \qquad (T_t u)'(x) = \int_{\mathbb{R}} \delta(x-y-t\, B'(u(y)))\, u'(y)\, dy,$$

wich leads to :

$$(1.11) \qquad T_t u(x) = \int_{\mathbb{R}} Y(x-y-t\, B'(u(y)))\, u'(y)\, dy \qquad \text{(Y=Heavyside func-tion)}$$

taking into account that u is compact supported. It is quite easy to check that (1.11) defines a consistent time discretization for (1.1), that is :

$$(1.12) \qquad T_0\, u(x) = u(x) \; ; \; \frac{d}{dt} T_t u(x) = -(B(u(x)))_x \quad \text{for } t = 0,$$

since :

$$(1.13) \qquad T_0\, u(x) = \int_{\mathbb{R}} Y(x-y)u'(y)dy \qquad \text{and}$$

$$\frac{d}{dt} T_t\, u(x) = \int_{\mathbb{R}} \delta(x-y)\, B'(u(y))\, u'(y)\, dy \qquad \text{for } t = 0$$

* INRIA, Centre de Rocquencourt, 78153 Le Chesnay Cedex, France

Moreover in the linear case :

(1.14) $\quad B(w) = b \cdot w,$

where b is some constant matrix with distinct real eigenvalues, $T_t u$ is the exact solution of (1.1) at time t. We have just defined $T_t u$ when u is smooth. Now we consider the case when u may be discontinuous. Nevertheless, we assume for simplicity that u has a compact support. Moreover we assume that there exists a Lipschitzian parametrization $s \in \mathbb{R} \to (X(s), U(s)) \in \mathbb{R} \times \mathbb{R}^m$ of the graph of u (which is reasonable when u has a bounded variation). Then, a natural generalization of (1.11) is given by :

(1.15) $$T_t u(x) = \int_{\mathbb{R}} Y(x - X(s) - t\, B'(U(s)))\, U'(s)\, ds$$

(there is no difference between (1.11) and (1.15) when u is smooth). For instance, if u is piecewise constant :

(1.16) $$u(x) = U_j \quad \text{for} \quad x_{j-\frac{1}{2}} < x < x_{j+\frac{1}{2}}, \quad j \in \mathbb{Z}$$

where

$$-\infty < x_{j-\frac{1}{2}} < x_{j+\frac{1}{2}} < +\infty,$$

a natural parametrization of its graph is given by :

(1.17) $$\begin{cases} X(s) = x_{j-\frac{1}{2}} + (S-2j)(x_{j+\frac{1}{2}} - x_{j-\frac{1}{2}}), \ U(s) = U_j & \text{for } 2j<s<2j+1 \\ X(s) = x_{j-\frac{1}{2}}, \ U(s) = W_{j-\frac{1}{2}}(s-2j+1) & \text{for } 2j-1<s<2j \end{cases}$$

where $C_{j-\frac{1}{2}} = (W_{j-\frac{1}{2}}(r),\ 0 < r < 1)$ is some smooth curve in $\mathbb{R}^m$

between U_{j-1} and U_j : $W_{j-\frac{1}{2}}(0) = U_{j-1}$, $W_{j-\frac{1}{2}}(1) = U_j$.

Clearly this curve is not unique (except in the scalar case when m=1 for which it is natural to set

$$W_{j-\frac{1}{2}}(r) = U_{j-1} + (U_j - U_{j-1})r, \qquad 0 \le r \le 1.$$

From (1.15, 17) we deduce :

$$(1.18) \qquad T_t u(x) = \sum_{j\in Z} \int_{C_{j-1/2}} Y(x-x_{j-1/2}-t\, B'(w))\, dw.$$

So we see that $T_t u$ depends on the choice of the different paths $C_{j-1/2}$ at the discontinuity points $x_{j-1/2}$.

2. Space discretization. In this section, we deduce several numerical schemes (many of them are already known) from the time discretization previously introduced. To achieve this, we introduce a uniform grid :

$$(2.1) \qquad x_{j+1/2} = x_{j-1/2} + h, \qquad j \in \mathbb{Z}; \quad h > 0,$$

and we consider the set of all piecewise constant functions of the form :

$$(2.2) \qquad u(x) = U_j, \qquad x_{j-1/2} < x < x_{j+1/2}, \quad j \in \mathbb{Z},$$

with

$$(2.3) \qquad U_j = 0 \text{ for } |j| \text{ large enough}; \quad \sum_{j\in\mathbb{Z}} \|U_j - U_{j-1}\| < +\infty,$$

where $\|\cdot\|$ denotes the Euclidian norm on $\mathbb{R}^m$.

From (1.18) we get :

$$(2.4) \qquad T_t\, u(x) = \sum_{j\in\mathbb{Z}} \int_{C_{j-1/2}} Y(x-x_{j-1/2}-t\, B'(w))\, dw$$

where $C_{j-1/2}$ is a smooth path in $\mathbb{R}^m$ connecting U_{j-1} to U_j and that is yet to be chosen.

We may define a new function of the type (2.2,3) by averaging $T_t u(x)$ on each cell $]x_{j-1/2}, x_{j+1/2}[$.

We set :

$$(2.5) \qquad \tilde{U}_j = \int_0^1 (T_t u)\, (x_{j-1/2} + rh)\, dr$$

From (2.4) we deduce :

$$h \sum_{j<i} \tilde{U}_j = \int_{-\infty}^{x_{i-1/2}} T_t u(x)dx = \sum_{j\in Z} \int_{C_{j-1/2}} \left(\int_{-\infty}^{x_{i-1/2}} Y(x-x_{j-1/2}-tB'(w))dx\right) dw$$

$$= \sum_{j\in Z} \int_{C_{j-1/2}} \max(0,\ x_{i-1/2}-x_{j-1/2}-t\ B'(w))\ dw$$

Since $T_0\ u = u$, we also have (by setting t=0 and substituting U_j for $\tilde{U}_j$ in the previous equation).

$$h \sum_{j<i} U_j = \sum_{j\in Z} \int_{C_{j-1/2}} \max\ (0, x_{i-1/2} - x_{j-1/2})\ dw.$$

So we have :

$$(2.6) \qquad h \sum_{j<i} (\tilde{U}_j - U_j) =$$

$$= \sum_{j\in Z} \int_{C_{j-1/2}} [\max(0,x_{i-1/2}-x_{j-1/2}-tB'(w))-\max(0,x_{i-1/2}-x_{j-1/2})]dw$$

Formula (2.6) defines a conservative numerical scheme which may be put in the classical form

$$(2.7) \qquad (\tilde{U}_j - U_j)\cdot h + (F_{j+1/2} - F_{j-1/2})\cdot t = 0$$

and we can easily see that the numerical flux $F_{i-1/2}$ is given by :

$$(2.8) \qquad t\ F_{i-1/2} = \text{const.}$$

$$- \sum_{j\in Z} \int_{C_{j-1/2}} [\max(0,x_{i-1/2}-x_{j-1/2}-tB'(w))-\max(0,x_{i-1/2}-x_{j-1/2})]\ dw$$

Now let us introduce the Courant number :

$$(2.9) \qquad q = 1 + \text{integer part of } h^{-1}\ t\ \rho(B)$$

where $\rho(B) = \sup_{\substack{w\in\mathbb{R}^m \\ k=1,m}} |\lambda_k(w)|$ is assumed to be finite.

We have for each $w \in \mathbb{R}^m$,

$$\max(0,x_{i-1/2}-x_{j-1/2}-tB'(w))-\max(0,x_{i-1/2}-x_{j-1/2})$$

$$= \max(0,(i-j)h-t\,B'(w)) - \max(0,(i-j)h) \qquad \text{(by (2.1))}.$$

$$= \begin{cases} -t\,B'(w) & \text{if } i-j \geq q \quad \text{that is } j \leq i-q \\ 0 & \text{if } i-j \leq -q \text{ that is } j \geq i+q \end{cases}$$

Thus, since $U_j = 0$ for $|j|$ large enough, it follows from (2.8), after a short computation,

$$\text{(2.10)} \qquad t\,F_{i-1/2} = \text{const} + (B(U_{i-q}) - B(0))t$$

$$- \sum_{|j-i|<q} \int_{C_{j-1/2}} [\max(0,x_{i-1/2}-x_{j-1/2}-tB'(w))-\max(0,x_{i-1/2}-x_{j-1/2})]dw$$

Since $\max(0,a) = (a+|a|)/2$, we finally obtain

$$\text{(2.11)} \qquad t\,F_{i-1/2} = (B(U_{i+q-1})+B(U_{i-q}))\,t/2$$

$$-\frac{1}{2} \sum_{|j-i|<q} \int_{C_{j-1/2}} (|x_{i-1/2}-x_{j-1/2}-tB'(w)|-|x_{i-1/2}-x_{j-1/2}|)dw$$

In particular when the Courant number is less than 1 :

$$\text{(2.12)} \qquad h^{-1}\,t\,\rho(B) < 1, \quad \text{that is } q = 1,$$

we get

$$\text{(2.13)} \quad F_{i-1/2} = \frac{1}{2}\left\{B(U_i)+B(U_{i-1}) - \int_{C_{j-1/2}} |B'(w)|dw\right\} + \text{const.}$$

Formula (2.13) defines precisely the Osher scheme [14] provided that for each j we choose the path $C_{j-1/2}$ connecting U_{j-1} to U_j in the same way as Osher does : $C_{j-1/2}$ is piecewise parallel to the right eigenvector $r_k(w)$ of $B'(w)$ in each point w. Consequently scheme (2.7, 11) can be seen as a generalization of the Osher scheme for large Courant numbers.

Now in order to simplify the computation of the numerical fluxes we are tempted to modify formulas (2.4) and (2.11) by substituting on each path $C_{j-1/2}$ a constant matrix $b_{j-1/2}$ with distinct real eigenvalues for $B'(w)$.

For simplicity, this matrix depends only on U_{j-1} and U_j. In other words we consider an application $(v,w) \to b(v,w)$ such that

(2.14) $$b_{j-1/2} = b(U_{j-1}, U_j)$$

It is natural to assume that :

(2.15) b is smooth; for each $(v,w) \in \mathbb{R}^m \times \mathbb{R}^m$, $b(v,w)$ has distinct real eigenvalues.

(2.16) $$b(v,v) = B'(v)$$

(2.17) $$b(v,w)(w-v) = B(w)-B(v)$$

Assumption (2.17) quaranties that :

(2.18) $$\int_{C_{j-1/2}} b_{j-1/2}\, dw = \int_{C_{j-1/2}} B'(w)\, dw.$$

The existence of such an application b is discussed in [10]. If such a modification is possible, we get approximations to (2.4) and (2.11), that are respectively,:

(2.19) $$\tilde{T}_t u(x) = \sum_{j \in Z} Y(x-x_{j-1/2} - tb_{j-1/2})(U_j - U_{j-1})$$

(2.20) $$t\tilde{F}_{i-1/2} = (B(U_{i+q-1}) + B(U_{i-q}))t/2$$

$$- \frac{1}{2} \sum_{|j-i|<q} (|x_{i-1/2} - x_{j-1/2} - tb_{j-1/2}| - |x_{i-1/2} - x_{j-1/2}|)(U_j - U_{j-1})$$

In particular if $q = 1$ (that is if (2.12) holds) :

(2.21) $$t\,\tilde{F}_{i-1/2} = \frac{t}{2}\left\{B(U_i) + B(U_{i-1}) - |b_{i-1/2}|(U_j - U_{j-1})\right\}$$

Formula (2.21) defines precisely Roe's scheme (with assumptions (2.14, 15, 16, 17) made) [16], and more generally the upstream schemes in the sense of [10] when assumption (2.17) is abondoned.

So, (2.20) defines a generalization of the Roe scheme for large Courant numbers. Scheme (2.20) was previously obtained by Le Veque [13] and defined as a variant of his generalization of the Godunov scheme.

Finally, formula 2.11), (2.13), (2.20) and (2.21) define four schemes, three of them are already known. We can summarize this in the following table :

	Exact Numerical Flux	Approximate Flux
Courant number < 1	Osher's scheme (2.13)	Roe's scheme (2.21
Courant number > 1	The author's generalization of Osher's scheme (2.11)	Le Veque's generalization of Roe's scheme (2.20)

3. <u>A particle scheme</u>. In the section we describe a different way to obtain a numerical scheme from the time discretization operator T_t introduced in the first section. Wen we defined T_t in the first section, we used a continuous distribution of Dirac masses to discribe the first derivative $u'(x)$ of the initial value u. Now replace this continuous distribution by a discrete one. This leads to the following approximation of the initial value and its first derivative :

$$(3.1)\quad u'(x) \simeq \sum_{j=1,P} \delta(x-x_{j-1/2})W_{j-1/2};\; u(x) \simeq \sum_{j=1,P} Y(x-x_{j-1/2})\, W_{j-1/2}$$

where $x_{j-1/2}$ denotes the position of the $j^{\underline{th}}$ Dirac mass and $W_{j-1/2} \in \mathbb{R}^m$ its weight.

In fact, this corresponds to a piecewise constant approximation of the initial value exactly as in the previous section. The main difference is that the assumption of a distribution of Dirac masses on a fixed grid as in (2.1) is no longer made, but instead.

$$(3.2)\qquad \|W_{j-1/2}\| \le \delta u, \quad j = 1,\dots,P$$

where δu is some well chosen small parameter. Nevertheless, we have:

$$(3.3)\quad \begin{cases} u(x)=\text{const}= U_i = \sum_j W_{j-1/2}\, Y(x_{i-1/2}-x_{j-1/2}) \text{ for } x_{i-1/2}<x<x_{i+1/2} \\ W_{j-1/2} = U_j - U_{j-1} \end{cases}$$

and we deduce from (1.18) :

$$T_t u(x) = \sum_j \int_{C_{j-1/2}} Y(x-x_{j-1/2}-t\,B'(w))\,dw$$

Making the same simplification as in section 2, we replace on $C_{j-1/2}$ $B'(w)$ by a constant matrix $b_{j-1/2}$ satisfying (2.14, 15, 16, 17). Then we get :

(3.4) $$\tilde{T}_t u(x) = \sum_j \int Y(x-x_{j-1/2}-tb_{j-1/2})(U_j-U_{j-1})$$

that is

(3.5) $$\tilde{T}_t u(x) = \sum_j Y(x-x_{j-1/2}-tb_{j-1/2})\,W_{j-1/2}$$

and

(3.6) $$(\tilde{T}_t u)'(x) = \sum_j \delta(x-x_{j-1/2}-tb_{j-1/2})\,W_{j-1/2}$$

Denoting by $\lambda_{k,j-1/2}$ the $k^{\underline{th}}$ eigenvalue of $b_{j-1/2}$ and by $P_{k,j-1/2}$ the corresponding projector, we have (utilizing the notation of (1.4)):

(3.6) $$(T_t u)'(x) = \sum_{j,k} \delta(x-x_{j-1/2}-t\lambda_{k,j-1/2})\,P_{k,j-1/2}W_{j-1/2}$$

So, starting from the initial distribution of particles (Dirac masses)

(3.7) $$(x_{j-1/2} \in \mathbb{R},\ W_{j-1/2} \in \mathbb{R}^m,\ j = 1,\dots,P)$$

we have obtained a new distribution :

(3.8) $$(x_{k,j-1/2}\ ;\ W_{k,j-1/2}\ ;\ j = 1,..,P\ ;\ k = 1,\dots,m)$$

defined by :

(3.9) $$x_{k,j-1/2} = x_{j-1/2} + t\,\lambda_{k,j-1/2}$$

(3.10) $$W_{k,j-1/2} = P_{k,j-1/2}\,W_{j-1/2}$$

This defines a particles scheme. Unfortunately, except in the case $m = 1$, we see that the amount of particles is multiplied m times at each time step, which is numerically unacceptable. So, to completely define a numerical scheme, we must introduce a procedure to reduce the number of particles ("merging" procedure). Moreover, in order to control the weight of the particles (which is not guaranteed by (3.10)),

(3.10)), we also have so define a breaking procedure to break the particles which become too large.

It is not clear to us yet, which "merging-breaking" procedure is the best. Let us just notice that an efficient merging procedure consists in replacing the new distribution of particles (3.8, 9, 10) by an equivalent one located on a uniform grid. This corresponds to project $\tilde{T}_t u$ (defined by (3.4, 5)) on a fixed grid and is equivalent to the Le Veque scheme considered in section 2.

Nevertheless, it seems of interest to avoid any projection on a fixed grid in order to keep a grid free particle method, capable to reduce dramatically the numerical diffusion phenomena, that are typical of fixed-grid methods. Finally let us notice that this particle method was suggested to us by the works of Chorin (especially [7]) and introduced in [2] for the scalar case (numerical results are given in [2] showing a great efficiency of the method).

4. Boltzmann-type schemes. In [10], following [11, 15], a class of time discretization operators was introduced for (1.1) :

$$T_{G,t}\, u(x) = \int_{\mathbb{R}} G(q,u(x-qt))dq \tag{4.1}$$

where G is a given application from $\mathbb{R}\times\mathbb{R}^m$ onto $\mathbb{R}^m$ such that :

$$\int_{\mathbb{R}} G(q,u)\, dq = u \;;\; \int_{\mathbb{R}} q\, G(q,u)\, dq = B(u) \tag{4.2}$$

which enforces the consistency conditions :

$$T_{G,0}\, u(x) = u(x) \;;\; \frac{d}{dt} T_{G,t}\, u(x) = -(B(u(x)))_x \text{ for } t = 0. \tag{4.3}$$

In fact, the time discretization operator T_t introduced in section 1 is closely related to this class of Boltzmann-type schemes. From (4.1) we deduce :

$$(T_{G,t})'(x) = \int_{\mathbb{R}} g(q,u(x-qt))\, u'(x-qt)dq \tag{4.4}$$

where g is the application (or more generally the distribution) from $\mathbb{R}\times\mathbb{R}^m$ onto $\mathbb{R}^{m\times m}$ defined by :

$$g(q,u) = D_2 G(q,u) \tag{4.5}$$

and satisfying (by (4.2))

$$\int_{\mathbb{R}} g(q,u)\, du = 1 \text{ (1=identity matrix) } ;\; \int_{\mathbb{R}} qg(q,u)\, dq = B'(u) \tag{4.6}$$

Using the transformation $y = x - qt$, $dy = -tdq$, we get :

(4.7) $$(T_{G,t}u)(x) = t^{-1} \int g((x-y)t^{-1}, u(y))\, u'(y)\, dy$$

Also, from (1.10) we get :

(4.8) $$(T_t u)'(x) = t^{-1} \int \delta((x-y)t^{-1} - B'(u(y)))\, u'(y) dy$$

since δ satisfies :

(4.9) $$\delta(tx) = t^{-1}\delta(x), \quad \forall\, t > 0, \ \forall\, x \in \mathbb{R}$$

Thus we get $T_t = T_{G,t}$ with :

(4.10) $$g(q,u) = \delta(q-B'(u))$$

However, strictly speaking, T_t does not define a Boltzmann type scheme (4.1, 2), since a priori, there is no application $G(q,u)$ such that :

(4.11) $$D_2\, G(q,u) = \delta(q-B'(u)),$$

except in the case $m = 1$ (scalar case).

5. The Scalar case. It follows from the previous section that in the scalar case $m = 1$, we have :

(5.1) $$T_t u(x) = \int_{\mathbb{R}} G(q,u(x-qt))\, dq$$

with :

(5.2) $$G(q,u) = \int_0^u \delta(q-B'(w))\, dw$$

After a short calculation, an other formulation of T_t may be obtained :

(5.3) $$T_t u(x) = \int_{\mathbb{R}} K(u(x-tB'(w)), w)\, dw$$

where

(5.4) $$K(v,w) = \begin{cases} +1 & \text{if } 0 < w < v \\ -1 & \text{if } 0 > w > v \\ 0 & \text{else} \end{cases}$$

To check that (5.1) and (5.3) are equivalent, multiply each of them by f(x), where f is same test function, and integrate with respect to x. From (5.1) we get :

$$\int_{\mathbb{R}} T_t\, u(x)f(x)dx = \int\int_{\mathbb{R}\times\mathbb{R}} G(q,u(x))f(x+tq)\ dw\ dq =$$

$$= \iint \left(\int_0^{u(x)} \delta(q-B'(w))\ f(w+tq)\ dw\right) dx\ dq$$

$$= \int_{\mathbb{R}} \left(\int_0^{u(x)} f(x+tB'(w))\ dw\right) dx$$

which does not differ from what we obtain from (4.3). Formulation (5.3,4) was extensively studied in [1,3] and [8]. It may be proved that T_t is monotone, conservative, L_1 non expansive, total variation diminishing. Moreover, if B is Lipschitzian, we have :

(5.5) $$S_t u = \lim_{n\to\infty} (T_{t/n})^n u \quad (\text{for the } L_1(\mathbb{R}) \text{ norm})$$

for each $u \in L_1(\mathbb{R})$, where $S_t u$ denotes the solution at time t of equation (1.1), satisfying the Kruẑkov entropy condition [12].

REFERENCES

[1] Y. BRENIER, Une application de la symétrisation de Steiner aux équations hyperboliques, C.R.Acad. Sc. Paris, Série I, 298 (1981) 563-566.

[2] Y. BRENIER, Calcul de lois de conservation scalaire par la méthode de transport-écroulement, Rapport INRIA 53(1981).

[3] Y. BRENIER, Résolution d'équations d'évolution quasilinéaires en dimension N d'espace à l'aide d'équations linéaires en dimension N+1, J.Diff. Equ., 50 (1983) 375-390.

[4] Y. BRENIER, Averaged Multivalued Solutions for scalar conservation laws, SIAM J. on Num. Analysis (in press).

[5] A.J. CHORIN, Numerical study of slightly viscous flows, J.Fluids Mech. 57 (1973) 785-796.

[6] A.J. CHORIN, P.S. BERNARD, Discretization of a vortex sheet, with an example of roll-up, J.Comp. Phys. 13 (1973) 423-428.

[7] A.J. CHORIN, Numerical methods for use in combustion theory modeling, Proc. 4th International IRIA Symposium on Computing Methods in Science and Engineering, Versailles Déc. 1979, R. Glowinski and J.L.Lions eds, North Holland, Amsterdam, 1981.

[8] Y.GIGA, T.MIYAKAWA, A kinetic construction of global solutions of first order quasilinear equations, Duk Math. J.50 (1983) 505-515.

[9] J. GLIMM, Solutions in the large for nonlinear hyperbolic systems of equations, C.P.A.M 18 (1965) 695-715.

[10] A. HARTEN, P.D.LAX, B.VAN LEER, On upstream differencing and Godunov-type schemes for hyperbolic conservation laws, SIAM Review 25 (1983) 35-61

[11] S. KANIEL, J. FALCOVITZ, Transport approach for compressible flow, Proc. 4th International IRIA Symposium on Computing Meho Methods in Science and Engineering, Versailles, Dec. 1979, R. Glowinski and J.L.Lions eds. North Holland, Amsterdam,1981,

[12] S.N. KRUZKOV, First order quasilinear equations in several independent variables, Math. USSR Sb. 10 (1970) 217-243.

[13] R. LE VEQUE, Convergence of a large time step generalization of Godunov's method for conservation laws, to appear.

[14] S. OSHER, Riemann solvers the entropy condition, and difference approximations, SIAM J. on Num. Analysis, (1984)

[15] D.J. PULLIN, Direct simulation methods for compressible inviscid ideal gas flow, J.Comp.Phys. 34 (1980) 231-244.

[16] P.L. ROE, Approximate Riemann solvers, parameter vectors, and differences schemes, J.Comp.Phys. 43 (1981) 357-372.

ON NUMERICAL SCHEMES FOR SOLVING THE EULER EQUATIONS OF GAS DYNAMICS

A. DERVIEUX* AND G. VIJAYASUNDARAM**

First order upwind schemes of Godunov-Van Leer, Steger-Warming, Godunov, Roe, Osher and Glimm ; Godunov type scheme I ; second order upwind schemes of Van Leer, Fromm-Van Leer, Hancock-Van Leer and Moretti ; and the second order centered schemes of Richtmyer, Mac Cormack, Lerat-Peyret and Jameson are described. Their performance for the shock tube problem proposed by Sod is compared. Godunov-Van Leer scheme, Glimm scheme, Hancock-Van Leer scheme and Fromm-Van Leer scheme produced better results.

There was a great interest in the construction of upwind schemes for the Euler equations of gas dynamics in the last decade. Since the compressible flow equations are nonlinear hyperbolic system of conservation laws, upwind schemes are preferred over the centered ones, as they reflect the physics of the problem. They are robust and have better dispersive and dissipative properties. Here an attempt is made to compare several upwind schemes and some centered ones to have a global picture of the state of the art. The shock tube problem proposed by Sod [18] is taken as the test problem.

0. Governing Equations. The flow of an inviscid, compressible fluid is governed by the Euler equations

$$W_t + F(W)_x = 0 \qquad (1)$$

where

$$W = \begin{pmatrix} \rho \\ \rho u \\ e \end{pmatrix}, \quad F(W) = \begin{pmatrix} \rho u \\ \rho u^2 + p \\ (e+p)u \end{pmatrix}, \qquad (2)$$

$$p = (\gamma-1)(e - \tfrac{1}{2}\rho u^2), \; \gamma = 1.4.$$

ρ denotes the density, u velocity, p pressure and e total energy per

* INRIA, Centre de Sophia Antipolis, Route des Lucioles 06560 VALBONNE (France).

** T.I.F.R. Centre, P.O. Box 1234, BANGALORE 560 012, (India).

unit volume, of the fluid. Equation (1) is in conservation form with W as the conserved variable and F as the flux term. Equations of motion in the physical variables $\tilde{W}$ is

(3) $$\tilde{W}_t + \tilde{A}(\tilde{W})\,\tilde{W}_x = 0$$

(4) $$\tilde{W} = \begin{pmatrix} \rho \\ u \\ p \end{pmatrix}, \quad \tilde{A}(\tilde{W}) = \begin{pmatrix} u & \rho & o \\ o & u & 1/\rho \\ o & \gamma p & u \end{pmatrix}$$

which is not in conservation form. Let $A(W) = F'(W)$ be the Jacobian matrix and $\lambda_1(W)$, $\lambda_2(W)$, $\lambda_3(W)$ be the eigenvalues of $A(W)$. A and $\tilde{A}$ are similar and diagonalisable. The following notations will be useful subsequently.

$$A = T\,\Lambda\,T^{-1}, \quad A^{\pm} = T\,\Lambda^{\pm}\,T^{-1},$$
$$\Lambda = \mathrm{diag}\,\{\lambda_j\}, \quad \Lambda^{\pm} = \mathrm{diag}\,\{\lambda_j^{\pm}\},$$
$$\lambda_j^{+} = \max\,(\lambda_j, o), \quad \lambda_j^{-} = \min\,(\lambda_j, o).$$

Flux vector F(W) is a homogeneous function of degree one.

(5) $$F(\alpha W) = \alpha F(W), \ \forall \alpha \in \mathbb{R}, \ W \in \mathbb{R}^3.$$

(6) As a consequence $F(W) = F'(W)W$.

(7) Let $F^{\pm}(W) = A^{\pm}(W)W$.

The Cauchy problem for (1) is the equation (1) with the initial condition

(8) $$W(x,o) = W_o(x)$$

The Riemann problem is a special initial value problem with the initial data

(9) $$W_o(x) = \begin{cases} W_L & \text{if } x < 0 \\ W_R & \text{if } x > 0 \end{cases}$$

where W_L and W_R are constant states. The shock Tube problem is a special Riemann problem with initial velocities equal to zero :

(10) $$u_L = 0 = u_R$$

This problem has a rarefaction wave, a contact discontinuity and a shock. Exact solution of this problem is known and it can be found by solving a nonlinear functional relation and expressing all the unknowns in terms of the initial data and the shock strength. A detailed description can be found in [30].

Let $\{x_i\}$ denote the grid points of a uniform mesh with mesh width Δx. Basic cell at x_i is the interval with x_i as centre and of length Δx. The interface of the cells at x_i and at x_{i+1} is denoted by $x_{i+1/2}$. Let t^n be a generic time level and Δt be the time step. The following notations

$$\lambda = \frac{\Delta t}{\Delta x}, \quad \nu = \lambda \max_{1 \le i \le 3} \sup_{W} |\lambda_i(W)|$$

$$W_i^n = W(x_i, t^n), \quad F_i^n = F(W_i^n)$$

are useful.

1. Godunov-Van Leer Scheme. The Godunov-Van Leer scheme introduced in [27] is an upstream centered matrix splitting scheme which uses the homogenity of the flux vector F. A version of the scheme for general F can be found in [22]. In conservation form, it is written as

$$(11) \qquad W_i^{n+1} = W_i^n - \lambda\{(\phi_F\ Gv)_{i+1/2}^n - (\phi_F\ Gv)_{i-1/2}^n\}$$

$$(12) \qquad (\phi_F\ Gv)_{i+1/2}^n = (A^+)_{i+1/2}^n W_i^n + (A^-)_{i+1/2}^n W_{i+1}^n$$

$$(13) \qquad (A^\pm)_{i+1/2}^n = A^\pm(W_{i+1/2}^n), \quad W_{i+1/2}^n = \frac{1}{2}(W_i^n + W_{i+1}^n)$$

2. Steger-Warming Scheme. By splitting the flux vector into positive flux and negative flux, Steger-Warming [19] constructed an upwind scheme which is

$$(14) \qquad W_i^{n+1} = W_i^n - \lambda\{(\phi_F\ SW)_{i+1/2}^n - (\phi_F\ SW)_{i-1/2}^n\}$$

$$(15) \qquad (\phi_F\ SW)_{i+1/2}^n = (F^+)_i^n + (F^-)_{i+1}^n$$

3. Godunov Scheme. Using the exact solution $W(x/t\ ; W_L, W_R)$ of the Riemann problem (1), (8), (9), Godunov [5] designed an ingeneous method generalizing the upwind scheme for nonlinear hyperbolic systems. As a finite difference scheme it has the form

$$(16) \qquad W_i^{n+1} = W_i^n - \lambda\{(\phi_F G)_{i+1/2}^n - (\phi_F G)_{i-1/2}^n$$

$$(17) \qquad (\phi_F G)_{i+1/2}^n = F(W(o, W_i^n, W_{i+1}^n))$$

Godunov gave an iterative procedure to compute the exact solution of the Riemann problem for Euler equations. The solution of the Riemann problem in the x-t plane consists of four constant states separated by the lines $x = u_i t$, $i = 1,3$. The lines $x = u_1 t$ and $x = u_3 t$ correspond to either a shock or a rarefaction wave. The line $x = u_2 t$ is the contact discontinuity. In [28], a characterization of the different cases, bounds and initial guess for the shock strength are given. Godunov's procedure involves the iteration of a 3×3 nonlinear system whereas the new method needs the solution of just one scalar nonlinear functional equation.

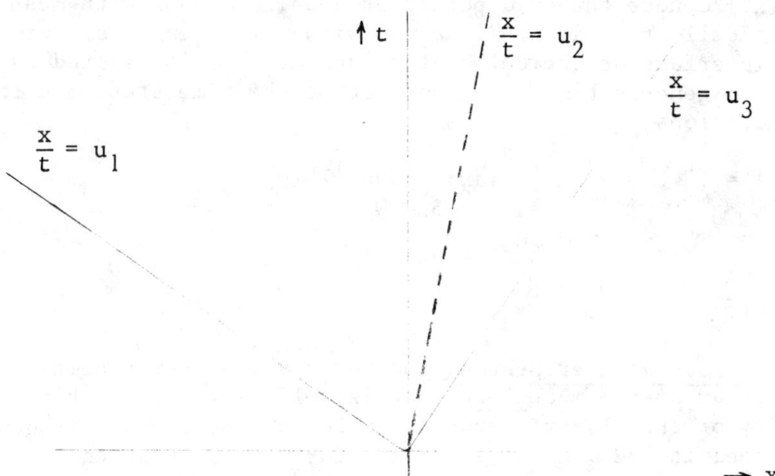

Fig. 1 : x-t diagram for the Riemann problem

4. Roe's Method. With a simplistic approach to the problem, Roe [17] could convert a scheme for the scalar linear equation into a scheme for the nonlinear hyperbolic systems. A finite difference scheme

$$(18) \qquad u_i^{n+1} = \sum_k c_k u_{i+k}^n$$

for the linear scalar problem

$$(19) \qquad u_t + cu_x = 0$$

can be put in the increment form

$$(20) \qquad u_i^{n+1} - u_i^n = -\lambda \sum_k \gamma_k (u_{i+k}^n - u_{i+k-1}^n)$$

where

$$(21) \qquad \nu = c\,\frac{\Delta t}{\Delta x}, \quad \gamma_k = \gamma_k(\nu, c)$$

The scheme is extended to linear systems

$$(22) \qquad u_t + Au_x = 0$$

with a diagonalisable matrix $A = T \Lambda T^{-1}$ by reducing the system into equations of the type (19) in the variables $(T^{-1}u)_j$. The scheme takes the form

$$(23) \qquad u_i^{n+1} - u_i^n = -\lambda \sum A_{\gamma_k} (u_{i+k}^n - u_{i+k-1}^n)$$

$$(24) \qquad A_{\gamma_k} = T \Lambda_{\gamma_k} T^{-1}, \quad \Lambda_{\gamma_k} = \text{diag}\,\{\lambda_j \gamma_{k,j}\}$$

$$(25) \qquad \gamma_{k,j} = \gamma_k(\nu_j, \lambda_j)$$

Extensions to nonlinear systems are carried out through a special linearization technique. Given two state vectors W_L, W_R, a new state vector

$$(26) \qquad \bar{W} = \bar{W}(W_L, W_R)$$

is found such that

$$(27) \qquad A(\bar{W})(W_R - W_L) = F_R - F_L$$

with

$$(28) \qquad F_R = F(W_R), \ F_L = F(W_L) .$$

Roe computed one such $\bar{W}$ for the Euler equations. Nonlinear version of the scheme is

$$(29) \qquad W_i^{n+1} - W_i^n = -\lambda \sum_k A_{\gamma_k}(W_{i,k}^n)(W_{i+k}^n - W_{i+k-1}^n)$$

$$(30) \qquad W_{i,k}^n = \bar{W}(W_{i+k-1}^n, W_{i+k}^n)$$

As

$$(31) \qquad \bar{W}(W_L, W_R) = \bar{W}(W_R, W_L)$$

Roe's method admists entropy violating shocks. In [7]an artificial viscosity term is given to eliminate such shocks. Roe's method offers some generality in handling schemes with ease. It has the limitation of not identifying nonlinear variants of the same linear scheme.

5. Osher Scheme. Osher generalized the first order upwind scheme to nonlinear systems by defining a path of integration in state space

$$(32) \qquad W_i^{n+1} = W_i^n - \lambda\left\{ \int_{W_{i-1}^n}^{W_i^n} A^+(W)\,dW + \int_{W_i^n}^{W_{i+1}^n} A^-(W)\,dW \right\}$$

where the path Γ_j^n connecting W_{j-1}^n and W_j^n is given by

$$(33) \qquad \Gamma_j^n = \bigcup_{k=1}^{3} \Gamma_{kj}^n$$

$$(34) \quad \Gamma_{kj}^n \begin{cases} \dfrac{dW^{(k)}}{ds} = r_k(W^{(k)}) \text{ for either } 0 \le s \le s_j^{(k)} \\ \qquad\qquad\qquad\qquad \text{or } 0 \ge s \ge s_j^{(k)} \\ W^{(k)}(0) = W^{(k+1)}(s_j^{(k+1)}) \end{cases}$$

$$(35) \qquad W^{(m+1)}(s_j^{(m+1)}) = W_{j-1}^n, \ W^{(1)}(s_j^{(1)}) = W_j^n$$

$r_k(W)$ being the right eigenvectors of $A(W)$.

Riemann invariants are used to compute the integrals over Γ^n_{kj} exactly. In [15] analytical computations of the integrals over the path Γ^n_{kj} are given. Numerical flux of Osher's scheme are smoother than that of the other schemes. This makes the scheme a better candidate for the implicit version.

6. Godunov Type Scheme I. Harten-Lax-Van Leer [8] proposed Godunov type schemes using approximate Riemann solvers. Let $\tilde{W}(x/t\ ;\ W_L,\ W_R)$ be an approximate Riemann solver satisfying the conditions (36) and (37) :

$$(36)\quad \begin{cases} \displaystyle\int_{-\frac{\Delta x}{2}}^{\frac{\Delta x}{2}} \tilde{W}(x/\Delta t\ ;\ W_L,\ W_R)dx = \frac{\Delta x}{2}(\tilde{W}_L + \tilde{W}_R) - \Delta t(\tilde{F}_R - \tilde{F}_L) \\ \qquad\qquad \text{for } \lambda \max |a_k| \leq \frac{1}{2} \end{cases}$$

(consistency with the integral form of conservation law)

$$(37)\quad \begin{cases} \displaystyle\int_{-\frac{\Delta x}{2}}^{\frac{\Delta x}{2}} \eta(\tilde{W}(x/\Delta t\ ;\ W_L,\ W_R))dx \leq \frac{\Delta x}{2}(\tilde{\eta}_L + \tilde{\eta}_R) - \Delta t(\tilde{q}_R - \tilde{q}_L) \\ \qquad\qquad \text{for } \lambda \max |a_k| \leq \frac{1}{2} \end{cases}$$

(consistency with the integral form of the entropy condition).

where η is an entropy function with entropy flux q and a_k are the signal speeds.

Godunov-type scheme for an approximate Riemann solver $\tilde{W}$ is defined as

$$(38)\quad \begin{aligned} W_i^{n+1} &= W_i^n - \lambda\{(\Phi_F\ \text{HLV})^n_{i+1/2} - (\Phi_F\ \text{HLV})^n_{i-1/2}\} \\ (\Phi_F\ \text{HLV})^n_{i+1/2} &= F_i^n - \frac{1}{\Delta t}\int_{-\frac{\Delta x}{2}}^{o} \tilde{W}(x/\Delta t\ ;\ W_i^n,\ W_{i+1}^n)dx + \frac{1}{2}\lambda\, W_i^n \end{aligned}$$

$$(39)\quad = F_{i+1}^n + \frac{1}{\Delta t}\int_o \tilde{W}(x/\Delta t;\ W_i^n,\ W_{i+1}^n)dx - \frac{1}{2}\lambda\, W_{i+1}^n$$

In [8] two approximate Riemann solvers are presented with one intermediate state and with two intermediate states. Godunov type scheme I has the approximate Riemann solver

$$(40)\quad W(x/t;W_L,W_R) = \begin{cases} W_L & \text{if } x/t < a_L \\ W_{LR} & \text{if } a_L < x/t < a_R \\ W_R & \text{if } a_R < x/t \end{cases}$$

where a_L, a_R are respectively minimum and maximum of the signal speeds. The formulae for W_{LR} and $\Phi_F(W_L, W_R)$ are

$$(41) \qquad W_{LR} = \frac{W_R a_R - W_L a_L}{a_R - a_L} - \frac{F_R - F_L}{a_R - a_L}$$

$$(42) \qquad (\Phi_F \text{ HLV})(W_L, W_R) = \begin{cases} F_L & \text{for } 0 < a_L \\ F_{LR} & \text{for } a_L < 0 < a_R \\ F_R & \text{for } a_R < 0 \end{cases}$$

$$(43) \qquad F_{LR} = \frac{-a_L}{a_R - a_L} F_R + \frac{a_R}{a_R - a_L} F_L + \frac{a_R a_L}{a_R - a_L}(W_R - W_L)$$

7. Glimm Scheme. Glimm's constructive proof for the existence of solutions for nonlinear hyperbolic conservation laws [4] was made into an efficient numerical tool by Chorin [2]. Sod [18] gave a flow chart for the method. In [28], an efficient exact Riemann solver from a new viewpoint was developed and was also utilised for Glimm scheme.

Let $W(x/t\ ;\ W_L, W_R)$ be the exact Riemann solver and $\xi^n_{i+1/2}$ be an equi-distributed random variable, in the interval $(-\frac{1}{2}, \frac{1}{2})$ of the Lebesgue measure. Then Glimm scheme is

$$(44) \qquad W^{n+1/2}_{i+1/2} = W\left(\frac{2\xi^n_{i+1/2}}{\lambda}\ ;\ W^n_i, W^n_{i+1}\right)$$

A similar construction allows one to define the values W^{n+1}_i. Proper choice of ξ is crucial to the success of the method. Chorin gave one such choice.

The upwind schemes so far described are characteristic upwind schemes. There are also convective upwind schemes [1], [14],[18] which are not described here for want of space.

8. Second Order Upwind Scheme. The second order upwind scheme is a five point scheme constructed by approximating the first and second derivatives in terms of the upstream values. The scheme can be put in conservation form by the upstream-centering technique of Van Leer.

$$(45) \qquad W^{n+1}_i = W^n_i - \lambda\{(\Phi_F\ SU)^n_{i+1/2} - (\Phi_F\ SU)^n_{i-1/2}\}$$

$$(46) \qquad \begin{cases} (\Phi_F\ SU)^n_{i+1/2} = (A^+)^n_{i+1/2} W^n_i + (A^-)^n_{i+1/2} W^n_{i+1} \\ + \frac{1}{2}(A^+)^n_{i-1/2}((W^n_i - W^n_{i-1}) - \lambda(F^n_i - F^n_{i-1})) \\ - \frac{1}{2}(A^-)^n_{i+3/2}((W^n_{i+2} - W^n_{i+1}) + \lambda(F^n_{i+2} - F^n_{i+1})). \end{cases}$$

The scheme is linearly stable for $\nu \leq 2$. There are other ways of obtaining a second order upwind scheme [22].
Warming and Beam [29] modified the Mac Cormack scheme into a second order upwind scheme. The results obtained are nearly same as that of (45)-(46).

9. Fromm-Van Leer Scheme. Second order centered schemes are known for their predominantly lagging phase error while second order upwind schemes exhibit predominantly leading phase error. Fromm [3] devised an ingeneous method, which he termed "zero average phase error method" by averaging out an upwind scheme and a centered one both of second order accuracy for the equation governing the convection of vorticities. Van Leer [22] had put them in conservation form for nonlinear hyperbolic conservation laws.

$$(47)\qquad W_i^{n+1} = W_i^n - \lambda\{(\Phi_F\ FV)_{i+1/2}^n - (\Phi_F\ FV)_{i-1/2}^n\}$$

$$(48)\qquad (\Phi_F\ FV)_{i+1/2}^n = \frac{1}{2}(\Phi_F\ SU)_{i+1/2}^n + \frac{1}{2}(\Phi_F\ LW)_{i+1/2}^n$$

$$(49)\qquad (\Phi_F\ LW)_{i+1/2}^n = \frac{1}{2}(F_{i+1}^n + F_i^n) - \frac{1}{2}\lambda|A|_{i+1/2}^n(F_{i+1}^n - F_i^n)$$

10. Hancock-Van Leer Scheme. Van Leer [24] extended the first order method of Godunov to second order accuracy by considering piecewise linear functions for the Lagrangian form of inviscid, compressible flow equations. Hancock [25] gave a two-step formulation which is much simpler than that of Van Leer.

Piecewise linear functions for the physical variables $\tilde{W}$ are taken as the space of approximate test functions.

$$(50)\qquad \tilde{W}^n(x) = \tilde{W}_i^n + \frac{(x-x_i)}{\Delta x}(\delta\tilde{W})_i^n \ \text{ in } x_{i-1/2} < x < x_{i+1/2}$$

The two degrees of freedom correspond to average and slope. The relations

$$(51)\qquad \left(\frac{\partial\tilde{W}^n}{\partial x}\right)_i = \frac{(\delta\tilde{W})_i^n}{\Delta x},\quad \left(\frac{\partial\tilde{W}}{\partial t}\right)_i^n = -\tilde{A}_i^n\frac{(\delta\tilde{W})_i^n}{\Delta x}$$

derived from (50), (3) are used to advance the cell averages to the time level n+1/2 and then the boundary values are calculated

$$(52)\qquad \tilde{W}_i^{n+1/2} = \tilde{W}_i^n + \frac{\Delta t}{2}\left(\frac{\partial\tilde{W}}{\partial t}\right)_i^n$$

$$(53)\qquad \tilde{W}_{(i\pm 1/2)\mp}^{n+1/2} = \tilde{W}_i^{n+1/2} \pm \frac{1}{2}(\delta\tilde{W})_i^n$$

$$(54)\qquad W_{(i\pm 1/2)\mp}^{n+1/2} = W(\tilde{W}_{(i\pm 1/2)\mp}^{n+1/2})$$

The time centered fluxes at the cell boundary i + 1/2 are computed from $W^{n+1/2}_{(i+1/2)-}$ and $W^{n+1/2}_{(i+1/2)+}$ by any upwind biased numerical flux formula Φ_F.

$$(\Phi_F\ HV)^{n+1/2}_{i+1/2} = \Phi_F(W^{n+1/2}_{(i+1/2)-},\ W^{n+1/2}_{(i+1/2)+}) \tag{55}$$

The scheme defining the cell averages is

$$W^{n+1}_i = W^n_i - \lambda\{(\Phi_F\ HV)^{n+1/2}_{i+1/2} - (\Phi_F\ HV)^{n+1/2}_{i-1/2}\} \tag{56}$$

The slope values $(\delta\tilde{W})^{n+1}_i$ are calculated from

$$(\delta u)^{n+1}_i = c^{n+1}_i \ \text{ave}\ (\frac{u^{n+1}_{i+1} - u^{n+1}_i}{c^{n+1}_i}),\ \frac{u^{n+1}_i - u^{n+1}_{i-1}}{c^{n+1}_i}) \tag{57}$$

$$(\delta\rho)^{n+1}_i = \rho^{n+1}_i \ \text{ave}\ (2.\frac{\rho^{n+1}_{i+1} - \rho^{n+1}_i}{\rho^{n+1}_{i+1} + \rho^{n+1}_i},\ 2\frac{\rho^{n+1}_i - \rho^{n+1}_{i-1}}{\rho^{n+1}_i + \rho^{n+1}_{i-1}}) \tag{58}$$

and a similar expression for the pressure. The averaging function is a special average defined by

$$\text{ave}\ (a,b) = \frac{(b^2+\varepsilon^2)a + (a^2+\varepsilon^2)b}{a^2 + b^2 + 2\varepsilon^2} \tag{59}$$

with the small positive bias ε^2 of the order of $(\Delta x)^3$.

11. The λ-Scheme. The λ-scheme of Moretti [13] is a two-step second order scheme which, for the model problem (19) with $c > 0$, reads as

$$\tilde{u}^{n+1}_i = u^n_i - \nu(2u^n_i - 3u^n_{i-1} + u^n_{i-2}) \tag{60}$$

$$u^{n+1}_i = \frac{1}{2}(u^n_i + \tilde{u}^{n+1}_i) - \frac{1}{2}\nu(\tilde{u}^{n+1}_i - \tilde{u}^{n+1}_{i-1}) \tag{61}$$

Roe's method is used to extend this to nonlinear hyperbolic conservation laws.

12. Two-step Lax-Wendroff Scheme. Two-step Lax-Wendroff scheme [10] introduced by Richtmyer [16] is

$$W^{n+1/2}_{i+1/2} = \frac{1}{2}(W^n_i + W^n_{i+1}) - \frac{1}{2}\lambda(F^n_{i+1} - F^n_i) \tag{62}$$

$$W^{n+1}_i = W^n_i - \lambda(F^{n+1/2}_{i+1/2} - F^{n+1/2}_{i-1/2}) \tag{63}$$

13. Mac Cormack Scheme. Mac Cormack scheme [12] is

$$\bar{W}_i^{n+1} = W_i^n - \lambda(F_i^n - F_{i-1}^n) \tag{64}$$

$$W_i^{n+1} = \frac{1}{2}(W_i^n + \bar{W}_i^{n+1}) - \frac{1}{2}\lambda(\bar{F}_{i+1}^{n+1} - \bar{F}_i^{n+1}) \tag{65}$$

14. S_β^α - Scheme. The S_β^α - scheme of Lerat-Peyret [11] is

$$W_{i+\beta}^{n+\alpha} = (1-\beta)W_i^n + \beta W_{i+1}^n - \alpha\lambda(F_{i+1}^n - F_i^n) \tag{66}$$

$$\left\{\begin{array}{l} W_i^{n+1} = W_i^n - \dfrac{\lambda}{2\alpha}\{(\alpha-\beta)F_{i+1}^n + (2\beta-1)F_i^n \\ \quad + (1-\alpha-\beta)F_{i-1}^n + (F_{i+\beta}^{n+\alpha} - F_{i+\beta-1}^{n+\alpha})\} \end{array}\right. \tag{67}$$

Equations (66)-(67) define a two-parameter family of schemes with parameters α and β. Two-step Lax-Wendroff scheme and Mac Cormack scheme are members of this family. The optimal scheme of Lerat-Peyret is the one with

$$\alpha = 1 + \frac{\sqrt{5}}{2}, \quad \beta = \frac{1}{2} \tag{68}$$

15. Jameson Scheme. Jameson [9] constructed a fourth order Runge-Kutta time stepping scheme with centered differencing for space derivatives. An artificial viscosity term was added to damp the oscillations. A semi-discrete second order approximation for (1) is

$$\frac{dW_i(t)}{dt} + \frac{F_{i+1}^n - F_{i-1}^n}{2\Delta x} = 0 \tag{69}$$

As centered differencing is not stable a diffusion term $\frac{1}{\Delta x} DW_i^n$ is added to make it stable.

$$DW_i^n = (dW)_{i+1/2}^n - (dW)_{i-1/2}^n \tag{70}$$

$$\begin{array}{l} (dW)_{i+1/2}^n = \lambda\{\varepsilon_{i+1/2}^{(2)}(W_{i+1}^n - W_i^n) - \\ \varepsilon_{i+1/2}^{(4)}(W_{i+2}^n - 3W_{i+1}^n + 3W_i^n - W_{i-1}^n)\} \end{array} \tag{71}$$

$$\varepsilon_{i+1/2}^{(2)} = K^{(2)}\max(\nu_i^n, \nu_{i+1}^n), \quad \varepsilon_{i+1/2}^{(4)} = \max(0, K^{(4)} - \varepsilon_{i+1/2}^{(2)})$$

$$\nu_i^n = \left|\frac{p_{i+1}^n - 2p_i^n + p_{i-1}^n}{p_{i+1}^n + 2p_i^n + p_{i-1}^n}\right|, \quad K^{(2)} = \frac{1}{4}, \quad K^{(4)} = \frac{1}{256}.$$

The ODE so obtained is solved using the fourth order Runge-Kutta scheme

$$(72)\quad \begin{cases} W^{(0)} = W^n \\ W^{(1)} = W^{(0)} - \frac{\Delta t}{2} PW^{(0)} \\ W^{(2)} = W^{(0)} - \frac{\Delta t}{2} PW^{(1)} \\ W^{(3)} = W^{(0)} - \Delta t . PW^{(2)} \\ W^{(4)} = W^{(0)} - \frac{\Delta t}{2} (PW^{(0)} + 2PW^{(1)} + 2PW^{(2)} + PW^{(3)}) \\ W^{n+1} = W^{(4)} \end{cases}$$

where

$$PW_i^n = \frac{F_{i+1}^n - F_{i-1}^n}{2\Delta x} - \frac{1}{\Delta x} DW_i^n .$$

Since it is expensive to compute the diffusion term for each fractional time step the diffusion term is frozen.

Artificial Viscosity of Lapidus :

As the second order centered difference schemes introduce spurious oscillations in the solution, an artificial viscosity term has to be added to reduce such oscillations. The artificial viscosity of Lapidus is of third order and can easily be added to a scheme. If $\tilde{W}_i^{n+1}$ is the solution obtained by any one of the schemes, then the scheme with artificial viscosity is given by

$$(73)\quad W_i^{n+1} = \tilde{W}_i^{n+1} + \chi.\lambda \, \{(d\tilde{W})_{i+1/2}^{n+1} - (d\tilde{W})_{i-1/2}^{n+1} \}$$

$$(74)\quad (d\tilde{W})_{i+1/2}^{n+1} = |\tilde{W}_{i+1}^{n+1} - \tilde{W}_i^{n+1}| . (\tilde{W}_{i+1}^n - \tilde{W}_i^n)$$

where χ is the coefficient of viscosity.

Test Problem and Numerical Experiments :

The shock tube problem proposed by Sod [18] is taken to test the performance of the schemes described above. It is the problem (1), (8), (9) with

$$(75)\quad W_L = \begin{cases} 1.0 \\ 0.0 \\ 2.5 \end{cases} , \quad W_R = \begin{cases} 0.125 \\ 0.0 \\ 0.25 \end{cases}$$

and initial discontinuity at x = 0.5.

Distributions of density, pressure, velocity and internal energy at time t = 0.14 obtained using the schemes are compared with the exact one. The computational domain is taken as (0,1) and is divided into

50 uniform intervals.
Figure 2 indicates the exact solution of the problem. All the schemes are linearly stable for $\nu \leq 1$. In most of the cases the courant number ν is chosen as 0.9. At each time level Δt is chosen satisfying the CFL stability condition.

Schemes 1-6 are first order accurate upwind schemes and the results obtained using them are shown in Figures 3-8. Contact discontinuity is smeared out to a considerable extent by the first order schemes, total smearing occuring with Steger-Warming scheme and Godunov type scheme I. As the first order schemes are more diffusive, they are free from oscillations. The rarefaction zone is approximated less accurately. Shock spreads over three to five zones ; shock is sharper with Godunov-Van Leer scheme. Upwind schemes are more complicated to program . Godunov type scheme I and Godunov-Van Leer scheme take less CPU time whereas Osher scheme is very expensive. The CPU time taken to reach the instant t = 0.14 by schemes 1-7 are listed in Table I.

Table I

CPU time taken to reach t = 0.14 on PRIME 450

S.No.	Scheme	No. of iterations	CPU time in seconds
1	Godunov-Van Leer	17	05
2	Steger-Warming	17	06
3	Godunov	17	06
4	Roe	17	07
5	Osher	17	11
6	Godunov type I	17	05
7	Glimm	17	07

Glimm scheme produced very sharp shock profiles with no point on the shock and no oscillations near the shock. The position of the shock is displaced by one mesh width. Contact discontinuity is also sharp and is located at the correct position. The constant state between rarefaction wave and contact discontinuity is extended by 3 mesh width. Due to randomness the rarefaction zone is not smoothly approximated.

The efficient Riemann solver used reduced the CPU time considerably ; The Glimm scheme presented here is much more cost effective than the one presented by Sod [18] .

Figures 10-13 indicate approximate solutions resulted with second order accurate upwind schemes. With these schemes smearing of contact discontinuity is reduced to a great extent. As they are second order accurate dissipative errors are less. Second order upwind scheme and Moretti scheme show predominantly leading phase error. Dispersive errors are reduced remarkably for the Fromm-Van Leer scheme. Hancock-Van Leer scheme has slight overshoots. Rarefaction zone is better approximated.
Second order upwind schemes are as complicated to programme as first order upwind schemes but they take more CPU time (Table II)

Table II

S.No.	Scheme	No. of iterations	CPU time in seconds
8	Second order upwind	18	08
9	Fromm-Van Leer	18	09
10	Hancock-Van Leer	17	11
11	Moretti	17	11

Figures 14-17 show the outcome of second order centered schemes. Artificial viscosity of Lapidus is added to Richtmyer scheme, Mac Cormack scheme and Lerat-Peyret scheme, with viscosity coefficient $\chi = 0.125$. Smearing of contact discontinuity is less. These schemes have predominantly lagging phase error. Oscillations are least with Richtmyer scheme and greatest with Lerat-Peyret scheme. Rarefaction zone is more accurately approximated than the first order schemes. Shock spreads over three to four zones. Centered schemes are simpler to program and take less CPU time than their counterparts. Jameson scheme is the most economical of all.

Table III

S.No.	Scheme	No. of iterations	CPU time in seconds
12	Richtmyer	18	05
13	Mac Cormack	18	06
14	Lerat-Peyret	20	06
15	Jameson	6	04

Conclusion :

Of all the schemes, the schemes of Godunov-Van Leer, Glimm, Fromm-Van Leer and Hancock-Van Leer scheme produced the best results. All the first order upwind schemes, Glimm scheme, Jameson scheme and Hancock-Van Leer scheme can be extended to two dimensions in the finite element setting. Extensions of other second order upwind schemes remain open.

Acknowledgement :

The authors acknowledge with thanks T.I.F.R. Centre at Bangalore for providing excellent computer facility.

References :

[1] F. ANGRAND, A. DERVIEUX, Some explicit triangular finite element schemes for the Euler equations, in I.J. for Num. Methods in Fluids (1984).

[2] A.J. CHORIN, Random choice solution of hyperbolic systems, J. Comp. Phys., 22, pp. 517, 533(1976).

[3] J.E. FROMM, A method for reducing dispersion in convective difference schemes, J. Comp. Phys. 3, pp. 176-189 (1968).

[4] J. GLIMM, Solutions in the large for nonlinear hyperbolique systemes of equations, Comm. Pure Appl. Math., 18, pp. 697-715 (1965).

[5] S.K. GODUNOV, Mat. Sb. 47 (1959), 271 also Cornell Aeronautical Lab. Translation.

[6] A. HARTEN, On second order accurate Godunov-type schemes, to appear.

[7] A. HARTEN, J.M. HYMAN, A self-ajusting grid for the computation of weak solutions of hyperbolic conservation laws, J. Comp. Phy , 50, pp. 235-269 (1983).

[8] A. HARTEN, P.D. LAX, B. VAN LEER, On Upstream differencing and Godunov type schemes for hyperbolic conservation laws, SIAM Review, Vo.25, pp. 35-61 (1983).

[9] A. JAMESON, Steady-state solution of the Euler equations for transonic flow ; transonic shock and multi-dimensional flows : Advances in Scientific Computing, EJ. R.E. MEYER, Proceedings of a symposium conducted by the Mathematics Research Center, The University of Wisconsin-Madison, pp. 37-70 (1981).

[10] P.D. LAX, B. WENDROFF, Systems of conservation laws, Comm. Pure Appl. Math., 13, pp. 217-237 (1960).

[11] A. LERAT, Sur le calcul des solutions faibles des systèmes hyperboliques de lois de conservation à l'aide de schémas aux différences, Thèse à l'Université de Paris VI (1981).

[12] R.W. MAC CORMACK, The effect of viscosity in Hypervelocity impact cratering, AIAA paper 69 - 354 Cincinnati, Ohio, 1969.

[13] G. MORETTI, The λ-scheme, Computers and Fluids, Vol. 7, pp. 191-205, (1979).

[14] T. NAGAYAMA, T. ADACHI, Numerical analysis of flow through turbine cascade by the modified FLIC method, Comm. to Tokyo Joint Gas Turbine Congress, May 22-27, 1977, Tokyo (Nagasaki Technical Institute, Mitsubishi Heavy Industries,LTD, Nagasaki, Japan).

[15] S. OSHER, F. SOLOMON, Upwind difference schemes for Hyperbolic systems of conservation laws, Math. Computation (1982).

[16] R. RICHTMYER, A survey of difference methods for non-steady fluid dynamics, NCAR Technical Note 63-2, National Center for Atmospheric Research, Boulder, Colo, 1962.

[17] P.L. ROE, The use of Riemann problem in finite difference schemes, Proceedings of the 7th Int. Conf. on Num. Methods in Fluid Dynamics, Stanford 1980, Springer Verlag 1981.

[18] G.A. SOD, A survey of several finite difference methods for systems of nonlinear hyperbolic conservation laws, Review, J. Comp. Phys. 27, 1-31 (1978).

[19] J. STEGER, R.F. WARMING, Flux vector splitting for the inviscid gas dynamic equations with applications to finite difference methods, J. Comp. Phys., Vol. 40, No.2, pp. 263-293 (1981).

[20] B. VAN LEER, Towards the ultimate conservative difference scheme, I. The quest of Monotonicity, Lecture Notes in Physics, pp 163-168 (1973).

[21] B. VAN LEER, II. Monotonicity and conservation combined in a second-order scheme, J. Comp. Phys. 14, pp. 361-370 (1974).

[22] B. VAN LEER, III. Upstream-centered finite difference schemes for ideal compressible flow, J. Comp. Phys. 23, pp. 263-275 (1977).

[23] B. VAN LEER, IV. A new approach to numerical convection, J. Comp. Phys. 23, pp 276-299 (1977).

[24] B. VAN LEER, V. A second-order sequel to Godunov's method, J. Comp. Phys. 32, pp. 101-136 (1979).

[25] B. VAN LEER, Computational methods for ideal, compressible flow ; Computational Fluid Dynamics, Von Karman Institute for Fluid Dynamics, Lecture Series 1983-04.

[26] G. VIJAYASUNDARAM, Résolution numérique des équations d'Euler pour des écoulements transsoniques avec un schéma de Godunov en éléments finis, Thèse Université P. et M. Curie, Paris VI (1982).

[27] G. VIJAYASUNDARAM, Transonic flow simulations using an upstream-centered scheme of Godunov in finite elements, to appear.

[28] G. VIJAYASUNDARAM, An efficient Riemann solver for Euler equations of gas dynamics, in preparation.

[29] R.F. WARMING, R.M. BEAM, Upwind second-order difference schemes and applications in aerodynamic flow, AIAA Journal, Vol. 14, No.9, pp. 1241-1249 (1976).

[30] G.B. WHITHAM, Linear and Nonlinear Waves, John Wiley Interscience Publications (1974).

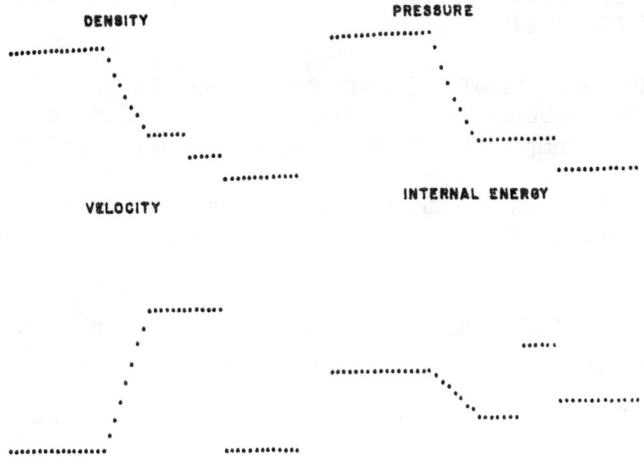

FIG.2.EXACT SOLUTION

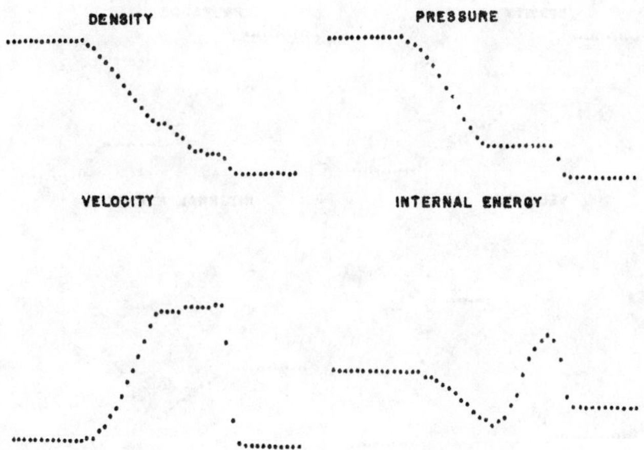

FIG.3. GODUNOV-VAN LEER SCHEME

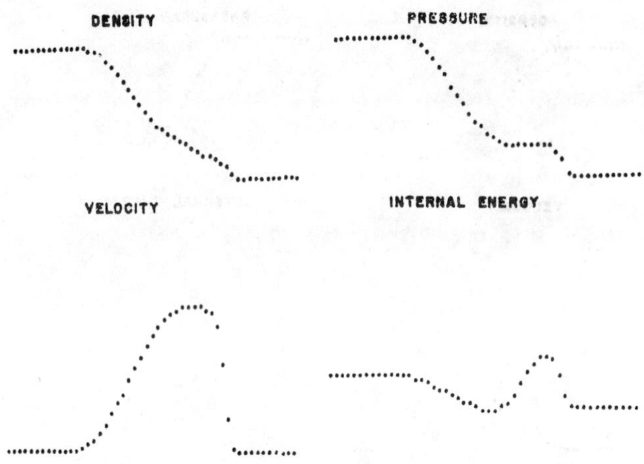

FIG.4. STEGER-WARMING SCHEME

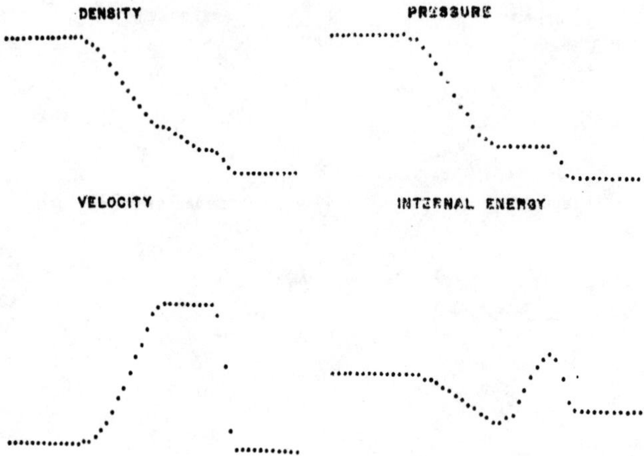

FIG. 5. GODUNOV SCHEME

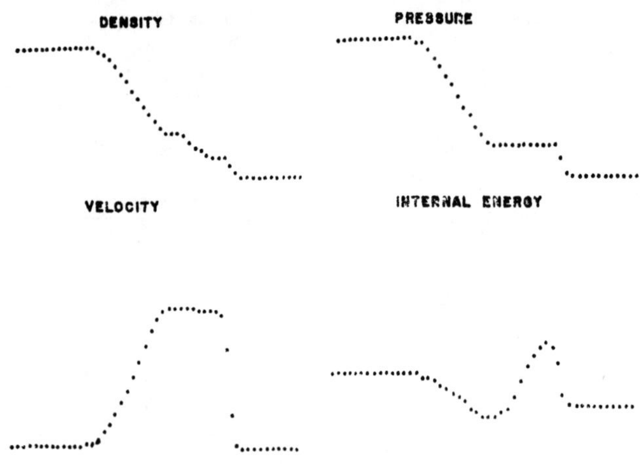

FIG. 6. ROE SCHEME

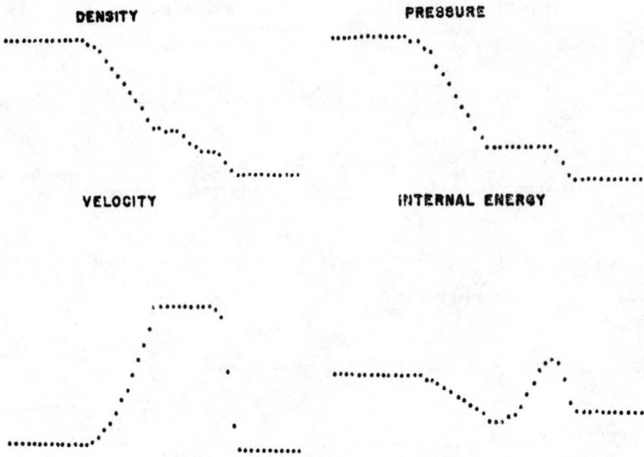

FIG.7. OSHER SCHEME

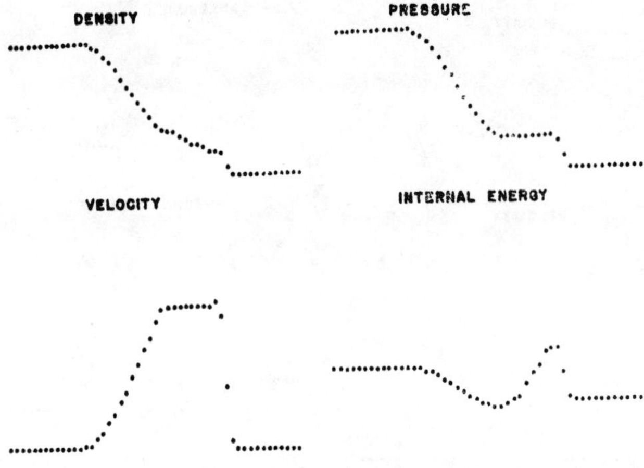

FIG.8. GODUNOV-TYPE SCHEME I

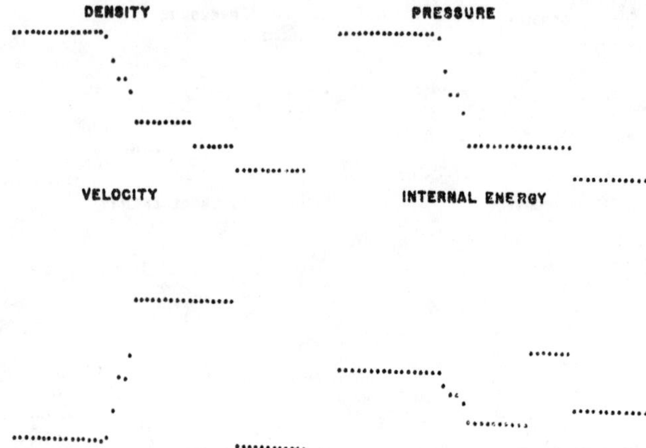

FIG. 9. GLIMM SCHEME

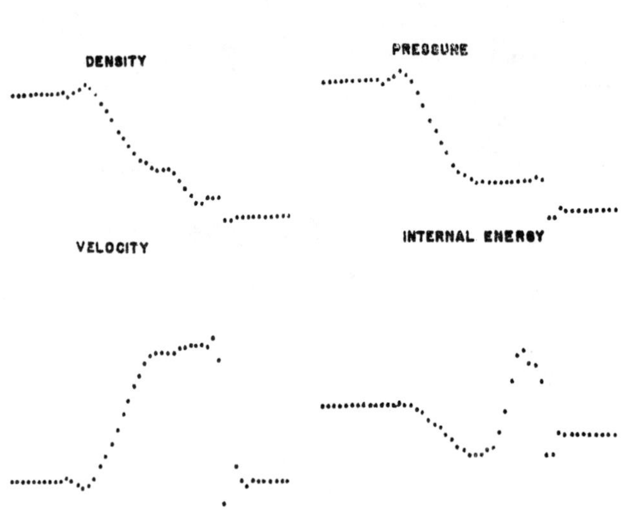

FIG. 10. SECOND ORDER UPWIND SCHEME

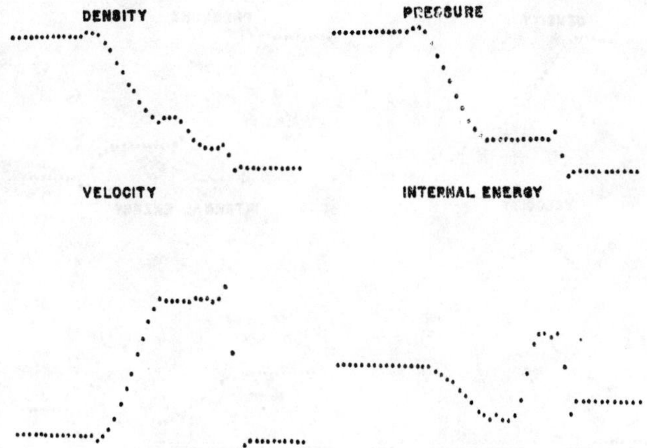

FIG.11. FROMM-VAN LEER SCHEME

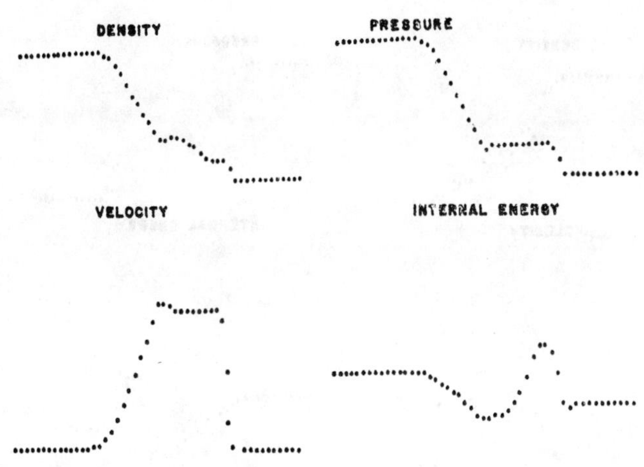

FIG.12. HANCOCK-VAN LEER SCHEME

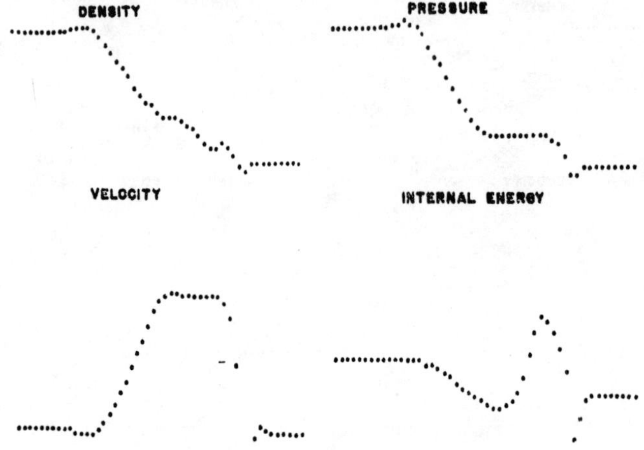

FIG.13. MORETTI SCHEME

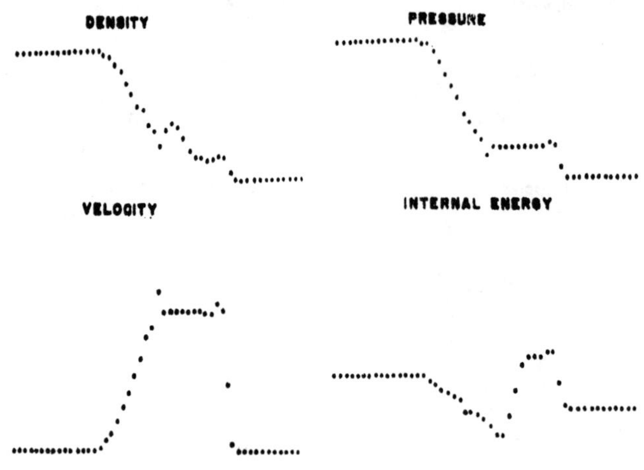

FIG.14. RICHTMYER SCHEME

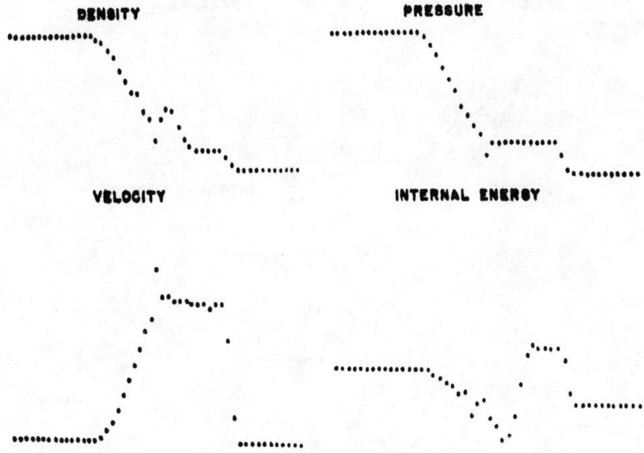

FIG.15. MAC CORMACK SCHEME

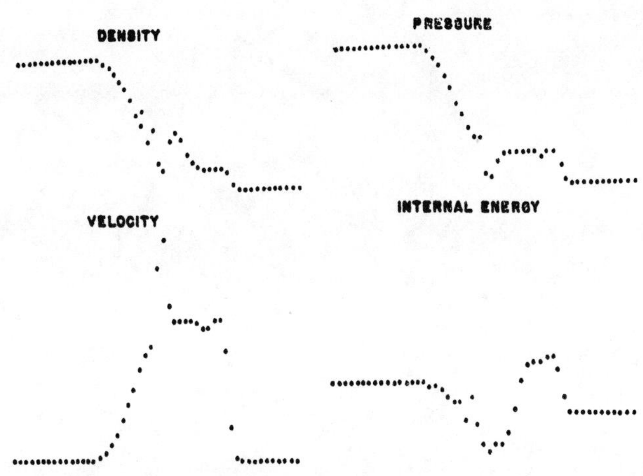

FIG.16. LERAT-PEYRET SCHEME

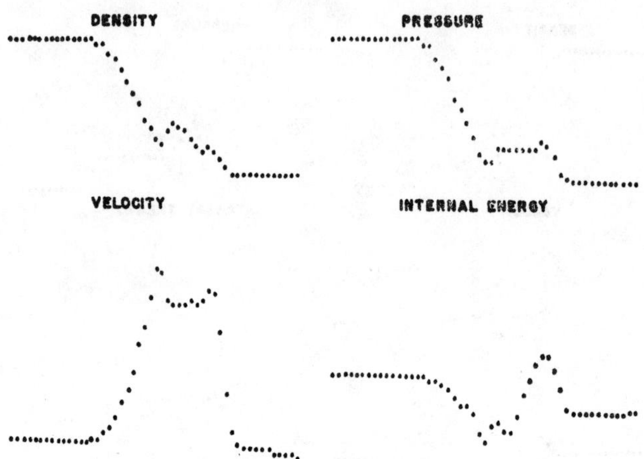

FIG.17. JAMESON SCHEME

THE CONVECTION-PROJECTION APPROACH TOWARDS THE SOLUTION OF EULER EQUATIONS

PHILIPPE MORICE*

Abstract. Some aspects of the convection-projection approach are presented to contribute to the construction of numerical schemes of upwind type. Links of this concept with a weak formulation based on characteristic variables are explored and allow the generation of various schemes for scalar conservation laws. In particular, a large time step method consisting of a generalization of the method of Godunov is presented. In one and two dimensions, the solution of Euler equations by Lagrangian-Eulerian schemes is proposed. The interpretation of these schemes in terms of the convection-projection approach may lead to them being better understood and to new developments.

1. Introduction. Among the guiding principles in the development of numerical methods for solving unsteady flow problems modelled by the hyperbolic system of Euler equations, one can roughly distinguish two main points of view.

In one approach, the first motivation is to get a high accuracy in smooth flow regions (subsonic or supersonic) and centered difference schemes often emerge after a study of the mathematical stability and the minimization of truncations errors. The necessity of using more or less complex forms of artificial viscosity is the rule for these schemes in the presence of discontinuities.

A second approach consists in including in numerical schemes some amount of upwinding more or less related to the characteristics of the hyperbolic system. From the first order method of Godunov [1] to higher-order upwind-differenced schemes recently published [2]-[6], many methods are based on sets of physical or mathematical principles like conservation, proper domain of influence, and monotonicity properties among many others. It seems that, at least for unsteady flows with strong shocks, this second family of methods could provide better answers than the first one.

*Office National d'Etudes et de Recherches Aérospatiales
BP 72, 92322 CHATILLON CEDEX, France

The purpose of this paper is to present a contribution to the understanding of the convection-projection approach as a general tool for generating numerical schemes with some built-in upwinding.

We begin in § 2 with a scalar conservation law and show the connection between the convection-projection concept and a weak formulation with characteristic variables. The choice of various continuous or discontinuous approximation functions leads to several numerical schemes. Of special interest for their accuracy are results with discontinuous piecewise linear and continuous piecewise quadratic approximation in the presence of shocks. These schemes have been proposed by van Leer [7]. The piecewise constant approximation gives an explicit upwind first order scheme. However a new technique taking into account the proper domain of influence of the initial solution allows one to take arbitrarily large time steps with proper treatment of shocks. The method presented here is valid for a conservation law under a convexity condition, and the numerical results show the strong improvement due to the large CFL values.

In § 3, an interpretation of the convection-projection approach for its application to the system of Euler equations leads to Lagrangian-Eulerian schemes. Various first-order and second-order schemes are presented in one dimension for the shock-tube problem in order to illustrate their respective merits.

We have developed in two dimensions some Lagrangian-Eulerian schemes which are based on a piecewise constant approximation and avoid the use of splitting. They share with the fluid-in-cell method [8] and some of its variants the property of simplicity, and with the recent method proposed by Woodward and Colella [5] the property of accuracy.

2. Convection-projection approach for a scalar conservation law

2.1. Generalities

We consider the quasi-linear, one-dimensional conservation law :

$$(2.1)\quad \left|\begin{array}{l} \dfrac{\partial u}{\partial t} + \dfrac{\partial}{\partial x} f(u) = 0 \quad , (x,t) \in [x_{min}, x_{max}] \times [t_0, T] \\ u(x, t_0) = u_0(x) \ , \end{array}\right.$$

with suitable boundary conditions.

We make the simplifying assumption that $f(u)$ is a strictly convex, C^2 function whose derivative $f'(u) = a(u)$ is monotone. A typical example is the inviscid Burgers equation where $f(u) = u^2$.

For smooth u(x,t), (2.1) takes the form :

$$\frac{\partial u}{\partial t} + a(u)\frac{\partial u}{\partial t} = 0 \; ; \quad u(x,0) = u_0(X) \, , \tag{2.2}$$

expressing that u is constant along characteristics defined by $dx - adt = 0$. Therefore, a(u) also is constant on characteristic curves which are straight lines.

Assuming $u_o(x)$ differentiable, it can be shown that the value of the spatial derivative of u on a characteristic curve C_ζ passing through a point (ζ, t_o) is given by :

$$\left.\frac{\partial u}{\partial x}\right|_{C_\zeta} = \left[\frac{du_0}{dx} \Big/ \left(1+(t-t_0)\frac{d}{dx}a(u_0(x))\right) \right]_{(x=\zeta)} \tag{2.3}$$

and the time t_{shock} at which the solution u(x,t) loses its smoothness is such that :

$$t_{shock} - t_0 = \min_\zeta \left(1 \Big/ \max\left[0, -\frac{d}{dx}a(u_0)\Big|_\zeta\right] \right)$$

With $t_1 < t_{shock}$, one can define, on the time interval (t_o, t_1), a change of variables from the cartesian frame (x,t,u) to a curvilinear one (X, τ, U) :

$$(\mathbb{T}) : \; X = x + a(u)(t_1 - t) \; ; \; \tau = t \; ; \; U = u \, .$$

For any sufficiently smooth function w(x,t) which takes the form $W(X, \tau)$ we have :

$$\begin{bmatrix} \frac{\partial w}{\partial x} \\ \\ \frac{\partial w}{\partial t} \end{bmatrix} = \begin{bmatrix} 1 & 0 \\ \\ -a(w) & 1 \end{bmatrix} \begin{bmatrix} \frac{\partial W}{\partial X} \\ \\ \frac{\partial W}{\partial \tau} \end{bmatrix} \, . \tag{2.4}$$

Then a smooth solution u(x,t) of (2.1) or (2.2) satisfies :

$$\frac{\partial U}{\partial \tau}(X, \tau) = 0 \, . \tag{2.5}$$

This clearly results from the links between $(\mathbb{T})$ and the properties of characteristics which permit a geometric construction for the solution of the initial value problem.

The necessary and sufficient condition for such a construction on the time interval $]t_o, t_1]$ is the following :

(H) : The application $x \longrightarrow \overline{X}(x) = x + \Delta t \; a(u_0(x))$ is a strictly increasing function, where $\Delta t = (t_1 - t_0)$

It must be noted that (H) is equivalent to :

(H_1) $\quad 1 + \Delta t \; a'(u_0) \, u_0'(x) > 0 \quad$ for continuous $u_o(x)$, and to :

(H_2) $a(u_o(x_D - o)) < a(u_o(x_D + o))$ for u_o discontinuous at x_D.

Such a discontinuity of u_o disappears just after t_o and thus is "non-physical".

The assumption (H) entails that the surface representing the solution $u^*(x,t)$ of (2.1) on $[x_{min}, x_{max}] \times [t_o, t_1]$ define also a function $U^*(X, \tau)$ in the coordinate system $\mathbb{T}$. In particular, the graph of the function $x \to u_o(x)$ (possibly supplemented with the set of values $[u_o(x_D - o), u_o(x_D + o)]$ if x_D is a non physical discontinuity point) defines a continuous mapping $X \to U_o(X)$. And (2.5) becomes :

$$(2.6) \qquad U^*(X, \tau) = U_o(X) \qquad \forall \tau \in [t_o, t_1] .$$

A precise definition of the function $U_o(X)$ is easily found if assumption (H) is fulfilled.

We denote by "b" the inverse function of $u \to a(u)$; ($a \to b(a)$ is strictly increasing by the convexity assumption on $f(u)$).

The following equation in x :

$$(2.7) \qquad \overline{X}(x) = \zeta$$

has a unique solution $\overline{x}(\zeta)$ for any ζ (assuming that if x_D is a non physical discontinuity point for u_o -and $\overline{X}$ - this value is taken as the solution when $\overline{X}(x-o) \leq \zeta \leq \overline{X}(x+o)$).

Taking :

$$(2.8) \qquad U_o(\zeta) = b\left(\frac{\zeta - \overline{x}(\zeta)}{\Delta t}\right)$$

gives a precise meaning to the function $X \to U_o(X)$.

For continuous u_o with the (H_1) assumption it suffices to set :

$$(2.9) \qquad U_o(X) = u_o(\overline{x}(X))$$

or :

$$(2.10) \qquad U_o(\overline{X}(x)) = u_o(x) .$$

An illustration of the change of variable $\mathbb{T}$ with an example of an application for the construction of the solution $u^*(x,t)$ from $u_o(x)$ is given in Figures 1 and 2. In Figure 1 is represented the coordinate surface $X = 0$. In Figure 2, the surface represents the solution u^* over the time interval (t_o, t_1) with an initial function $u_o(x)$ for which (H) holds.

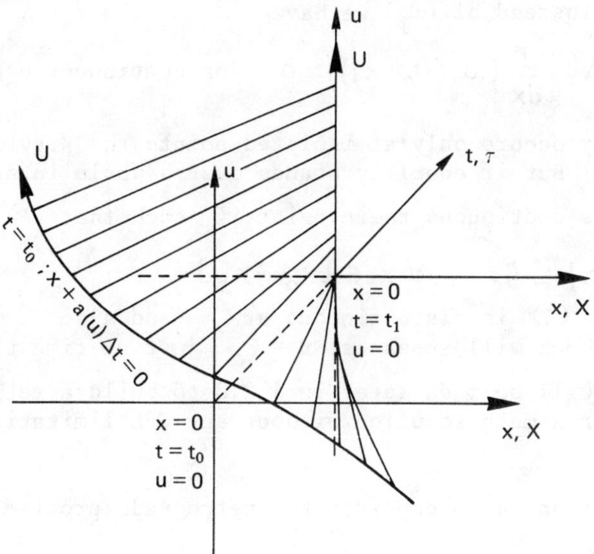

Figure 1. Scalar conservation law. Change of variables over a time interval Δt.

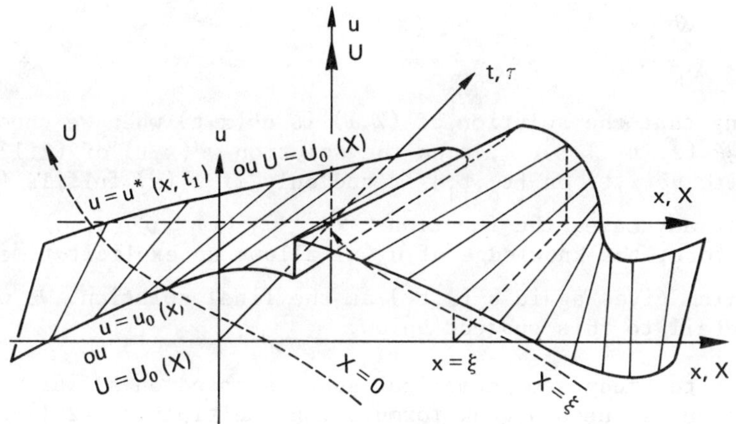

Figure 2. Scalar conservation law. Explicit construction of a solution.

When the hypothesis (H) is no longer satisfied, shocks must appear during the time step $[t_o, t_1]$ or are already present at the initial time t_o.

An interesting special case is provided by weakening (H) to :

$(\tilde{H})$ $\quad x \longrightarrow \overline{X}(x) = x + \Delta t\, a(u_o(x))$ is non decreasing.

Therefore instead of (H_1) we have

$$(\tilde{H}_1) \qquad 1+\Delta t \frac{d}{dx}(a(u_o(x))) \geq 0 \quad \text{for continuous } u_o(x).$$

If equality occurs only at isolated points the previous construction remains valid. But if equality stands over a whole interval $[x_a, x_b]$ where $u_o(x)$ is continuous there exists ζ_c such that :

$$(2.11) \qquad \overline{X}(x) = \zeta_c \quad , \quad \forall x \in [x_a, x_b] .$$

Thus $\zeta \to x(\zeta)$ is discontinuous at ζ_c (and also $\zeta \to U_o(\zeta)$). Therefore a shock will appear at $x = \zeta_c$ just at time t_1. Special use of this case will be made later in § 2.3 to build a method for calculating an approximate solution without any CFL limitation on the time step.

The basic idea is to consider the retrograde problem :

$$(2.12) \qquad \left| \begin{array}{ll} \dfrac{\partial w}{\partial t} + \dfrac{\partial}{\partial x} f(w) = 0 & (x,t) \in [x_{min}, x_{max}] \times [t_o, t_1] \\ \\ w(x, t_1) = w_1(x) & . \end{array} \right.$$

Assuming that the solution of (2.1) is $u^*(x,t)$ when we choose $w_1(x) = u^*(x, t_1)$, then the solution $w^*(x,t)$ of (2.12) coincides with $u^*(x,t)$ on $[t_o, t_1]$ if and only if $u_o(x)$ fulfils $(\tilde{H})$. However, in any case, the function $x \to \tilde{u}_o(x) \equiv w^*(x, o)$ fulfils $(\tilde{H})$. Therefore, the knowledge of $\tilde{u}_o(x)$ allows an explicit time integration which gives again $u^*(x,t_1)$ as the final solution. We will come back in detail to this subject below.

In order to study some numerical schemes based on a finite element approximation, we need a weak formulation. Multiplying (2.1) by a test function $\varphi(x,t)$ and then integrating by parts the integral defined on a space-time domain Ω we get :

$$(2.13) \qquad \iint_\Omega \left(u \frac{\partial \varphi}{\partial t} + f(u) \frac{\partial \varphi}{\partial x}\right) dx\,dt - \int_{\partial\Omega} (u\,\nu_t + f(u)\,\nu_x)\,\varphi\,d\gamma = 0$$

And, if $\Omega = K_i \times [t_o, t_1]$ with $K_i = [x_i, x_{i+1}]$ and $\varphi \equiv 1$ on K_i :

$$(2.14) \qquad \int_{x_i}^{x_{i+1}} u\,dx \Bigg|_{t_o}^{t_1} + \int_{t_o}^{t_1} f(u)\,dt \Bigg|_{x_i}^{x_{i+1}} = 0$$

This is the starting point for finite-volume schemes but several other choices remain to be made to complete a numerical scheme.

Another weak formulation can be derived from (2.5) using the new variables defined in $\mathbb{T}$. With $I = [X_m, X_M]$:

$$\iint_{I\times[t_0,t_1]} U \frac{\partial \Phi}{\partial \tau} \, dX \, d\tau - \int_I \Phi U \, dX \Big|_{t_0}^{t_1} = 0 \quad \forall \, \Phi(X,\tau) . \tag{2.15}$$

The choice of test functions satisfying $\partial\Phi/\partial\tau = 0$ leads to :

$$\int_I \Phi(X)\left(U(X,t_1) - U(X,t_0)\right) dX = 0 \quad \forall \, \Phi(X) . \tag{2.16}$$

Another interesting formula results from (2.13) where, for a given $u_0(x) = u(x,t_0)$, we set $\varphi \equiv 1$ over a triangle in the (x,t) plane formed by the characteristic line passing through the points (x_0,t_1) and $(\bar{x}(x_0),t_0)$, and the two axis $t = t_0$ and $x = x_0$. We get :

$$\int_{t_0}^{t_1} f(u(x_0,t))\,dt = \int_{\bar{x}(x_0)}^{x_0} u_0(x)\,dx - \Delta t \left[a(u)u - f(u)\right]_{u = u_0(\bar{x}(x))} \tag{2.17}$$

Then (2.13) becomes : $\displaystyle \frac{1}{\Delta x}\int_{x_i}^{x_{i+1}} [u(x,t_1) - u_0(x)]\,dx = \dots$

$$\frac{1}{\Delta x}\left[\int_{\bar{x}(x_i)}^{x_i} u_0 \, dx - \int_{\bar{x}(x_{i+1})}^{x_{i+1}} u_0 \, dx\right] + \frac{\Delta t}{\Delta x}\left[a(u)u - f(u)\right]_{u_0(\bar{x}(x_i))}^{u_0(\bar{x}(x_{i+1}))} . \tag{2.18}$$

The right hand side of (2.18) depends only on the knowledge of $u_0(x)$ and can be calculated if u_0 fulfils $(\tilde{H})$.

2.2. Numerical schemes

We discretise $[x_{min}, x_{max}]$ with nodes x_i delimiting cells for a piece-wise polynomial approximation of functions. In order to simplify the presentation, we have chosen a uniform spacing with $\Delta x = x_{i+1} - x_i$, $\forall i$.

Let us define the discontinuous and continuous piecewise polynomial approximation spaces :

$$P_{kd} = \left\{ u(x) \;\middle|\; u\big|_{(x_i,x_{i+1})} = \text{polynomial of } k^{th} \text{ degree} \right\}$$

$$P_{kc} = P_{kd} \cap C^0([x_{min}, x_{max}]) \qquad (k \geq 1) .$$

We shall denote such an approximation space by V_h if x is the independent variable and by W_h if it is X as defined by $\mathbb{T}$. Then for $t = t_1$, $V_h \equiv W_h$. At $t = t_o$, u_h in V_h can be considered as a function $U_h(X)$ if u_h fulfils ($\tilde{H}$) but in general $U_h \notin W_h$.

As proposed by van Leer [7] we shall use the Legendre polynomial decomposition for representing polynomials in P_{kd} or P_{kc}.

(2.19)

$$u_h(x)\Big|_{[x_i,x_{i+1}]} = \bar{u}_{i+1/2} + \left(\frac{x-x_{i+1/2}}{\Delta x}\right)\bar{\Delta}_{i+1/2}u + \frac{1}{2}\left[\left(\frac{x-x_{i+1/2}}{\Delta x}\right)^2 - \frac{1}{12}\right]\bar{\Delta}^2_{i+1/2}u + \dots$$

where only the (k + 1) first terms must be considered. This representation is useful because of the orthogonality properties of the Legendre polynomials and of the meaning of $\bar{u}$, $\bar{\Delta}$, $\bar{\Delta}^2$... :

(2.20)

$$\bar{u}_{i+1/2} = \frac{1}{\Delta x}\int_{x_i}^{x_{i+1}} u_h(x)\,dx$$

$$\frac{\bar{\Delta}u_{i+1/2}}{\Delta x} = \frac{1}{\Delta x}\int_{x_i}^{x_{i+1}} \frac{du_h(x)}{dx}\,dx = \frac{12}{\Delta x^2}\int_{x_i}^{x_{i+1}} \left(\frac{x-x_{i+1/2}}{\Delta x}\right)u_h(x)\,dx$$

$$\frac{\bar{\Delta}^2 u_{i+1/2}}{\Delta x^2} = \frac{1}{\Delta x}\int_{x_i}^{x_{i+1}} \frac{d^2u_h(x)}{dx^2}\,dx = \frac{360}{\Delta x^3}\int_{x_i}^{x_{i+1}}\left[\left(\frac{x-x_{i+1/2}}{\Delta x}\right)^2 - \frac{1}{12}\right]u_h(x)\,dx\,.$$

Discontinuous approximation

Knowing an initial function $u_{oh}(x) \in V_h = P_{kd}(x)$ we want to find $u_{1h}(x) \in V_h$, an approximate solution at time $t = t_1$. Assuming that u_{oh} fulfils (H), the function $X \to U_{oh}(X)$ can be derived and the weak formulation (2.16) gives :

(2.21)
$$\int_{X_i}^{X_{i+1}} v_h(X)\left[u_{1h}(X) - U_{oh}(X)\right]dX = 0 \quad, \quad \forall\, v_h(X) \in W_h\,,\ \forall i$$

Then, $u_{1h}(X)$ is defined through its (k + 1) first moments :

(2.22)
$$\left|\begin{aligned} \bar{u}^{i+1/2} &= \frac{1}{\Delta x}\int_{x_i}^{x_{i+1}} U_{oh}(X)\,dX \\ \frac{\overline{\Delta u}_{i+1/2}}{\Delta x} &= \frac{12}{\Delta x^2}\int_{x_i}^{x_{i+1}}\left(\frac{X-x_{i+1/2}}{\Delta x}\right) U_{oh}(X)\,dX \\ \frac{\overline{\Delta^2 u}_{i+1/2}}{\Delta x^2} &= \frac{360}{\Delta x^2}\int_{x_i}^{x_{i+1}}\left(\left(\frac{X-x_{i+1/2}}{\Delta x}\right)^2-\frac{1}{12}\right) U_{oh}(X)\,dX \end{aligned}\right.$$

It must be noticedthat (2.21) corresponds to the definition of u_{1h} as the solution of :

(2.23)
$$\min_{u_h\in V_h}\int_{x_{min}}^{x_{max}}\left[u_h(x)-U_{oh}(x)\right]^2 dx \,.$$

Therefore u_{1h} is seen to be the L^2-projection into V_h of the function $x\to U_{oh}(x)$ which is obtained by the transport of $x\to u_{oh}(x)$ according to the rules given above.

One could choose another norm for this projection instead of the L^2 norm as was done by Morton [9].

Another way to get simple schemes is to use numerical integration formulas in (2.22).

For $V_h = P_{od}$, a direct application of (2.21) gives the classical method of Godunov which can be described by the following general formulas.

(2.24)
$$\left|\begin{aligned} \bar{u}^{i+1/2} &= \frac{1}{\Delta x}\int_{x_i}^{x_{i+1}} U_{oh}(x)\,dx = \bar{u}_{i+1/2} - \frac{\Delta t}{\Delta x}\left[\langle f_{i+1}\rangle - \langle f_i\rangle\right] \\ \Delta t\,\langle f_i\rangle &= \int_{\bar{x}(x_i)}^{x_i} u_{oh}(x)\,dx - \Delta t\,\big(a(u)u - f(u)\big)\Big|_{u=u_{oh}(\bar{x}(x_i))} \end{aligned}\right.$$

with

(2.25)
$$\Delta t \le \Delta t_{max} = \Delta x \Big/ \max_i \left|a\left(\bar{u}_{i+1/2}\right)\right| .$$

For $V_h = P_{1d}$, exact calculation of the first moment of U_{oh} in (2.22) leads to a scheme proposed in van Leer [7] as his scheme III whereas the use of Simpson's quadrature rule to evaluate this integral gives his scheme II. All these schemes have been numerically tested on Burgers equation and results will be presented below.

Continuous approximation

The weak formulation (2.16) can be written as :

$$(2.26)\qquad \int_{x_{i-1}}^{x_{i+1}} (u_{1h}(X) - U_{oh}(X))\, v_h(X)\, dX = 0 \ .$$

for all $v_h(X)$ such that $Supp(V_h) \subset [x_{i-1}, x_{i+1}]$.

Exact integration of (2.26) gives an implicit scheme even in the simplest case of continuous piecewise linear approximation ($V_h \in P_{1c}$) :

$$(2.27)\qquad u^{i-1} + 4u^{i} + u^{i+1} = \frac{6}{\Delta x}\left[\int_{x_{i-1}}^{x_i}\left(\frac{x - x_{i-1}}{\Delta x}\right) u_{oh}(\bar{X}(x))\,dx + \int_{x_i}^{x_{i+1}}\left(\frac{x_{i+1} - x}{\Delta x}\right) u_{oh}(\bar{X}(x))\,dx\right].$$

This implicit scheme is very similar to a scheme proposed by Morton [10] with several others under the generic name of Euler characteristic Galerkin (ECG) schemes.

Having the same number of unknowns as the scheme resulting from a piecewise constant approximation, its implementation and extension to several dimensions seems more difficult. Moreover, the use of some mass lumping in order to suppress the implicitness in (2.27) would also compromise its better accuracy.

An other interesting way of finding explicit schemes with continuous approximation appears for polynomials of higher degree (P_k, $k \geqslant 2$).

It consists, with u_{1h} belonging to P_{kc}, of using in (2.16) test functions chosen in $P_{(k-2)d}$, and of enforcing continuity of u_{1h} at nodes by :

$$(2.28)\qquad u_{1h}(x_i) = u_{oh}(\bar{X}(x_i)) \ .$$

In the case of a parabolic approximation (2.28) is combined with :

$$(2.29)\qquad \bar{u}^{i+1/2} = \frac{1}{\Delta x}\int_{x_i}^{x_{i+1}} u_{oh}(\bar{X}(x))\,dx \ .$$

This is exactly the scheme V proposed in van Leer [7].

In our description of various schemes we have assumed that the initial data satisfies the hypothesis (H) or ($\tilde{H}$). This is clearly too severe a restriction on u_{oh} or Δt. Indeed, however smooth a continuous initial solution may be, there exists a maximum value of Δt for which (H) can hold. Moreover, once a shock has been created, no value of Δt would allow (H) to be satisfied, and for that reason the use of a discontinuous approximation leads us to weaken (H) or ($\tilde{H}$). For example, in the piecewise constant case the assumption ($\tilde{H}$) becomes :

(2.30) $$1 + \frac{\Delta t}{\Delta x} \left(a(\bar{u}_{i+1/2}) - a(\bar{u}_{i-1/2}) \right) \geq 0 \quad .$$

If u_{oh} does not fulfil such a condition an obvious remedy is to limit Δt. If $a(u_{oh})$ is a non-negative function the constraint (2.30) is equivalent to CFL $\leq$ 1. In the case of arbitrary sign for $a(u_{oh})$, the restriction can at worst become CFL $\leq$ 1/2. Such a limitation may be found too severe ; then it is better to keep the usual condition CFL $\leq$ 1 and to define an adhoc treatment of shocks by using any available knowledge about discontinuous solutions. This approach has been used for discontinuous piecewise linear approximation where the position of shocks at the end of the time step is explicitly calculated in order to evaluate u_{1h} by means of its two first moments as described in (2.22).

A similar choice has been made for the continuous piecewise parabolic approximation. With CFL $\leq$ 1, the formation and displacement of a shock occurs near the nodal points, so that they can easily be taken into account in (2.28) and (2.29). Indeed, this latter formula can be transformed according to (2.24) and (2.17), since it is possible to recognize in the time integral of (2.17) the algebraic value of the area bounded by the two curves $x = x_o$ and $X = x_o$ and by the initial curve $x \to u_o(x)$. This fact permits one to decide on which side of a node a shock has been created if any.

Some numerical results are presented in Figures 3 and 4. They illustrate the good behaviour of schemes with $V_h = P_{1d}$ and P_{2c} which correspond to van Leer's schemes III and V. The problem solved is the propagation of a wave made up of two arcs of parabolas in the case of Burgers equation ($f(u) = u^2/2$).

The results are shown at various times. For both of these approximation functions, the cell mean values exhibit a superconvergent behaviour. Either with discontinuous piecewise linear or continuous piecewise quadratic functions the shocks are fairly well represented.

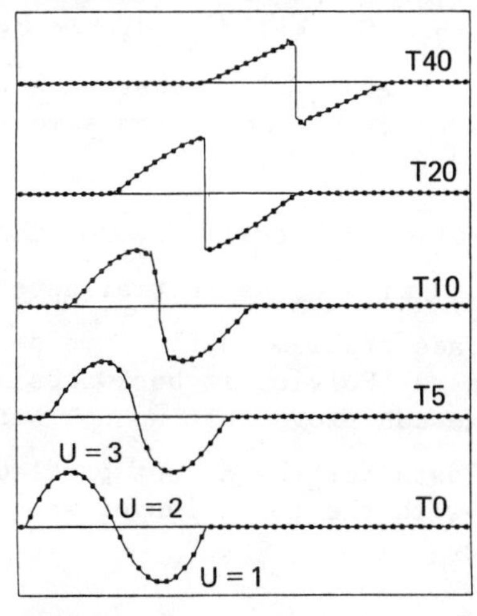

Fig. 3. Burgers equation
Shock wave formation and propagation by a scheme with a discontinuous piecewise linear approximation.

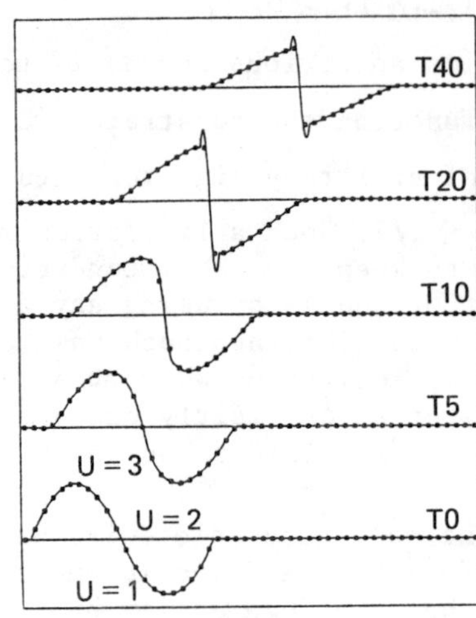

Fig. 4. Burgers equation. Shock wave formation and propagation by a scheme with a continuous piecewise quadratic approximation.

2.3. A large time step method

It is clear that if u_{oh} is continuous and very smooth the condition (H) holds for very large values of Δt, say $\Delta t \leqslant \Delta t_M$. The definition of the various schemes presented here is valid eventhough $\Delta t_M >> \frac{\Delta x}{a}$. In other words, very large values of CFL may be reached in this case. However, the general situation is that, due to the shock formation process, (H) does not hold for a small value of the time step Δt.

We present here a way to get rid of the time step limitation attached to the formation or propagation of shocks.

The basic idea is to replace the initial data $u_o(x)$ which, for the chosen Δt, violates (H) and $(\tilde{H})$ by another function $\tilde{u}_o(x)$ which satisfies $(\tilde{H})$ precisely for the given value of Δt. The substitute function $\tilde{u}_o$ is so defined that it provides us with exactly the same solution u_1 at time $t_1 = t_o + \Delta t$ as does u_o.

Indeed, assume that we know the solution u_1 at time t_1 for the problem (2.1) on the time interval $[t_o, t_1]$ with u_o as initial data violating $(\tilde{H})$. Then consider the retrograde problem on the same period of time with u_1 given as the final time data. Solving it backwards in time gives us at time t_o a function $\tilde{u}_o$. One can prove that $\tilde{u}_o$ now satisfies $(\tilde{H})$ and that the use of $\tilde{u}_o$ as initial data for the direct problem (forwards in time) restores u_1 at time t_1 with the formation of shocks taking place exactly at the final time.

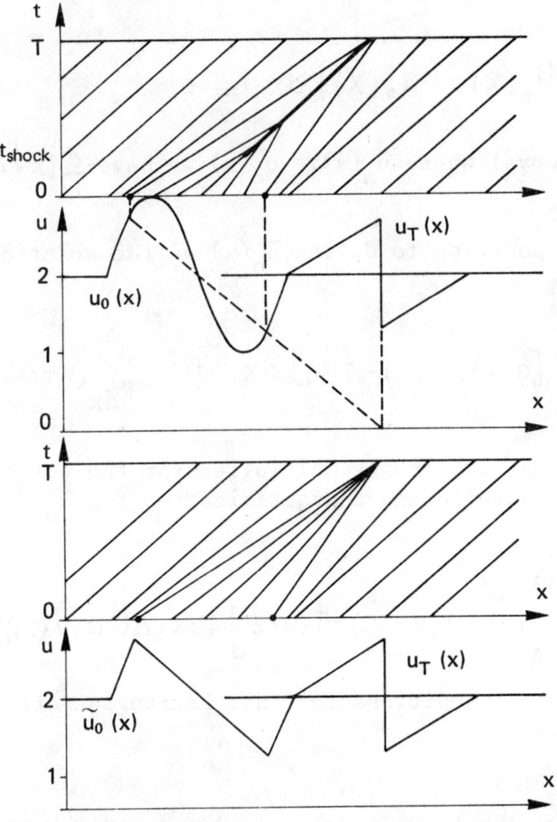

Figure 5. Regularization of initial data (Burgers equation).

This construction rests on the fact that f(u) is convex and therefore any non physical discontinuity resolves instantaneously and fully into a rarefaction wave. This would not be the same with an arbitrary f(u) which could give both a rarefaction and a shock wave by resolution of a non physical discontinuity.

This property is illustrated in Figure 5 which represents the characteristics with u_o and $\widetilde{u}_o$ as initial data for the same problem as in Figures 3 and 4.

Inspection of the characteristics of the problems with u_o and $\widetilde{u}_o$ as initial data shows that the function $\widetilde{u}_o$ possesses the following properties :

- The function $\widetilde{u}_o(x)$ can be different from $u_o(x)$ only over bounded intervals where ($\mathcal{P}_H$) is violated on compact subsets of I.

- On such an interval I, we have :

(2.31) $$\frac{d}{dx}\left(x + \Delta t\, a(\tilde{u}_o(x))\right) = 0 \qquad \text{a.e on } I$$

(2.32) $$\int_I (\tilde{u}_o(x) - u_o(x))\, dx = 0 .$$

- On a small interval where $\tilde{u}_o(x) = u_o(x)$ we have $\frac{d}{dx}(x + \Delta t\, a(u_o(x))) > 0$ a.e.

We are now in position to define $\tilde{u}_o(x)$ as the solution of the problem :

(2.33) $$\min_{v(x)} \int_{-A}^{A} [a(v)(v-u_o) - f(u)]\, dx \quad \text{with} \quad \frac{d}{dx}(x + \Delta t\, a(v(x))) \geq 0 \text{ a.e.}$$

Indeed, introducing a Lagrange multiplier for the positivity constraint we get the following characterization for $\tilde{u}_o$:

$P_{v,\lambda}$:

$$\max_{\lambda(x)\geq 0}\ \min_{v(x)} \int_{-A}^{A} \left[a(v)(v-u_o) - f(v) + \frac{d\lambda}{dx}(x + \Delta t\, a(v(x)))\right] dx = \mathcal{L}(v,\lambda).$$

The stationarity conditions for this Lagrangian are defined firstly by :

(2.34) $$\left(\frac{\partial \mathcal{L}}{\partial v}, \delta v\right) = 0 \iff v(\lambda) - u_o + \Delta t \frac{d\lambda}{dx} = 0 \ , \ \forall x \ ,$$

and if

(2.35) $$F(\lambda) \equiv \mathcal{L}(v(\lambda), \lambda) = \int_{-A}^{A} \left[x \frac{d\lambda}{dx} - f\left(u_o - \Delta t \frac{d\lambda}{dx}\right)\right] d\lambda \ ,$$

by

(2.36) $$\left|\begin{array}{l} \max\limits_{\lambda \geq 0} F(\lambda) \Rightarrow \displaystyle\int_{-A}^{A} (x + \Delta t\, a(v(\tilde{\lambda})))\frac{d}{dx}(\delta\lambda)\, dx = 0 \quad \forall \delta\lambda, \\ \tilde{u}_o = v(\tilde{\lambda}) . \end{array}\right.$$

Therefore $v(\tilde{\lambda}) = u_o$ on intervals where $\tilde{\lambda} \equiv 0$ so that $\frac{d}{dx}(x + \Delta t\, a(u_o)) > 0$.

On the contrary, if $\tilde{\lambda}$ has a compact support I where $\tilde{\lambda} > 0$, from (2.34)-(2.36), we can deduce :

(2.37) $$\frac{d}{dx}(x + \Delta t\, a(v(\tilde{\lambda}))) = 0 \text{ on } I \text{ and } \int_I \left(v(\tilde{\lambda}) - u_o + \Delta t \frac{d\lambda}{dx}\right) dx = 0 \Rightarrow \int_I (\tilde{u}_o - u_o)\, dx = 0.$$

To complete the characterization of $\tilde{u}_o$ as the solution of (2.33), it suffices to prove that the functional $\mathcal{L}^o(v, \lambda)$ is convex in v and concave in λ and therefore that there exists a unique saddle-point of $\mathcal{L}(v, \lambda)$ and a unique solution to (2.33). This result is guaranteed by the convexity of f(u).

The interest of these theoretical findings lies in the possibility of their application to the discrete problem, particularly in the lowest order of approximation.

From $u_{oh} \in P_o$ we wish to construct $\tilde{u}_{oh} \in P_o$ satisfying, for a given large Δt, the $(\tilde{H})$ assumption (2.30) and which can replace u_{oh} as the initial value for this time step, while allowing the use of (2.24) to find the discrete solution at time $t_1 = t_o + \Delta t$.

Such a function $\tilde{u}_{oh}$ is given by solving the P_{v_h, λ_h} problem :

$$P_{v_h,\lambda_h} : \left| \begin{array}{l} \max\limits_{\{\lambda_i \geq 0\}} \min\limits_{\{\bar{v}_{i+1/2}\}} \mathcal{L}_h(v_h, \lambda_h) \\ \mathcal{L}_h(v_h,\lambda_h) = \sum\limits_i \Big\{ a(\bar{v}_{i+1/2})(\bar{v}_{i+1/2} - \bar{u}_{i+1/2}) - f(\bar{v}_{i+1/2}) \ldots \\ \qquad\qquad + (\lambda_{i+1} - \lambda_i)\left(\dfrac{x_{i+1/2}}{\Delta t} + a(\bar{v}_{i+1/2})\right)\Big\}. \end{array} \right.$$

The P_{v_h, λ_h} problem follows from a direct discretization of the problem with $v_h \in P_o$ and $\lambda_h \in P_{1c}$.

The minimization of $\mathcal{L}_h(v_h, \lambda_h)$ with respect to v_h gives :

$$\bar{v}_{i+1/2}(\lambda_h) = \bar{u}_{i+1/2} - (\lambda_{i+1} - \lambda_i) \tag{2.38}$$

and the maximization of $\mathcal{L}_h(v_h(\lambda_h), \lambda_h)$ with respect to λ_h gives :

$$-\frac{\Delta x}{\Delta t} - a(\bar{v}_{i+1/2}(\lambda_h)) + a(\bar{v}_{i-1/2}(\lambda_h)) \equiv G_i = 0 . \tag{2.39}$$

A practical algorithm for solving P_{v_h, λ_h} is the following :

(2.40)

i/ Initialize λ_h by $\lambda_i^o = 0$,

Assuming that λ_i^n is know,

ii/ Set $\bar{v}^n_{i+1/2} = \bar{u}_{i+1/2} - (\lambda^n_{i+1} - \lambda^n_i)$ and $G^n_i = -\dfrac{\Delta x}{\Delta t} - a(\bar{v}^n_{i+1/2}) + a(\bar{v}^n_{i-1/2})$.

iii/ If $G^n_i \leq \varepsilon \dfrac{\Delta x}{\Delta t}$ Stop (ε very small positive constant).

Else,

iv/ Set $\lambda^{n+1}_i = \max\left(0, \lambda^n_i + \rho\, G^n_i\right)$ and go to ii/.

This algorithm converges to a solution $\tilde{u}_{oh}$ if $0 < \rho < \rho_{max}$. An estimation for ρ_{max} can be found from the problem without the positivity constraint on λ_h.

We have now to define in a precise manner how to calculate the solution u_{1h} at time $t_1 = t_o + \Delta t$ from $\tilde{u}_{oh}$ the initial data constructed to replace u_{oh} as just indicated above.

By strict application of (2.34) we can derive the following formulas :

$$(2.41)\quad \left| \begin{array}{l} \bar{u}^{i+1/2} = \bar{\tilde{u}}_{i+1/2} - \dfrac{\Delta t}{\Delta x}\left(< f_{i+1} > - < f_i >\right) \\ < f_i > = \max\left\{ \sum_{l=1}^{l_{max}} I^-_{i,l} \, , \, \sum_{l=1}^{l_{max}} I^+_{i,l} \right\} , \end{array} \right.$$

Where l_{max} is the smallest integer greater than $\eta \max_i |a(\bar{\tilde{u}}_{i+1/2})|$, $(\eta = \Delta t/\Delta x)$.

$$(2.42)\quad \left| \begin{array}{l} I^-_{i,l} = \left| \begin{array}{l} \left[g(\bar{\tilde{u}}_{i-l+1/2}, \min(\bar{\tilde{u}}_{i-l+1/2}, v)) \right]_{v=b(\frac{l-1}{\eta})}^{v=b(\frac{l}{\eta})} , \\ \quad \text{if } \bar{\tilde{u}}_{i-l+1/2} = \bar{u}_{i-l+1/2} \\ \max\left(0, \left[g(\bar{\tilde{u}}_{i-l+1/2}, v) \right]_{v=b(\frac{l-1}{\eta})}^{v=b(\frac{l}{\eta})} \right) \\ \quad \text{if } \bar{\tilde{u}}_{i-l+1/2} \neq \bar{u}_{i-l+1/2} \end{array} \right. \\ I^+_{i,l} = \left| \begin{array}{l} \left[g(\bar{\tilde{u}}_{i+l-1/2}, \max(\bar{\tilde{u}}_{i+l-1/2}, v)) \right]_{v=b(-\frac{l-1}{\eta})}^{v=b(-\frac{l}{\eta})} , \\ \quad \text{if } \bar{\tilde{u}}_{i+l-1/2} = \bar{u}_{i+l-1/2} \\ \max\left(0, \left[g(\bar{\tilde{u}}_{i+l-1/2}, v) \right]_{v=b(-\frac{l-1}{\eta})}^{v=b(-\frac{l}{\eta})} \right) . \\ \quad \text{if } \bar{\tilde{u}}_{i+l-1/2} \neq \bar{u}_{i+l-1/2} \end{array} \right. \end{array} \right.$$

with

$$(2.43)\quad g(u,w) = f(w) - f(u) + a(w)(u-w),$$

and $a \to b(a)$ is the inverse function of $u \to a(u)$.

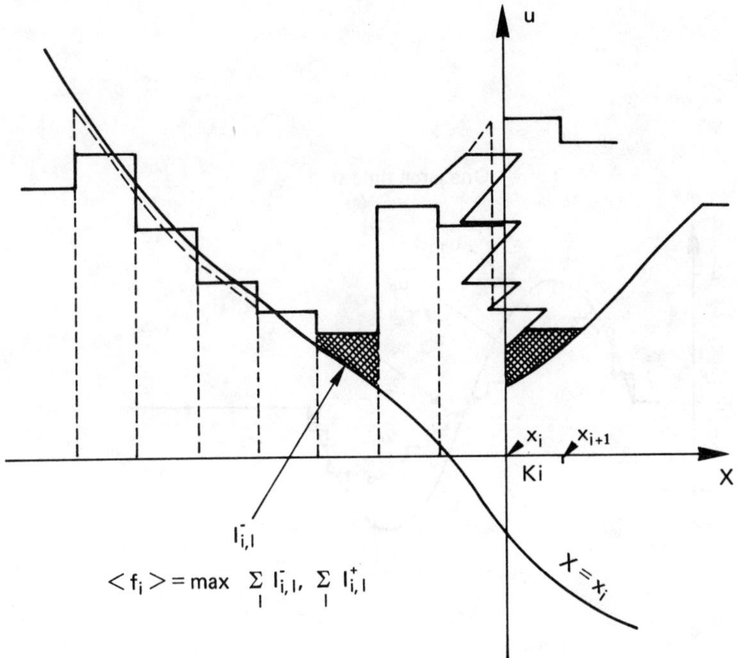

Figure 6. Schematic description of fluxes in the large time step method.

The quantity $I^-_{i,l}$ is the contribution of the cell $[x_{i-l-1}\ ,\ x_{i-l}]$ to the flux $\langle f_i \rangle$ at node x_i. An intermediate formulation of $I^-_{i,l}$ can be given after a slight transformation of (2.17) :

$$
(2.44)\qquad I^-_{i,l} = \left| \begin{array}{ll} \displaystyle\int_{x_{i-l-1}}^{x_{i-l}} \max\left(0, \tilde{u}_{i-l+1/2} - b\left(\frac{x_i - x}{\Delta t}\right)\right) dx & \text{if } \tilde{u}_{i-l+1/2} = \bar{u}_{i-l+1/2} \\[2ex] \displaystyle\max\left(0, \int_{x_{i-l-1}}^{x_{i-l}} \left(\tilde{u}_{i-l+1/2} - b\left(\frac{x_i - x}{\Delta t}\right)\right) dx\right) & \text{if } \tilde{u}_{i-l+1/2} \neq \bar{u}_{i-l+1/2}. \end{array} \right.
$$

An analogous formula holds for $I^+_{i,l}$. The graphical interpretation of this integral can be seen on Figure 6 where is depicted the convection of the shaded area from above the $X = x_i$ curve at time t_o to the right of node x_i at time t_1.

Figure 7 gives an interpretation of this large time step scheme as a convection projection scheme for the problem already solved in figures 3 and 4 with high order approximation but CFL $\leqslant 1$.

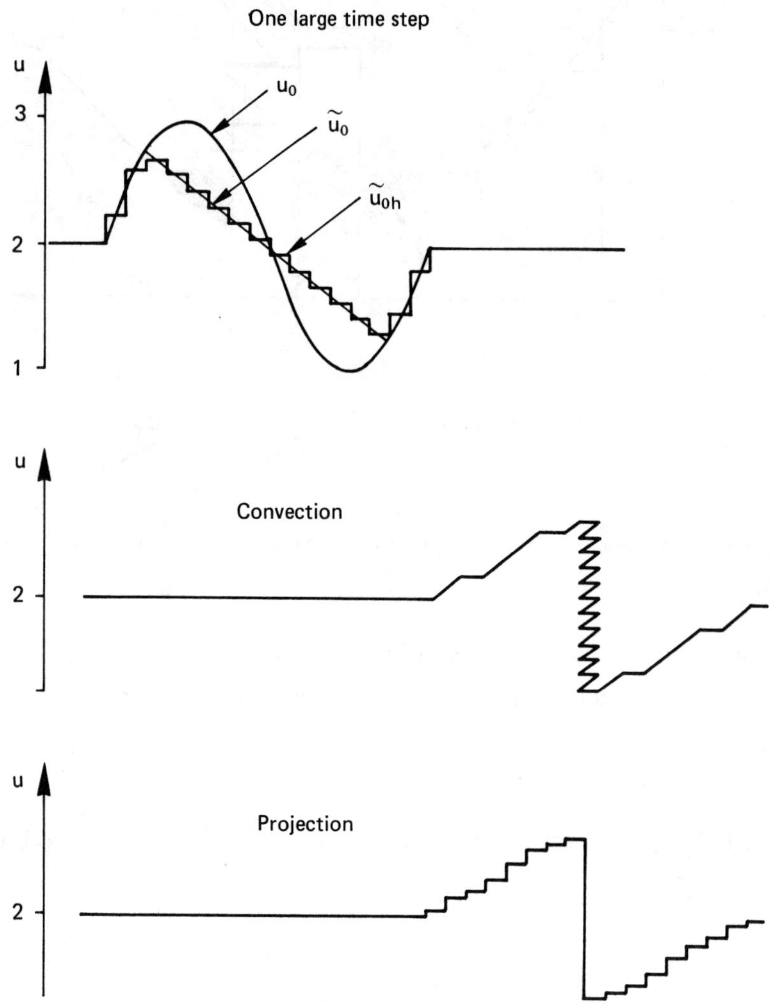

Figure 7. Burgers equation. Schematic description of the large time step method.

To be more specific, for the Burgers equation, the formulas (2.41) become :

$$g(u,w) = -\frac{1}{2}(u-w)^2 ,$$

$$(2.45)\quad \begin{cases} I^-_{i,l} = \begin{cases} \frac{1}{2}\left\{\left(\left[\tilde{\bar{u}}_{i-l+1/2}-\frac{l-1}{\eta}\right]^+\right)^2-\left(\left[\tilde{\bar{u}}_{i-l+1/2}-\frac{l}{\eta}\right]^+\right)^2\right\} \\ \qquad \text{if } \tilde{\bar{u}}_{i-l+1/2}=\bar{u}_{i-l+1/2} , \\ \frac{1}{\eta}\left[\tilde{\bar{u}}_{i-l+1/2}-\frac{l-1/2}{\eta}\right]^+ \\ \qquad \text{if } \tilde{\bar{u}}_{i-l+1/2}\neq\bar{u}_{i-l+1/2} \end{cases} \\ I^+_{i,l} = \begin{cases} \frac{1}{2}\left\{\left(\left[\tilde{\bar{u}}_{i+l-1/2}+\frac{l-1}{\eta}\right]^-\right)^2-\left(\left[\tilde{\bar{u}}_{i+l-1/2}+\frac{l}{\eta}\right]^-\right)^2\right\} \\ \qquad \text{if } \tilde{\bar{u}}_{i+l-1/2}=\bar{u}_{i+l-1/2} , \\ -\frac{1}{\eta}\left[\tilde{\bar{u}}_{i+l-1/2}+\frac{l-1/2}{\eta}\right]^- \\ \qquad \text{if } \tilde{\bar{u}}_{i+l-1/2}\neq\bar{u}_{i+l-1/2} \end{cases} \end{cases}$$

where $[y]^+ = \max(o,y)$, $[y]^- = \min(o,y)$.

The usual case where Δt is chosen sufficiently small for giving a CFL $\leqslant 1$ corresponds to $l_{max} = 1$. If moreover $\tilde{u}_{oh} \equiv u_{oh}$ (for example when $u_{oh} > 0 \ \forall x$) then formulas (2.45) boil down to :

$$(2.46)\quad \langle f_i \rangle = \max\left\{\frac{1}{2}\left(\left[\bar{u}_{i-1/2}\right]^+\right)^2, \frac{1}{2}\left(\left[\bar{u}_{i+1/2}\right]^-\right)^2\right\}$$

Numerical results obtained for the same problem as the one shown on Figures 3 - 4 are presented in Figure 8 both for small time steps (2.46) and for large time steps (2.45). Notice that we have taken a Δx twice as small as before in order to keep the same number of unknowns.

In Figure 8a, the various results are obtained with CFL < 1 and in a number of time steps proportional to the elapsed time since T_o. On the contrary, each one of the results in Figure 8b has been calculated from T_o in only one time step with a CFL proportional to this increasingly large time step.

First, it is clear that with CFL < 1 the simple scheme given by (2.46) is of low accuracy compared with those corresponding to higher order polynomial approximations presented in Figures 3 and 4. It is only first-order accurate and thus of little use when precise details of the solution are desired.

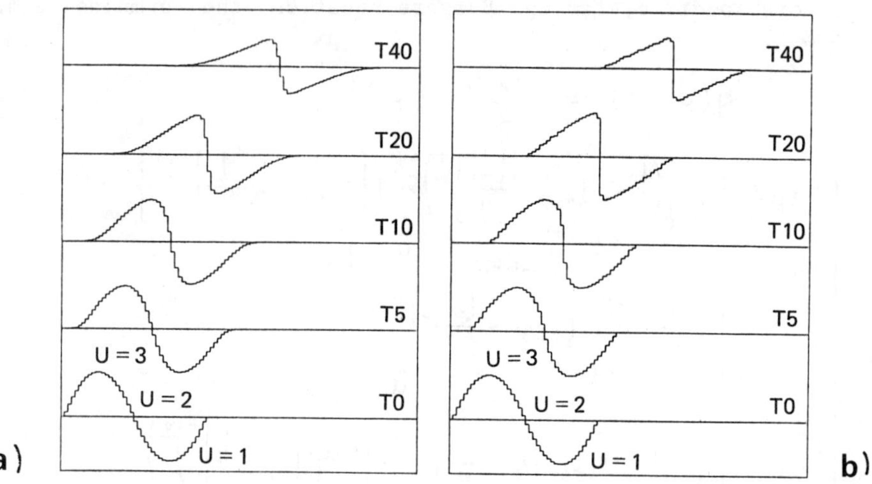

Figure 8. Burgers equation. Piecewise constant approximation a) standard results with CFL < 1. b) results for the large time step method.

The contribution of this large time step technique can be seen in Figure 8b. These results are now comparable to those of Figures 3 and 4. The errors at the various times in the evolution of the solution are due to the representation of the initial function $\tilde{u}_o$ by a piecewise constant approximation P_o and also to the projection of the convected solution at the final time into the same space P_o.

It is the accumulation of the repeated errors in projections which is responsible for the smearing of shocks and of slope discontinuities when small time steps are used.

This argues in favour of large time steps for the piecewise constant approximation. However, the limitation of this technique to convex (or concave) f(u) is a drawback which seems unavoidable. Also, the extension of these ideas to hyperbolic systems of genuinely non linear equations remains to be done.

Before concluding this chapter, we note that our treatment of large time steps for the Godunov method applied to scalar conservation laws brings clear and complete answers to the problems encountered by R.J. Leveque [11] in his efforts to define a large time step method by explicit handling of discontinuities.

3. Euler equations

3.1. Generalities

We have seen in the preceding paragraph how the convection-projection approach for constructing numerical schemes is related to a weak formulation based on characteristic variables. For systems of conservation laws one possible extension would consist in using first a characteristic decomposition of the system before applying scalar techniques of § 2. How successful this approach would be is not clear and much work remains to be done in this direction.

We have chosen a simpler way of extending some ideas presented above. It is based on the Lagrangian formulation in which, among all the characteristic directions, the fluid velocity plays a central role. As before, we have a coordinate transformation by which the Euler weak formulation can be changed into the Lagrangian weak formulation. Thus the convection phase will now be a Lagrangian evolution of the dependent variables and it will be followed by a projection step which is an Eulerian remapping at the final time.

To be more specific, we consider the equations of gaz dynamics in Eulerian form :

(3.1)
$$\left| \begin{array}{l} w_t + [F(w)]_x + [G(w)]_y = 0 \\ \\ w = \begin{pmatrix} \rho \\ \rho u \\ \rho v \\ \rho E \end{pmatrix}, \quad F(w) = \begin{pmatrix} \rho u \\ p + \rho u^2 \\ \rho u v \\ (p+\rho u)E \end{pmatrix}, \quad G(w) = \begin{pmatrix} \rho v \\ \rho u v \\ p + \rho v^2 \\ (p+\rho v)E \end{pmatrix}, \\ \\ E = e + \frac{1}{2}(u^2+v^2) \; ; \; e = \frac{1}{\gamma - 1}\frac{p}{\rho} \; ; \; A(w) = \frac{dF(w)}{dw} \; ; \; B(w) = \frac{dG(w)}{dw}, \end{array} \right.$$

where ρ, p, u, v, E, e are respectively the density, pressure, cartesian components of fluid velocity, specific total energy and specific internal energy. And γ is the ratio of specific heats.

For a unit vector $\vec{\nu}$ forming an angle ν with the positive x-axis we define the matrix :

(3.2) $$\mathcal{A}(\nu) = (\cos \nu) A + (\sin \nu) B .$$

Its eigenvalues are :

(3.3) $$\lambda_\nu^{(1)} = V_\nu - c \; ; \; \lambda_\nu^{(2)} = \lambda_\nu^{(3)} = V_\nu \; ; \; \lambda_\nu^{(4)} = V_\nu + c ,$$

where $V_\nu = \vec{V}.\vec{\nu}$, with $\vec{V}$ the absolute fluid velocity and c the speed of sound.

(3.4) $$c^2 = \gamma p / \rho .$$

A weak formulation for (3.1) is :

(3.5)
$$\forall \varphi_i \,, \quad i=1,\dots,4$$
$$\int_0^T\int_{K(t)} \Big(W_i \frac{\partial \varphi_i}{\partial t} + F_i \frac{\partial \varphi_i}{\partial x} + G_i \frac{\partial \varphi_i}{\partial y}\Big)\, dx\, dy\, dt - \Big[\int_{K(t)} W_i \varphi_i \, dx\, dy\Big]_0^T - \int_0^T\int_{\partial K(t)} \vec{\mathcal{F}}_i \cdot \vec{n}\, \varphi_i \, d\gamma\, dt = 0,$$

where $\mathcal{F}_i = (F_i, G_i)^t$ and $K(t)$ is a bounded part of the flow domain and $\vec{n}$ the unit outward normal to ∂K.

Choosing a test function φ_i independent of x,y,t gives the conservation laws :

$$(3.6) \quad \int_{K(t)} W_i \, dx\, dy \Big|_{t=T} = \int_{K(t)} W_i \, dx\, dy \Big|_{t=0} - \int_0^T \int_{\partial K(t)} \vec{\mathcal{F}}_i \cdot \vec{n}\, d\gamma\, dt \,, \quad i=1,\dots,4.$$

The introduction of the Lagrangian variables (a,b,τ) attached to the fluid particles according to :

$$(3.7) \quad x(a,b,\tau) = a + \int_{t_0}^{t} u(a,b,\tau)\,d\tau \,, \quad y(a,b,\tau) = b + \int_{t_0}^{t} v(a,b,\tau)\,d\tau \,, \quad t=\tau,$$

leads to the weak formulation in Lagrangian variables :

(3.8)
$$\int_0^T\int_{\tilde{K}} \Big(\tilde{W}_i \frac{\partial \tilde{\varphi}_i}{\partial \tau} + \tilde{F}_i \frac{\partial \tilde{\varphi}_i}{\partial a} + \tilde{G}_i \frac{\partial \tilde{\varphi}_i}{\partial b}\Big)\, da\, db\, d\tau - \Big[\int_{\tilde{K}} \tilde{W}_i \tilde{\varphi}_i \, da\, db\Big]_0^T - \int_0^T\int_{\partial \tilde{K}} \vec{\tilde{\mathcal{F}}}_i \cdot \vec{N}\, \tilde{\varphi}_i \, d\sigma\, d\tau = 0,$$

where K is the Lagrangian representation in variables a,b of K(t), and

$$(3.9) \quad \left|\; \begin{aligned} &\tilde{W} = \begin{pmatrix} J\rho \\ J\rho u \\ J\rho v \\ J\rho E \end{pmatrix}, \quad \tilde{F} = \begin{pmatrix} 0 \\ y_b p \\ -x_b p \\ (u y_b - v x_b) p \end{pmatrix}, \quad \tilde{G} = \begin{pmatrix} 0 \\ -y_a p \\ x_a p \\ (-u y_a + v x_a) p \end{pmatrix}. \\ &J = x_a y_b - x_b y_a \qquad \vec{N} = (y_\sigma, -x_\sigma)^t \end{aligned} \right.$$

The system of conservation laws in Lagrangian coordinates is :

$$(3.10) \quad \frac{\partial \tilde{W}}{\partial \tau} + \frac{\partial \tilde{F}}{\partial a} + \frac{\partial \tilde{G}}{\partial b} = 0 \;.$$

But, it must be supplemented by the kinematic relations :

$$(3.11) \quad x_\tau = u \;;\; y_\tau = v.$$

A useful equation in conservation form can be derived for J :

$$\frac{\partial J}{\partial \tau} - \frac{\partial}{\partial a}(u y_b - v x_b) - \frac{\partial}{\partial b}(u y_a - v x_a) = 0 . \tag{3.12}$$

When we consider the one-dimensional case, we can write (3.9) - (3.12) in the following manner :

$$\left| \begin{array}{l} \tilde{W} = \begin{pmatrix} J\rho \\ J\rho u \\ J\rho E \end{pmatrix} ; \quad \tilde{F} = \begin{pmatrix} 0 \\ p \\ pu \end{pmatrix} ; \quad J = x_a , \\ \dfrac{\partial \tilde{W}}{\partial \tau} + \dfrac{\partial \tilde{F}}{\partial a} = 0 \\ \dfrac{\partial J}{\partial \tau} - \dfrac{\partial u}{\partial a} = 0 . \end{array} \right. \tag{3.13}$$

The one dimensional Lagrangian equations appear to be somewhat simpler than their Eulerian counterpart. The simplification is not so impressive as in the scalar case since only one conservation law reduces to the constancy condition :

$$\frac{\partial}{\partial \tau}(J\rho) = 0 . \tag{3.14}$$

The other characteristic directions are now in a symmetric arrangement with eigenvalues $(-\rho c, 0, \rho c)$ for the Jacobian of the system (3.13). These two properties are preserved in the multidimensional case eventhough it is difficult to find (3.9-10) simpler than (3.1).

Thus the main difference between the present case and the scalar case is that here the Lagrangian change of variable no longer leads to an explicit solution of the problem. But the distinction in numerical schemes between a convection phase and a projection phase can be kept if we consider that the convection phase consists of numerically solving on a time interval T the Lagrangian formulation (3.9). The Lagrangian results at final time can then be projected into the initial grid to serve as new data for the next time interval T.

We have studied such numerical schemes both in one dimension and in two dimensions, and we will report here some of these results.

3.2. One dimensional Lagrangian-Eulerian schemes

For simplicity, we have considered only piecewise constant approximation. The notation is the same as in § 2 for the mean value on cells. A spatial index is written as a subscript for initial time values and as a superscript for final time values. From (3.8) we get, in one dimension, the general conservation law for the Lagrangian conservative variables :

$$\tilde{w}^{i+1/2} = \tilde{w}_{i+1/2} - \frac{\Delta t}{\Delta x}\left(\tilde{F}_{i+1} - \tilde{F}_i\right). \qquad (3.15)$$

The flux $\tilde{F}_i$ has to be estimated from the cells surrounding the interface node x_i. A choice consistant with the present context is the adoption of the Godunov scheme where $\tilde{F}_i$ is given by the solution of a Riemann problem between the two constant states $\tilde{w}_{i-1/2}$ and $\tilde{w}_{i+1/2}$. The contact discontinuity remains attached to the node $a = x_i$ and the corresponding values of pressure p_i^* and fluid velocity u_i^* are used directly to give $\tilde{F}_i = \tilde{F}(u_i^*, p_i^*)$.

The evolution from t_o to $t_o + \Delta t$ of the position of the ith interface node is defined by :

$$x^i = x_i + \Delta t\, u_i^* , \qquad (3.16)$$

where we have identical values for x and a at the intial time t_o and therefore :

$$J^{i+1/2} = (x^{i+1} - x^i)/\Delta x = 1 + \frac{\Delta t}{\Delta x}(u_{i+1}^* - u_i^*) . \qquad (3.17)$$

The projection of Lagrangian results onto the the given initial grid may be done at each time step or only after m Lagrangian steps.

In that case some book-keeping must be done in order to remap the Lagrangian results defined at $t_o + m\Delta t$ onto a grid which has been moved by several cell lengths since t_o. In contrast, if a projection step is done at the end of every Lagrangian step, one can easily express it in a flux formulation :

$$(3.18)\quad \left| \begin{array}{l} w^{i+1/2} = \tilde{w}^{i+1/2} - \dfrac{\Delta t}{\Delta x}\left[u_j^* \,(\tilde{w}/J)^{k(j)}\right]_{j=i}^{j=i+1} \\ \text{where} \quad k(j) = \left| \begin{array}{ll} j-1/2 & \text{if } u_j^* \geq 0 \\ j+1/2 & \text{if } u_j^* < 0 . \end{array} \right. \end{array} \right.$$

To be exact, this projection formula must be applied with the time step limitation : $\Delta t \max |u_i^*| < \Delta x$.

Exact projection is only first order accurate as in the Godunov scheme. Therefore a projection at every time step leads to a strong dissipation as can be seen in Figure 9a where results for the Sod's shock tube problem [12] are plotted. In Figure 9b, c, d, the results when a projection is done every 5th, 10th and 20th Lagrangian time steps are presented. Only two projections were done after the initialization in the latter case.

The benefit of spacing the projection steps even a little is evident for the discontinuities, however nothing is changed in the smearing of the slope discontinuities bounding the rarefaction fan.

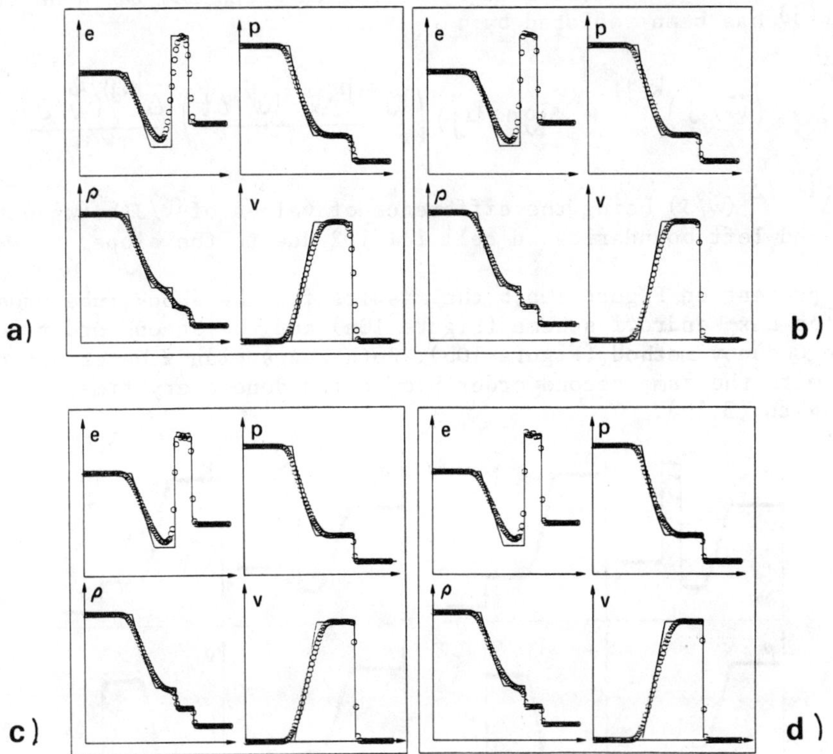

Figure 9. Shock tube problem : First order Lagrangian-Eulerian method with a projection step done at every m^{th} time step. a) m = 1. b) m = 5. c) m = 10. d) m = 20.

Second order accuracy is clearly a necessity for the Lagrangian scheme and also for the projection step if it is frequently carried out. Concerning the Lagrangian scheme we have used in our tests both a Lax-Wendroff scheme and a second order extension of the Godunov method. This extension was proposed by van Leer and Hancock [13]. Schematically, it consists of recovering slopes from the piecewise constant initial data. These slopes are used in a non conservative way to predict cell values at the mid time step. By associating these cell values with the previous slopes one can define left and right values from each side of an interface which in turn are used as data for a numerical evaluation of fluxes. In our tests the solution of a Riemann problem has been used. The recovery of slopes is done by a non linear averaging from three consecutive values as proposed by van Albada (see van Leer [13]).

The same ideas can lead to a second order accurate projection. First, the piecewise constant Lagrangian results calculated on the displaced grid are transformed into piecewise linear values while preserving conservation. Then, with the same time step restriction as before, the projection is done by (2.18) in which now the quantity $(\tilde{W}/J)^{k(j)}$ has been replaced by :

$$(3.19) \qquad (\tilde{W}/J)^{k(j)} + \operatorname{sign}(u_j^*) \left(\frac{J^{k(j)}\Delta x - |u^*|\Delta t}{2} \right) \frac{\bar{\Delta}^{k(j)}(\tilde{W}/J)}{J^{k(j)}\Delta x} \; ,$$

with $\bar{\Delta}^{i+1/2}(\tilde{W}/J)$ being the difference of values of $\tilde{W}/J$ between the right and left boundaries on cell $i + 1/2$ due to the slope.

We present on Figure 10a,b the results for the shock tube problem with the Lax-Wendroff scheme (Figure 10a) and the second order version of the Godunov method (Figure 10b). Both Lagrangian schemes are combined with the same second order projection done every time step according to (3.19).

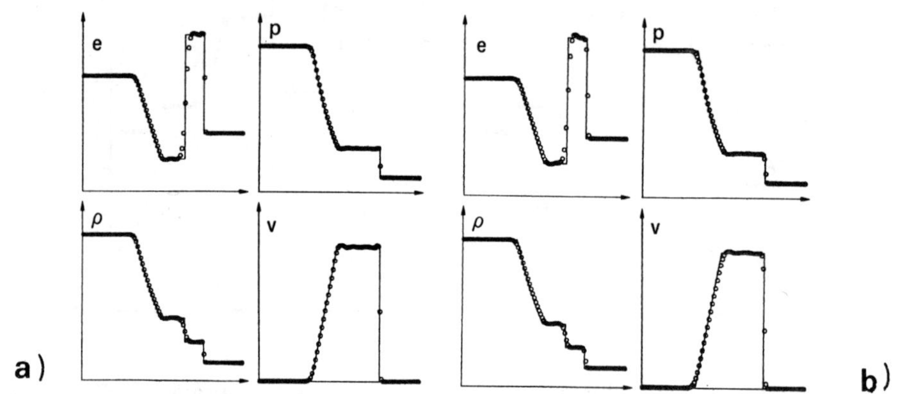

Figure 10. Shock tube problem : a) 2nd order Lax-Wendroff Lagrangian scheme with a 2nd order projection. b) 2nd order version of Godunov Lagrangian scheme with a 2nd order projection.

3.3. Lagrangian-Eulerian methods in two dimensions

As already mentioned, the Lagrangian equations in the multidimensional case do not appear to be simpler than the Eulerian ones. However, the Lagrangian-Eulerian approach keeps its inherent merits. Namely it has both a mathematically and physically sound basis and it gives robust schemes without the need of adding artificial viscosity. In fact, the Lagrangian phase is centered in nature whereas the projection of the Lagrangian results onto the Eulerian grid introduces a natural upwinding effect (see (3.18)).

To extend to two dimensions the schemes studied in one dimension, we have chosen the simple piecewise constant approximation on quadrilateral cells. We also decided to make this extension without resorting to a splitting algorithm.

Various first-order and second-order schemes have been studied. The simplest first order method was composed of the method of Godunov for the Lagrangian phase followed by an exact or approximate first order projection step. The method of Godunov can be described by the following formula expressing the evolution on a cell K_0 of the Lagrangian variables defined in (3.9) :

$$[\tilde{w}]_{K_0}^{n+1} = [\tilde{w}]_{K_0}^{n} - \frac{\Delta t}{|K_0|}\int_{\partial K_0} (\tilde{F} n_a + \tilde{G} n_b)\, d\sigma \tag{3.20}$$

where :

$$(\tilde{F} n_a + \tilde{G} n_b)\Big|_{\partial K_l} = \left|\begin{array}{l} 0 \\ p y_\sigma \\ -p x_\sigma \\ p \vec{V}\cdot\vec{N}_l \end{array}\right. \quad \text{with} \quad \vec{N}_l = \left|\begin{array}{l} y_\sigma \\ -x_\sigma \end{array}\right. \text{ on } \partial K_l . \tag{3.21}$$

Practically, the calculation of a Lagrangian step is done by first estimating the velocity of the grid at nodes. Then the solution of Riemann problems at each interface between two cells gives a pressure p^* and a normal velocity $(\vec{V}.\vec{n})^*$ which are used for calculation of fluxes according to (3.21) and (3.20).

It must be noticed that the Riemann problems are very easy to solve, and the Newton method converges towards a sufficiently accurate solution in at most 3 iterations so that "approximate Riemann solvers" are not truly a necessity.

The exact projecting of Lagrangian results after m time steps boils down to calculating the area of the polygonal intersection of Lagrangian and initial Eulerian cells in order to distribute in proportion mass, momentum and energy. This remapping technique is rather expensive and this is another argument in favour of less frequent projections. Some tests have been done on unsteady flows in channels and have been presented in [14] showing better shocks and contact discontinuities with less frequent projections.

However, the first order accuracy is not sufficient, and among many possibilities of extensions from one to two dimensions, we have chosen to develop first a method combining a Lagrangian scheme resembling the Richtmyer-Burstein scheme of second order of accuracy with an approximate projection of second order of accuracy done every Lagrangian step. Precise details can be found in [15] and we only recall here the main features of the method.

The Lagrangian scheme is of predictor-corrector type. The prediction is built with a finite element weak formulation and gives velocity and pressure at nodes at the end of a half time step. These quantities allow the calculation of the mesh displacement and the evaluation of fluxes at the middle of interfaces between cells. The corrector step is simply the classical finite volume formulation of the conservation laws over a time step.

The second order accurate projection step is a direct extension in two dimensions of the method described above in the one-dimensional case.

Some results obtained with this method are presented in Figure 11. They illustrate the time evolution of isomach lines in the Emery test problem. At the initial time a uniform flow with M = 3. is disturbed by the sudden application of the impervious boundary conditions.

Many variants could be considered with modifications concerning either the Lagrangian phase or the projection step. The present method is not far from the BBC scheme quoted by Woodward and Collela [5]. In a previous paper [16] they presented a recently developed high order and quite sophisticated method (the Piecewise-Parabolic-Method : PPM) of Lagrangian Eulerian type which could be considered as an optimal scheme as far as accuracy is concerned.

In conclusion, the convection-projection approach is shown to give a sound basis for the construction or at least for the interpretation of various upwind numerical schemes with useful applications from Burgers equation to multidimensional Euler equations. Further exploration of this concept especially towards the generation of flux splitting methods is an attractive subject for future work.

Acknowledgements. The author wishes to thank Michel Borrel for his very hepful cooperation in the course of this research work.

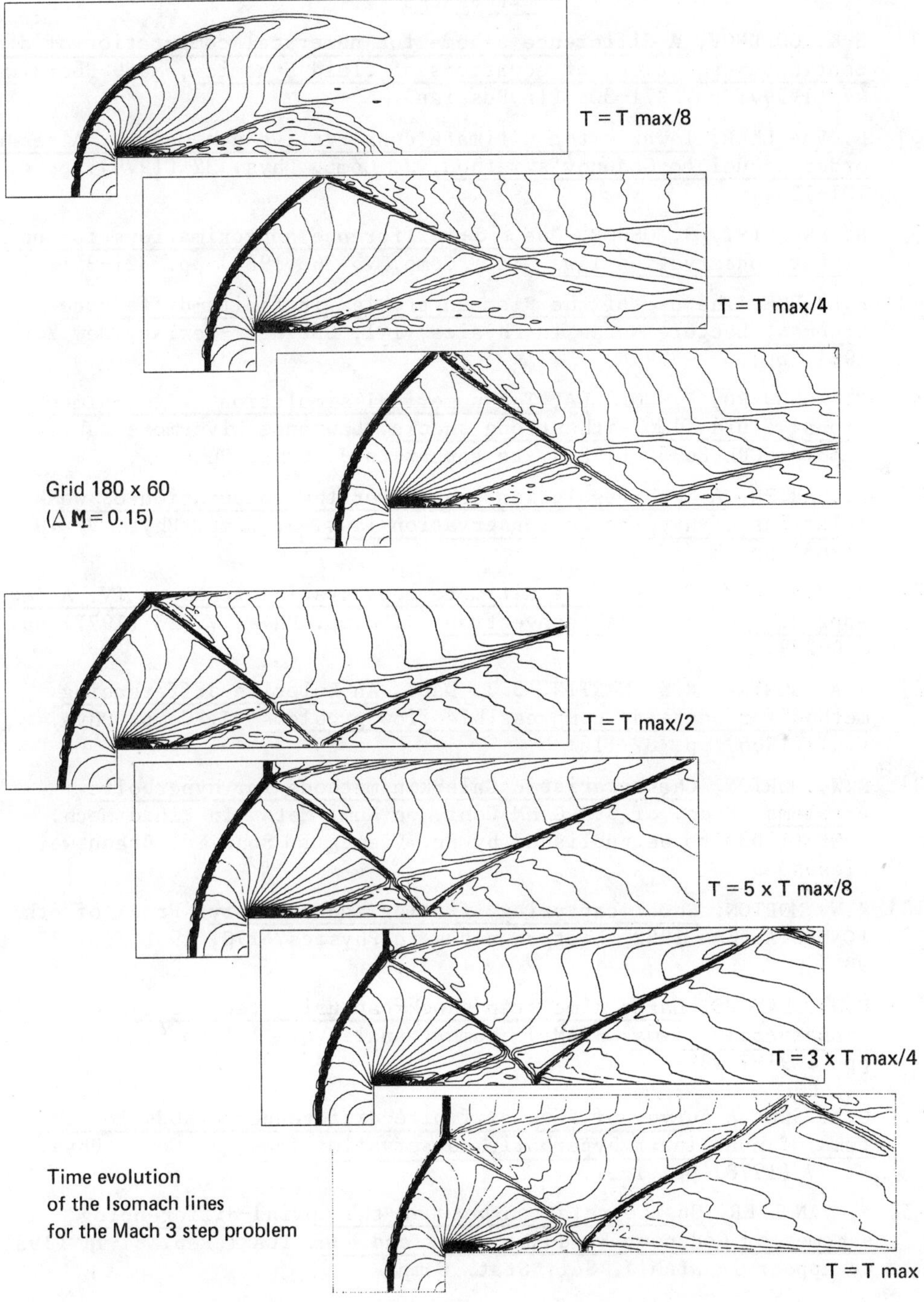

Figure 11. Time evolution of the Isomach limes for the Emery test problem.

REFERENCES

[1] S.K. GODUNOV, A difference scheme for numerical computation of discontinuous solutions of equations of fluid dynamics, Mat. Sbornik, 47 (1959), pp. 271-306 (In Russian).

[2] B. VAN LEER, Towards the ultimate conservative scheme : V. A second order sequel to Godunov's method, J. Comp. Phys. 32 (1979), pp. 101-136.

[3] B. ENGQUIST, S. OSHER, One sided difference approximations for non linear conservation laws. Math. Comp. V. 36 (1981) pp. 321-351.

[4] P.L. ROE, The use of the Riemann problem in finite-difference schemes. Lecture Notes in Physics, 141, Springer-Verlag, New York 1981, pp. 354-359.

[5] P. WOODWARD, P. COLELLA, The numerical simulation of two-dimensional fluid flow with strong shocks. Lawrence Livermore Lab. Report UCRL 86952 (1983), to appear in J. Comp. Phys.

[6] A. HARTEN, A high resolution scheme for the computation of weak solutions of hyperbolic conservation laws. J. Comp. Phys. V. 49 (1983) pp. 357-393.

[7] B. VAN LEER, Towards the ultimate conservative scheme : IV. A new approach to numerical convection. J. Comp. Phys. V. 23 (1977) pp. 276-299.

[8] R.A. GENTRY, R.E. MARTIN, B.J. DALY, An Eulerian differencing method for unsteady compressible flow problems. J. Comp. Phys. V. 1 (1966) pp. 87-118.

[9] K.W. MORTON, Characteristic Galerkin methods for hyperbolic problems, Proc. of 5th GAMM Conf. on Num. Meth. In Fluid Mech. ROME (1983) to be published by Fr. Vieweg und Sohn Ed. Braunsweig/ Wiesbaden.

[10] K.W. MORTON, Shock capturing, fitting and recovery. Proc. of 8th ICNMFD Aachen 1982, Lecture notes in Physics, 170, 1982, pp. 77-93.

[11] R.J. LEVEQUE, Large time step shock-capturing techniques for scalar conservations laws, SIAM J. Numer. Anal. Vol. 19, no 6, Dec. 1982, pp. 1091-1109.

[12] G.A. SOD, A survey of several finite-difference methods for systems of non linear hyperbolic conservation laws. J. Comp. Phys. V. 27 (1978) pp. 1-31.

[13] B. VAN LEER, On the relation between the upwind-differencing schemes of Godunov, Engquist-Osher and Roe. ICASE Rep. 81-11 (1981) to appear in SIAM J. Sci. Stat. Comp..

[14] M. BORREL, Ph. MORICE, A Lagrangian-Eulerian approach to the computation of unsteady transonic flows. Proc. of the 4th GAMM Conf. on Num. Meth. in Fl. Mech. Paris (1981). Fr. Vieweg und Sohn Ed. Braunschweig/Wiesbaden.

[15] M. BORREL, Ph. MORICE, A second order Lagrangian-Eulerian method for computation of two-dimensional unsteady transonic flows. Proc. of the 5th GAMM Conf. on Num. eth. in Fluid Mech. ROME (1983) to be published by Fr. Vieweg und Sohn Ed. Braunschweig/Wiesbaden.

[16] P. COLELLA, P. WOODWARD, The Piecewise-Parabolic-Method (PPM) for gas-dynamical simulations. LBL Rep. no 14661, July 1982.

A TWO-DIMENSIONAL LAGRANGE-EULER TECHNIQUE FOR GAS DYNAMICS

ERWAN LE GRUYER* AND ALAIN YVES LE ROUX**

Abstract. We describe a numerical method for the equations of isentropic gas dynamics where the boundary condition V.n = 0 is replaced by a simple continuity property. This is done by using some particular paths in a splitting technique and by using an Antidiffusive Lagrange Euler method. This method is broken into steps adapted to each term of the equations, and works with a weakened stability condition. We also report a few numerical experiments.

1. Introduction. Lagrange methods are never easy to implement near the boundary since with the usual alternating direction techniques, the mesh length may decrease or even vanish. In order to be able to go on with the computation, the approximate solution is first projected on the Lagrange mesh and then on the fixed Euler mesh. However this technique introduces a lot of diffusion, which is difficult to correct near the boundary. This problem becomes harder in the case of a reflection against this boundary since the wave has to travel twice across a zone of high diffusion.

In this paper, the space discretisation is chosen in such a way as to overcome these difficulties. It will present also another advantage, which is to weaken the stability condition. This is mainly true in the cases where the speed of sound presents locally small variations, for example when the density itself does.

The method is first described for a half-plane problem. Other domains will be considered in the last section. We choose the axes in such a way that the domain is defined by $\{x+y > 0\}$. The two components of the velocity $\vec{V}$ are denoted by u and v, ρ is the density and p is the pressure. In fact p is a given increasing function of the density $\rho > 0$, which belongs to $C^1([0,+\infty[)$ and, without any loss of generality satisfies

$$p(0) = 0.$$

In practice p is often a convex function.

* Laboratoire d'Analyse Numérique, INSA, 20, avenue des Buttes de Coësmes 35043 RENNES CEDEX

** UER Math. et Informatique. Université de Bordeaux 33405 TALENCE

The dynamical equations are

$$\rho(u_t + u\,u_x + v\,u_y) + p_x = 0 \quad , \tag{1}$$

$$\rho(v_t + u\,v_x + v\,v_y) + p_y = 0 \quad , \tag{2}$$

and the continuity equation expressing mass conservation is

$$\rho_t + (\rho u)_x + (\rho v)_y = 0 \ . \tag{3}$$

The boundary condition is

$$\vec{V}.\vec{n} = 0 \quad , \tag{4}$$

where $\vec{n}$ is a unit vector normal to the boundary. Here this condition reduces to

$$u + v = 0 \quad , \tag{5}$$

on the boundary. Initial values u_o, v_o, and $\rho_o > 0$ are prescribed in $L^\infty(\mathbb{R} \times \mathbb{R}^+)$ for the velocity and the density.

2. The splitting formulation on the half plane

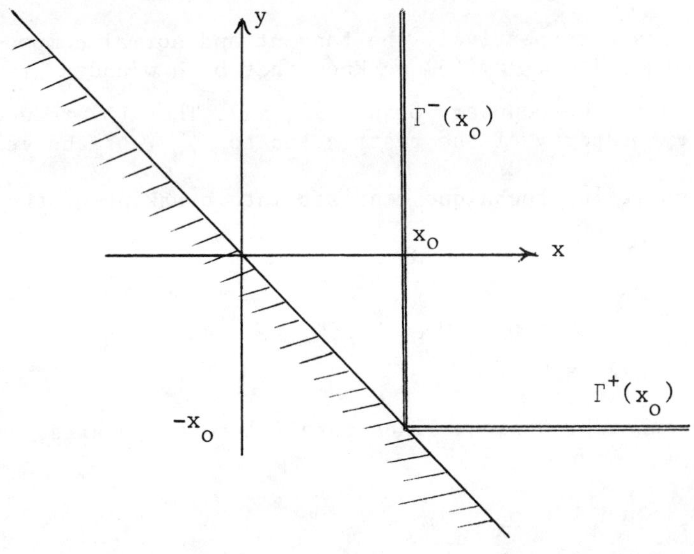

For any point $(x_o,-x_o)$ on the boundary we define the sets

$$\Gamma^+_{(x_o)} = \{ (x,y),\ y = -x_o\ ,\ x \geq x_o \}\ ,$$

and

$$\Gamma^-_{(x_o)} = \{ (x,y),\ x = x_o\ ,\ y \geq -x_o \}\ .$$

Then, we define

$$\Gamma_{(x_o)} = \Gamma^+_{(x_o)} \cup \Gamma^-_{(x_o)}\ .$$

We also introduce a new variable s, equal to $x-x_o$ on $\Gamma^+_{(x_o)}$ and equal to $-y-x_o$ on $\Gamma^-_{(x_o)}$. Thus $s \in \mathbb{R}$, with $s \leq 0$ on $\Gamma^-_{(x_o)}$ and $s \geq 0$ on $\Gamma^+_{(x_o)}$.

We also have $s = 0$ at the boundary point $(x_o, -x_o)$.

Next, we define new components of the velocity $\vec{V}$ along $\Gamma_{(x_o)}$ by setting

$$w = \begin{cases} u & \text{on } \Gamma^+_{(x_o)}\ , \\ -v & \text{on } \Gamma^-_{(x_o)}\ , \end{cases}$$

$$z = \begin{cases} -v & \text{on } \Gamma^+_{(x_o)}\ , \\ u & \text{on } \Gamma^-_{(x_o)}\ . \end{cases}$$

Then w and z are respectively the tangent and normal components of $\vec{V}$ along the path $\Gamma_{(x_o)}$. From (5) we know that both w and z are continuous functions at the boundary point $(x_o,-x_o)$. Thus (5) reduces to this continuity property of the restriction to $\Gamma_{(x_o)}$ of the velocity.

The usual splitting technique consists into breaking up the system in two parts

$$\text{(6)} \qquad \begin{cases} u_t + v\,u_y = 0\ , \\ \rho(v_t + v\,v_y) + p_y = 0\ , \\ \rho_t + (\rho v)_y = 0\ , \end{cases}$$

which is only considered along lines parallel to the y-axis, and similarly

$$\text{(7)} \qquad \begin{cases} \rho(u_t + u\,u_x) + p_x = 0\ , \\ v_t + u\,v_x = 0\ , \\ \rho_t + (\rho u)_x = 0\ , \end{cases}$$

along the x-axis. Following this idea the computations are carried using (6) along $\Gamma^-_{(x_o)}$ and (7) along $\Gamma^+_{(x_o)}$. This is done for some values of x_o.

By using the notations s and (w,z) we get the system

(8) $\quad z_t + w z_s = 0$,

(9) $\quad \rho(w_t + w w_s) + p_s = 0$,

(10) $\quad \rho_t + (\rho w)_s = 0$,

on both $\Gamma^-_{(x_o)}$ and $\Gamma^+_{(x_o)}$, that is on the path $\Gamma_{(x_o)}$, since the continuity of w and z is prescribed by (5) at the boundary point $(x_o,-x_o)$.

Let h be a given real positive number. Our discretisation technique consists in solving the system (8), (9), (10) on each path,

$$\Gamma_j = \Gamma(jh) ,$$

where $j \in \mathbb{Z}$.

This is done by an approximate method using h as the space mesh length on each Γ_j. As a matter of fact we put for any $i \in \mathbb{Z}$

$$s_{i+\frac{1}{2}} = (i + \tfrac{1}{2})\, h$$

on a given Γ_j. We write for any i and j in $\mathbb{Z}$,

$$\{a_{ij}\} = \Gamma_i \cap \Gamma_j \qquad \text{if} \quad i \neq j$$

and for i = j

$$a_{ii} = \{ih,-ih\}$$

which lies on the boundary. Then for any i and j the points

$$a_{ij} \ , \ a_{i+1,j} \ , \ a_{i,j+1} \ , \ a_{i+1,j+1}$$

are the four vertices of a square of side h. Thus, by considering all paths Γ_j, for all $j \in \mathbb{Z}$, each vertex $a_{k\ell}$, for $k \in \mathbb{Z}$, $\ell \in \mathbb{Z}, k < \ell$, is considered twice : the first time during the computations performed along Γ_k, and the second time along Γ_ℓ . Since these paths are orthogonal the usual splitting technique is performed.

Now, let us fix some $j \in \mathbb{Z}$. We are concerned with the system (8), (9), (10) on Γ_j. From Equations (9) and (10), we derive the momentum conservation law, that is

(11) $\quad q_t + (qw + p)_s = 0$,

where $\quad q = \rho w$.

Thus (10) becomes

$$\rho_t + q_s = 0 \ . \tag{12}$$

The term $(qw)_s$ in (11) expresses the convection by the velocity field w of the momentum q, in a conservation form. This term will be approximated by using the same Lagrange meshes as in (8). Now, in order to get the wave propagation, the remaining term, namely $q_t + p_s$, is set equal to zero as in any classical splitting technique, and coupled to Equation (12).

Thus we get the system

$$q_t + p_s = 0 \ ,$$
$$\rho_t + q_s = 0 \ ,$$

which describes the propagation of waves. We easily find that the characteristics speeds are $\pm c$, denoting by

$$c = \sqrt{p'(\rho)}$$

the speed of sound. To get the Riemann invariants R and S, we define the function Φ on $[0,+\infty[$, by

$$\Phi(\rho) = \int_0^\rho \sqrt{p'(r)}dr \ . \tag{13}$$

We obtain

$$R = q + \Phi(\rho) \ , \tag{14}$$

$$S = q - \Phi(\rho) \ , \tag{15}$$

which satisfy

$$R_t + c\,R_s = 0 \ , \tag{16}$$

$$S_t - c\,S_s = 0 \ . \tag{17}$$

The Riemann invariants will be used to give the shape of the waves which travel with the speed of sound c. They will give the pressure at any step of the computation. However the continuity equation will no longer be used in the form given by (12), but by (10). It will be effectively approximated on the same Lagrange mesh as was used for the convective term in (11) ; this will also be the case in the treatment of Equation (8).

The space mesh length is still h. For the time discretisation, we introduce an increasing sequence $(t_n)_{n\in\mathbb{N}}$, with $t_o = 0$. We denote by r_n the ratio

$$r_n = \frac{1}{n}(t_n - t_{n-1}) \ .$$

The space

$$V_h = \left\{ w \in L^1(\mathbb{R}) \ , \ w = \text{constant on } \left]s_{i-1/2}, s_{i+1/2}\right] \text{ for any } i \in \mathbb{Z} \right\}$$

and the associated L^2-projection operator P_o will be used in the Euler step.

3. The convective terms, on the half plane

This section deals with the approximation of following two equations

$$(18) \qquad \begin{cases} q_t + (qw)_s = 0 \\ \rho_t + (\rho w)_s = 0 \end{cases}$$

which are coupled by

$$q = \rho w.$$

Next a similar approximation will be adapted to (8), which is not in conservative form. The approximate value we shall obtain for q will be regarded as the contribution of the convective term in (11) at each time step. The remaining term in (11) corresponds to the propagation of waves, and will be added later.

Letting

$$m = \int_0^s \rho(\sigma,t)d\sigma$$

and noting that $q = \rho w$, we see that w and m satisfy

$$(19) \qquad \begin{cases} w_t + \left(\frac{w^2}{2}\right)_s = 0 \\ m_t + w\, m_s = 0 \end{cases}$$

Therefore m verifies the same equation as z in (8). This formulation, (19), allows us to get solutions w and m of bounded variation on $\mathbb{R} \times \left]0,T\right[$ for any $T > 0$. Hence the solutions of (18) have to be considered as Radon measures. However, with a minor extension, the averaging operators apply on both q and ρ . A convergence proof of Lagrange methods adapted to (19) may be found in [6].

We assume that we are starting with a known approximate solution, for $t = t_{n-1}$

$$q^n(.,t_{n-1}) \in V_h \quad , \qquad \rho^n(.,t_{n-1}) \in V_h \quad ,$$

where, of course, density is positive. They are easily obtained by projecting the restrictions to Γ_j of the initial data at $t = 0$.

We set

$$w^n(.,t_{n-1}) = \frac{q^n(.,t_{n-1})}{\rho^n(.,t_{n-1})} ,$$

which belongs to V_h. Then we denote by

$$q_i^n \quad , \quad \rho_i^n \quad , \quad w_i^n$$

the constant values of q^n, ρ^n and w^n on each space interval $]s_{i-\frac{1}{2}}, s_{i+\frac{1}{2}}[$, for $i \in \mathbb{Z}$, and $t = t_{n-1}$. Also, we denote by q^n and ρ^n the solutions of (18), as defined from (19), on the strip $\mathbb{R} \times]t_{n-1}, t_n[$. The value of w^n on any characteristic

(20) $$s = ih + (t-t_n)w_i^n$$

is a constant, as long as this characteristic does not intersect any other one. The value of s is given for $t = t_n$ by

(21) $$s_i^n = ih + r_n h w_i^n \quad .$$

This sequence, $(s^n_{i+\frac{1}{2}})_{i \in \mathbb{Z}}$, is increasing if the following condition is fulfilled,

(22) $$1 + r_n(w_{i+1}^n - w_i^n) \geq 0 \quad .$$

We can now introduce the space corresponding to the Lagrange mesh

$$W_h^n = \left\{ w \in L^1(\mathbb{R}), \ w = \text{constant on }]s^n_{i-\frac{1}{2}}, s^n_{i+\frac{1}{2}}[, \ i \in \mathbb{Z} \right\} ,$$

and the associated L^2-orthogonal projection Q_n. The scheme is the following

(23) $$\begin{cases} \tilde{q}^n = P_o Q_n \, q^n(.,t_n) & , \\ \tilde{\rho}^n = P_o Q_n \, \rho^n(.,t_n) & . \end{cases}$$

Here $\tilde{q}^n$ and $\tilde{\rho}^n$ correspond to the provisional values of the solutions at $t = t_n$, prior to the computation of the pressure field. This scheme, (23), may be detailed as a transport step across the strip $\mathbb{R} \times]t_{n-1}, t_n[$, followed by a Lagrange projection on W_h^n, and then by another projection on V_h, i.e. the fixed Euler mesh. Integrating (18) on each quadrilateral having as vertices

$$(ih, t_{n-1}) \ , \quad ((i+1)h, t_{n-1}) \ , \quad (s_{i+1}^n, t_n) \ , \ (s_i^n, t_n) \quad ,$$

we get for any $i \in \mathbb{Z}$ and for $s_i^n < s < s_{i+1}^n$

$$(24)\quad \begin{cases} Q_n\, q^n(s,t_n) = \dfrac{1}{2}\, \dfrac{q^n_{i+1} + q^n_i}{1 + r_n (w^n_{i+1} - w^n_i)} \\[2ex] Q_n\, \rho^n(s,t_n) = \dfrac{1}{2}\, \dfrac{\rho^n_{i+1} + \rho^n_i}{1 + r_n (w^n_{i+1} - w^n_i)} \end{cases}$$

Since we have for any $s \in \left] s^n_i,\, s^n_{i+1} \right[$

$$w^n(s,t_n) = \frac{q^n_{i+1} + q^n_i}{\rho^n_{i+1} + \rho^n_i} = \frac{\rho^n_{i+1}}{\rho^n_{i+1} + \rho^n_i}\, w^n_{i+1} + \frac{\rho^n_i}{\rho^n_{i+1} + \rho^n_i}\, w^n_i$$

the scheme defined by (23) obviously preserves the L^∞-norm and the total variation of w^n at each step.

Note that we find no flux across the characteristics bounding the quadrilateral in (23). This comes from the conservative form of (18).

The functions $\tilde{q}^n$ and $\tilde{\rho}^n$ are computed by another projection on the Euler mesh. This is done very easily since $Q_n q^n(.,t_n)$ and $Q_n \rho^n(.,t_n)$ are piecewise constant functions.

However the scheme (23) needs to be improved, for it gives overly diffused results. This diffusion may be easily expressed by inserting in (24)

$$w^n_{i+1/2} = 0$$

for any $i \in \mathbb{Z}$.

Here we forget for a moment that q is ρw. After the Euler-projection step, we get

$$\tilde{q}^n_i = \frac{1}{4}\left[q^n_{i+1} + 2q^n_i + q^n_{i-1} \right]$$

instead of the exact value which is equal to q^n_i .

By writing, for any $i \in \mathbb{Z}$, the flux

$$\delta^n_{i+1/2} = q^n_{i+1} - q^n_i$$

we get

$$\tilde{q}^n_i = q^n_i + \frac{1}{4}\,(\delta^n_{i+1/2} - \delta^n_{i-1/2})$$

The antidiffusion technique consists in removing these fluxes from $\tilde{q}^n_{i+1/2}$, by using the last computed values $(\tilde{q}^n_{j+1/2})_{j\in\mathbb{Z}}$ instead of $(q^n_{j+1/2})_{j\in\mathbb{Z}}$. In fact only parts of these fluxes will be used so as not to degrade the stability, as it would happen if new local extrema were created. Hence we introduce for any $i\in\mathbb{Z}$ the following corrected flux

$$(25)\qquad a^n_{i+1/2} = sg^n_{i+1/2}\ \mathrm{Max}\left\{0,\min\left(sg^n_{i+1/2}\,\Delta^n_{i+3/2},\ sg^n_{i+1/2}\,\Delta^n_{i-1/2},\ \frac{1}{2}\,|\Delta^n_{i+1/2}|\right)\right\}$$

where

$$sg^n_{i+1/2} = \mathrm{sign}(\Delta^n_{i+1/2})$$

and for any $j\in\mathbb{Z}$,

$$\Delta^n_{j+1/2} = \frac{1}{2}\,(\tilde{q}^n_{j+1} - q^n_j)$$

Then for any $i\in\mathbb{Z}$ we compute

$$(26)\qquad \overset{*}{q}{}^n_i = \tilde{q}^n_i - a^n_{i+1/2} + a^n_{i-1/2}\ ,$$

to which the contribution of the pressure still has to be added.

Similarly, for the density, we set for any $j\in\mathbb{Z}$,

$$D^n_{j+1/2} = \frac{1}{2}\,(\tilde{\rho}^n_{j+1} - \tilde{\rho}^n_j)$$

and for any $i\in\mathbb{Z}$, we compute the corrected flux

$$(27)\qquad b^n_{i+1/2} = sg^n_{i+1/2}\ \mathrm{Max}\left\{0,\min\left(sg^n_{i+1/2}\,D^n_{i+3/2},\ sg^n_{i+1/2}\,D^n_{i-1/2},\ \frac{1}{2}\,|D^n_{i+1/2}|\right)\right\}$$

where now

$$sg^n_{i+1/2} = \mathrm{sign}\,(D^n_{i+1/2})\ .$$

Finally we get, for any $i\in\mathbb{Z}$,

$$(28)\qquad \rho^{n+1}_i = \tilde{\rho}^n_i - b^n_{i+1/2} + b^n_{i-1/2}\ .$$

Using (26) and (28) we have computed $\overset{*}{q}{}^{n+1}(.,t_n)$ and $\rho^{n+1}(.,t_n)$ belonging to V_h. The antidiffusion steps (25) and (27) can be performed by the same subroutine. In fact the coefficients of this flux correction are independent of the function to apply, and of any other part of the method. The effect of the diffusion are significantly reduced. The shapes of the shock waves are notably improved. The convergence was proved in the case of quasilinear equations in [8].

Antidiffusion methods of this type having a higher order of accuracy can be found in [3],[4],[5],[10] or [12].

4. The propagation of waves on the half plane

We have seen in Section 2 that this propagation is described by

$$(29)\qquad \begin{aligned} q_t + p_s &= 0 \quad , \\ \rho_t + q_s &= 0 \quad , \end{aligned}$$

which is a strictly hyperbolic system. The associated Riemann invariants R and S were given in (14), (15). Our aim in this section is to build the pressure gradient in order to be able to compute its contribution in (11). The second equation in (29) was already solved by the method described in section 3. However it must be coupled with the first equation to get a correct wave propagation.

We propose two methods for the computation of the pressure gradient in this section. The first one uses a Lagrange projection, and the second one an interpolation on a Lagrange mesh, which directly gives the gradient p_s. We assume for both methods that we are able to compute Φ and the speed of sound. However we propose an approximation of Φ at the end of this section, since in practice the exact computation is sometimes difficult.

Then we can compute R and S at any point where ρ and q are known. This is the case at time $t = t_{n-1}$ or whenever the values computed in the previous section are available. These two Riemann invariants are associated with two waves travelling respectively with the velocity c and -c. This results from Equations (16) and (17).

Let us now describe the first method. For any given $i \in \mathbb{Z}$ we define two approximate characteristics

$$\sigma = \sigma_{i+1/2}(t) \qquad \text{and} \qquad \tau = \tau_{i+1/2}(t)$$

that satisfy

$$(30)\qquad \begin{aligned} \sigma'_{i+1/2}(t) &= c\Big(\rho(\sigma_{i+1/2}(t), t_{n-1})\Big) \quad , \\ \tau'_{i+1/2}(t) &= -c\Big(\rho(\tau_{i+1/2}(t), t_{n-1})\Big) \quad , \end{aligned}$$

with the starting values

$$\sigma_{i+1/2}(t_{n-1}) = \tau_{i+1/2}(t_{n-1}) = s_{i+1/2}\Big(= (i+\tfrac{1}{2})h\Big)$$

Next we write

$$(31)\qquad \sigma^n_{i+1/2} = \sigma_{i+1/2}(t_n) \quad , \quad \tau_{i+1/2} = \tau^n_{i+1/2}(t_n) \quad .$$

The equations in (30) enforce that density should remain constant on each mesh between t_{n-1} and t_n. Therefore $c\left(\rho(\sigma_{i+1/2}(t),t_{n-1})\right)$ and $-c\left(\rho(\tau_{i+1/2}(t),t_{n-1})\right)$ are piecewise constant functions and both differential equations in (30) can be solved exactly. This permits us to get $\sigma^n_{i+1/2}$ and $\tau^n_{i+1/2}$ easily from (31). This linearization that we get by using in (16) and (17) the speed of sound as in (30), looks like a hypothesis of superposition of waves. Obviously $(\sigma^n_{i+1/2})_{i\in\mathbb{Z}}$ and $(\tau^n_{i+1/2})_{i\in\mathbb{Z}}$ are monotone increasing, since they are both obtained from integrating the differential equations with an increasing sequence of starting values. This implies that this step is free of stability limitation.

Now, from (16) and (17), R and S are constant along the characteristics. We use the same argument by following the approximate characteristics (30). More precisely we define two functions, denoted by $R^n(t.,t_n)$ and $S^n(.,t_n)$, such that

$$R^n(s,t_n) = q^n_i + \Phi(\rho^n_i) \qquad \text{for} \qquad \sigma^n_{i-1/2} < s < \sigma^n_{i+1/2} \quad ,$$

and

$$S^n(s,t_n) = q^n_i - \Phi(\rho^n_i) \qquad \text{for} \qquad \tau^n_{i-1/2} < s < \tau^n_{i+1/2} \quad .$$

Next these functions are projected on the translated Euler space,

$$V_{h,1/2} = \left\{ v \in L^1(\mathbb{R}),\ v(.+\tfrac{h}{2}) \in V_h \right\} \quad ,$$

by using the associated L^2-projection operator $P_{1/2}$. We get

$$R^{n+1}(.,t_n) = P_{1/2}\, R^n(.,t_n) \quad ,$$

$$S^{n+1}(.,t_n) = P_{1/2}\, S^n(.,t_n) \quad .$$

Since the function Φ is often difficult to invert, we shall use the formula

$$(32) \qquad p_s = c(\rho)\,\frac{\partial}{\partial s}\,\Phi(\rho) \quad ,$$

to compute the pressure gradient. We assume that each ρ^{n+1}_i has been already calculated by the method of Section 3. Then for any $i \in \mathbb{Z}$, we write

$$(33) \qquad h(p_s)^n_i = \frac{1}{2}\, c(\rho^{n+1}_i)\left(R^{n+1}_{i+1/2} - R^{n+1}_{i-1/2} - S^{n+1}_{i+1/2} + S^{n+1}_{i-1/2}\right)$$

However the speed of sound in (33) may be computed from the formula

$$(34) \qquad c(\rho) = \Phi' \circ \Phi^{-1}\left(\frac{R-S}{2}\right)$$

in the few cases for which this can be done easily. Now we add the contributions already computed in the preceding section, by writing for any $i \in \mathbb{Z}$,

$$q_i^{n+1} = \overset{*}{q}{}_i^{n+1} - r_n h(p_s)_i^n \quad . \tag{35}$$

The second method we propose for the pressure term computation is the following. The approximate characteristics are drawn by starting from the middles of the mesh intervals. This is done by integrating the differential equations (30), but with the initial data

$$\sigma_{i+1/2}(t_{n-1}) = \tau_{i+1/2}(t_{n-1}) = ih \quad , \qquad i \in \mathbb{Z} \; . \tag{36}$$

Thus we define for every $i \in \mathbb{Z}$

$$\sigma_i^n = \tau_{i+1/2}(t_n) \; , \qquad \tau_i^n = \tau_{i+1/2}(t_n) \quad .$$

Now we build two continuous piecewise-linear functions

$$R^n(.,t_n) \qquad \text{and} \qquad S^n(.,t_n)$$

such that

$$R^n(\sigma_i^n,t_n) = q_i^n + \Phi(\rho_i^n)$$

and

$$S^n(\tau_i^n,t_n) = q_i^n - \Phi(\rho_i^n)$$

The derivatives

$$\frac{\partial R^n}{\partial s}(.,t_n) \qquad \text{and} \qquad \frac{\partial S^n}{\partial s}(.,t_n)$$

are piecewise constant functions. We compute them and project them on the Euler space V_h. Writing for any $i \in \mathbb{Z}$

$$DR_i^n = P_o \frac{\partial R^n}{\partial s}(s,t_n) \qquad \text{for} \qquad (i-\tfrac{1}{2}) < s < (i+\tfrac{1}{2})h \quad ,$$

and

$$DS_i^n = P_o \frac{\partial S^n}{\partial s}(s,t_n) \qquad \text{for} \qquad (i-\tfrac{1}{2}) < s < (i+\tfrac{1}{2})h \quad ,$$

we get from (32),

$$(p_s)_i^n = \frac{1}{2}\, c(\rho_i^n) \left(DR_i^n - DS_i^n \right) . \tag{37}$$

Here the value of c may also be computed using (34) if it is possible. Newt we compute q_i^{n+1} as in (35).

Strictly speaking, these two methods apply only if the function ϕ can be computed exactly. This is not always true in practice. We now suggest a way to approximate this function Φ, the speed of sound c and then the Riemann invariants R and S. Before the first time step we choose a sample increasing sequence

$$0 \leqslant \rho_o < \rho_1 \quad .... < \rho_j < \rho_{j+1} \quad < ... < \rho_J \quad ,$$

that should be somehow related to the specific problem considered. Then we compute the sequences, for $0 \leqslant j \leqslant J-1$

$$c_{j+1/2} = \sqrt{\frac{p(\rho_{j+1}) - p(\rho_j)}{\rho_{j+1} - \rho_j}} \quad , \tag{38}$$

and

$$\Phi_{j+1} = \Phi_j + \sqrt{\left(p(\rho_{j+1}) - p(\rho_j)\right)(\rho_{j+1} - \rho_j)} \quad , \tag{39}$$

with

$$\Phi_o = 0 \quad .$$

Next we compute the continuous piecewise-linear interpolates $\tilde{c}$ and $\tilde{\Phi}$ at these points ρ_j, $0 \leqslant j \leqslant J$. Hence we proceed with them instead of c and Φ as before. Note that the formula in (39) is similar to the equation of the shock curves that we follow by solving a Riemann problem. It is well known that a shoch curve of equation,

$$q = \pm \sqrt{\left(p(\rho) - p(\rho_j)\right) \quad (\rho - \rho_j)} + q_j \quad ,$$

for a fixed $j \in \{0,...,J\}$, where q_j is some real constant, has the same slope as

$$q = \pm \left(\Phi(\rho) - \Phi(\rho_j)\right) + q_j$$

for $\rho = \rho_j$, and has the same convexity. This last formula is the equation of rarefaction waves, corresponding to hold constant a Riemann invariant.

By both methods, we are able to compute the propagation of the waves without any stability condition. Obviously the advantage will be the greatest for problems with a speed of sound of locally small variations. This is the case of large scale ocean circulation models, for which

$$c = \sqrt{g(A + h)} \quad ,$$

where g is the gravity constant, A is the depth (a few thousands meters) and h is the distance between the surface and its rest position (less than a few meters). Such examples also occur in gas dynamics when locally ρ does not have a too large variation, and in some models arising in elasticity theory. This situation is met in physics of cold plasmas, where c is exactly a constant ; see [6]. For these models the speed of sound is rather large and would have placed very strong restrictions on the choice of the timestep if the usual Courant-Friedrichs-Lewy stability condition had been enforced.

A variant of these methods is to take the expression

$$\frac{1}{r_n h}\left[q(.,t_{n-1}) - \frac{1}{2}\left(R^n(.,t_n) + S^n(.,t_n)\right)\right]$$

as the gradient of pressure. We can compute for instance

$$(p_s)_i^n = \frac{1}{r_n h}\left[q_i^n - \left\{P_o\,\frac{R^n(.,t_n) + S^n(.,t_n)}{2}\right\}_{s=ih}\right],$$

for any $i \in \mathbb{Z}$, as the approximate gradient of pressure. This last technique seems to give better results in examples with a strong pressure and a relatively small velocity.

In place of the methods detailed in this section, one can compute the pressure gradient by any other scheme of the Euler type. In this case a stability condition must be specified, where only the speed of sound is involved ; see [10], [4] or [11].

5. The general method on the half plane

For any i and j in $\mathbb{Z}$, we associate the set

$$K_{ij} = \left\{\{a_{ij}\} + \left[-\frac{h}{2},\frac{h}{2}\right] \times \left[-\frac{h}{2},\frac{h}{2}\right]\right\} \cap \left\{(x+y),\ x+y \geq 0\right\},$$

with the point a_{ij}

This set is a square of side h for $i \neq j$, and is a right-angled triangle for $i = j$. We define

$$\mathcal{V}_h = \left\{W \in L^1\left(\{x+y > 0\}\right),\ W = \text{constant on } K_{ij} \text{ for } i,j \in \mathbb{Z}\right\},$$

and the associated L^2-projection operator $\mathcal{P}$. The initial data are projected by this operator. Thus we get the averages of the velocity $V_o = (u_o,v_o)$,

$$V_h(x,y;0) = \frac{1+\delta_{ij}}{h^2}\iint_{K_{ij}} V_o(\xi,\eta)\,d\xi\,d\eta \quad \text{for } (x,y) \in K_{ij}, \tag{40}$$

and a similar formula for the density

$$\rho_h(x,y;0) = \frac{1+\delta_{ij}}{h^2}\iint_{K_{ij}} \rho_o(\xi,\eta)\,d\xi\,d\eta \quad \text{for } (x,y) \in K_{ij}, \tag{41}$$

where δ_{ij} is the Kronecker symbol. Hence for any $j \in \mathbb{Z}$, both restrictions to Γ_j of $u_h(.,0)$ and of $\rho_h(.,0)$ belong to V_h. Note that (40) also gives the starting values of w and z.

Conversely for any $j \in \mathbb{Z}$, we denote by $V_h(\Gamma_j)$ the set of the piecewise constant functions belonging to V_h when they are expressed by means of the curvilinear abscissa s of Γ_j. Then for the family $(W_j)_{j \in \mathbb{Z}}$ with $W_j \in V_h(\Gamma_j)$ for any $j \in \mathbb{Z}$, we can build a function $W \in \mathcal{V}_h$ as follows. We set for $(x,y) \in K_{ij}$, with $i < j$,

$$W(x,y) = \frac{1}{2}\left\{W_i\big((j-i)h\big) + W_j\big((i-j)h\big)\right\}$$

and for $i = j$

$$W(x,y) = \frac{1}{2} W_i(o) .$$

We shall denote by B this operator defined on $\bigcup_{i \in \mathbb{Z}} V_h(\Gamma_j)$, with values in $\mathcal{V}_h$.

Let $n \in \mathbb{N}^*$. We assume we know, for any $j \in \mathbb{Z}$,

$$q_h^{(j)}(.,t_{n-1}) \in V_h(\Gamma_j) , \rho_h^{(j)}(.,t_{n-1}) \in V_h(\Gamma_j) .$$

Then we compute

$$\overset{*}{q}{}_h^{n+1}(.,t_n) \in V_h \quad \text{and} \quad \rho_h^n(.,t_n) \in V_h$$

by the antidiffusive Lagrange Euler method of Section 3. The values are given in (26) and (28), for each path Γ_j, $j \in \mathbb{Z}$. Next we compute the propagation of waves as in Section 4. Thus we get $q_h^{n+1}(.,t_n)$ whose values are given by (35) on each path. Note that these values are computed by starting with the data of $q_h^{(j)}$ and $\rho_h^{(j)}$ at $t = t_{n-1}$. However, we can proceed, in this calculation, by using the values of $\overset{*}{q}{}_h^{n+1}$ and ρ_h^{n+1} just calculated, on the same path.

Now it remains to compute z satisfying (8). This is done by a Lagrange method similar to the second one of Section 4. It is however easier to apply in this case since the characteristics are straight lines. On any path Γ_j, with $j \in \mathbb{Z}$, we assume we have

$$z_h(.,t_{n-1}) \in V_h ,$$

whose values are denoted by z_i^n, with $i \in \mathbb{Z}$. We use the Lagrange meshes computed by (21) and set for any $i \in \mathbb{Z}$,

$$\tilde{z}_h(s,t_n) = \frac{1}{2}(z_{i+1}^n + z_i^n) \quad \text{if} \quad s_i^n < s < s_{i+1}^n .$$

Then $\tilde{z}_h$ is projected on V_h by using P_o, and is antidiffused by means of the same method as in (25), (26). Thus we obtain the values of

$$z_h^{(j)}(.,t_n) \in V_h ,$$

and the computation of (8), (9), (10) is performed for one time step.

Before starting again the next timestep, we have to combine the values that we have just obtained on each path Γ_j, $j \in \mathbb{Z}$. This may be done after the computation on each path by setting the last computed values to the values of a function of $\mathcal{V}_h$. This means that the values calculated at $t = t_n$ on the path Γ_j should be taken in account as the starting values on $K_{jk} \cap \Gamma_k$ for any $k > j$ at $t = t_{n-1}$. This technique is very easy to implement and does not give any additional diffusion. However we observe a very fast dispersion of errors. As a matter of fact the finite speed of propagation of the date is numerically violated in this case. The experiments show a concentration of errors along the boundary.

A second technique consists in using only the data given at $t = t_{n-1}$ on any path, and to combine the values computed at $t = t_n$ by the projection on $\mathcal{V}_h$ with the operator B. We notice for this method that the time ratio r_n must be doubled for the approximation to be consistant with (1), (2), (3). The dispersion of data is no longer observed but a certain amount of diffusion still remains. It may be reduced significantly by one additionnal call of the antidiffusion subroutine as in (25), (26), after proceeding with the operator B has proceeded. However this antidiffusion technique still needs to be improved.

To get the starting values at $t = t_n$ on each $V_h(\Gamma_j)$, $j \in \mathbb{Z}$, from the function we have just built in $\mathcal{V}_h$, we only have to identify their values. In fact, for each path Γ_j, $j \in \mathbb{Z}$, the curvilinear interval $\left](i-\frac{1}{2})h,(i+\frac{1}{2})h\right[$ exactly corresponds to $\Gamma_j \cap K_{ij}$. Therefore the computation on $]t_n, t_{n+1}[$ is performed by the same technique and so on.

6. Conclusion

We have tested some parts of this method. The antidiffusion Lagrange Euler techniques perform quite well. The accuracy is rather good, even for shock waves and with time steps far greater than those allowed by the Courant-Friedrichs-Lewy condition. The propagation of a density by a linear scalar equation as (16) for the Riemann invariants was also tested with a given constant velocity on the same family of paths. The starting values were equal to one on a square of side 8h, and zero elsewhere. The velocity was parallel to the boundary. We obtained the results of Figure 1 after 90 timesteps (i.e. $r_n = 1/3$) and of Figure 2 after 24 timesteps (i.e. $r_n = 5/4$). Both results are given for the same value of t. We see that the diffusion in the second case is more important. We also notice that this diffusion is mainly spreading in directions which are orthogonal to the velocity. In both examples, the antidiffusion was performed before B operates. In fact, it works better for the smaller value of the ratio r_n. However the difference between these results before the antidiffusion step was not so large, since for both sets of results the solution has a rather bad shape. This remark seems to show that a coefficient depending on the size of r_n should be inserted to control this antidiffusion step. Other examples can be found in [8].

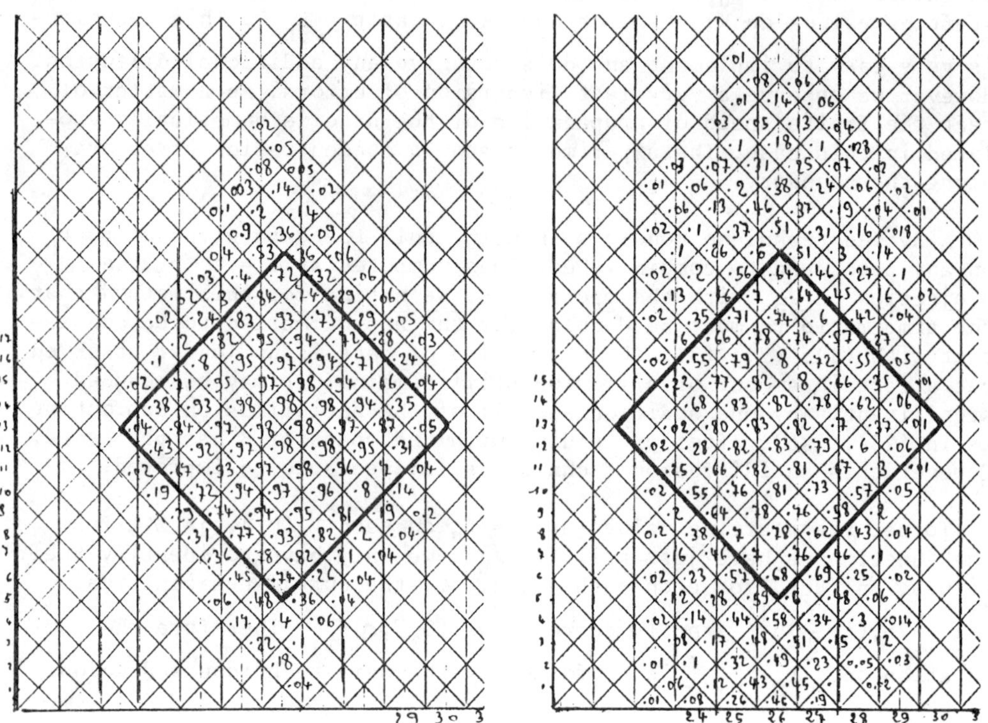

Distribution of the density for $r_n = \frac{1}{3}$

Figure 1

Distribution of the density for $r_n = \frac{5}{4}$

Figure 2

In these two cases the exact solution is still one in the square and zero elsewhere.

These examples have shown the stability of the method, and that the data exactly move with the given velocity. The following example was tested to verify the stability of the Lagrange method of Section 3. It dealt with Burgers equation in the case of a continuous initial data on ℝ. A shock wave appears for t = 3. We give the results obtained with the Antidiffusive Lagrange Euler method after three timesteps compared with the results of Godunov's method after 30 timesteps as required to satisfy the C.F.L. condition if the meshsize is the same. The method of Section 3 was also tested with only one timestep as allowed by the C.F.L. condition. It gives a sharp shock in this case, but the accuracy need to be improved for the rarefaction wave ; see [1], [7].

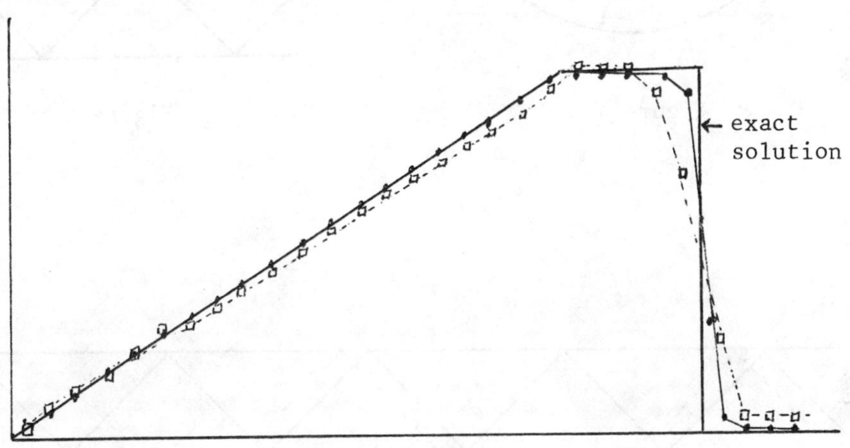

Antidiffusive Lagrange Euler scheme with h=0.2, at $t_3 = 3$ (after $t_1=1.2$, $t_2=2.7$)

Godunov's scheme with h=0.2, $\Delta t=0.2$

Figure 3

This method can be adapted to many other domains than the half plane. We give three examples : a square, a circle and a strip. The case of the square was studied in [7] with a velocity computed by a vortex method, by means of a mixed finite element technique described in [9], [10]. The triangulation and the set of paths were perfectly adapted to one another. As a matter of fact, the finite element technique gave exactly the normal components of the velocity at the end of each element of each path. Then the computation of the convection of the vortex on these paths allows to calculate a new velocity field for the next time step.

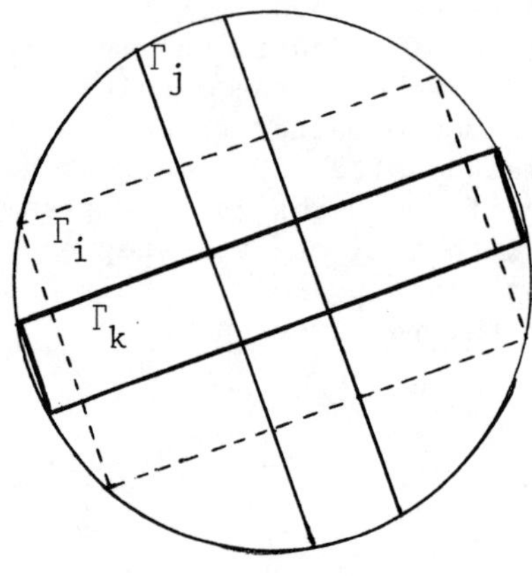

Figure 4

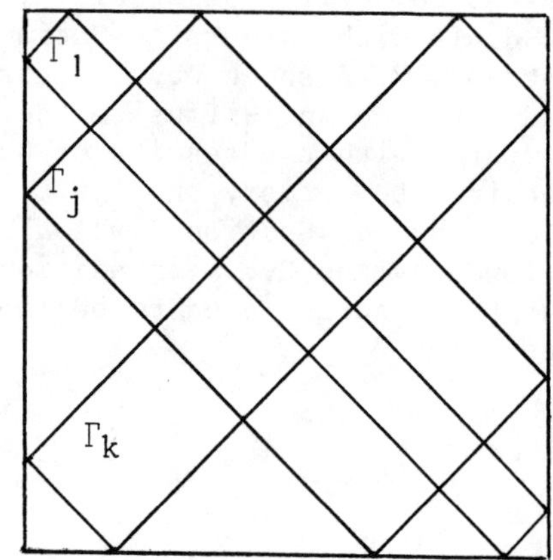

Figure 5

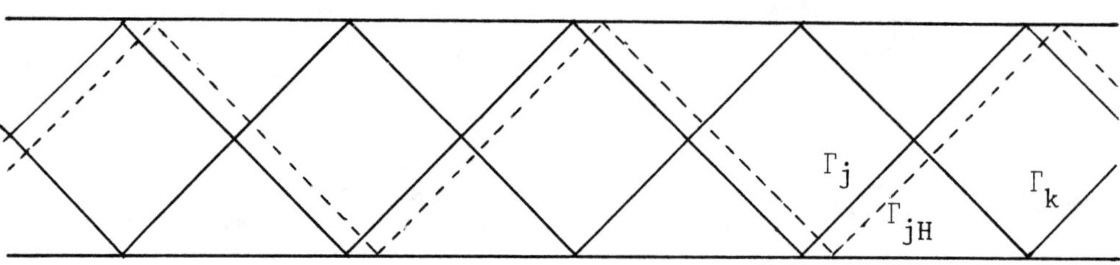

Figure 6

The construction of such a family of paths can be difficult for a general domain. It is theorically possible for connected domains with a polygonal boundary since they can be transformed into the half plane by means of the well known technique of conformal mapping, which preserves the angles between the paths. We can also get around some difficulties by using paths of nonconstant width. For closed paths the data must be extended to $\mathbb{R}$ by periodicity.

This study and the forecoming research on this subject are partially supported by a convention with the Centre d'Etudes de Gramat.

REFERENCES

[1] J. LAVIGNAT, Th. LEHNER, A.Y. LE ROUX, One dimension beam-plasma interaction using finite element methods (to appear).

[2] A.Y. LE ROUX, P. QUESSEVEUR, Convergence of an antidiffusion Lagrange Euler Scheme for quasilinear equations (to appear in SIAM Numerical Analysis).

[3] B. VAN LEER, Towards the ultimate conservative difference scheme. III & IV, J. Comp. Physics - 23 (1977), pp. 263-279.

[4] C. COSTE, B. MELTZ, J. OVADIA, E.A.D. Un nouvel algorithme en hydrodynamique multifluide. Computing Method in Applied Sciences and Engineering V.R. Glowinski and J.L. Lions Editors. North Holland (1982).

[5] D.L. BOOK, J.P. BORIS, K. HAIN, Flux corrected transport II : generalisation of the method. J. of Comp. Physics. 18 (1975).

[6] A. MARTEN, On a class of high resolution total variation stable finite difference schemes. Nasa report N° NCA2 - OR - 525-201 (1982).

[7] P. QUESSEVEUR, Thèse de 3e cycle - Bordeaux (april 1984).

[8] E. LE GRUYER, Traitement de la convection par des techniques Lagrangiennes en dimension deux - Thèse de 3e cycle - Bordeaux (January 1984).

[9] M.N. LE ROUX, A mixed finite element method for a weighted elliptic problem. R.A.I.R.O. Analyse Numérique, Vol. 16, N° 3, (1982), pp. 243-273.

[10] B. BEN BARKA, E. LE GRUYER, M.N. and A.Y. LE ROUX, P. QUESSEVEUR, Modélisation océanique avec topographie. Rapport de Recherche/ CNEXO, Rennes (1982).

[11] B. BEN BARKA, E. LE GRUYER, A.Y. LE ROUX, P. QUESSEVEUR, Quelques techniques numériques du type Lagrange Euler. Rapport de contrat C.E. GRAMAT, Rennes (1983).

[12] J. OVADIA, P.A. RAVIART, Nouvelles méthodes numériques pour le traitement de la convection. Report CEA - N - 2052 (1978).

PART II: STEADY STATE CALCULATIONS

NUMERICAL SOLUTION OF THE EULER EQUATION FOR COMPRESSIBLE INVISCID FLUIDS

ANTONY JAMESON*

1. Objective and Guiding Principles

The objective of this work is to develop a satisfactory numerical method for the calculation of steady solutions of the Euler equations for inviscid compressible gas flows. The intended application is the prediction of the aerodynamic properties of airplanes flying at transonic speeds. This will ultimately call for solutions of the viscous equations. The Reynolds numbers prevailing in full scale flight are typically very large, with the result that the boundary layers become turbulent, forcing recourse to statistical averaging and the introduction of turbulence models. We may regard the solution of the Euler equations as a way station on the route to this longer term objective. My emphasis is on steady flow.

Some of the principal difficulties of the problem are:

(1) The equations of gas dynamics are nonlinear.

(2) Solutions in the transonic range will ordinarily be discontinuous: Cathleen Morawetz has shown that shock free solutions are isolated points [1]. Thus we may expect to find shock waves. The solutions will also generally contain contact surfaces in the form of vortex sheets (shed both by wings in three-dimensional flow, and also by profiles in two-dimensional flow in the event that shock waves of differing strength produce different amounts of entropy on the upper and lower surfaces).

(3) There are regions of the flow in the neighborhood, for example, of stagnation points, the wing trailing edge or the wing tip, where the derivatives may become very large or even unbounded, leading to large discretization errors.

(4) The equations are to be solved in an unbounded domain.

(5) We are generally interested in calculating flows over bodies of extreme geometric complexity (including cases where the

domain is multiply connected).

In order to find one's way through this thicket of difficulties and pitfalls, Peter Lax has suggested the need for "design principles" to guide the construction of a numerical scheme appropriate to the problem*. The present work follows the general plan of attack which I proposed in 1981 [2]. A steady solution is obtained as the asymptotic state of a time dependent problem. Since the unsteady problem is used only as a vehicle for reaching the steady state, alternative iterative schemes might be comtemplated. An example is the least squares method, which has reached a high level of sophistication at the hands of Glowinski and his co-workers [3]. I believe, however, that the time marching formulation is worth considering for some of the following reasons:

(1) Its simplicity (with consequent reduction of the risk of programming error).

(2) It offers the possibility of providing a dual purpose computer program for both steady and unsteady problems.

(3) Where there is a possibility of a non-unique steady state or perhaps a non-physical solution, containing, for example, a discontinuous expansion, one may rely on the modeling of a true physical process to make an appropriate selection.

(4) Algorithms can rather easily be devised which take maximum advantage of vector computers.

None of these virtues could be considered decisive if the convergence of the time marching scheme to a steady state were excessively slow in comparison with competing methods. I hope to show, however, that the time marching method can be modified in such a way that this need not be the case.

Within the framework of the time marching formulation the design principles which have guided this work are:

(1) The conservation laws of gas dynamics will be satisfied in discrete conservation form by the numerical approximation. (We may then rely on the theorem of Lax and Wendroff that the correct shock jump conditions will be satisfied by the solution if it converges in the limit of decreasing mesh width [4].)

(2) Unphysical solutions are to be excluded by the introduction of appropriate dissipative terms in the discrete approximation.

*Baejter Seminar, Princeton, October 1983.

(3) The final steady state should be independent of the time marching procedure. (I do not wish to exclude the possibility that the final steady state will depend on the initial state, although there is evidence provided by numerous flying objects suggesting the repeatability of a substantial class of steady solutions). This requirement does, however, exclude the use of a number of popular difference schemes, including schemes with fractional steps, and the Lax Wendroff and MacCormack schemes.

(4) If a quantity is known to be invariant in the true solution, it should also be invariant in the numerical solution. In particular, the total enthalpy should be constant in the steady state solution.

(5) Uniform flow should be an exact solution of the difference equations on an arbitrary mesh.

In order to meet requirement (3) I believe that it is helpful to separate the space discretization procedure entirely from the time marching procedure by applying first a semi-discretization. This has the advantage of allowing the problems of spatial discretization error, artificial dissipation and shock modeling to be studied independently of the problems of time marching stability and convergence acceleration. I believe that my own studies of convergence acceleration procedures will be found to be essentially complementary to the large body of recent work that has been devoted to improving the modeling of shock waves and contact discontinuities.

The main measures for accelerating the convergence to steady state which have been applied in this work are:

(1) Modification of the differential equations for faster convergence to a steady state.

(2) The use of a hybrid multistage time stepping scheme with distinct and separately optimized treatment of the hyberbolic and parabolic terms.

(3) The use of residual averaging to permit a larger time step without violating stability restrictions.

(4) Parallel time stepping on multiple grids.

The use of all these measures in conjunction has made it possible to obtain satisfactory solutions of the Euler equations for two dimensional flows in 25-50 steps.

Since I first proposed the use of multistage time stepping schemes for solving the Euler equations [2], numerous modifications of both the time stepping and space discretization schemes have been introduced, and the treatment of the boundary condition has also been altered. In this text I therefore describe the entire scheme as it now stands. Modifications of the Euler equations for faster convergence to a steady state are reviewed in the next section. Section 3 discusses the spatial discretization scheme which is derived from the integral form of the conservation laws. Section 4 discusses a relatively simple form of adaptive dissipation which leads to reasonably satisfactory modeling of shock waves in steady flows. The following section discusses schemes designed to improve shock resolution. These rely on more complicated constructions of artificial dissipation based on the concept of diminishing total variation for scalar conservation laws, following the ideas of Lax, Harten, van Leer and Osher [5,6]. In the light of this discussion it can be seen that the adaptive dissipation can be designed to yield a scheme which locally behaves like a TVD scheme in the neighborhood of a shock wave. The optimal construction, however, requires a decomposition into characteristic fields along the lines proposed by Roe [7,8]. Section 6 reviews the treatment of boundary conditions. Section 7 presents a class of hybrid multi-stage time stepping schemes, while Sections 8 and 9 review the residual averaging and multiple grid schemes. In order to test the various ingredients in a less complex setting, Section 10 presents some results of tests with Burgers' equation. Section 11 presents some representative results for the Euler equations.

2. Modification of the Equations to Improve Convergence to a Steady State

The motion of an inviscid compressible gas is governed by the Euler equations. Let p, ρ, u, v, E, H and c denote the pressure, density, Cartesian velocity components, total energy, total enthalpy and speed of sound. For a perfect gas

$$E = \frac{p}{(\gamma-1)\rho} + \frac{1}{2}\,(u^2 + v^2),\quad H = E + \frac{p}{\rho}, \tag{2.1}$$

$$c^2 = \frac{\gamma p}{\rho}$$

where γ is the ratio of specific heats. Using Cartesian space coordinates x and y, the Euler equations for a two dimensional time dependent flow are

$$\frac{\partial w}{\partial t} + \frac{\partial}{\partial x}\, f(w) + \frac{\partial}{\partial y}\, g(w) = 0 \tag{2.2}$$

where t is the time coordinate and

$$w = \begin{bmatrix} \rho \\ \rho u \\ \rho v \\ \rho E \end{bmatrix}, \qquad f = \begin{bmatrix} \rho \\ \rho u^2 + p \\ \rho uv \\ \rho uH \end{bmatrix}, \qquad g = \begin{bmatrix} \rho v \\ \rho vu \\ \rho v^2 + p \\ \rho vH \end{bmatrix}. \tag{2.3}$$

These equations are to be solved for a steady state $\frac{\partial w}{\partial t} = 0$.

Aside from trying to construct the most efficient possible numerical method, it is natural to consider the possibility of modifying the Euler equations to improve the rate of convergence to a steady state. Three main approaches have been tried in this work.

The first is to use locally varying time steps such that the difference scheme operates everywhere in the flow field close to its stability limit. This is equivalent to solving

$$\frac{\partial w}{\partial t} + \alpha\{ \frac{\partial}{\partial x} f(w) + \frac{\partial}{\partial y} g(w)\} = 0 \tag{2.4}$$

where α is a variable scaling factor. This method ensures that disturbances will be propagated to the outer boundary in a fixed number of steps, proportional to the number of mesh intervals between the body and the outer boundary.

The second approach is based on the observation that if the enthalpy has a constant value H_∞ in the far field, it is constant everywhere in a steady flow, as can be seen by comparing the equations for conservation of mass and energy. If we set the value $H = H_\infty$ everywhere in the flow field throughout the evolution, then the pressure can be calculated from the equation

$$p = \frac{\gamma-1}{\gamma} \rho \left(H_\infty - \frac{u^2+v^2}{2}\right) .$$

This eliminates the need to integrate the energy equation. The resulting three equation model still constitutes a hyperbolic system, which approaches the same steady state as the original system. The use of this model has been advocated by Veuillot and Viviand [9].

The third approach is to retain the energy equation, and to add forcing terms proportional to the difference between H and H_∞ [10]. In a subsonic irrotational flow one can introduce a velocity potential ϕ and set

$$u = \phi_x , \quad v = \phi_y .$$

The Euler equations then reduce to the unsteady potential flow equation

$$\phi_{tt} + 2u\phi_{xt} + 2v\phi_{yt}$$
$$= (c^2 - u^2)\,\phi_{xx} - 2uv\phi_{xy} + (c^2 - v^2)\phi_{yy}\ . \qquad (2.6)$$

This is simply the equation for undamped wave motion in a moving coordinate frame. Damping could be introduced by adding a term $\alpha\phi_t$ to this equation. Such a term cannot directly be added to the Euler equations. However, the Bernoulli equation for unsteady potential flow is

$$\phi_t + H = H_\infty$$

Thus one can simulate the addition of a term $\alpha\phi_t$ to (2.6) by adding terms proportional to $H-H_\infty$ to the Euler equations. Provided that the space discretization scheme is constructed in such a way that $H = H_\infty$ is consistent with the steady state solution of the difference equations, these terms do not alter the final steady state. Numerical experiments have confirmed that they do assist convergence. The terms added to the mass and momentum equations are $\alpha\rho(H-H_\infty)$, $\alpha\rho u(H-H_\infty)$ and $\alpha\rho v(H-H_\infty)$, while that added to the energy equation is $\alpha\rho(H-H_\infty)$. In calculations using multiple grids an effective strategy is to include these terms only on the fine grid, and to increase the parameter α.

3. Finite Volume Formulation

The space discretization scheme is developed by writing the Euler equations in integral form

$$\frac{\partial}{\partial t}\iint_S w\,dS + \int_{\partial S} (f\,dy - g\,dx) = 0 \qquad (3.1)$$

for a domain S with boundary ∂S. The computational domain is divided into quadrilateral cells denoted by the subscripts i,j as sketched in Figure 1. Assuming that the dependent variables are known at the center of each cell, a system of ordinary differential equations is obtained by applying equation (3.1) separately to each cell. These have the form

$$\frac{d}{dt}\left(S_{i,j}w_{i,j}\right) + Q_{i,j} = 0 \qquad (3.2)$$

where $S_{i,j}$ is the cell area, and $Q_{i,j}$ is the net flux out of the cell. This may be evaluated as

$$\sum_{k=1}^{4} (\Delta y_k f_k - \Delta x_k g_k) \tag{3.3}$$

where f_k and g_k denote values of the flux vectors f and g on the kth edge, Δx_k and Δy_k are the increments of x and y along the edge with appropriate signs, and the sum is over the four sides of the cell. The flux vectors are evaluated by taking the average of the values in the cells on either side of each edge. For example

$$f_2 = \frac{1}{2} (f_{i+i,j} + f_{i,j}) \tag{3.4}$$

where $f_{i,j}$ denotes $f(w_{i,j})$. Alternatively one may evaluate first the flux velocity

$$Q_k = \frac{\Delta y_k (\rho u)_k - \Delta x_k (\rho u)_k}{\rho_k}$$

on each edge. Then the flux for the x momentum component, for example, is

$$\sum_{k=1}^{4} \{Q_k (\rho u)_k + \Delta y_k \, p_k\} \tag{3.5}$$

Schemes constructed in this manner reduce to central difference schemes on Cartesian meshes, and are second order accurate if the mesh is sufficiently smooth. They also satisfy the design principle (5) that uniform flow should be an exact solution of the difference equations. Provided that they are augmented by appropriate dissipative terms, they have been found to give quite accurate results, and they can easily be extended to three dimensional flows [11,12].

4. Adaptive Dissipation

The finite volume scheme (3.2) is not dissipative, allowing undamped oscillations with alternate sign at odd and even mesh points. In order to eliminate spurious oscillations, which will be triggered by discontinuities in the solution, one can follow either of two strategies. The first is to begin with a non-dissipative scheme such as (3.2), or a fourth order scheme, and to add just enough dissipation where it is needed to control the tendency to produce spurious oscillations. The second approach, which can be traced to the work of Boris, Book and Zalesak [13,14], is to begin by adding enough dissipation everywhere to guarantee the absence of unwanted oscillations. This leads to a first order accurate scheme which excessively smears discontinuities. A correction is then added to cancel the first order error, but the correction is limited to prevent the introduction of overshoots near discontinuities. This idea is the basis of an ingenious method recently proposed by Harten [15] for the construction of schemes which promise to give sharp resolution of shock waves. In this section I describe an adaptive scheme for adding dissipation which has proved effective in practice, and in the next section, I shall consider the more complex high resolution schemes. The idea of the adaptive scheme is to add third order dissipative terms throughout the domain to provide a base level of dissipation sufficient to prevent nonlinear instability, but not sufficient to prevent oscillations in the neighborhood of shock waves. In order to capture shock waves additional first order dissipative terms are added locally by a sensor designed to detect discontinuities.

With the addition of dissipative terms $D_{i,j}$, the semi-discrete equations (3.2) take the form

$$\frac{d}{dt}(S_{i,j}\, w_{i,j}) + Q_{i,j} - D_{i,j} = 0. \tag{4.1}$$

To preserve conservation form the dissipative terms are generated by dissipative fluxes. The dissipation for the density equation, for example, is

$$D_{i,j}(\rho) = d_{i+1/2,j} - d_{i-1/2,j} + d_{i,j+1/2} - d_{i,j-1/2} \tag{4.2}$$

where the dissipative flux $d_{i+1/2,j}$ is defined by

$$d_{i+1/2,j} = \varepsilon^{(2)}_{i+1/2,j}\, R_{i+1/2,j}\,(\rho_{i,j} - \rho_{i-1,j})$$

$$- \varepsilon^{(4)}_{i+1/2,j}\, R_{i+1/2,j}\,(\rho_{i+2,j} - 3\rho_{i+1,j} + 3\rho_{i,j} - \rho_{i-1,j}). \tag{4.3}$$

Here $\varepsilon^{(2)}_{i+1/2,j}$ and $\varepsilon^{(4)}_{i+1/2,j}$ are adaptive coefficients, and $R_{i+1/2,j}$ is a coefficient chosen to give the dissipative terms the proper scale. An appropriate scaling factor is

$$R_{i+1/2,j} = \lambda_{i+1/2,j} \tag{4.4}$$

where λ is the spectral radius of the Jacobian matrix

$$\Delta y \frac{\partial f}{\partial w} - \Delta x \frac{\partial g}{\partial w}$$

for the flux across the cell face. This can be estimated as

$$\lambda = |\Delta y\, u - \Delta x\, v| + c\sqrt{\Delta x^2 + \Delta y^2} \tag{4.5}$$

where c is the speed of sound, and Δx and Δy are the displacements of the face in the x and y directions.

If one is using an explicit time stepping scheme one needs an estimate of the time step limit. A conservative estimate for a nominal Courant number of unity is

$$\Delta t^*_{i,j} = \frac{S_{i,j}}{\lambda_{i,j} + \mu_{i,j}}$$

where $\lambda_{i,j}$ and $\mu_{i,j}$ are the average spectral radii of the Jacobian matrices in the i and j directions. It is then convenient to avoid duplicate calculations by taking

$$R_{i+1/2,j} = \frac{1}{2}\left(\frac{S_{i+1,j}}{\Delta t^*_{i+1,j}} + \frac{S_{i,j}}{\Delta t^*_{i,j}}\right). \tag{4.4*}$$

An effective sensor of the presence of a shock wave can be constructed by taking the second difference of the pressure. Define

$$\nu_{i,j} = \left| \frac{p_{i+1,j} - 2p_{i,j} + p_{i-1,j}}{p_{i+1,j} + 2p_{i,j} + p_{i-1,j}} \right| \tag{4.6}$$

Set

$$\bar{\nu}_{i+1/2,j} = \max\,(\nu_{i+2,j}, \nu_{i+1,j}, \nu_{i,j}, \nu_{i-1,j})$$

Then we take

$$\varepsilon^{(2)}_{i+1/2,k} = \min\,(\tfrac{1}{2}, k^{(2)}\, \bar{\nu}_{i+1/2,j})$$

and

$$\varepsilon^{(4)}_{i+1/2,j} = \max\,(0, k^{(4)} - \alpha\, \bar{\nu}_{i+1/2,j}) \tag{4.7}$$

where $k^{(2)}$, $k^{(4)}$ and α are constants. Typically,

$$k^{(2)} = 1, \quad k^{(4)} = 1/32 , \quad \alpha = 2.$$

In a smooth region of the flow $\nu_{i,j}$ is proportional to the square of the mesh width, with the result that $\varepsilon^{(2)}_{i+1/2,j}$ is also proportional to the square of the mesh width, while $\varepsilon^{(4)}_{i+1/2,j}$ is of order one, and the dissipative fluxes are of third order in comparison with the convective fluxes. In the neighborhood of a shock wave $\nu_{i,j}$ is of order 1, so that the scheme behaves locally like a first order scheme. The term $\alpha \bar{\nu}_{i+1/2,j}$ is subtracted from $k^{(4)}$ to cut off the fourth differences which would otherwise cause an oscillation in the neighborhood of the shock wave.

The dissipative terms for the momentum and energy equations are constructed in the same way as those for the mass equation, substituting ρu, ρv or ρH for ρ in equation (4.3). The purpose of using differences of ρH rather than ρE in the dissipative terms for the energy equation is to produce difference equations which admit the solution $H = H_\infty$ in the steady state. With this choice the energy equation reduces to the mass equation multiplied by H_∞ when the time derivatives vanish.

5. Schemes Designed to Improve the Resolution of Shock Waves

There is by now a rather extensive theory of difference schemes for the treatment of a scalar conservation law

$$\frac{\partial u}{\partial t} + \frac{\partial}{\partial x} f(u) = 0 \tag{5.1}$$

This theory stems from the mathematical theory of shock waves, as it has been formulated by Lax [16]. It is known that the total variation

$$TV = \int_{-\infty}^{\infty} \left|\frac{\partial u}{\partial x}\right| dx$$

of a solution of (5.1) can never increase. Correspondingly it seems desirable that the discrete total variation

$$TV = \sum_{i=-\infty}^{\infty} |u_{i+1} - u_i|$$

of a solution of a difference approximation to (5.1) should not increase. A semi-discretization will have this property (now generally called total variation diminishing or TVD) if it can be cast in the form

$$\frac{du_i}{dt} = c^+_{i+1/2}(u_{i+1} - u_i) - c^-_{i-1/2}(u_i - u_{i-1}) \tag{5.2}$$

where the coefficients $c^+_{i+1/2}$ and $c^-_{i-1/2}$ are non-negative [17,18].

This can be proved as follows. The total variation can be expressed as

$$TV = \sum_{i=-\infty}^{\infty} s_{i+1/2}(u_{i+1} - u_i)$$

where

$$s_{i+1/2} = \begin{cases} 1 \text{ if } u_{i+1} - u_i \geq 0 \\ -1 \text{ if } u_{i+1} - u_i < 0 \end{cases}$$

Thus

$$\begin{aligned} \frac{d}{dt} TV &= \sum_{i=-\infty}^{\infty} s_{i+1/2} \frac{d}{dt}(u_{i+1} - u_i) \\ &= \sum_{i=-\infty}^{\infty} s_{i+1/2} \{c^+_{i+3/2}(u_{i+2} - u_{i+1}) \\ &\qquad - (c^+_{i+1/2} + c^-_{i+1/2})(u_{i+1} - u_i) \\ &\qquad + c^-_{i-1/2}(u_i - u_{i-1})\} \\ &= -\sum_{i=-\infty}^{\infty} v_{i+1/2}(u_{i+1} - u_i) \end{aligned}$$

where

$$\begin{aligned} v_{i+1/2} &= (c^+_{i+1/2} + c^-_{i+1/2})\, s_{i+1/2} \\ &\quad - c^+_{i+1/2}\, s_{i-1/2} - c^-_{i+1/2}\, s_{i+3/2}, \end{aligned}$$

If $c^+_{i+1/2}$ and $c^-_{i+1/2}$ are non-negative it follows that $v_{i+1/2}$ is either zero, or else it has the same sign as $u_{i+1} - u_i$. Hence

$$\frac{d}{dt} TV \leq 0 .$$

This is a special case of a more general result for multi-point schemes [17].

Since the total variation will increase if an initially monotone profile ceases to be monotone, TVD schemes preserve monotonicity. The TVD property does not, however, guarantee fulfillment of the entropy condition that the only admissible discontinuous solutions are those for which the characteristics of (5.1) converge on the discontinuities from both sides [16]. Some TVD schemes admit stationary expansion shock waves, for example. Monotone time stepping schemes of the form

$$u_i^{n+1} = H(u_{i-\ell}^n, u_{i-\ell+1}^n, \ldots, u_{i+\ell}^n)$$

where

$$\frac{\partial H}{\partial u_\alpha} \geq 0 \qquad \text{for all } \alpha$$

are known to exclude entropy violating solutions [18,19,20]. Recently Osher has shown that semi-discrete approximations to (5.1) of the form

$$\frac{du_i}{dt} + \frac{1}{\Delta x}(h_{i+1/2} - h_{i-1/2}) = 0 \tag{5.3}$$

exclude entropy violating solutions if the numerical flux $h_{i+1/2}$ satisfies the condition

$$\text{sgn}(u_{i+1} - u_i)(h_{i+1/2} - f(u)) \leq 0$$

for all u between u_i and u_{i+1}. Osher calls schemes which satisfy this condition E schemes [18].

It can be shown that monotone and E schemes are at best first order accurate [18,19]. Harten devised a method of constructing TVD schemes which are second order accurate almost everywhere [15]. His construction can be adapted to a semi-discrete scheme as follows. Denoting $f(u_i)$ by f_i, define the numerical flux in equation (5.3) as

$$h_{i+1/2} = \frac{1}{2} (f_{i+1} + f_i) - \alpha_{i+1/2}(u_{i+1} - u_i) . \tag{5.4}$$

Now suppose that

$$\alpha_{i+1/2} = \frac{1}{2} k|a_{i+1/2}| \tag{5.5}$$

where

$$a_{i+1/2} = \begin{cases} \dfrac{f_{i+1}-f_i}{u_{i+1}-u_i} & \text{if } u_{i+1} \neq u_i \\ \left.\dfrac{\partial f}{\partial u}\right|_{u=u_i} & \text{if } u_{i+1} = u_i \,. \end{cases} \tag{5.6}$$

Then it follows that

$$h_{i+1/2} = f_i - \frac{1}{2}(k|a_{i+1/2}| - a_{i+1/2})(u_{i+1} - u_i)$$

and

$$h_{i-1/2} = f_i - \frac{1}{2}(k|a_{i-1/2}| + a_{i-1/2})(u_i - u_{i-1}).$$

This gives (5.2) with

$$c^+_{i+1/2} = \frac{1}{2}(k|a_{i+1/2}| - a_{i+1/2})$$

and

$$c^-_{i-1/2} = \frac{1}{2}(k|a_{i-1/2}| + a_{i-1/2})$$

Then $c^+_{i+1/2} \geq 0$, $c^-_{i-1/2} \geq 0$, provided that $k \geq 1$. In order to achieve a second order accurate scheme which satisfies the TVD condition, one need only apply the same sequence of operations defined by equations (5.3–5.6) to a corrected flux

$$F_i = f_i + g_i$$

in which g_i is an anti-diffusive flux which approximates

$$\Delta x \, \alpha \frac{\partial u}{\partial x} \,,$$

and thus cancels the first order error. It is necessary to limit the anti-diffusive flux, however, to prevent the possibility of $g_{i+1} - g_i)/(u_{i+1} - u_i)$ becoming unbounded. For this purpose define

$$\tilde{g}_{i+1/2} = \alpha_{i+1/2}(u_{i+1} - u_i) \tag{5.7}$$

and

$$g_i = B(\tilde{g}_{i+1/2}, \tilde{g}_{i-1/2}) \tag{5.8}$$

where B is an averaging function which limits the magnitude attainable by $(g_{i+1} - g_i)/(u_{i+1} - u_i)$ and satisfies the condition

$$B(r,r) = r.$$

A simple choice is the min mod function

$$B(r,s) = \begin{cases} 0 \text{ if } r \text{ and } s \text{ have opposite signs} \\ r \text{ if } |r| \le |s| \text{ and } r \text{ and } s \text{ have the same sign} \\ s \text{ if } |s| \le |r| \text{ and } r \text{ and } s \text{ have the same sign} \end{cases} \tag{5.9}$$

The numerical flux is now

$$h_{i+1/2} = \frac{1}{2}(f_{i+1} + f_i) + \frac{1}{2}(g_{i+1} + g_i) \tag{5.10}$$

$$-\alpha_{i+1/2}(u_{i+1} - u_i) - \frac{1}{2}k|g_{i+1} - g_i| \operatorname{sign}(u_{i+1} - u_i)$$

In most of the domain the second term cancels the third term to order Δx^2, and the last term is of order Δx^2, yielding the desired second order accuracy.

Alternative flux limiting functions have been studied by Roe [21], and Sweby [22]. Let r_i be the ratio of consecutive gradients

$$r_i = \frac{\tilde{g}_{i-1/2}}{\tilde{g}_{i+1/2}}$$

One can set

$$B(\tilde{g}_{i+1/2}, \tilde{g}_{i-1/2}) = \phi(r_i)\tilde{g}_{i+1/2}$$

where the function ϕ satisfies the symmetry condition

$$\phi(r) = r\,\phi(1/r)$$

and

$$\phi(1) = 1.$$

Then denoting $\phi(r_i)$ by ϕ_i,

$$\frac{g_{i+1} - g_i}{u_{i+1} - u_i} = \left(\frac{\phi_{i+1}}{r_{i+1}} - \phi_i\right)\alpha_{i+1/2}$$

which is bounded if $\phi(r)/r$ and $\phi(r)$ are bounded for all r. The resulting flux is

$$h_{i+1/2} = f_i + g_i - (\alpha_{i+1/2} - \frac{1}{2} a_{i+1/2})(u_{i+1} - u_i)$$
$$- 1/2 \{k \left| \frac{\phi_{i+1}}{r_{i+1}} - \phi_i \right| - (\frac{\phi_{i+1}}{r_{i+1}} - \phi_i)\}(u_{i+1} - u_i)$$

and it can be seen that the TVD condition (5.2) is satisfied.

The min mod function (5.9) is obtained by setting

$$\phi(r) = \begin{cases} 0, & r \le 0 \\ r, & 0 \le r \le 1 \\ 1, & r \ge 1 \end{cases}.$$

A less stringent limiter which may be expected to improve the accuracy is

$$\phi(r) = \frac{|r| + r}{1 + r}.$$

Roe [21], Sweby [22], and Osher and Chakravarthy [6] have constructed TVD schemes which omit the last term of equation (5.10) by suitably restricting $\phi(r)$.

The concept of converting a first order monotone scheme to a second order scheme by adding an anti-diffusive flux which is limited to prevent spurious oscillations may already be found in the work of Boris, Book and Zalesak [13,14]. Harten's advance was to develop a mathematical formulation of the idea which allowed the TVD property to be proved for a second order scheme. The concept of flux limiting was independently introduced by van Leer [23].

There are difficulties in extending these ideas to systems of equations, and also to equations in more than one space dimension. Firstly the total variation of the solution of a system of hyperbolic equations may increase. Secondly it has been shown by Goodman and Leveque that a TVD scheme in two space dimensions is no better than first order accurate [24]. One might add dissipative terms by applying the same construction to the complete system of equations, using for $\alpha_{i+1/2}$ the spectral radius of the Jacobian matrix $\partial f/\partial u$, evaluated for an average value $u_{i+1/2}$. My own numerical experiments indicate that this leads to an excessively dissipative scheme. It can be seen however, that with the scaling (4.4), the adaptive dissipation proposed in the previous section leads to a scheme which will behave locally like a TVD scheme if the constants are chosen to make sure that the coefficient $\varepsilon^{(2)}_{i+1/2,j}$ = 1/2 in the neighborhood of a shock wave. This will be illustrated in Section 10 for the case of Burgers' equation.

The discrimination needed for sharp resolution of shock waves in a fluid flow can be obtained by first decomposing the fluctuations into characteristic fields, following the suggestion set forth by Roe in a very striking paper [7]. The TVD construction is then separately applied to each field. For this purpose each flux difference is expressed in terms of components in the basis defined by the eigenvectors of $\partial f/\partial u$, evaluated at the interface. The dissipative terms are then defined separately for each field by equations (5.5)-(5.11), taking for $a_{i+1/2}$ the eigenvalues q, q, q+c and q-c of the different fields. The dissipative terms are finally recombined to form dissipative fluxes corresponding to the original variables.

Roe recommends that $\left.\frac{\partial f}{\partial u}\right|_{i+1/2}$ be represented by a matrix $A(u_{i+1},u_i)$ with the property that

$$A(u_{i+1},u_i)(u_{i+1} - u_i) = f_{i+1} - f_i$$

and has proposed a method of constructing such a matrix [8]. This has the advantage that a stationary shock wave can be resolved with a single interior point. Otherwise a non-oscillatory scheme will produce a smeared out shock wave with an extended tail.

These properties are not obtained without a cost. Firstly there is a large increase in the number of arithmetic operations required in the realization of the scheme. Secondly it is no longer possible to satisfy the design principle (4), stated in Section 1, that the total enthalpy of the steady state solution should be constant. Because the dissipative terms entering the mass and energy equations are independently constructed, these two equations are not consistent with each other in the steady state when the total enthalpy is constant.

There remains the pitfall of perhaps generating an unphysical solution which violates the entropy condition. The scheme defined by equations (5.3)-(5.10) allows a stationary discontinuous expansion. The flux is conserved across a stationary discontinuity of a solution of (5.1). Thus if u_{i+1} and u_i lie on either side of a stationary discontinuity, $f_{i+1} = f_i$, and according to equation (5.6) $a_{i+1/2} = 0$, with the result that the dissipative coefficient $\alpha_{i+1/2}$ defined by equation (5.5) also vanishes, and the difference equations are satisfied whether or not the discontinuity is an expansion or a compression. This difficulty can be corrected by modifying the definition of $\alpha_{i+1/2}$ to make sure that it cannot vanish in this situation.

6. Boundary Conditions

At a solid boundary the only contribution to the flux balance (3.3) comes from the pressure. The normal pressure gradient $\frac{\partial p}{\partial n}$ at the wall can be estimated from the condition that $\frac{\partial}{\partial t}(\rho q_n) = 0$, where q_n is the normal velocity component. The pressure at the wall is then estimated by extrapolation from the pressure at the adjacent cell centers, using the known value of $\frac{\partial p}{\partial n}$.

The rate of convergence to a steady state will be impaired if outgoing waves are reflected back into the flow from the outer boundaries. The treatment of the far field boundary condition is based on the introduction of Riemann invariants for a one dimensional flow normal to the boundary. Let subscripts ∞ and e denote free stream values and values extrapolated from the interior cells adjacent to the boundary, and let q_n and c be the velocity component normal to the boundary and the speed of sound. Assuming that the flow is subsonic at infinity, we introduce fixed and extrapolated Riemann invariants

$$R_\infty = q_{n_\infty} - \frac{2c_\infty}{\gamma-1}$$

and

$$R_e = q_{n_e} + \frac{2c_e}{\gamma-1}$$

corresponding to incoming and outgoing waves. These may be added and subtracted to give

$$q_n = \frac{1}{2}(R_e + R_\infty)$$

and

$$c = \frac{\gamma-1}{4}(R_e - R_\infty)$$

where q_n and c are the actual normal velocity component and speed of sound to be specified in the far field. At an outflow boundary, the tangential velocity component and entropy are extrapolated from the interior, while at an inflow boundary they are specified as having free stream values. These four quantities provide a complete definition of the flow in the far field. If the flow is supersonic in the far field, all the flow quantities are specified at an inflow boundary, and they are all extrapolated from the interior at an outflow boundary.

7. Hybrid Multistage Time Stepping Schemes

Multi-stage schemes for the numerical solution of ordinary differential equations are usually designed to give a high order of accuracy. Since the present objective is simply to obtain a steady state as rapidly as possible, the order of accuracy is not important. This allows the use of schemes selected purely for their properties of stability and damping. For this purpose it pays to distinguish the hyperbolic and parabolic parts stemming respectively from the convective and dissipative terms, and to treat them differently. This leads to a new class of hybrid multi-stage schemes.

Since the cell area S_{ij} is independent of time, equation (4.1) can be written as

$$\frac{dw}{dt} + R(w) = 0 \tag{7.1}$$

where $R(w)$ is the residual

$$R_{i,j} = \frac{1}{S_{i,j}} (Q_{i,j} - D_{i,j}) . \tag{7.2}$$

Let w^n be the value of w after n time steps. Dropping the subscripts i,j the general m stage hybrid scheme to advance a time step Δt can be written as

$$\begin{aligned}
w^{(0)} &= w^n \\
w^{(1)} &= w^{(0)} - \alpha_1 \quad \Delta t \ R^{(0)} \\
&\dots \\
w^{(m-1)} &= w^{(0)} - \alpha_{m-1} \Delta t \ R^{(m-2)} \\
w^{(m)} &= w^{(0)} - \quad \Delta t \ R^{(m-1)} \\
w^{n+1} &= w^{(m)}
\end{aligned} \tag{7.3}$$

where the residual in the q+1st stage is evaluated as

$$R^{(q)} = \frac{1}{S} \sum_{r=0}^{q} \{\beta_{qr} Q(w^{(r)}) - \gamma_{qr} D(w^{(r)})\} \tag{7.4}$$

subject to the consistency constraint that

$$\sum_{r=0}^{q} \beta_{qr} = \sum_{r=0}^{q} \gamma_{qr} = 1. \tag{7.5}$$

In order to assess the properties of these schemes it is useful to consider the model problem

$$u_t + u_x + \mu \Delta x^3 u_{xxxx} = 0 . \tag{7.6}$$

In the absence of the third order dissipative term this equation describes the propagation of a disturbance without distortion at unit speed. With centered differences the residual has the form

$$\Delta t R_i = \frac{\lambda}{2}(u_{i+1} - u_{i-1}) + \lambda\mu(u_{i+2} - 4u_{i+1} + 6u_i - 4u_{i-1} + u_{i-2})$$

where $\lambda = \Delta t/\Delta x$ is the Courant number. If we consider a Fourier mode $\hat{u} = e^{ipx}$ the discretization in space yields

$$\Delta t \frac{d\hat{u}}{dt} = z\hat{u}$$

where z is the Fourier symbol of the residual. Setting $\xi = p\Delta x$, this is

$$z = -\lambda i \sin\xi - 4\lambda\mu(1 - \cos \xi)^2 . \tag{7.7}$$

A single step of the multistage scheme yields

$$\hat{u}^{n+1} = g(z)\hat{u}^n$$

where g(z) is the amplification factor. The stability region of the scheme is given by those values of z for which $g(z) \leq 1$.

A simple procedure is to recalculate the residual at each stage using the most recently updated value of w. Then (7.4) becomes

$$R^{(q)} = \frac{1}{S}\{Q(w^{(q)}) - D(w^{(q)})\} \tag{7.8}$$

Schemes of this subclass have been analyzed in a book by van der Houwen [25], and more recently in papers by Sonneveld and van Leer [26], and Roe and Pike [27]. They are second order accurate in time for both linear and nonlinear problems if $\alpha_{m-1} = 1/2$. An efficient 4 stage scheme, which is also fourth order accurate for linear problems, has the coefficients

$$\alpha_1 = 1/4 \quad , \alpha_2 = 1/3 \quad , \alpha_3 = 1/2 . \tag{7.9}$$

The amplification factor of this scheme is given by the polynomial

$$g(z) = 1 + z + \frac{z^2}{2} + \frac{z^3}{6} + \frac{z^4}{24} .$$

The stability region of this scheme is shown in Figure 2(a), which displays contour lines for $|g| = 1, .9, .8, \ldots$. The Figure also shows the locus of z as the wave number is varied between 0 and 2π for a Courant number $\lambda = 2.8$, and a dissipation coefficient $\mu = 1/32$. The corresponding variation of $|g|$ with ξ is shown in Figure 2(b). The intercept of the stability region with the imaginary axis is $2\sqrt{2}$; the corresponding bound on the time step is $\Delta t \le 2\sqrt{2}\,\Delta x$.

The maximum stability interval along the imaginary axis attainable by an m stage scheme is m-1. This has been proved for the case when m is odd by van der Houwen, who also gave formulas for the coefficients α_q [25]. The case of m even has recently been solved by Sonneveld and van Leer [26]. For example, the 4 stage scheme with coefficients

$$\alpha_1 = 1/3, \ \alpha_2 = 4/15, \ \alpha_3 = 5/9$$

has a stability region extending to 3 along the imaginary axis, at the expense of a slight reduction along the real axis.

A substantial saving in computational effort can be realized by evaluating the dissipative part only once. This leads to a class of schemes in which (7.4) becomes

$$R^{(q)} = \frac{1}{S}\{Q(w^{(q)}) - D(w^{(0)})\} . \tag{7.10}$$

The amplification factor can no longer be represented as a polynomial, but it can easily be calculated recursively. The stability region for a 4 stage scheme of this class with the coefficients (7.9) is shown in Figure 3. This also shows that this scheme is stable for the model problem (7.6) with a Courant number $\lambda = 2.6$, and dissipation coefficient $\mu = 1/32$. Schemes in the subclass defined by

(7.10) are clearly more efficient than the more conventional schemes defined by (7.8), in which the dissipative terms are repeatedly evaluated.

If the time stepping scheme is to be used in a multigrid procedure it is important that it should be effective at damping high frequency modes. One can fairly easily devise 3 and 4 stage schemes in the class defined by (7.10) which meet this requirement. An effective 3 stage scheme is given by the coefficients

$$\alpha_1 = .6, \qquad \alpha_2 = .6 \quad . \tag{7.11}$$

Its stability region is shown in Figure 4.

Additional flexibility is provided by a class of schemes in which the dissipative terms are evaluated twice. This may be used to make a further improvement in the high frequency damping properties, or else to extend the stability region along the real axis to allow more margin for the dissipation introduced by an upwind or TVD scheme of the type described in Section 5. In this class of schemes

$$\begin{aligned} R^{(0)} &= \frac{1}{S}\{Q(w^{(0)}) - D(w^{(0)})\}, \\ R^{(q)} &= \frac{1}{S}\{Q(w^{(q)}) - \beta D(w^{(1)}) - (1-\beta)D(w^{(0)})\}, \quad q \geq 1 \end{aligned} \tag{7.12}$$

In the case of pure dissipation ($Qw = 0$), the amplification factor reduces to

$$g = 1 + z + \alpha_1 \beta\, z^2 .$$

Thus if β is chosen such that $\alpha_1\beta = 1/4$, the stability region will contain a double zero at $z = -2$ on the real axis. A maximum stability interval of 8 can be attained along the real axis by choosing β such that $\alpha_1\beta = 1/8$. Figures 5 and 6 show stability regions of 4 and 5 stage schemes in this class with the coefficients

$$\alpha_1 = 1/4, \quad \alpha_2 = 1/3, \quad \alpha_3 = 1/2, \qquad \beta = 1 \tag{7.13}$$

and

$$\alpha_1 = 1/4, \quad \alpha_2 = 1/6, \quad \alpha_3 = 3/8, \quad \alpha_4 = 1/2, \quad \beta = 1 \tag{7.14}$$

The improvement of Figure 5 over Figure 2 is clear. The 5 stage scheme combines van der Houwen's optimal coefficients with two evaluations of the dissipative terms to attain a stability interval of 4 along both the imaginary and the real axes.

8. Residual Averaging

A simple method of increasing the time step is to replace the residuals by weighted averages of the neighboring residuals [28]. Consider the general multi-stage scheme (7.3). In a one dimensional case one might replace the residual R_i by the average

$$\bar{R}_i = \varepsilon R_{i-1} + (1-2\varepsilon)\, R_i + \varepsilon R_{i+1} \tag{8.1}$$

at each stage of the scheme. This smooths the residuals and also increases the support of the scheme, thus relaxing the restriction on the time step imposed by the Courant-Friedrichs-Lewy condition. If $\varepsilon \geq 1/4$, however, there are Fourier modes such that $\bar{R}_i = 0$ when $R_i \neq 0$. To avoid this restriction one may perform the averaging implicitly by setting

$$-\varepsilon\bar{R}_{i-1} + (1+2\varepsilon)\bar{R}_i - \varepsilon\bar{R}_{i+1} = R_i \text{ ,} \tag{8.2}$$

For an infinite interval this equation has the explicit solution

$$R_i = \frac{1-r}{1+r} \sum_{q=-\infty}^{\infty} r^{|q|} R_{i+q} \tag{8.3}$$

where

$$\varepsilon = \frac{r}{(1-r)^2} \text{ , } r < 1 \text{ .} \tag{8.4}$$

Thus the effect of the implicit smoothing is to collect information from residuals at all points in the field, with an influence coefficient which decays by factor r at each additional mesh interval from the point of interest.

Consider the model problem (7.6). According to equation (8.4) the Fourier symbol (7.7) will be replaced by

$$z = -\lambda\, \frac{i \sin \xi + 4\mu(1-\cos\xi)^2}{1 + 2\varepsilon(1-\cos\xi)} \text{ .}$$

In the absence of dissipation one now finds that stability can be maintained for any Courant number λ, provided that the smoothing parameter satisfies

$$\varepsilon \ge \frac{1}{4}\left\{\frac{\lambda^2}{\lambda^{*2}} - 1\right\}$$

where λ^* is the stability limit of the unsmoothed scheme. In terms of the decay parameter r the stability condition becomes

$$\frac{1+r}{1-r} \ge \frac{\lambda}{\lambda^*} \text{ .}$$

Suppose that the time step is increased to the limiting value permitted by a given value of r. Then according to (8.3) the Fourier symbol in the absence of dissipation is

$$z = -\lambda^* \{1 + 2\sum_{q=1}^{\infty} r^q \cos\xi q\}\, i\sin\xi$$

$$= -\lambda^* \frac{1-r^2}{1-2r\cos\xi+r^2}\, i\sin\xi$$

and it may be verified that $|z| \le \lambda^*$.

In the case of a finite interval with periodic conditions equation (8.3) still holds, provided that the values R_{i+q} outside the interval are defined by periodicity. In the absence of periodicity one can also choose boundary conditions such that (8.3) is a solution of (8.2), with $R_{i+q} = 0$ if $i + q$ lies outside the interval. This solution can be realized by setting

$$\tilde{R}_1 = R_1 ,$$

$$\tilde{R}_i = R_i - r(R_i - \tilde{R}_{i-1}) \text{ for } 2 \le i \le n$$

and then

$$\bar{R}_n = \frac{1}{1+r}\tilde{R}_n ,$$

$$\bar{R}_i = \tilde{R}_i - r(\tilde{R}_i - \bar{R}_{i+1}) \text{ for } 1 \le i \le n-1 \text{ .}$$

In the two dimensional case the implicit residual averaging is applied in product form

$$(1-\varepsilon_x\delta_x^2)(1-\varepsilon_y\delta_y^2)\,\bar{R}_{ij} = R_{ij}$$

where δ_x^2 and δ_y^2 are the second difference operators in the x and y directions, and ε_x and ε_y are the corresponding smoothing parameters.

In practice the best rate of convergence is generally obtained by using a value of λ about three times λ^*, and the smallest possible amount of smoothing to maintain stability. The idea of residual averaging may be compared with Lerat's implicit scheme [29]. This may be interpreted as a method of smoothing the corrections calculated by a 2 stage scheme. In the general case of an m stage scheme it is not sufficient to smooth the final corrections after the mth stage. It is possible, however, to devise stable schemes by smoothing the residuals at alternate stages, starting with the 1st stage if m is odd, and the 2nd stage if m is even.

9. Multigrid Scheme

While the available theorems in the theory of multigrid methods generally assume ellipticity, it seems that it ought to be possible to accelerate the evolution of a hyperbolic system to a steady state by using large time steps on coarse grids, so that disturbances will be more rapidly expelled through the outer boundary. The interpolation of corrections back to the fine grid will introduce errors, however, which cannot be rapidly expelled from the fine grid, and ought to be locally damped if a fast rate of convergence is to be attained. Thus it remains important that the driving scheme should have the property of rapidly damping out high frequency modes.

In a novel multigrid scheme proposed by Ni [30], the flow variables are stored at mesh nodes, and the rates of change of mass, momentum and energy in each mesh cell are estimated from the flux integral appearing in equation (3.1). The corresponding change δw in the flow variables is then distributed unequally between the nodes at its four corners. The distribution rule is such that the accumulated corrections at each node correspond to the first two terms of a Taylor series in time, like a Lax Wendroff scheme. As it stands, this scheme does not damp oscillations between odd and even points. Ni introduces artificial viscosity by adding a further correction proportional to the difference between the value at each corner and the average of the values at the four corners of the cell. Residuals on the coarse grid are formed by taking weighted averages of the corrections at neighboring nodes of the fine grid, and corrections are then assigned to the corners of coarse grid cells by the same distribution rule. When several grid levels are used, the distribution rule is applied once on each grid down to the coarsest grid, and the corrections are then interpolated back to the fine grid. Using 3 or 4 grid levels, Ni obtained a mean rate of convergence of about .95, measured by the average reduction of the residuals in a multigrid cycle. The performance of the scheme seems to depend critically on the presence of the added dissipative terms to provide the necessary damping of high frequency nodes. In Ni's published version of the scheme these introduce an error of first order.

The flexibility in the formulation of the hybrid multi-stage schemes allows them to be designed to provide effective damping of high frequency modes with higher order dissipative terms. This makes it possible to devise rapidly convergent multigrid schemes without any need to compromise the accuracy through the introduction of excessive levels of dissipation [31].

In order to adapt the multi-stage scheme for a multigrid algorithm, auxiliary meshes are introduced by doubling the mesh spacing. Values of the flow variables are transferred to a coarser grid by the rule

$$w_{2h}^{(0)} = (\sum S_h w_h)/S_{2h} \tag{9.1}$$

where the subscripts denote values of the mesh spacing parameter, S is the cell area, and the sum is over the 4 cells on the fine grid composing each cell on the coarse grid. This rule conserves mass, momentum and energy. A forcing function is then defined as

$$P_{2h} = \sum R_h(w_h) - R_{2h}(w_{2h}^{(0)}) \tag{9.2}$$

where R is the residual of the difference scheme. In order to update the solution on a coarse grid, the multistage scheme (7.3) is reformulated as

$$\begin{aligned} w^{(1)} &= w_{2h}^{(0)} - \alpha_1 \Delta t (R_{2h}^{(0)} + P_{2h}) \\ &\cdots \\ w_{2h}^{(q+1)} &= w_{2h}^{(0)} - \alpha_q \Delta t \; (R_{2h}^{(q)} + P_{2h}) \\ &\cdots \end{aligned} \tag{9.3}$$

where $R^{(q)}$ is the residual at the q+1st stage. In the first stage of the scheme, the addition of P_{2h} cancels $R_{2h}(w_{2h}^{(0)})$ and replaces it by $\sum R_h(w_h)$, with the result that the evolution on the coarse grid is driven by the residuals on the fine grid. This process is repeated on successively coarser grids. Finally the correction calculated on each grid is passed back to the next finer grid by bilinear interpolation.

Since the evolution on a coarse grid is driven by residuals collected from the next finer grid, the final solution on the fine grid is independent of the choice of boundary conditions on the coarse grids. The surface boundary condition is treated in the same way on every grid, by using the normal pressure gradient to extrapolate the surface pressure from the pressure in the cells adjacent to the wall. The far field conditions can either be transferred from the fine grid, or recalculated by the procedure described in Section 6.

It turns out that an effective multigrid strategy is to use a simple saw tooth cycle (as illustrated in Figure 7), in which a transfer is made from each grid to the next coarser grid after a single time step. After reaching the coarsest grid the corrections are then successively interpolated back from each grid to the next finer grid without any intermediate Euler calculations. On each grid the time step is varied locally to yield a fixed Courant number, and the same Courant number is generally used on all grids, so that progressively larger time steps are used after each transfer to a coarser grid. In comparison with a single time step of the Euler scheme on the fine grid, the total computational effort in one multigrid cycle is

$$1 + 1/4 + 1/16 + \ldots \leq 4/3$$

plus the additional work of calculating the forcing functions P, and interpolating the corrections.

10. Trials with Burgers' Equation

The simplest nonlinear conservation law is Burgers' equation without viscosity:

$$\frac{\partial u}{\partial t} + \frac{\partial}{\partial x}\left(\frac{u^2}{2}\right) = 0 . \tag{10.1}$$

This provides a useful text of the properties of numerical schemes which avoids other complicating factors such as boundary conditions and complex geometry. If the boundary conditions are given as

$$u(0) = v, \quad u(1) = -v, \quad v \geq 0 \tag{10.2}$$

equation (10.1) admits a steady entropy satisfying solution with a shock wave at a location which depends on the initial state. This section presents the results of some numerical experiments in which the multigrid multi-stage time stepping scheme was applied to initial data satisfying (10.2), to test its ability to recover a steady solution. The semi-discretization has the form

$$\frac{du_i}{dt} + \frac{1}{\Delta x} \left(f_{i+1/2} - f_{i-1/2} - d_{i+1/2} + d_{i-1/2}\right) = 0$$

where $f_{i+1/2}$ is the convective flux

$$f_{i+1/2} = 1/4(u_{i+1}^2 + u_i^2)$$

and $d_{i+1/2}$ is the artificial dissipative flux.

Several alternative forms of the dissipative flux were tested:

1). <u>Adaptive Dissipation (Section 4)</u>

$$d_{i+1/2} = \varepsilon_{i+1/2}^{(2)} \alpha_{i+1/2} (u_{i+1} - u_i)$$

$$- \varepsilon_{i+1/2}^{(4)} \alpha_{i+1/2} (u_{i+2} - 3u_{i+1} + 3u_i - u_{i-1})$$

where

$$\alpha_{i+1/2} = \frac{1}{4} \left| u_{i+1} + u_i \right| \qquad \text{(scheme 1a)}$$

or

$$\alpha_{i+1/2} = \frac{1}{4} \{ | u_{i+1} | + | u_i | \} \qquad \text{(scheme 1b)}$$

Also $\varepsilon_{i+1/2}^{(2)}$ and $\varepsilon_{i+1/2}^{(4)}$ are determined by defining a sensor

$$\nu_i = 2 \frac{| u_{i+1} - 2u_i + u_{i-1} |}{| u_{i+1} | + 2 | u_i | + | u_{i-1} |} .$$

Set

$$\bar{\nu}_{i+1/2} = \max (\nu_{i+2}, \nu_{i+1}, \nu_i, \nu_{i-1})$$

Then

$$\varepsilon_{i+1/2}^{(2)} = \min (k^{(2)}, \bar{\nu}_{i+1/2})$$

and

$$\varepsilon^{(4)}_{i+1/2} = \max\,(0,\, k^{(4)} - \bar{\nu}_{i+1/2}) \,.$$

By choosing $k^{(2)} = 1$ the scheme is locally TVD wherever $\nu_{i+1/2} \geq 1$.

2). <u>TVD Scheme (Section 5)</u>

$$d_{i+1/2} = \alpha_{i+1/2}(u_{i+1} - u_i) - \frac{1}{2}\,(g_{i+1} + g_i)$$

$$+ \frac{1}{2}\,|g_{i+1} - g_i|\,\mathrm{sign}(u_{i+1} - u_i)$$

where

$$\alpha_{i+1/2} = \frac{1}{4}\,|u_{i+1} + u_i| \qquad \text{(scheme 2a)}$$

or

$$\alpha_{i+1/2} = \frac{1}{4}\,\{|u_{i+1}| + |u_i|\} \qquad \text{(scheme 2b)}$$

Also the anti-diffusive flux g_i is determined by

$$\tilde{g}_{i+1/2} = \alpha_{i+1/2}(u_{i+1} - u_i)$$

and

$$g_i = B(\tilde{g}_{i+1/2}, \tilde{g}_{i-1/2})$$

where $B(r,s)$ is the min mod function (5.9). The choice of $\alpha_{i+1/2}$ in schemes 1a and 2a allows a stationary shock with at most one interior point, but also allows a stationary expansion shock. The alternative form used in schemes 1b and 2b excludes the stationary expansion shock, and still allows a discrete shock with no tail, but now with 2 interior points, as may be verified solving a quadratic equation.

Calculations were performed with a 3 stage time stepping scheme in which the dissipative terms were evaluated twice, as defined by equations (7.3) and (7.12), with the coefficients

$$\alpha_1 = .6\ ,\ \alpha_2 = .6,\quad \beta = .35$$

The stability region of this scheme is shown in Figure 8. The grid contained 128 cells, and 5 grid levels were used in the multigrid scheme.

Figure 9 shows a typical evolution using scheme 1a. Figure 9(a) shows the first 10 time steps, starting from the initial data at the bottom of the figure. It can be seen that by the completion of the 6th step the solution is indistinguishable from the final steady solution, illustrated in Figure 9(b). In this case the symmetry of the initial data results in a shock wave with no interior points. The convergence history measured by the average value of $|\partial u/\partial t|$ is shown in Figure 9(c). Figure 10 shows a similar evolution with shifted initial data. The shock wave in the final steady state is displaced to the right and contains a single interior point, as a consequence of the choice of the viscosity coefficient $\alpha_{i+1/2}$. Figure 11 shows the first calculation repeated with scheme 1b. The final solution now has a shock wave with 2 interior points because of the additional dissipation introduced by the choice of an entropy satisfying viscosity coefficient. The TVD schemes 2a and 2b produced results which were essentially identical to those produced by schemes 1a and 1b, but with a slower rate of convergence to a steady state.

11. Results for the Euler Equations

This section presents some typical results of multigrid calculations of the Euler equations for two-dimensional flow. Results are first presented for the standard scheme with adaptive dissipation as defined in Section 4. The 5 stage scheme defined by equation (7.14) was used in all these examples, and residual averaging was also used to allow time steps corresponding to a Courant number of 7.5.

The first example is the transonic flow past an NACA 64A410 airfoil at Mach .720 and an angle of attack $\alpha = 0^{o}$. Figure 6(a) shows the inner part of the grid, which contained 160 cells in the circumferential direction and 32 cells in the direction normal to the profile. The outer boundary of the grid is located at a distance of about 50 chords. Figure 6(b) displays the final pressure distribution in terms of the pressure coefficient

$$c_p = (p - p_\infty)/1/2\ \rho_\infty\ q_\infty^2 .$$

There is a moderately strong shock wave at about 60% of the chord on the upper surface. This result was obtained with 25 multigrid cycles on a 40 x 8 mesh, followed by 25 cycles on an 80 x 16 mesh, and finally 25 cycles on the 160 x 32 mesh. The calculation on the first mesh was started from an initial condition of uniform flow, and subsequently the result on each mesh was interpolated to provide the initial condition of the next mesh. 2 grid levels were used in the multigrid scheme on the 40 x 8 mesh, 3 grid levels on the 80 x

16 mesh, and 4 grid levels on the 160 x 32 mesh. Figure 6(c) shows the convergence history on the 160 x 32 mesh. Two measures are displayed. One curve shows the decay of the logarithm of the error (measured by the root mean square root of change of the density). The other curve shows the build-up of the number of grid points in the supersonic zone, which is a useful measure of global convergence. It can be seen that the flow field is already fully developed.

Two other typical examples are shown in the next two figures. Figure 13 shows the result for the NACA 0012 at Mach .8 and an angle of attack $\alpha = 1.25^{\circ}$, and Figure 14 shows the result for the Korn airfoil at Mach .75 and $\alpha = 0^{\circ}$. Each result was obtained with 50 cycles on an 80 x 16 mesh, followed by 50 cycles on a 160 x 32 mesh. This is sufficient for full development of the flow field. Such a calculation can be performed in 10 seconds with one processor of a Cray X/MP. The flow past the NACA 0012 airfoil contains a fairly strong shock wave on the upper surface, which is resolved in about four mesh cells, and a weak shock wave on the lower surface, which is quite smeared. The Korn airfoil is designed to be shock free at the given Mach number and angle of attack. The result of the Euler calculation is in close agreement with the result of the design calculation [32]. The drag should be zero in a shock free flow, and the calculated value of the drag coefficient CD = .0005 is an indication of the level of discretization error. Another measure of error is the entropy, which should also be zero. In this calculation the entropy, measured by $\frac{p}{\rho^{\gamma}} - \frac{p_{\infty}}{\rho_{\infty}^{\gamma}}$, is less than .001 everywhere in the flow field.

The final example shows the effect of replacing the scheme with adaptive dissipation by the TVD scheme for the case of the NACA 0012 airfoil at Mach .8 and $\alpha = 1.25^{\circ}$. The result is shown in Figure 15. It can be seen that sharper shock waves are obtained, with one interior cell. The entropy level ahead of the shock waves is much higher than in the calculation shown in Figure 12, however, and the stagnation enthalpy is no longer constant.

These results clearly demonstrate that the convergence of a time dependent hyperbolic system to a steady state can be substantially accelerated by the introduction of multiple grids. With the formulation here proposed, the steady state is entirely determined by the space discretization scheme on the fine grid. It is independent of both the time stepping scheme, and the discretization procedure used on the coarse grids. Either of these could be modified in any way which would improve the rate of convergence or reduce the computational effort. The existing scheme captures shocks fairly well using the adaptive dissipation of Section 4. The shock resolution is improved by the use of the TVD scheme of Section 5, at the expense of larger errors in the smooth part of the flow.

REFERENCES

(1) Morawetz, C.S., On the Non-Existence of Continuous Transonic Flows Past Profiles, Comm. Pure. Appl. Math., 9, 1956, pp. 45-48.

(2) Jameson, Antony, Transonic Aerofoil Calculations Using the Euler Equations, Proc. IMA Conference on Numerical Methods in Aero-Nautical Fluid Dynamics, edited by P. L. Roe, Academic Press, 1982, pp. 289-308.

(3) Bristeau, M.O., Glowinski, R., Periaux, J., Perrier, P., Pironneau, O., and Poirier, G., Transonic Flow Simulations by Finite Elements and Least Squares Methods, Proc. of 3rd International Conference on Finite Elements in Flow Problems, Banff, 1980, Vol. 1, edited by D.H. Norrie, pp. 11-29.

(4) Lax, Peter and Wendroff, Burton, Systems of Conservation Laws, Comm. Pure. Math., 13, 1960, pp. 217-237.

(5) Harten, Amiram, Lax, Peter and van Leer, Brum, On Upstream Differencing and Godunov Type Schemes for Hyperbolic Conservation Laws, SIAM Review, 25, 1983, pp. 35-61.

(6) Osher, Stanley and Chakravarthy, Sukumar, High Resolution Schemes and the Entropy Condition, ICASE Report NASA CR 172218, 1983.

(7) Roe, P.L., The Use of the Riemann Problem in Finite Difference Schemes, Proc. 7th International Conference on Numerical Methods in Fluid Dynamics, Stanford 1980, edited by W.C. Reynolds and R.W. MacCormack, Springer, 1981, pp. 354-359.

(8) Roe, P.L., Approximate Riemann Solvers, Parameter Vectors, and Difference Schemes, J. Computational Physics, 43, 1981, pp. 357-372.

(9) Veiullot, J.P. and Viviand, H., Methodes Pseudo-Instationnaire pour l'Calcul d' Ecoulements Transsoniques, ONERA Publication 1978-4, 1978.

(10) Jameson, Antony, Steady-State Solution of the Euler Equations for Transonic Flow, Proc. Symposium on Transonic, Shock and Multidimensional Flows, Madison, 1981, edited by R.E. Meyer, Academic Press, 1982, pp. 37-70.

(11) Jameson, A., Schmidt, W., and Turkel, E., Numerical Solution of the Euler Equations by Finite Volume Methods Using Runge-Kutta Time Stepping Schemes, AIAA Paper 81-1259, 1981.

(12) Schmidt, W., and Jameson, A., Recent Developments in Finite Volume Time Dependent Techniques for Two and Three Dimensional Transonic Flows, Von Karman Institute Lecture Series 1982-04, 1982.

(13) Boris, J.P., and Book, D.L., Flux Corrected Transport, 1, SHASTA, A Fluid Transport Algorithm that Works, J. Computational

Physics, 11, 1973, pp. 38-69.

(14) Zalesak, Stephen, Fully Multidimensional Flux Corrected Transport Algorithm for Fluids, J. Computational Physics, 31, 1979, pp. 335-362.

(15) Harten, A., High Resolution Schemes for Hyperbolic Conservation Laws, New York University Report DOE/ER 03077-175, 1982.

(16) Lax, P.D., Hyperbolic Systems of Conservation Laws and the Mathematical Theory of Shock Waves, CBMS Regional Conference Series on Applied Mathematics 11, SIAM, 1973.

(17) Jameson, Antony, and Lax, Peter D., Conditions for the Construction of Multi-Point Total Variation Diminishing Difference Schemes, Princeton University, Report MAE 1650, 1984.

(18) Osher, S., Riemann Solvers, The Entropy Condition, and Difference Approximations, SIAM J. Numer Anal., 21, 1984, pp. 217-235.

(19) Harten, A., Hyman, J.M., Lax, P.D., and Keyfitz, B., On Finite Difference Approximations and Entropy Conditions for Shocks, Comm. Pure Appl. Math, 29, 1976, pp. 297-322.

(20) Crandall, M., and Majda, A., Monotone Difference Approximations for Scalar Conservation Laws, Math. Comp., 34, 1980, pp. 1-21.

(21) Roe, P.L., Some Contributions to the Modelling of Discontinuous Flows, Proc. AMS/SIAM Summer Seminar on Large Scale Computation in Fluid Mechanics, San Diego, 1983.

(22) Sweby, P.K., High Resolution Schemes Using Flux Limiters for Hyperbolic Conservation Laws, to appear in J. Computational Physics.

(23) van Leer, Bram, Towards the Ultimate Conservative Difference Scheme. II. Monotonicity and Conservation Combined in a Second Order Scheme, J. Computational Physics, 14, 1974, pp. 361-370.

(24) Goodman, J.B., and Leveque, R.J., On the Accuracy of Stable Schemes for 2-D Scalar Conservation Laws, to appear in Math. Comp.

(25) Van der Houwen, Construction of Integration Formulas for Initial Value Problems, North Holland, 1977.

(26) Sonneveld, P., and van Leer, B., Towards the Solution of Van der Houwen's Problem, to appear in Nieuw Archief voor Wiskunde.

(27) Roe, P.L., and Pike, J., Restructuring Jameson's Finite Volume Scheme for the Euler Equations, RAE Report, 1983.

(28) Jameson, A., and Baker, T.J., Solution of the Euler Equations for Complex Configurations, Proc. AIAA 6th Computational Fluid Dynamics Conference, Danvers, 1983, pp. 293-302.

(29) Lerat, A., Sides, J., and Daru, V., An Implicit Finite Volume Scheme for Solving the Euler Equations, Proc. 8th International Conference on Numerical Methods in Fluid Dynamics, Aachen, 1982, edited by E. Krause, Springer, 1982, 343-349.

(30) Ni, R.H., A Multiple Grid Scheme for Solving the Euler Equations, Proc. AIAA 5th Computational Fluid Dynamics Conference, Palo Alto, 1981, pp. 257-264.

(31) Jameson, Antony, Solution of the Euler Equations by a Multigrid Method, Applied Mathematics and Computation, 13, pp. 327-356.

(32) Bauer, F., Garabedian, P., Korn, D., and Jameson, A., Supercritical Wing Sections II, Springer, 1975.

ACKNOWLEDGMENTS

This work has been carried out with the support of the Office of Naval Research, under Grant N00014-81-K-0379, and the NASA Langley Research Center, under Grant NAG-1-186. The method has been rather widely used, and many people have contributed to its improvement. Wolfgang Schmidt has been my principal collaborator in developing the method as a useful tool for aerodynamic analysis. More recently Tim Baker has assumed a key role in the extension of the method to treat complex three dimensional configurations. Manuel Salas made an important improvement in the accuracy of the prediction of lift. I have also benefited from continuing exchanges of ideas with Eli Turkel. Discussions with Stan Osher, Bram van Leer and Paul Woodward have been a great help to me in preparing Section 5, on TVD schemes.

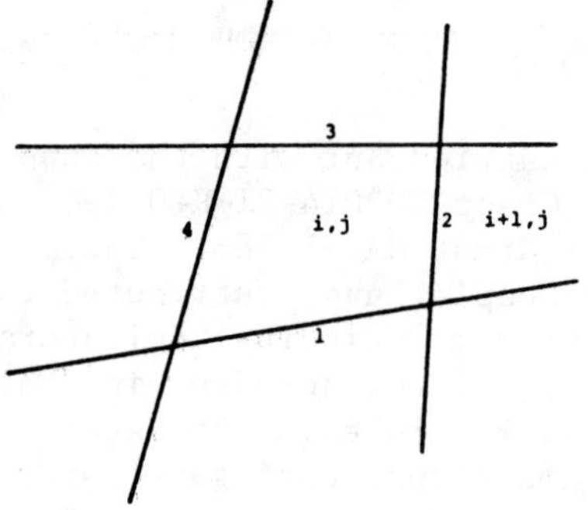

Figure 1

Finite Volume Scheme

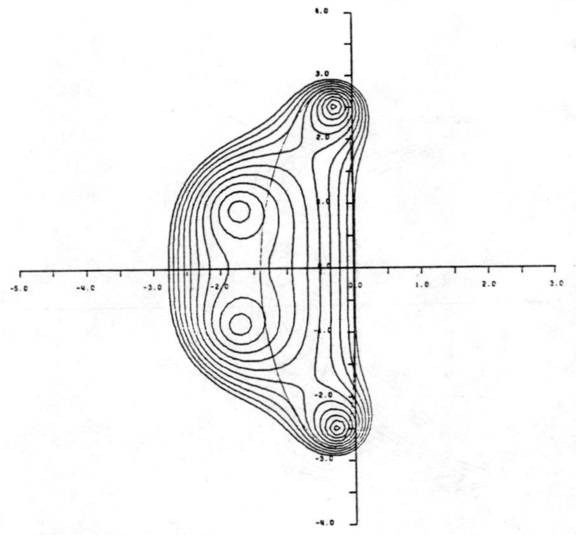

Figure 2(a)

Stability region of standard 4 stage scheme
Contour lines $|g|$ = 1., .9, .8, ...
and locus of $z(\xi)$ for $\lambda = 2.8$, $\mu = 1/32$
Coefficients $\alpha_1 = 1/4$, $\alpha_2 = 1/3$, $\alpha_3 = 1/2$

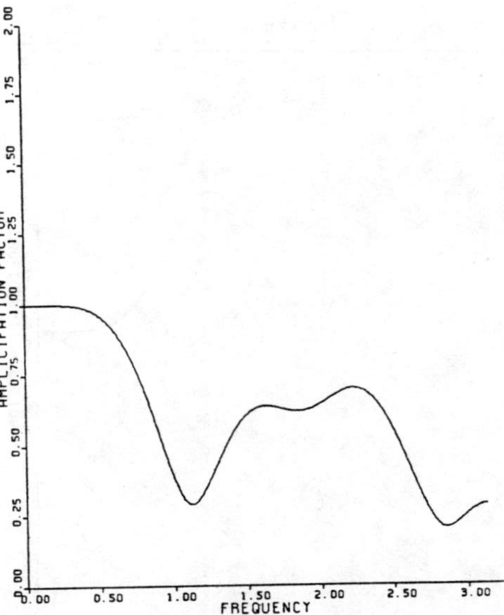

Figure 2(b)

Amplification factor $|g|$ of standard 4 stage scheme
for $\lambda = 2.8$, $\mu = 1/32$
Coefficients $\alpha_1 = 1/4$, $\alpha_2 = 1/3$, $\alpha_3 = 1/2$

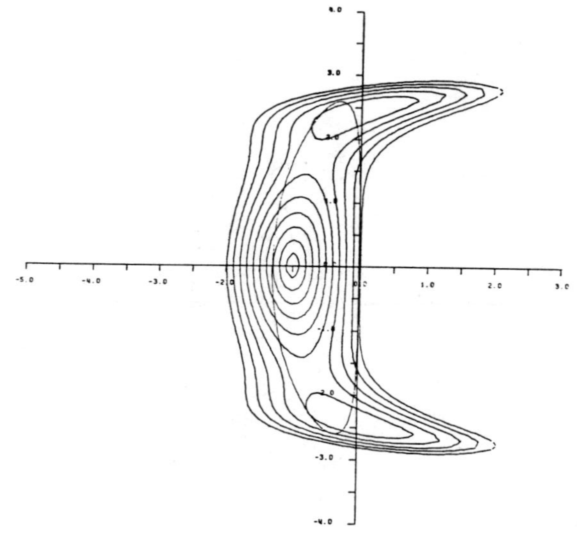

Figure 3(a)

Stability region of 4 stage scheme with single evaluation of dissipation
Contour lines $|g| = 1., .9, .8, \ldots$
and locus of $z(\xi)$ for $\lambda = 2.6$, $\mu = 1/32$
Coefficients $\alpha_1 = 1/4$, $\alpha_2 = 1/3$, $\alpha_3 = 1/2$

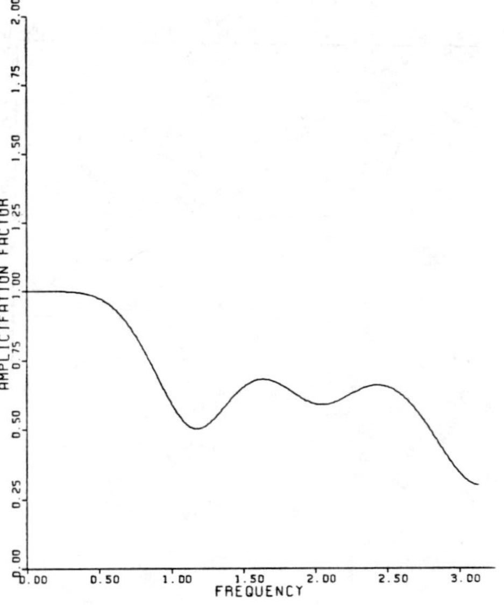

Figure 3(b)

Amplification factor $|g|$ of 4 stage scheme with single evaluation of dissipation
for $\lambda = 2.6$, $\mu = 1/32$
Coefficients $\alpha_1 = 1/4$, $\alpha_2 = 1/3$, $\alpha_3 = 1/2$

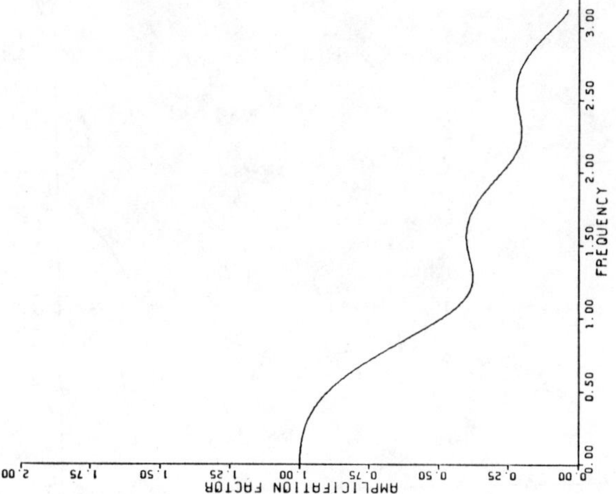

Figure 4(b)

Amplification factor $|g|$ of 3 stage scheme with single evaluation of dissipation for $\lambda = 1.5$, $\mu = .04$
Coefficients $\alpha_1 = .6$, $\alpha_2 = .6$

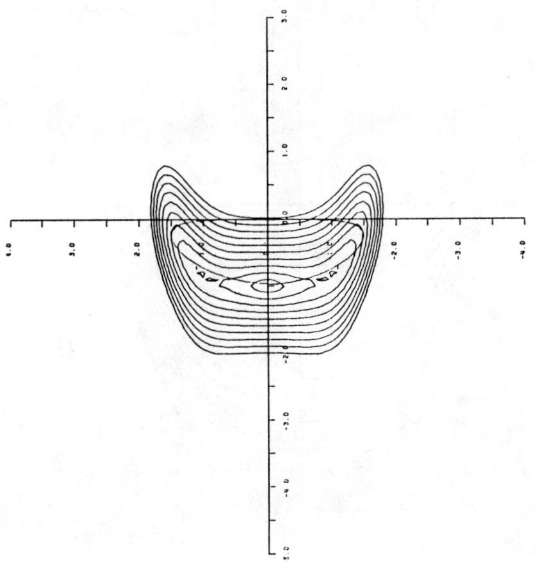

Figure 4(a)

Stability region of 3 stage scheme with single evaluation of dissipation
Contour lines $|g| = 1., .9, .8, \ldots$
and locus of $z(\xi)$ for $\lambda = 1.5$, $\mu = .04$
Coefficients $\alpha_1 = .6$, $\alpha_2 = .6$

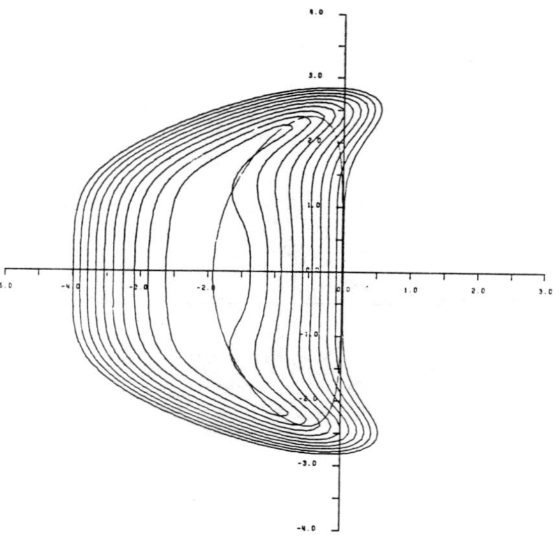

Figure 5(a)

Stability region of 4 stage scheme with two evaluations of dissipation
Contour lines $|g| = 1., .9, .8, \ldots$
and locus of $z(\xi)$ for $\lambda = 2.4$, $\mu = .05$
Coefficients $\alpha_1 = 1/4$, $\alpha_2 = 1/3$, $\alpha_3 = 1/2$, $\beta = 1$

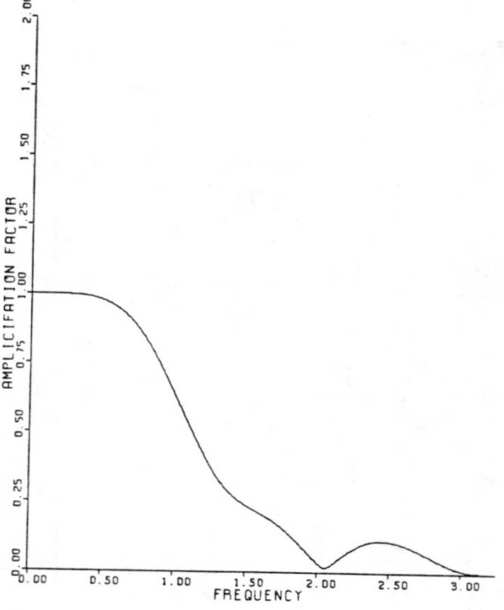

Figure 5(b)

Amplification factor 4 stage scheme with two evaluations of dissipation
for $\lambda = 2.4$, $\mu = .05$
Coefficients $\alpha_1 = 1/4$, $\alpha_2 = 1/3$, $\alpha_3 = 1/2$, $\beta = 1$

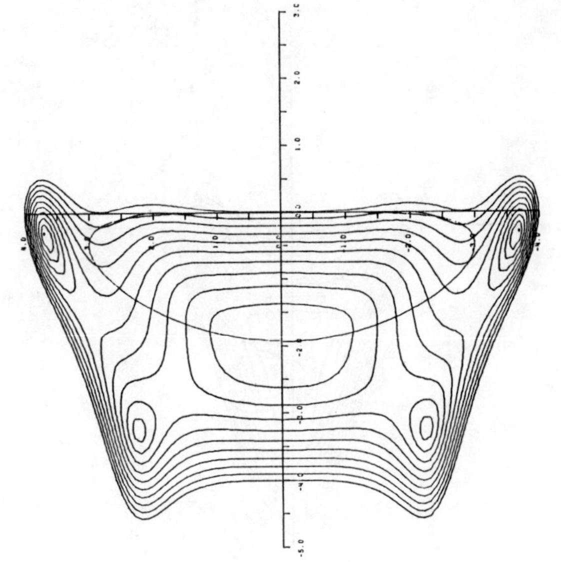

Figure 6(b)

Amplification factor $|g|$ of 5 stage scheme with two evaluations of dissipation for $\lambda = 3$, $\mu = .04$

Coefficients $\alpha_1 = 1/4$, $\alpha_2 = 1/6$, $\alpha_3 = 3/8$, $\alpha_4 = 1/2$, $\beta = 1$

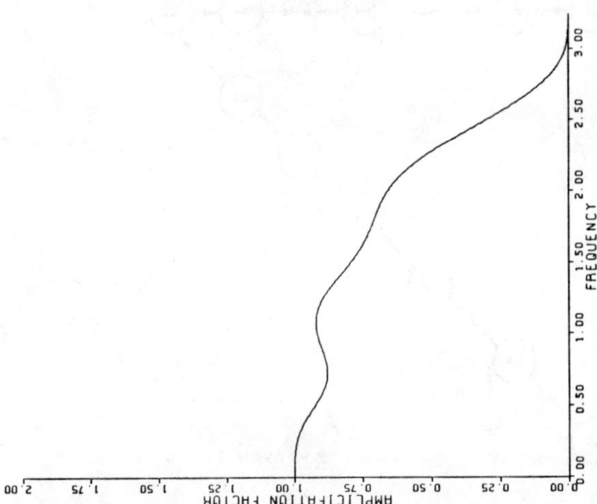

Figure 6(a)

Stability region of 5 stage scheme with two evaluations of dissipation

Contour lines $|g| = .9, .8, .7, \ldots$ and locus of $z(\xi)$ for $\lambda = 3.$, $\mu = .04$

Coefficients $\alpha_1 = 1/4$, $\alpha_2 = 1/6$, $\alpha_3 = 3/8$, $\alpha_4 = 1/2$, $\beta = 1$

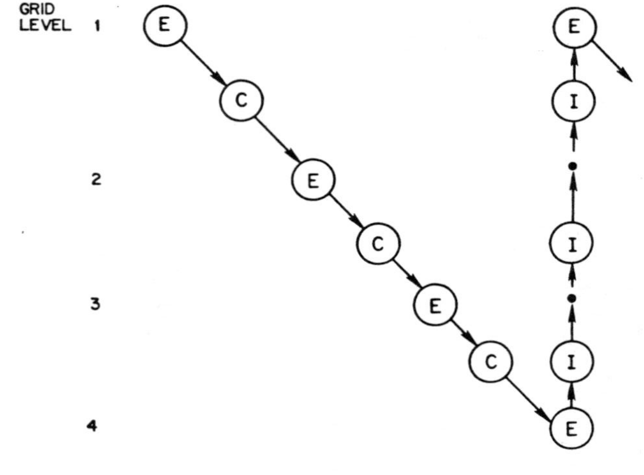

Figure 7

Saw Tooth Multigrid Cycle

(E) Euler Calculation

(C) Residual Collection

(I) Interpolation

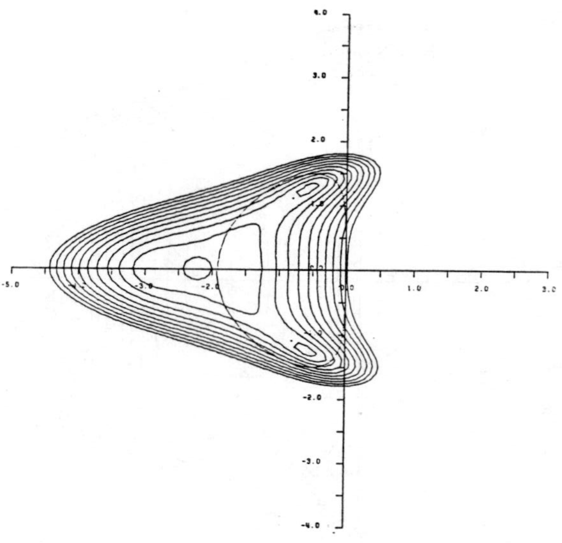

Figure 8

Stability region of 3 stage scheme with two evaluations of dissipation
Contour lines $|g| = 1., .9, .8, \ldots$
and locus of $z(\xi)$ for $\lambda = 1.5$, $\mu = .08$
Coefficients $\alpha_1 = .6$, $\alpha_2 = .6$, $\beta = .625$

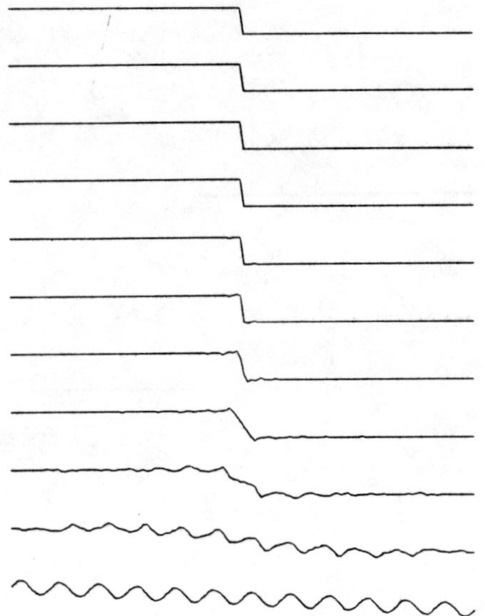

Figure 9(a)

Initial state and first 10 cycles in evolution of Burgers' equation (reading upwards)
Adaptive dissipation (scheme 1a)
128 cells 5 grids $\lambda = 2.0$

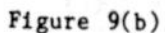

Figure 9(b)

Final State of Burgers' equation after 20 cycles of the multigrid scheme
Residual .5327 10-8

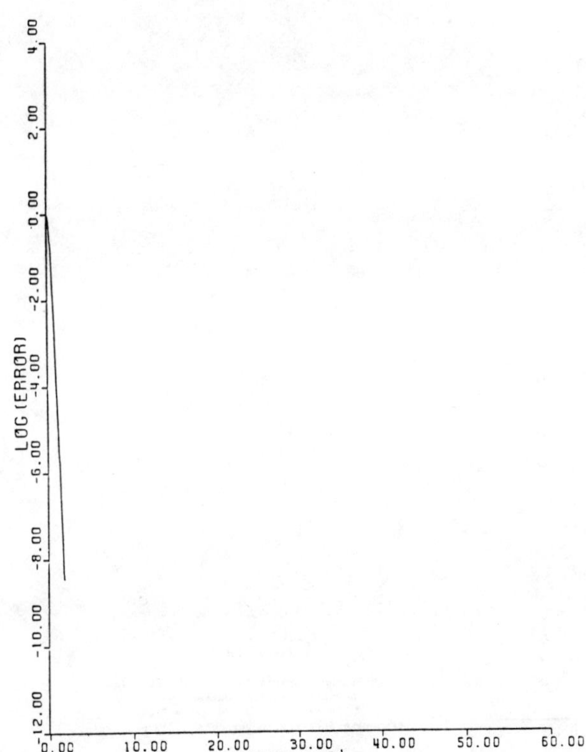

Figure 9(c)

Convergence history for Burgers' equation
Adaptive Dissipation (scheme 1a)
128 cells 5 grids $\lambda = 2.0$
Mean rate of error reduction .3587 per cycle.

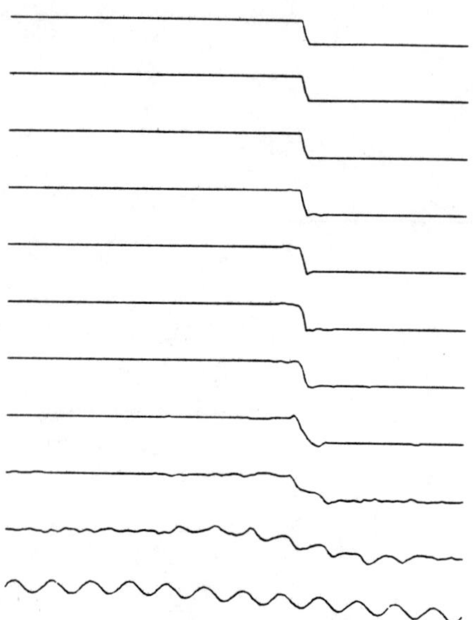

Figure 10(a)

Initial state and first 10 cycles in evolution of Burgers' equation (reading upwards)
Adaptive dissipation (scheme 1a)
128 cells 5 grids $\lambda = 2.0$

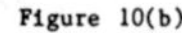

Figure 10(b)

Final State of Burgers' equation after 20 cycles of the multigrid scheme
Residual .2476 10^{-6}

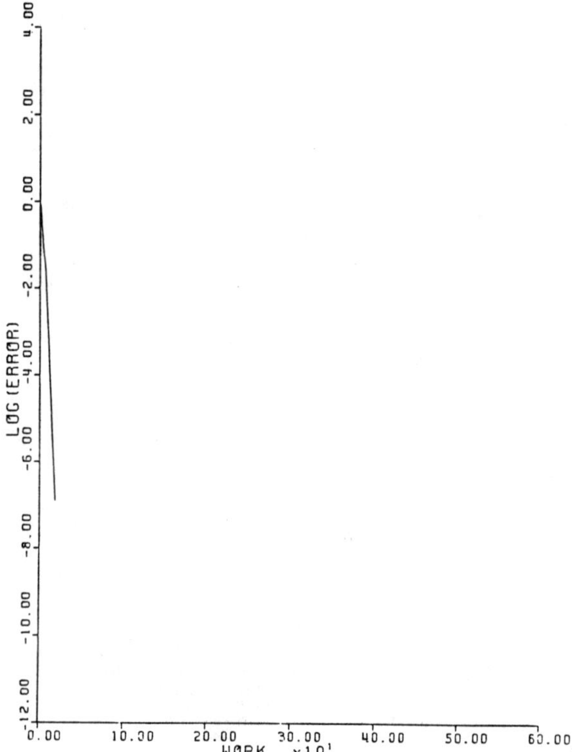

Figure 10(c)

Convergence history for Burgers' equation
Adaptive Dissipation (scheme 1a)
128 cells 5 grids $\lambda = 2.0$
Mean rate of error convergence .4339 per cycle.

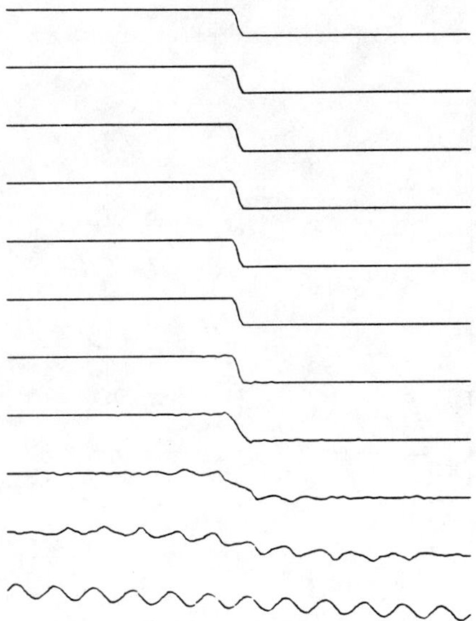

Figure 11(a)

Initial state and first 10 cycles in evolution of Burgers' equation (reading upwards)
Adaptive dissipation (scheme 1b)
128 cells 5 grids $\lambda = 2.0$

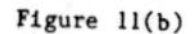

Figure 11(b)

Final State of Burgers' equation after 20 cycles of the multigrid scheme
Residual $.9028\ 10^{-8}$

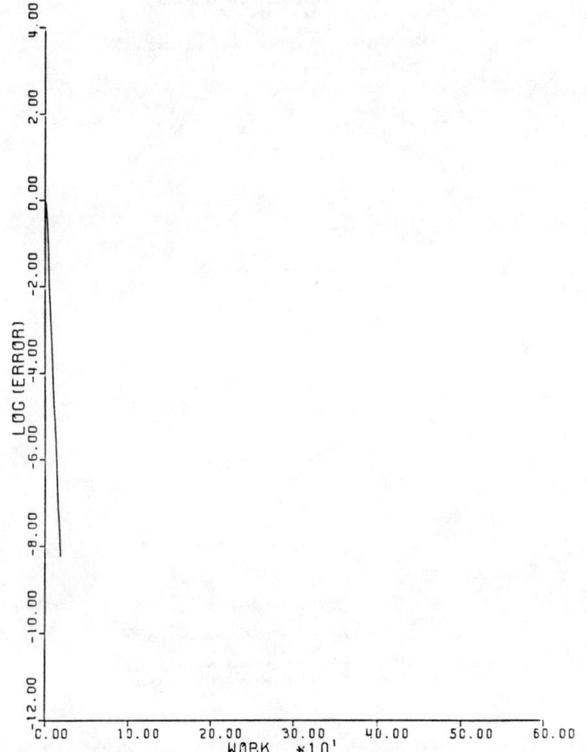

Figure 11(c)

Convergence history for Burgers' equation
Adaptive Dissipation (scheme 1a)
128 cells 5 grids $\lambda = 2.0$
Mean rate of convergence .3688 per cycle.

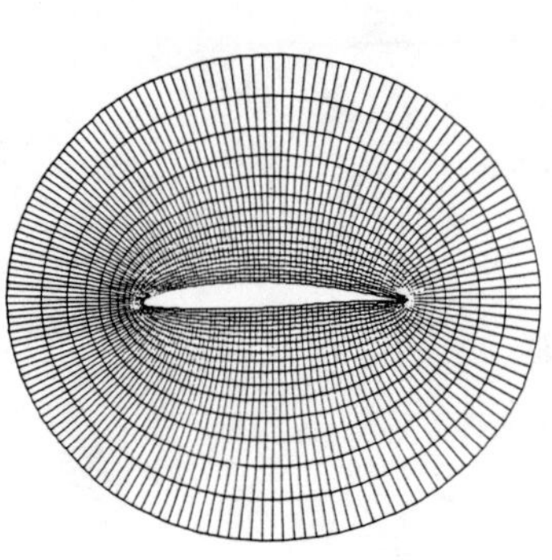

Figure 12(a)

Inner part of grid for NACA 64A410
160 x 32 cells

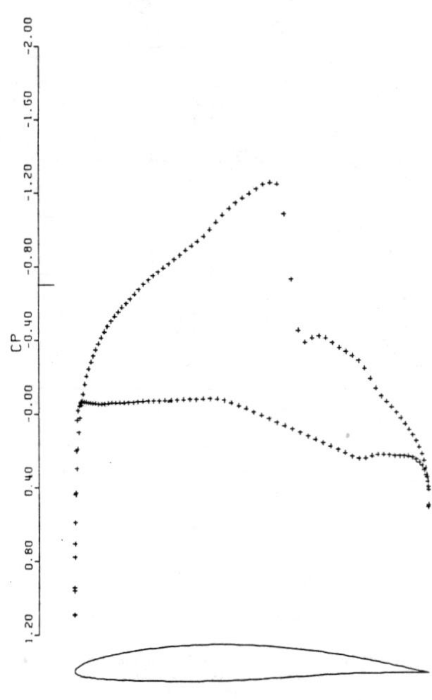

Figure 12(b)

Pressure distribution for NACA 64A410
Mach .450 α^{0}
CL .6292 CD .0036
160 x 32 grid 25 cycles Residual .248 10^{-3}

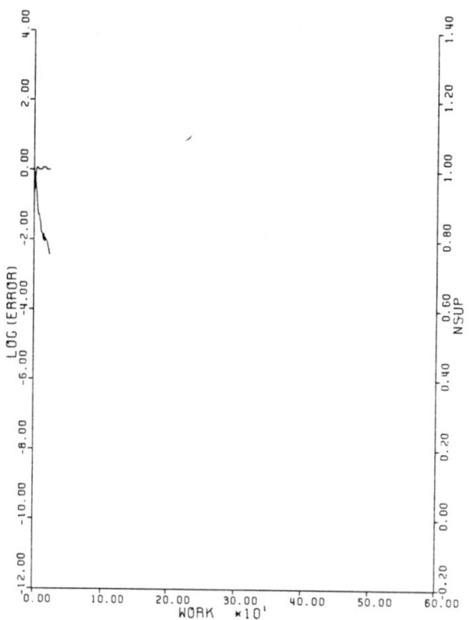

Figure 12(c)

Convergence history for NACA 64A410
Mach .720 $\alpha 0^{0}$
160 x 32 grid 25 cycles Residual .248 10^{-3}
Mean rate of convergence .7912 per cycle

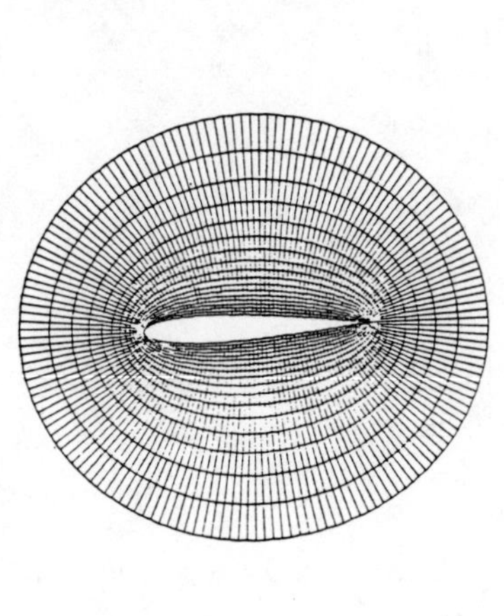

Figure 13(a)

Inner part of grid for NACA 0012
160 x 32 cells

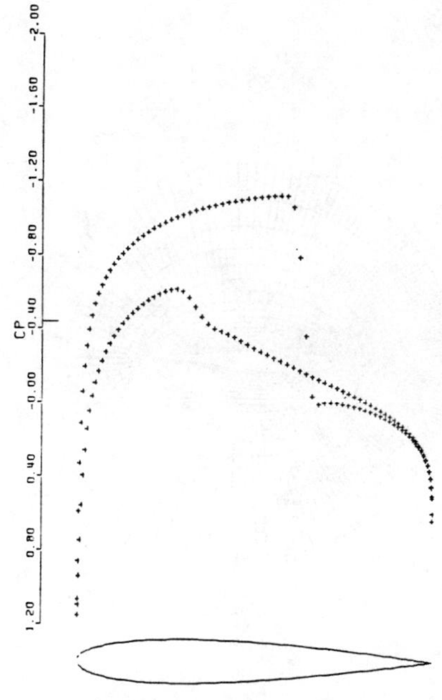

Figure 13(b)

Pressure distribution for NACA 0012
Mach .800 α 1.25°
CL .3504 CD .0227
160x32 grid 50 cycles Residual .152 10^{-3}

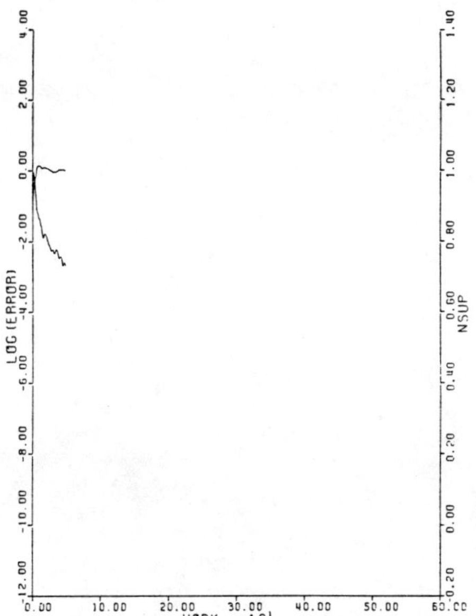

Figure 13(c)

Convergence history for NACA 0012
Mach .800 α 1.250
160 x 32 grid 50 cycles Residual .152 10^{-3}
Mean rate of convergence .8817 per cycle

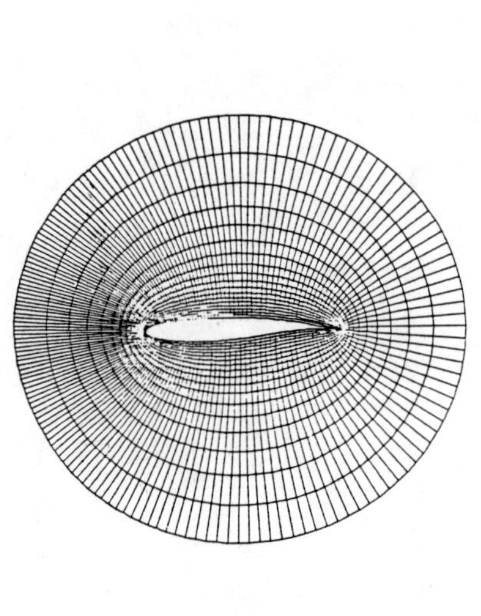

Figure 14(a)

Inner part of grid for Korn airfoil
160 x 32 cells

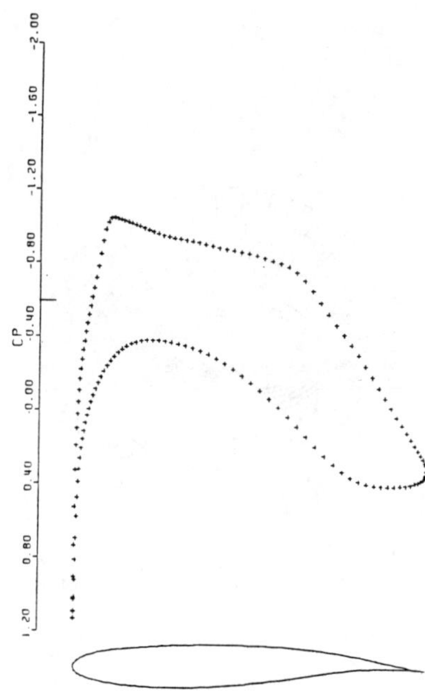

Figure 14(b)

Pressure distribution for Korn airfoil
Mach .750 α 0°
CL .6254 CD .0005
160 x 32 grid 50 cycles Residual .112 10^{-3}

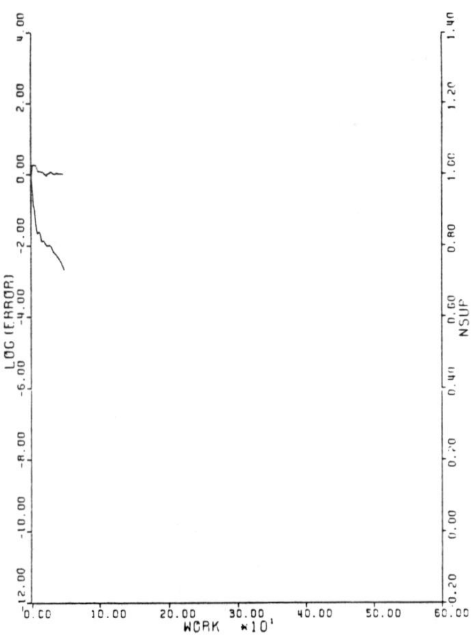

Figure 14(c)

Convergence history for Korn airfoil
Mach .750 α 0°
160 x 32 50 cycles Residual .112 10^{-3}
Mean rate of convergence .8820 per cycle

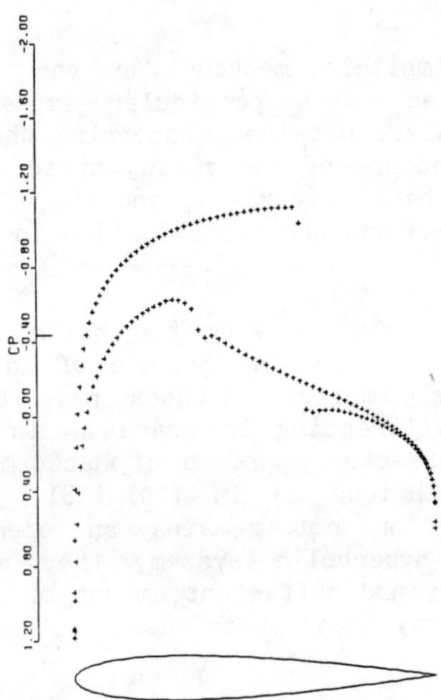

Figure 15

Pressure distribution for NACA 0012
Result using the TVD scheme.
Mach. .800 α 1.25°
CL .3349 CD .0262
160 x 32 grid 500 cycles Residual .869 10^{-4}

ANALYSIS OF AN IMPLICIT EULER SOLVER

V. DARU* AND A. LERAT†

Abstract. An implicit method for the time-dependent Euler equations is analyzed, with particular emphasis on steady-state calculations. Results are obtained concerning the nonlinear stability properties, the dependence of the steady-state on the time step, the convergence rate to the steady-state, and the effects on stability of the boundary conditions and the factorization in space.

1. Introduction. When solving the Euler equations, a way for reducing the computing costs is the use of an implicit method. In recent years, several implicit methods have been developed using either a centered differencing in space as in [1]-[6], or the non centered predictor-corrector approach of MacCormack [7]-[9], or else a real upwinding technique as in [10]-[18]. In our opinion, the choice of upwinding or not remains an open question. In the approximation of a hyperbolic system, they are several ways to generate the necessary and sufficient amount of numerical dissipation and upwinding is one way among other ones.

Here, we consider the centered implicit method proposed by Lerat [4], [5], [19] which has been applied to steady and unsteady transonic flows by Sidès and the present authors in [6], [20]. This method is second-order accurate in time and space, always linearly stable and dissipative, the numerical dissipation resulting from the implicit time-differencing and not from the addition of an artificial viscosity term. The application of the method to transonic aerodynamics has led to important reductions in computing time over explicit methods of the same accuracy, in particular for unsteady flows. However, for steady flows, further advances are needed for

+ONERA, 29 av. Division Leclerc 92320 Châtillon (FRANCE). Now at ENSAM

++ENSAM, 151 bd de l'Hôpital. 75013 PARIS. Consultant at ONERA.

treating complex flow configurations. In the present paper, we return to the analysis of the basic method and we describe new developments for improving the efficiency of steady-state calculations.

In the following section, we briefly review the implicit method and we point out what must be revised. Then, in Section 3 we consider the scalar case and we analyze the nonlinear stability of the time differencing scheme, the dependence of the steady-state on Δt and the convergence rate to the steady-state. In section 4, we come to the Euler equations and we study the effect of the boundary conditions on the linear stability. Finally in Section 5, we discuss the convergence rate of the implicit method for the one-dimensional Euler equations and for a model equation involving two space variables.

2. Review of the implicit method. In the one-dimensional case, the Euler equations form a hyperbolic system which can be written as:

$$w_t + f(w)_x = 0 \tag{1}$$

with

$$w = \begin{bmatrix} \rho \\ \rho u \\ \rho E \end{bmatrix} \qquad f(w) = \begin{bmatrix} \rho u \\ \rho u^2 + p \\ (\rho E + p) u \end{bmatrix} \tag{2}$$

where ρ is the density, p the pressure, u is the fluid velocity and E is the specific total energy which is the sum : $e + u^2/2$, the specific internal energy e being related to ρ and p by an equation of state.

For the approximation on a regular mesh of hyperbolic systems of conservation laws such as (1), the basic implicit method [4] is expressed in Δ-form as follows :

$$\Delta \check{w} = -\sigma \bar{\delta} f(w) + (1-2\alpha) \frac{\sigma^2}{2} \delta[A^2(\bar{w}) \delta w] \tag{3}$$

$$\Delta w + \alpha\sigma \bar{\delta}[A(w)\Delta w] + \beta \frac{\sigma^2}{2} \delta[A^2(\bar{w}) \delta(\Delta w)] + \frac{\gamma}{2} \delta^2(\Delta w) = \Delta \check{w} \tag{4}$$

In these expressions, Δw is the time variation of the numerical solution :

$$(\Delta w)_i^n = w_i^{n+1} - w_i^n$$

where w_i^n denotes the numerical solution at the mesh point $(i\Delta x, n\Delta t)$

and $\Delta \check{w}$ is a provisional time variation of w given by the explicit step (3).
The spatial differencing makes use of the operators :

$$(5) \quad (\delta\phi)_i^n = \phi_{i+1/2}^n - \phi_{i-1/2}^n \quad , \quad (\bar{\phi})_i^n = (\phi_{i+1/2}^n + \phi_{i-1/2}^n)/2$$

for any mesh function ϕ, so that :

$$(\bar{\delta}\phi)_i^n \equiv (\overline{\delta\phi})_i^n = (\phi_{i+1}^n - \phi_{i-1}^n)/2$$

$$\{\delta[A^2(\bar{w})\,\delta\phi]\}_i^n = A^2(\tfrac{1}{2}w_{i+1}^n + \tfrac{1}{2}w_i^n)(\phi_{i+1}^n - \phi_i^n) - A^2(\tfrac{1}{2}w_i^n + \tfrac{1}{2}w_{i-1}^n)(\phi_i^n - \phi_{i-1}^n)$$

The matrix A (w) stands for the derivative df (w)/dw, the real numbers α, β and γ are three parameters and

$$\sigma = \Delta t/\Delta x \,.$$

The implicit schemes (3), (4) involve only two time-levels and three points in space. Further, they are (see [4] or [19]):

- in conservation form
- second-order accurate in time and space
- space-centered
- linearly implicit (Δw is obtained by the solution of a linear algebraic system with a block-tridiagonal structure).

Now, if

$$(6) \qquad \alpha < \tfrac{1}{2} \quad , \quad \beta \leq \alpha - \tfrac{1}{2} \qquad \text{and} \qquad \gamma < \tfrac{1}{2}$$

the linear schemes obtained for a constant matrix A are ALWAYS

- uniquely solvable

- stable in L_2

- dissipative in the sense of Kreiss, whenever A is non-singular, for the pure initial value problem.

Furthermore, if

(7) $$\beta < - \frac{\alpha^2}{4(1-\gamma)}$$

the linear algebraic system is always strictly diagonally dominant for a single conservation law (no block).

The question is now : how should the parameters α, β and γ be selected in the range defined by (6) and (7) ? For a truly UNSTEADY PROBLEM, it is natural to minimize the evolutionary truncation error for large CFL numbers, where

$$CFL = \sigma \operatorname*{Max}_i \rho_A(w_i^n)$$

in which $\rho_A(w)$ denotes the spectral radius of the matrix A (w). With the constraints (6), (7), this choice leads to (see [19]) :

(8) $$\beta = \alpha - 1/2$$

For simplicity of the algorithm, we can choose $\gamma = 0$. Then, inequality (7), which is related to diagonal dominance, requires that:

$$-2 - \sqrt{6} < \alpha < -2 + \sqrt{6}$$

that is, α should be between about -4.449 and 0.449. The simplest possible choice is :

(9) $$\alpha = \gamma = 0 \quad \text{and} \quad \beta = -1/2$$

Now for the calculation of a STEADY PROBLEM with the unsteady method (3), (4), it turns out that the convergence to the steady-state can be importantly accelerated by taking :

(10) $$\beta = 2\alpha - 1$$

We will justify this point in Section 3 for a single conservation law and in Section 5 for the Euler equations. Choosing again $\gamma = 0$ for simplicity, the inequality (7) requires that :

$$-4 - 2\sqrt{5} < \alpha < -4 + 2\sqrt{5}$$

that is, α should be between -8.472 and 0.472. The simplest possible choice is now :

$$\alpha = \gamma = 0 \quad \text{and} \quad \beta = -1 \tag{11}$$

Thus, both the unsteady and steady problems can be treated with $\alpha = 0$, that is, without the implicit term due to Beam and Warming [1]. In this case, the explicit step (3) reduces to the original Lax and Wendroff scheme [21], which is by itself second-order accurate, and the implicit step can be considered as an implicit correction of the truncation error. To see this, one makes $\alpha = \gamma = 0$ to get :

$$\frac{\Delta w}{\Delta t} = \frac{\Delta \check{w}}{\Delta t} - \beta \frac{\Delta t^2}{2} \left[A^2(w)\, w_{tx} \right]_x + O(\Delta t^3)$$

This fact allows a further simplification : in the implicit term with coefficient β, we can replace the matrix $A\,(\bar{w})$ by its spectral radius, so that the blocks reduce to scalar coefficients in the algebraic linear system to be solved. The implicit step (4) with $\alpha = \gamma = 0$ becomes :

$$\Delta w + \beta \frac{\sigma^2}{2} \delta \left[\rho_A^2(\bar{w})\, \delta(\Delta w) \right] = \Delta \check{w} \tag{12}$$

The above simplification does not degrade the properties of solvability, stability and dissipation of the scheme and the strict diagonal dominance is maintained for any hyperbolic system. On the other hand, for unsteady problems the truncation error is slightly increased (see [19]) and for steady problems the optimal convergence rate due to the relation in (10) is lost, as we will see latter in Section 5.

The simplified scheme with the spectral radius has been applied to transonic aerodynamics in [6] and [20]. In these applications, the implicit step of the method has been factored in space by using the ADI technique. For the approximation of the hyperbolic system

$$w_t + f(w)_x + g(w)_y = 0$$

on a regular cartesian mesh, the ADI formulation reads :

$$\Delta w^* + \beta \sigma_x^2/2 \cdot \delta_x [\rho_A^2(\bar{w}^x)\, \delta_x(\Delta w^*)] = \Delta \check{w}$$

$$(13) \qquad \Delta w + \beta \sigma_y^2/2 \cdot \delta_y [\rho_B^2(\bar{w}^y)\, \delta_y(\Delta w)] = \Delta w^*$$

where $\Delta \check{w}$ is given by an explicit step of second-order accuracy and the spatial difference operators are the analogue of (5) in the x and y directions. The symbols ρ_A and ρ_B denote the spectral radii of $A = df/dw$ and $B = dg/dw$, respectively and

$$\sigma^x = \Delta t/\Delta x \quad , \quad \sigma^y = \Delta t/\Delta y$$

When the explicit step is of Lax and Wendroff type, it has been proven that a sufficient condition for the stability of the factored implicit method in the linear scalar case is $\beta \leq -1$.

Accurate numerical results have been obtained with (13) for various steady and unsteady transonic flows over airfoils. The computing time has been reduced by a factor of 3 to 4 over an explicit calculation for steady problems and by greater factors for unsteady problems. The limitation of the efficiency for steady flows seems due mainly to the use of the spectral radii ρ_A and ρ_B instead of the matrices A and B and to the present ADI factorization. These two techniques lead to a simple programming but they reduce the convergence rate of the method and therefore alternative approaches must be investigated. Other features of the method must also be considered such as the effects of the nonlinearities and of the boundary conditions. Finally, if very large time steps are allowable, the dependence of the steady state on the time-step must be analyzed.

Before discussing the above points, let us consider the implicit scheme involving blocks, (3)-(4), with the parameters being selected according to (11), as is suitable for steady problems. The algorithm is written in Δ-form, but its formulation can be simplified by returning to the w-form, since then, the Lax and Wendroff term disappears from the explicit step and the whole scheme reduces to

$$(14) \qquad w^{n+1} - \frac{\sigma^2}{2}\, \delta [A^2(\bar{w}^n)\, \delta w^{n+1}] = w^n - \sigma\, \bar{\delta} f(w^n)$$

where w^n is the numerical solution at time $n\Delta t$.

3. Scalar equation. In this Section, we consider the single conservation law :

$$w_t + f(w)_x = 0 \quad , \quad f(w) = w^2/2 \tag{15}$$

The derivative of the flux f is defined here by A (w) = w.

3.1 Nonlinear stability

It is well known that for smooth solutions of (15) with compact support, the L_2-norm in space

$$\| w(.,t) \| = \left(\int | w(x,t) |^2 \, dx \right)^{1/2}$$

is conserved during the time evolution.

Similarly, we have studied the evolution of the L_2-norm for the implicit scheme (14) applied to equation (15) . We have obtained a partial result which concerns only the time-differencing.

RESULT : The semi-discrete scheme

$$w^{n+1} - \frac{\Delta t^2}{2} \left[\left(\frac{w^n + w^{n+1}}{2} \right)^2 w_x^{n+1} \right]_x = w^n - \Delta t \, f(w^n)_x \tag{16}$$

is unconditionally stable in the sense that for any smooth solution of (16) with compact support

$$\| w^{n+1} \| \leqslant \| w^n \| \tag{17}$$

Proof : Let us compute the difference of the squares of both sides of (17) :

$$\| w^{n+1} \|^2 - \| w^n \|^2 = \int [(w^{n+1})^2 - (w^n)^2] \, dx$$

$$= I_1 - I_2$$

where

$$I_1 = 2 \int (w^{n+1} - w^n) w^{n+1} dx \quad \text{and} \quad I_2 = \int (w^{n+1} - w^n)^2 dx$$

To calculate the integral I_1, we express $w^{n+1} - w^n$ from the scheme defined by (16), multiply by $2w^{n+1}$ and integrate over space. We obtain :

$$I_1 = -2\Delta t\, I_3 + \Delta t^2 I_4$$

with

$$I_3 = \int f(w^n)_x\, w^{n+1}\, dx \quad , I_4 = \int \left[\left(\frac{w^n + w^{n+1}}{2}\right)^2 w_x^{n+1} \right]_x w^{n+1}\, dx$$

Integrating I_3 by parts yields :

$$I_3 = -\int f(w^n)\, w_x^{n+1}\, dx = -\frac{1}{2}\int (w^n)^2\, w_x^{n+1}\, dx$$

and we note that for a similar term :

$$\frac{1}{2}\int (w^{n+1})^2\, w_x^{n+1}\, dx = \frac{1}{6}\int \left[(w^{n+1})^3\right]_x dx = 0$$

Adding the above term to I_3, one finds :

$$I_3 = \frac{1}{2}\int \left[(w^{n+1})^2 - (w^n)^2\right] w_x^{n+1}\, dx = \int (w^{n+1} - w^n)\left(\frac{w^{n+1} + w^n}{2}\right) w_x^{n+1}\, dx$$

On the other hand, we integrate I_4 by parts :

$$I_4 = -\int \left(\frac{w^n + w^{n+1}}{2}\right)^2 (w_x^{n+1})^2\, dx$$

Therefore, the integral I_1 can be written as :

$$I_1 = -\int \left[2(w^{n+1} - w^n)\left(\Delta t\, \frac{w^n + w^{n+1}}{2}\, w_x^{n+1}\right) + \left(\Delta t\, \frac{w^n + w^{n+1}}{2}\, w_x^{n+1}\right)^2 \right] dx$$

Finally, substracting I_2 to I_1, we find a remarkable equality :

$$\|w^{n+1}\|^2 - \|w^n\|^2 = -\left\| w^{n+1} - w^n + \Delta t\, \frac{w^n + w^{n+1}}{2}\, w_x^{n+1} \right\|^2$$

which yields the result.

Of course, this theoretical result is not directly applicable. In practice, the time average $(w^n+w^{n+1})/2$ in the implicit term is replaced by w^n and the space derivatives are approximated by centered differences. Nevertheless the scheme given by (14) itself is very robust for steady-state calculations. In our numerical experiments with the nonlinear equation (15), the scheme (14) corresponding to $\alpha = \gamma = 0$ and $\beta = -1$ remains stable for very large CFL numbers. On the contrary, the scheme (3)-(4) with $\alpha = \gamma = 0$ and $\beta = -1/2$ (devised for unsteady problems) exhibits nonlinear unstabilities for large CFL numbers, even when it is used for a steady-state calculation.

3.2 Dependence of the steady-state on Δt

A shortcoming of the implicit method (3), (4) is a possible dependence of the steady state on Δt. Here we want to study what really occurs especially for a large Δt.

First, we fix Δx and Δt and we consider a steady solution w of (3), (4). This solution satisfies :

$$-\sigma\,\delta\,\overline{f(w)} + (1-2\alpha)\frac{\sigma^2}{2}\,\delta\left[A^2(\bar{w})\,\delta w\right] = 0 \tag{18}$$

For the single conservation law (15),

$$A(\bar{w}) = \bar{w} \quad , \quad A(\bar{w})\,\delta w = \delta f(w)$$

and equation (18) can be written as :

$$\delta\,\overline{f(w)} - (1-2\alpha)\frac{\sigma}{2}\,\delta\left[\bar{w}\,\delta f(w)\right] = 0$$

so that :

$$\overline{f(w)} - (1-2\alpha)\frac{\sigma}{2}\,\bar{w}\,\delta f(w) = \text{const.} \tag{19}$$

The expression (19) shows that in general, the numerical steady state depends on Δt. However, this does not necessarily mean that the numerical accuracy is low for large Δt.

Let us study the behaviour of the solutions of Eq. (19) for large Δt, with boundary conditions

$$(20) \qquad \begin{array}{ll} w = a > 0 & \text{on the left boundary} \\ w = -a & \text{on the right boundary} \end{array}$$

which are compatible with a non-trivial steady state (on both sides, the characteristics are ingoing and the fluxes are the same). Let us assume that the steady solution is constant near one of the boundaries, say the right one, so that:

$$(21) \qquad \delta f = 0 \qquad \text{on the right boundary}$$

From Eq. (20) and (21), we deduce that the constant on the right hand side of Eq. (19) is equal to f(a), which is independent of σ. Now we suppose that a bounded steady solution exits when $\sigma \to \infty$ and we denote it by w_∞. Then Eq. (19) gives:

$$\bar{w}_\infty \, \delta f\,(w_\infty) = 0$$

The solutions of the above equation with boundary conditions (20) satisfy:

$$(22) \qquad f(w_\infty) = f(a)$$

even if $\bar{w}_\infty = 0$ at some mesh points. They represent the weak solutions of the exact problem (15), (20).

Therefore, the implicit method can give the EXACT solutions for very large Δt . This is confirmed by the numerical experiments conducted with scheme (14).

However, various weak solutions satisfying (22) are possible. Some of them do not satisfy the entropy condition. These nonphysical solutions have not been found in the calculations, except when they are introduced as initial data+.

+ the authors are grateful to T. Hughes and S. Osher for valuable discussions on the question of non uniqueness.

3.3. Convergence rate to a steady state

We are now interested in the efficiency of the implicit method for large Δt. Let us carry out a linear study of the convergence rate based on a Fourier analysis. For the linear flux $f(w) = cw$ where c is a constant, the amplification factor of the implicit method (3), (4) can be written as :

$$g(\xi;\sigma) = h_0(\xi;\sigma) / h_1(\xi;\sigma) \tag{23}$$

where

$$h_0(\xi;\sigma) = 1-(1-\cos\xi)\left[(1-2\alpha+\beta)\eta^2+\gamma\right]+\underline{i}\sin\xi\,(\alpha-1)\eta$$

$$h_1(\xi;\sigma) = 1-(1-\cos\xi)(\beta\eta^2+\gamma)+\underline{i}\sin\xi\,\alpha\eta$$

in which $\xi = k\Delta x$, k is the wave number, $\underline{i}^2 = -1$ and $\eta = \sigma w$.

For a given $\xi \in\,]0,2\pi[$, it is apparent that the convergence rate increases as the modulus of g decreases from 1 to 0. For large-time-steps :

$$\lim_{\sigma\to\infty} |g(\xi;\sigma)| = \left|1+\frac{1-2\alpha}{\beta}\right| \leq 1 \qquad \text{for } \xi \neq 0 \tag{24}$$

the limit being 1 for $\xi = 0$, as required for consistency.

Now, in the case where equation (10) applies, and only in this case, the following result holds :

$$\lim_{\sigma\to\infty} |g(\xi;\sigma)| = 0 \tag{25}$$

that is, a very rapid convergence is obtained for large time steps.

Furthermore, the choice given by (10) and the assumption of a finite Δt being made, we get :

$$|g(\xi;\sigma)| = 1 - \frac{1}{8}\theta(\Gamma+\theta)\xi^4 + O(\xi^6)$$

for long wavelengths ($\xi \to 0$) and

$$|g(\pi;\sigma)| = \Gamma / (\Gamma + 2\theta)$$

for short wavelengths ($\xi = \pi$), where

$$\theta = (1-2\alpha)\eta^2 \quad \text{and} \quad \Gamma = 1-2\gamma$$

Due to the constraints (6), θ and Γ are positive and we can conclude that the convergence speed increases with Δt for both long and short wavelengths.

3.4. Numerical results

The implicit scheme (14) (corresponding to $\alpha = \gamma = 0$ and $\beta = -1$) has been applied to the single conservation law (15) for $0 < x < 1$ and $t > 0$, with the boundary conditions :

$$w(0,t) = 1 \quad \text{and} \quad w(1,t) = -1 \ , \ t > 0$$

and the initial condition :

$$w(x,0) = 1-2x \quad , \ 0 \leq x \leq 1$$

which is a linear interpolation between the boundary values.

The convergence to the steady-state has been terminated at the first time level $n = n_\infty$ for which :

$$R = \max_i |w_i^{n+1} - w_i^n| \leq 10^{-6}$$

Fig. 1 displays the initial data and the numerical steady-state solution in solid line for Δx = 1/50 and various CFL numbers from 0.5 to one million, where

$$CFL = \sigma \max_i |w_i^0| = \sigma$$

The number n_∞ is indicated on the figure, except for CFL = 4. For this Courant number, the residual cannot be reduced under the level $R = 10^{-4}$ (the calculation has been performed on a UNIVAC 1110 computer with simple precision). The exact solution appears in dashed line.

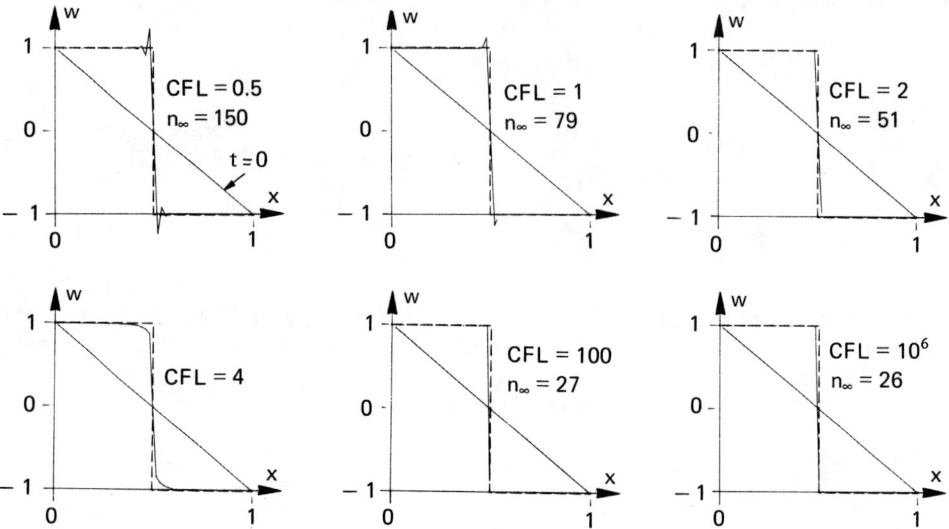

Fig. 1 — Computation of a steady shock for $w_t + (w^2/2)_x = 0$ by the implicit scheme (14) for various CFL numbers and $\Delta x = 1/50$. ---- Exact solution ; —— Numerical solution.

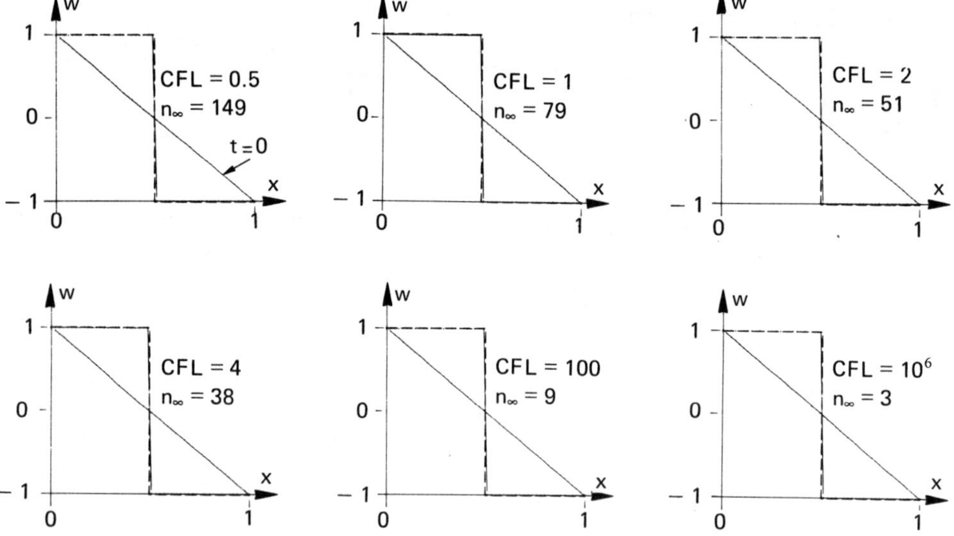

Fig. 2 — Computation of a steady shock for $w_t + (w^2/2)_x = 0$ by the implicit scheme (14) for various CFL numbers and $\Delta x = 1/49$. ---- Exact solution ; —— Numerical solution.

One can observe that :

- *The calculations are always stable*
- *the steady-state depends on Δt*
- *the steady state is very accurate for large Δt (the shock is spread over one mesh interval)*
- *the efficiency is maximum for large Δt.*

Other experiments have been made with the space increment Δx = 1/49. The corresponding mesh has no point at the exact shock location. The numerical steady solutions are shown on Fig. 2. They are independent of the time step. In this case we obtain the exact steady state as a solution of equation (19). The number of time-steps to reach the steady state decreases from 149 for CFL =.5 to only 3 for CFL = 10^6.

4. Boundary conditions and stability for the Euler equations.

4.1. Theory

We want now to examine the effect of the boundary conditions on the stability of the numerical approximation obtained when the implicit scheme is used at the interior points of the field. The study is made for the one-dimensional system of Euler equations. We first present an analysis based on the theory developed by Gustafsson, Kreiss and Sundström [22]. The results given by this analysis are then compared with numerical experiments conducted for the problem of the inviscid flow in a quasi-one-dimensional nozzle. The comparison shows a good agreement between numerical and theoretical results.

The theory of numerical stability for initial-boundary-value problems is now well developed, at least for the one-dimensional case, due to the works by Kreiss [23] [24], Osher [25] [26], and Gustafsson, Kreiss and Sundström I22] (see also [27] [28] [29]). More recently, several authors brought some complements to this theory. Among them, Yee, Beam and Warming [30] [31] [32], and also Gustafsson [33], have considered more precisely the stationary case. The report by Yee [34] contains an exhaustive bibliography of the various works concerning the subject up to 1981. Gottlieb and Turkel [35] pointed out that the results of a study in the scalar case cannot, in general, be extended to the case of a system, which needs a more complicated analysis, distinguishing clearly the role of the characteristic variables. In his thesis, Coughran [36] has studied the stability of several difference schemes with different kinds of boundary conditions, applied to a model system which is not written in the characteristic variables. He also gives a few results for the two-dimensional case.

A basic assumption of this theory is the stability of the difference scheme used at the interior point. The question of the stability of the discretized problem can be reduced to the examination of a set of resolvant equations, which are obtained by looking for solutions w_i^n of the form $z^n \psi_i$, where the complex number z is called an eigenvalue and ψ belongs to L_2 (on the spatial mesh). Then the approximation is stable if there exists no eigenvalue or generalized eigenvalue z with $|z| \geq 1$ and ψ in L_2 (see [22] for more details). In the case of a domain including two boundaries, Kreiss [23] has shown that the analysis reduces to the study of two "half-problems", which we get by removing one boundary to infinity : so it is sufficient to consider each boundary separately.

The Euler equations (1) can be rewritten as :

$$w_t + A(w)\, w_x = 0 \quad , x > 0,\ t > 0 \tag{26}$$

For a perfect gas with constant specific heats, the matrix A (w) can be expressed as :

$$A(w) = \begin{bmatrix} 0 & 1 & 0 \\ -\frac{3-\gamma}{2} u^2 & (3-\gamma) u & \gamma - 1 \\ -\left(\frac{2-\gamma}{2} u^2 + \frac{c^2}{\gamma-1}\right) u & \frac{3-2\gamma}{2} u^2 + \frac{c^2}{\gamma-1} & \gamma u \end{bmatrix}$$

where c is the sound speed defined by $(\gamma p/\rho)^{1/2}$ and γ is the ratio of specific heats.

We complete system (26) with initial conditions on w. For the well-posedness of the problem, we must give some "analytical" boundary conditions at x = 0, the number of which depends on the nature of the flow. Moreover, the numerical solution of the problem requires additional boundary conditions, which are generally called "numerical" boundary conditions. We examine here the case where these numerical boundary conditions are given by an extrapolation of the values at the interior points. We will only consider the case of a subsonic outflow at the boundary x = 0 ; the case of a subsonic inflow is analogous and leads to a similar analysis. When the flow is supersonic at the boundary, all the eigenvalues of the matrix A have the same sign, and the analysis is very simplified : the conclusion is that the global approximation is unconditionally stable if the implicit scheme is combined with the data of all the dependent variables at the boundary when the flow is ingoing, or an extrapolation of all these variables when the flow is outgoing.

For a subsonic outflow at x = 0, we have : $c > -u > 0$ at x = 0. Then, we know that the continuous problem is well posed if the pressure is given at $x = 0$ as the analytical boundary condition (see Yee [34]).

We make a linearized analysis by "freezing" the matrix A at $x = 0$ and $t = 0$. Let us rewrite the system (26) by taking p as one of the variables :

(27) $$W_t + \hat{A} W_x = 0$$

where
$$W = \begin{pmatrix} \rho \\ \rho u \\ p \end{pmatrix}, \quad \hat{A} = \begin{bmatrix} 0 & 1 & 0 \\ -\tilde{u}^2 & 2\tilde{u} & 1 \\ -\tilde{u}\tilde{c}^2 & \tilde{c}^2 & \tilde{u} \end{bmatrix}$$

the notation $\sim$ refers to the frozen values at the boundary. The matrices A and $\hat{A}$ are related by the similarity transformation :

$$\hat{A} = M^{-1} A M$$

where
$$M = \begin{bmatrix} 1 & 0 & 0 \\ 0 & 1 & 0 \\ -\frac{1}{2}\tilde{u}^2 & -(\gamma-1)\tilde{u} & \gamma-1 \end{bmatrix}$$

and
$$W_t = M^{-1} w_t, \quad W_x = M^{-1} w_x$$

The eigenvalues of matrix A are $\lambda_1 = \tilde{u}$, $\lambda_2 = \tilde{u}-\tilde{c}$, $\lambda_3 = \tilde{u}+\tilde{c}$ and we can diagonalize this matrix by the transformation :

$$\hat{A} = T \Lambda T^{-1}$$

where
$$T = \begin{bmatrix} 1 & 1 & 1 \\ \tilde{u} & \tilde{u}-\tilde{c} & \tilde{u}+\tilde{c} \\ 0 & \tilde{c}^2 & \tilde{c}^2 \end{bmatrix}$$

Λ being the diagonalized matrix.
System (27) can be written in terms of the vector of characteristic variables $U = T^{-1} w = (U_1, U_2, U_3)^t$. This gives :

$$U_t + \Lambda U_x = 0$$

So we have :

$$\begin{pmatrix} \rho \\ \rho u \\ p \end{pmatrix} = TU = \begin{pmatrix} U_1 + U_2 + U_3 \\ \tilde{u}\, U_1 + (\tilde{u} - \tilde{c})\, U_2 + (\tilde{u} + \tilde{c})\, U_3 \\ \tilde{c}^2 \,(U_2 + U_3) \end{pmatrix}$$

Now, we assume that the initial data are all zero and that the analytical boundary condition is :

$$p(0,t) = 0 \tag{28}$$

which is not a restriction to the generality as shown in [22]. For the numerical boundary conditions, we consider an extrapolation of order j-1 for ρ and ρu, that is :

$$D_+^j \, (\rho)_0^{n+1} = 0$$

$$D_+^j \, (\rho u)_0^{n+1} = 0 \tag{29}$$

At interior points, we use the implicit scheme, (3)-(4), with $\alpha = \gamma = 0$. In the characteristic variables, the scheme and the boundary conditions (28) and (29) become :

$$\left(I + \beta \frac{\sigma^2}{2} \Lambda^2 \delta^2 \right) \Delta U_i = \left(-\sigma \Lambda \bar{\delta} + \sigma^2 \Lambda^2 \delta^2 \right) \begin{pmatrix} U_1 \\ U_2 \\ U_3 \end{pmatrix}_i^n \tag{30.a}$$

$$\left(U_2 + U_3 \right)_0^{n+1} = 0 \tag{30.b}$$

$$D_+^j \left(U_1 + U_2 + U_3 \right)_0^{n+1} = 0 \tag{30.c}$$

$$D_+^j \left(\tilde{M} U_1 + (\tilde{M} - 1) U_2 + (\tilde{M} + 1) U_3 \right)_0^{n+1} = 0 \tag{30.d}$$

where $\beta \leq -1/2$ for stability of (30.a), $\tilde{M} = \tilde{u}/\tilde{c}$ and I is the identity matrix.

The characteristic equation of (30.a) is :

$$\det \left\{ \left[K I + \beta \frac{\sigma^2}{2} \Lambda^2 (K-1)^2 \right](z-1) + \frac{\sigma}{2} \Lambda (K^2-1) - \frac{\sigma^2}{2} \Lambda^2 (K-1)^2 \right\} = 0$$

where $\sigma = \Delta t / \Delta x$ and K, z are complex numbers. It gives :

$$(31) \quad \left[K_\ell + \beta \frac{\sigma^2}{2} \lambda_\ell^2 (K_\ell - 1)^2 \right](z-1) + \frac{\sigma}{2} \lambda_\ell (K_\ell^2 - 1) - \frac{\sigma^2}{2} \lambda_\ell^2 (K_\ell - 1)^2 = 0$$

for $\ell = 1, 2, 3$.

The fundamental solutions of the scheme are then :

$$\begin{pmatrix} \phi_1 \\ \phi_2 \\ \phi_3 \end{pmatrix} = \begin{pmatrix} K_1^i \\ K_2^i \\ K_3^i \end{pmatrix}$$

where the complex numbers K_ℓ ($\ell = 1, 2, 3$) are the roots of eq. (31) with modulus lower or equal to one.
This amounts to taking :

$$(32) \quad \begin{pmatrix} U_1^n \\ U_2^n \\ U_3^n \end{pmatrix} = z^n \begin{pmatrix} C_1 K_1^i \\ C_2 K_2^i \\ C_3 K_3^i \end{pmatrix}$$

where C_1, C_2, C_3 are complex constants.

Substituting (32) into (30) leads to the system :

$$(33) \quad \begin{aligned} & C_2 + C_3 = 0 \\ & C_1 (1-K_1)^j + C_2 (1-K_2)^j + C_3 (1-K_3)^j = 0 \\ & \tilde{M} C_1 (1-K_1)^j + (\tilde{M}-1) C_2 (1-K_2)^j + (\tilde{M}+1) C_3 (1-K_3)^j = 0 \end{aligned}$$

The condition for the existence of an eigenvalue or a generalized eigenvalue is obtained by setting the determinant of (33) to zero, which gives :

$$(K_1 - 1)^j \left[(K_2 - 1)^j + (K_3 - 1)^j \right] = 0$$

that is :

$$(K_1 - 1)^j = 0 \tag{34.a}$$

or

$$(K_2 - 1)^j + (K_3 - 1)^j = 0 \tag{34.b}$$

From a lemma of Kreiss [23], we know that since the scheme (30.a) is three points centered, second order accurate, Cauchy stable and dissipative and since $\lambda_1 < 0$, $\lambda_2 < 0$ and $\lambda_3 > 0$, then :
for $|z| \geq 1$, there exists positive constants δ_1 and δ_2 such that :

$$|K_1| \leq 1 - \delta_1 \tag{35.a}$$

$$|K_2| \leq 1 - \delta_2 \tag{35.b}$$

$$|K_3| \leq 1 \tag{35.c}$$

This shows that (34.a) cannot occur if $|z| \geq 1$. Moreover, in the case where j = 1 we can conclude that the global approximation, is stable, since :

$$|K_2 + K_3| \leq 2 - \delta_2 < 2$$

which proves that (34.b) is not verified either.

If j = 2 or 3, the analysis is more complicated. Let us rewrite the system which must be verified by K_2 and K_3:

$$\left[K_2 + \beta \frac{\tilde{\sigma}^2}{2} (\tilde{M}-1)^2 (K_2 - 1)^2\right](z-1) + \frac{\tilde{\sigma}}{2}(\tilde{M}-1)(K_2^2 - 1) - \frac{\tilde{\sigma}^2}{2}(\tilde{M}-1)^2 (K_2 - 1)^2 = 0 \tag{36.a}$$

$$\left[K_3 + \beta \frac{\tilde{\sigma}^2}{2} (\tilde{M}+1)^2 (K_3 - 1)^2\right](z-1) + \frac{\tilde{\sigma}}{2}(\tilde{M}+1)(K_3^2 - 1) - \frac{\tilde{\sigma}^2}{2}(\tilde{M}+1)^2 (K_3 - 1)^2 = 0 \tag{36.b}$$

(36.c) $(K_2-1)^{\gamma}+(K_3-1)^{\gamma}=0$ or $K_2=1+\omega(K_3-1)$

where $\tilde{\sigma}=\sigma\tilde{c}$ and $\omega^j=-1$.

The elimination of K_2 and z from these equations leads to the following polynomial equation for K_3

$$\frac{1}{2}\omega\beta\eta_2\eta_3[\omega(\eta_3-\eta_2)K_3+(2-\omega)\eta_3-\omega\eta_2](K_3-1)^2$$
$$\text{(37)}\quad -\omega[K_3^2[\eta_3(1+\eta_3)-\omega\eta_2(1+\eta_2)]-(1-\omega)\eta_3(1-\eta_3)$$
$$-K_3[\eta_3(1+(1-2\omega)\eta_3)-\omega\eta_2(2-\omega-\omega\eta_2)=0$$

where

$$\eta_2=\tilde{\sigma}(-\tilde{M}+1) \qquad \text{and} \qquad \eta_3=\tilde{\sigma}(-\tilde{M}-1)$$

This equation can be solved numerically. We explore a sample of numerical values of the parameters $\tilde{M}$ and of the CFL number which is here given by :

$$CFL=\sigma\,\mathrm{Max}\,(|\lambda_2|,|\lambda_3|)$$

with $|\tilde{M}|<1$ and $0<CFL\leq 100$

For each value of $\tilde{M}$ and CFL, we obtain a solution K_3 of (37), and then use (36.b) and (36.c) for the calculation of K_2 and z. Then. for every solution (K_2, K_3, z), one of the following situation occurs:

a) $|K|_\infty=\mathrm{Sup}(|K_2|,|K_3|)>1$. In this case, the solution is of no interest to the stability theory (it is not in L_2)

b) $|z|<1$ and $|K|_\infty<1$: This gives stability.

c) $|z|>1$ and $|K|_\infty<1$: z is an eigenvalue, and the approximation is unstable.

d) $|z|=1$ and $|K|_\infty=1$. This case needs a little more work before concluding whether or not z is a generalized eigenvalue. This is done by perturbing z so that it moves outside the unit circle and by looking if the corresponding K's give or not an L_2 solution.

The result are summarized in Table 1.

We can follow the same procedure to investigate the stability of the global approximation in the case where the simplified implicit scheme (3), (12) is used at the interior points of the domain, with the same boundary conditions. Again, we can conclude directly to stability if $j = 1$. For $j = 2$ and 3, we perform a numerical study. The results are displayed in Table 2.

Table 1
Stability of the scheme for a subsonic outflow
($\tilde{u}<0$ at $x = 0$) ; $j-1$ is the order of extrapolation
for ρ and ρu .

j=1	$\beta=-1/2$	stable
	$\beta=-1$	stable
j=2	$\beta=-1/2$	stable
	$\beta=-1$	stable if CFL < 5
j=3	$\beta=-1/2$	unstable
	$\beta=-1$	unstable

Table 2

Stability of the simplified scheme for a subsonic outflow

j=1	$\beta=-1/2$	stable
	$\beta=-1$	stable
j=2	$\beta=-1/2$	stable
	$\beta=-1$	stable
j=3	$\beta=-1/2$	unstable
	$\beta=-1$	unstable

The only difference in the results concerns the linear extrapolation and $\beta=-1$.

More details about this study can be found in [38].

4.2. Numerical application

We have conducted numerical experiments+ for the problem of the inviscid flow in a quasi-one-dimensional nozzle, which is governed by the following system (with the same notations as before) :

$$w_t + f(w)_x = h(w), \quad t \geq 0$$

where :

$$w = \begin{pmatrix} \rho S \\ \rho u S \\ \rho E S \end{pmatrix}, \quad f(w) = \begin{pmatrix} \rho u S \\ (\rho u^2 + p) S \\ (\rho E + p) u S \end{pmatrix}, \quad h(w) = \begin{pmatrix} 0 \\ p \frac{dS}{dx} \\ 0 \end{pmatrix}$$

$$\rho E = p/(\gamma-1) + 1/2\, \rho u^2$$

and S = S(x) is the normal section of the nozzle, here defined by :

$$S(x) = 1.398 + 0.347 \tanh(0.8x-4) \quad 0 \leq x \leq 10.$$

It is a divergent nozzle introduced in [37], and also considered by several authors.

Notice that the inhomogeneous term h(w) does not change the conclusions of the theoretical stability analysis, as shown in [22].

At the entry of the nozzle, the flow is supersonic and we fix the value of all the variables. At the boundary x = 10, there is a subsonic outflow, the pressure being fixed at the value 0.74664 (ρ and ρu are extrapolated). In order to compute these boundary conditions in a way as simple as possible, we use the relation :

$$(\rho E)_t = \frac{1}{\gamma-1} p_t + u(\rho u)_t - \frac{1}{2} u^2 \rho_t$$

and linearize it by :

$$\frac{\Delta(\rho E)_I}{\Delta t} = \frac{1}{\gamma-1} \frac{\Delta p_I}{\Delta t} + u_I^n \frac{\Delta(\rho u)_I}{\Delta t} - \frac{1}{2} u_I^{n^2} \frac{\Delta \rho_I}{\Delta t}$$

where the subscript I refers to values at the outflow boundary.

+No artificial viscosity is introduced in all the calculations.

Since $\Delta p_I=0$, we obtain a linear relation between ρ, ρu and ρE at the boundary :

$$\Delta(\rho E)_I = u_I^n (\Delta \rho u)_I - \frac{1}{2}(u_I^n)^2 \Delta \rho_I$$

Fig. 3 shows the numerical solution (here the density in the flow) obtained with the scheme with blocks and $\beta = -1$, for a CFL number equal to 4, which is a value lower than the theoretical limit for stability. It is clear that the constant extrapolation as well as the linear extrapolation are stable in this case. Figure 4 shows the same calculations, but for CFL = 10 : we see that the spurious waves created by the boundary are, in the case of linear extrapolation, amplified as time increases, and then an unstability occurs, according to the theoretical predictions. On the contrary, a constant extrapolation leads to stability.

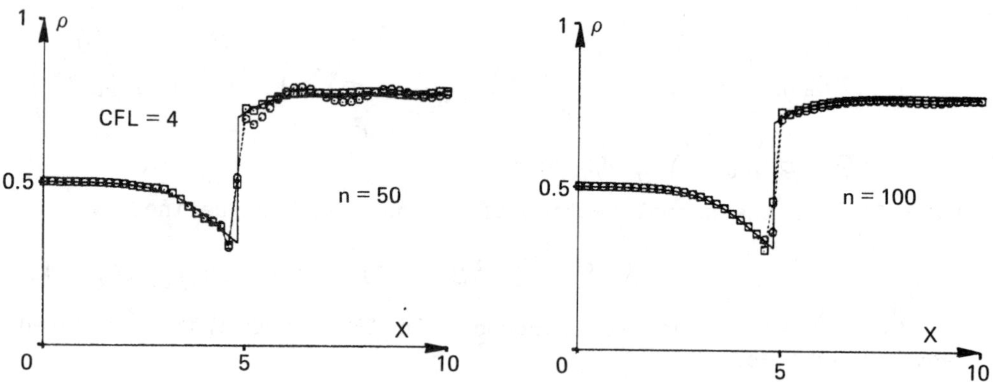

Fig. 3 – Effect of the boundary condition on the calculation of a flow with shock in a nozzle by the implicit scheme with blocks, for β = – 1 and CFL = 4.

—— Exact solution

---□--- Numerical solution with constant extrapolation of ρs and ρus at x = 10

---○--- Numerical solution with linear extrapolation of ρs and ρus at x = 10

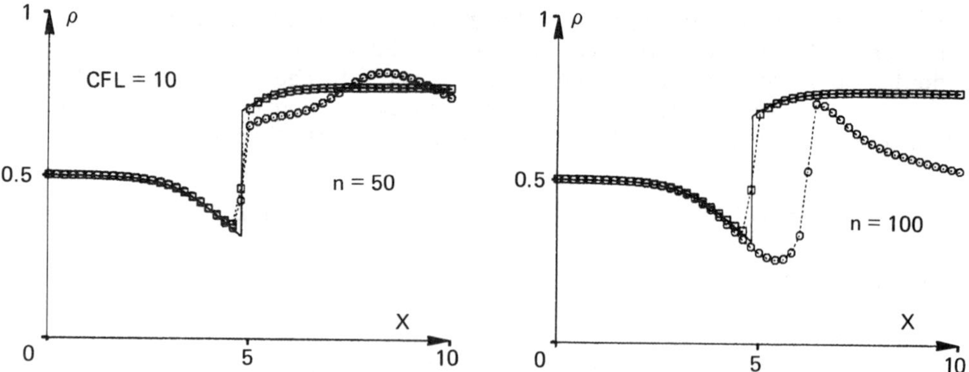

Fig. 4 – Effect of the boundary condition on the calculation of a flow with shock in a nozzle by the implicit scheme with blocks, for β = – 1 and CFL = 10.

—— Exact solution

---□--- Numerical solution with constant extrapolation of ρs and ρus at x = 10

---○--- Numerical solution with linear extrapolation of ρs and ρus at x = 10

In the other cases, the approximation is stable, as shown on Figure 5 (β = -1/2) or 6 (simplified scheme). So it is clear that the numerical experiments correlate very closely with the theoretical results.

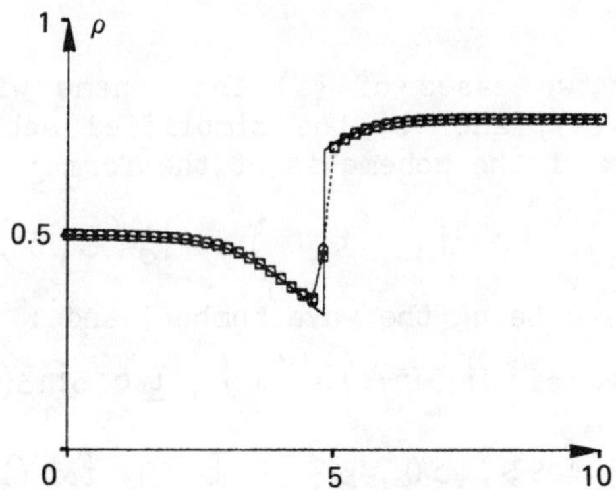

Fig. 5 – Effect of the boundary condition on the calculation of a flow with shock in a nozzle by the implicit scheme with blocks, for $\beta = -1/2$ and CFL = 10.

—— Exact solution

---□--- Numerical solution with constant extrapolation of ρs and $\rho u s$ at $x = 10$

---○--- Numerical solution with linear extrapolation of ρs and $\rho u s$ at $x = 10$.

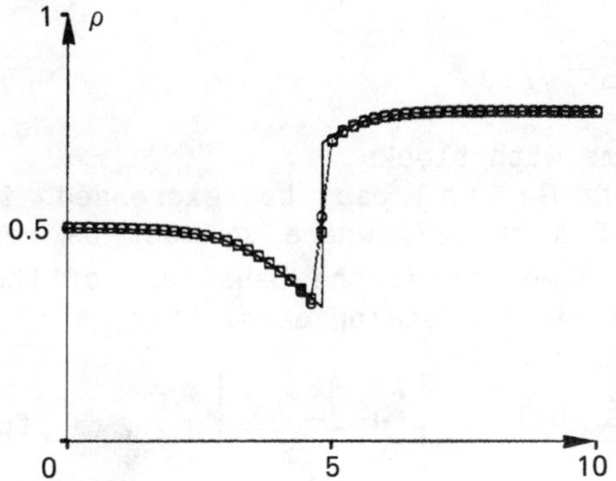

Fig. 6 – Effect of the boundary condition on the calculation of a flow with shock in a nozzle by the simplified implicit scheme, for $\beta = -1$ and CFL = 10.

—— Exact solution

--□-- Numerical solution with constant extrapolation of ρs and $\rho u s$ at $x = 10$

---○--- Numerical solution with linear extrapolation of ρs and $\rho u s$ at $x = 10$.

5. Convergence rate to the steady-state

5.1 The one-dimensional Euler equations

We consider here the implicit scheme when it is used as an iterative method to obtain a steady solution. As in the scalar case, we make a linear analysis of the amplification matrix of the scheme in order to study the rate of convergence to the steady state.

In the linear case, the system to be solved becomes :

$$w_t + A w_x = 0$$

where A is a constant matrix with real eigenvalues λ_i.

We consider the two cases of (a) the scheme with blocks in the implicit part, (3)-(4), and (b) the simplified scheme (3)-(12). The amplification matrix of the scheme is of the form :

$$G(\xi;\sigma) = H_1^{-1}(\xi;\sigma) \, . \, H_0(\xi;\sigma)$$

where $\xi = k\Delta x$, k being the wave number, and :

$$H_1(\xi;\sigma) = \begin{cases} I - (1-\cos\xi)[\beta(\sigma A)^2 + \gamma I] + i\,\alpha \sin\xi(\sigma A) & \text{for } (4) \\ [1 - (1-\cos\xi)\beta\sigma^2 \rho_A^2]\, I & \text{for } (12) \end{cases}$$

$$H_0(\xi;\sigma) = \begin{cases} H_1(\xi;\sigma) - (1-2\alpha)(1-\cos\xi)(\sigma A)^2 - i \sin\xi(\sigma A) & \text{for } (4) \\ H_1(\xi;\sigma) - (1-\cos\xi)(\sigma A)^2 - i \sin\xi(\sigma A) & \text{for } (12) \end{cases}$$

where

$$\sigma = \Delta t / \Delta x$$

a) Case of the scheme with blocks.
The eigenvalues of $G(\xi;\sigma)$ can be expressed in terms of the eigenvalues λ_q of A by (23) where η must be replaced by $\sigma\lambda_q$. Therefore for large time steps, the behaviour of the spectral radius of $G(\xi;\sigma)$ is the same as in the scalar case, that is :

$$\lim_{\sigma \to \infty} \rho[G(\xi;\sigma)] = \left| 1 + \frac{1-2\alpha}{\beta} \right| , \quad \text{for} \quad \xi \neq 0$$

In the present case, this limiting value depends only on the parameters of the scheme. As in the scalar case, the best choice to increase the convergence speed when σ is large is to satisfy (10), and the simplest scheme corresponds to (11).

b) Case of the simplified scheme.

The eigenvalues $g_q(\xi;\sigma)$ of $G(\xi;\sigma)$ are now :

$$g_q(\xi;\sigma) = h_{0,q}(\xi;\sigma) / h_{1,q}(\xi;\sigma)$$

where :

$$h_{0,q}(\xi;\sigma) = 1 - (1-\cos\xi)(\beta\rho^2(\sigma A) + (\sigma\lambda_q)^2) - i \sin\xi(\sigma\lambda_q)$$

$$\lambda_{1,q}(\xi;\sigma) = 1 - \beta\, \rho^2(\sigma A)(1-\cos\xi)$$

If σ tends to infinity, we find :

$$\lim_{\sigma\to\infty} |g_q(\xi;\sigma)| = \left|1 + \frac{1}{\beta}\left(\frac{\lambda_q}{\rho(A)}\right)^2\right| \qquad \text{for } \xi \neq 0$$

and therefore :

$$\lim_{\sigma\to\infty} \rho\left[G(\xi;\sigma)\right] = \operatorname{Max}\left(|1+\beta^{-1}|, |1+\beta^{-1}d^2|\right)$$

$$= \begin{cases} 1 & \text{if } \beta = -1/2 \\ |1-d^2| & \text{if } \beta = -1 \end{cases} \qquad \text{for } \xi \neq 0$$

where

$$d = \operatorname{Min}_q |\lambda_q| / \operatorname{Max}_q |\lambda_q|$$

Now, the limiting value of $\rho[G(\xi;\sigma)]$ may depend on the matrix A and in general we do not retain the good convergence properties of the scheme with blocks for $\beta = 2\alpha - 1$ (the value of d may be very close to zero, see [20]).

This shows that the use of large time steps in the simplified scheme is not a good strategy for improving the convergence speed .

<u>Numerical application</u>

We consider the same problem as in the previous section, that is the one-dimensional flow in a divergent nozzle, in the following two cases :

- a purely supersonic flow (all variables extrapolated at the exit of the nozzle)

- a flow with a shock (supersonic inflow, subsonic outflow) with the same boundary conditions as before.

For the supersonic flow, fig. 7 shows a comparison of the convergence histories obtained for :

- the explicit Lax-Wendroff scheme with CFL = 1
- the simplified implicit scheme with CFL = 100
- the implicit scheme with blocks with CFL = 100.

R_2 = Root mean square on the spatial mesh of $\Delta\rho/\Delta t$ in full line, $\Delta(\rho u)/\Delta t$ in dotted line and $\Delta(\rho E)/\Delta t$ in dashed line.

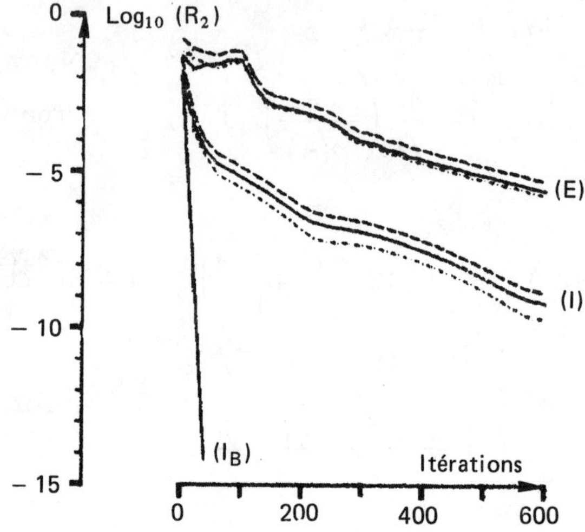

Fig. 7 — Convergence histories for a purely supersonic flow in a nozzle computed by
(E) the explicit Lax-Wendroff scheme, for CFL = 1
(I_B) the implicit scheme with blocks, for $\beta = -1$ and CFL = 100
(I) the simplified implicit scheme (no block), for $\beta = -1$ and CFL = 100.

One can appreciate the efficiency of the scheme with blocks which reaches a converged solution after less than 20 iterations. On the contrary, the simplified scheme is much less efficient. Let us also notice the good accuracy of the solution obtained with the implicit scheme for large values of the CFL number (Fig. 8).

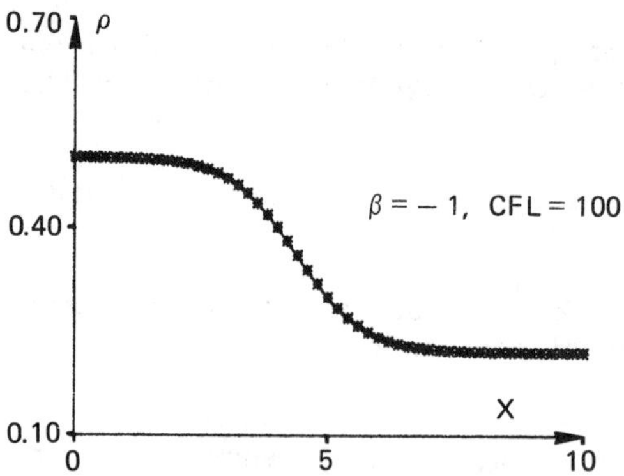

Fig. 8 — Calculation of a purely supersonic flow in a nozzle by the implicit schemes, for $\beta = -1$ and CFL = 100.
—— Exact solution
–x– Implicit scheme with blocks
–+– Simplified implicit scheme (no block).

For the flow with a shock, we compare on fig. 9 the convergence histories obtained for the same implicit schemes with CFL = 50. Again, for the scheme with blocks, the steady state is reached after less than 50 iterations. The shock structure of the numerical solution is monotone and spread over one mesh cell (fig. 10). Nevertheless, we remark a loss of accuracy behind the shock : we think that this is due to the discretization of the inhomogeneous term of the system, and are looking for possible improvements of this point.

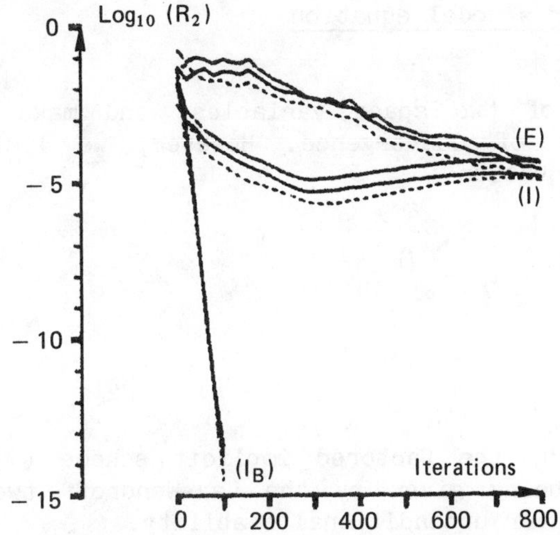

Fig. 9 – Convergence histories for a flow with shock in a nozzle computed by
(E) the explicit Lax-Wendroff scheme for CFL = 1
(I_B) the implicit scheme with blocks, for $\beta = -1$ and CFL = 50
(I) the simplified implicit scheme (no block), for $\beta = -1$ and CFL = 50.

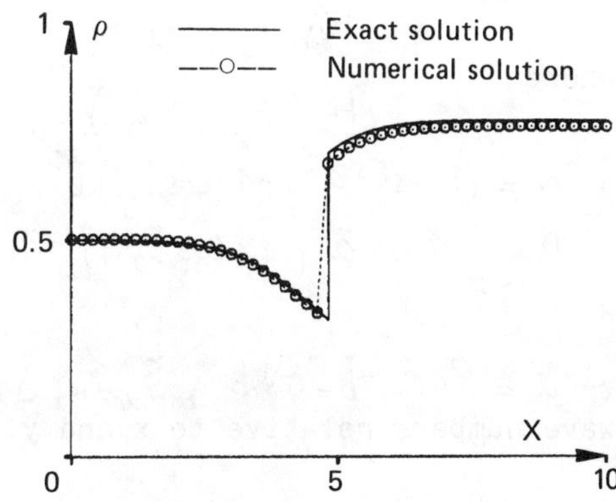

Fig. 10 – Calculation of a flow with shock in a nozzle by the implicit scheme with blocks, for $\beta = -1$ and CFL = 50.

All these results show that the best choice to make for steady calculations is certainly the implicit scheme with blocks and $\beta = -1$, with large CFL numbers, and this is in spite of the fact that the scheme with blocks needs about 3 times more computing time per iteration than the simplified scheme. We can also notice that, although the analysis of the rate of convergence of the schemes is based on a very simple argument, it provides a good estimation of the behaviour of the schemes, which is confirmed by the numerical experiments.

5.2 The two-dimensional case for a model equation

We now consider the case of two space variables, and make a similar analysis of the rate of convergence. However, we limit ourselves to the model scalar equation :

$$w_t + a w_x + b w_y = 0 \tag{38}$$

where a and b are constants.

First we approximate (38) by the factored implicit scheme (13) where $\rho_A = a$ and $\rho_B = b$, $\check{w}_i$ being given by the Lax-Wendroff two-dimensional scheme, and $\beta \leq -1$ for unconditional stability.

Making a Fourier transform of (13), we obtain the amplification factor g of the scheme :

$$g(\xi_x, \xi_y) = 1 - H(\xi_x, \xi_y) \tag{39}$$

where

$$H(\xi_x, \xi_y) = N/D$$

$$N = (1-\cos\xi_x)\bar{a}^2 + (1-\cos\xi_y)\bar{b}^2 + \sin\xi_x \sin\xi_y \,\bar{a}\bar{b} + i(\sin\xi_x \bar{a} + \sin\xi_y \bar{b})$$

$$D = [1-\beta\bar{a}^2(1-\cos\xi_x)][1-\beta\bar{b}^2(1-\cos\xi_y)]$$

with

$\bar{a} = \sigma^x a$, $\bar{b} = \sigma^y b$, $\xi_x = k_x \Delta x$ and $\xi_y = k_y \Delta y$, k_x and k_y being the wave numbers relative to x and y.

We study the behaviour of g when both σ^x and σ^y tend to infinity and we obtain :

$$\lim_{\sigma x, \sigma y \to \infty} |g(\xi_x, \xi_y)| = 1 \qquad \text{for } \xi_x \neq 0, \xi_y \neq 0$$

This means that we cannot expect an acceleration of convergence to steady state by using large time steps in the calculations. It is due to the factorization of the implicit part of the scheme which introduces a high-order term. This term destroys the convergence properties of the one-dimensional scheme. On the contrary, it is easy to see that the unfactored scheme would have the same behaviour than the one-dimensional scheme (also notice that it is stable under the same condition : $\beta \leq -1/2$ if $\alpha = \gamma = 0$).

The question which arises at this point is : is it possible to construct a two-dimensional scheme which retains at the same time a factored form of the implicit part and good steady- state convergence properties ?

In fact, we can see that it is the factorization <u>in the Δ-form</u> which destroys the convergence properties of the scheme. We are going to show now that if we make the factorization in the w-form, it is possible to retain a "one-dimensional" behaviour of the amplification factor. With this objective, we apply to (38) a two dimensional unfactored scheme corresponding to $\beta = -1$, that we write in a slightly modified form :

$$(40) \qquad w^{n+1} - \frac{1}{2}\left[\bar{a}^2 \delta_x^2 w^{n+1} + \bar{b}^2 \delta_y^2 w^{n+1}\right] = w^n - (\bar{a}\,\bar{\delta}_x w^n + \bar{b}\,\bar{\delta}_y w^n) + \frac{\Delta t^2}{2}\left(\frac{\bar{a}\bar{b}}{4}\,\bar{\delta}_x \bar{\delta}_y w^n\right)$$

Here the mixed derivatives are taken in an explicit way.

Now, if we factor the LHS of (40), we obtain :

$$(41) \qquad \begin{aligned} w^* - \frac{\bar{a}^2}{2}\delta_x^2 w^* &= w^n - (\bar{a}\,\bar{\delta}_x w^n + \bar{b}\,\bar{\delta}_y w^n) + \frac{\Delta t^2}{2}\frac{\bar{a}\bar{b}}{4}\bar{\delta}_x\bar{\delta}_y w^n \\ w^{n+1} - \frac{\bar{b}^2}{2}\delta_y^2 w^{n+1} &= w^* \end{aligned}$$

We have only added a fourth order term to (40). Let us write the amplification factor g' of (41), by using the same notations as before :

$$g'(\xi_x, \xi_y) = N'/D' \tag{42}$$

where

$$N' = (1 - i \sin\xi_x \bar{a})(1 - i \sin\xi_y \bar{b})$$

$$D' = (1 + (1-\cos\xi_x)\bar{a}^2)(1 + (1-\cos\xi_y)\bar{b}^2)$$

It is clear that (42) corresponds to the product of two one-dimensional amplification factors with $\alpha = \delta = 0$ and $\beta = -1$; this implies the following :

$$\lim_{\sigma^x, \sigma^y \to \infty} |g'(\xi_x, \xi_y)| = 0 \qquad \text{for } \xi_x \neq 0, \ \xi_y \neq 0$$

with :

$$|g'(\xi_x, \xi_y)| \leq 1$$

for all values of $\bar{a}$, $\bar{b}$, ξ_x and ξ_y.

We can extend the formulation (41) to the hyperbolic system :

$$w_t + f(w)_x + g(w)_y = 0$$

With the usual notations, we will write the scheme as :

$$\check{w} = w^n - [\sigma^x \bar{\delta}_x f(w^n) + \sigma^y \bar{\delta}_y g(w^n)] +$$

$$\frac{1}{2}\sigma^x \sigma^y \{\bar{\delta}_x [A(w^n)\bar{\delta}_y g(w^n)] + \bar{\delta}_y [B(w^n)\bar{\delta}_x f(w^n)]\},$$

$$w^* - \frac{\sigma^{x^2}}{2}\delta_x [A^2(\bar{w}^x)\delta_x w^*] = \check{w}$$

$$w^{n+1} - \frac{\sigma^{y^2}}{2}\delta_y [B^2(\bar{w}^y)\delta_y w^{n+1}] = w^*$$

where

$$A(w) = \frac{df}{dw}(w) \quad \text{and} \quad B(w) = \frac{dg}{dw}(w)$$

Of course this may not be the only possibility to extend the one-dimensional scheme without loosing its convergence properties (see, in the parabolic case, the work [39]).

6. Conclusion. An implicit Euler solver has been analyzed with the purpose of improving the calculation of steady-state solutions. The analysis has been first carried out for a single conservation law. For some value of the parameters ($\alpha = \gamma = 0$, $\beta = -1$) and large time steps, the implicit method has been shown to be robust (nonlinear stability property), accurate (exact solution obtained when $\Delta t \to \infty$) and efficient (optimal convergence rate to the steady state). Then for the Euler equations, the effect of the boundary conditions on the linear stability has been studied in one-space dimension and unconditional stability results have been obtained for the nozzle flow problem. Concerning the efficiency of the method for the Euler equations, it has been proven that it is necessary to keep the original version of the method involving a block-tridiagonal structure of the linear algebraic system and not to use the simplified scheme without the blocks. In addition, a new factorization has been proposed for the case of several space variables. A study of a model equation in two-space dimensions has shown that this factorization should be efficient. Research is in progress to obtain accurate solutions of the steady Euler equations with very large time steps similarly as is the case for a single conservation law.

REFERENCES

[1] R. BEAM and R.F. WARMING, An implicit finite-difference algorithm for hyperbolic systems in conservation law form, J. Comput. Phys., Vol. 22 (1976) pp. 87-110.

[2] R.F. WARMING and R. BEAM, On the construction and application of implicit factored schemes for conservation laws, SIAM-AMS Proceed., Vol. 11 (1978), pp. 85-120.

[3] J.L. STEGER, Implicit finite difference simulation of flow about arbitrary two-dimensional geometries, AIAA J., Vol. 16 (1978) pp. 679-686.

[4] A. LERAT, Une classe de schémas aux différences implicites pour les systèmes hyperboliques de lois de conservation, Comptes-rendus Acad. Sc. Paris t. 288A (1979), pp. 1033-1036.

[5] A. LERAT, Sur le calcul des solutions faibles des systèmes hyperboliques de lois de conservation à l'aide de schémas aux différences, Publication ONERA n° 1981-1 (1981). English transl. ESA TT 762.

[6] A. LERAT, J. SIDES and V. DARU, An implicit finite-volume method for solving the Euler equations, Lecture Notes in Physics, Vol. 170 (1982), pp. 343-349.

[7] R.W. MacCORMACK, Numerical solution of compressible viscous flows at high Reynolds numbers, Lecture Notes in Physics, Vol. 148 (1981), pp. 254-267.

[8] W. KORDULLA and R.W. MacCORMACK, Transonic flow computation using an explicit-implicit method, Lecture Note in Physics, Vol. 170 (1982), pp. 286-295.

[9] F. CASIER, H. DECONINCK and C. HIRSCH, A class of central bidiagonal schemes with implicit boundary conditions for the solution of Euler's equations, AIAA paper n° 83-0126 (1983).

[10] J.L. STEGER and R.F. WARMING, Flux Vector Splitting of the inviscid Gasdynamic Equations with Application to finite difference methods, J. Comput. Phys. Vol. 40 (1981), pp. 263-293.

[11] P.G. BUNING and J.L. STEGER, Solution of the Two-Dimensional Euler Equations with Generalized Coordinate Transformation Using Flux Vector Splitting, AIAA Paper n° 82-0971 (1982).

[12] T.H. PULLIAM, D.C. JESPERSEN and R.E. CHILDS, An Enhanced Version of an Implicit Code for the Euler Equations, AIAA Paper n° 83-0344 (1983).

[13] T.J. COAKLEY, Implicit Upwind Methods for the Compressible Navier-Stokes Equations, AIAA Paper n° 83-1958 (1983).

[14] A. DADONE and M. NAPOLITANO, An Implicit Lambda Scheme, AIAA J. Vol. 21 (1983), pp. 1391-1399.

[15] G. MORETTI, A fast Euler Solver for steady flows, AIAA Paper n° 83-1940 (1983).

[16] S.F. WORNOM, Implicit conservative characteristic modeling schemes for the Euler equations. A new approach, AIAA Paper n° 83-1939 (1983).

[17] W.A. MULDER and B. Van LEER, Implicit upwind methods for the Euler equations, AIAA Paper n°83-1930 (1983).

[18] H.C. YEE, R.F. WARMING and A. HARTEN, Implicit Total Variation Diminushing (TVD) schemes for steady-state calculations, AIAA Paper n° 83-1902 (1983).

[19] A. LERAT, Implicit methods of second order accuracy for the Euler Equations, AIAA Paper n° 83-1925 (1983).

[20] A. LERAT, J. SIDES and V. DARU, Efficient computation of steady and unsteady transonic flows by an implicit Euler solver, in "Advances in Computational Transonics", Habashi Ed., Pineridge Press (1984).

[21] P.D. LAX and B. WENDROFF, Systems of conservation laws, Comm. Pure Appl. Math., Vol. 13, (1960), pp. 217-237.

[22] B. GUSTAFSSON, H.O. KREISS and A. SUNDSTROM, Stability theory of difference approximations for mixed-initial-boundary value problems II, Math. Comp., Vol. 26 (1972), pp. 649-686.

[23] H.O. KREISS, Difference approximations for the initial-boundary value problem for hyperbolic differential equations, in "Numerical solutions of non-linear differential equations", Greenspan Ed., J. Wiley (1966), pp. 141-166.

[24] H.O. KREISS, Stability theory for difference approximations of mixed initial boundary value problems I., Math. Comp., Vol. 22 (1968), pp. 703-714.

[25] S. OSHER, Systems of difference equations with general homogeneous boundary conditions, Trans. Amer. Math. Soc., 137 (1969), pp. 177-201.

[26] S. OSHER, Stability of difference approximations of dissipative type for mixed initial-boundary value problems I, Math. Comp. Vol. 23 (1969), pp. 335-340.

[27] H.O. KREISS, Difference approximations for initial boundary value problems, Proc. Roy. Soc. London. A. 323 (1971), pp. 255-261.

[28] H.O. KREISS, Boundary conditions for difference approximations of hyperbolic differential equations. AGARD LS n° 64 (1973).

[29] B. GUSTAFSSON and H.O. KREISS, Boundary Conditions for time dependent problems with an artificial boundary, J. Comput. Phys. Vol. 30 (1979), pp. 333-351.

[30] H.C. YEE, R.M. BEAM and R.F. WARMING, Stable boundary approximations for a class of implicit schemes for the one-dimensional inviscid equations of gasdynamics, AIAA Paper n° 81-1009 (1981).

[31] R.M. BEAM, R.F. WARMING and H.C. YEE, Stability Analysis of Numerical Boundary Conditions and Implicit Difference Approximations for Hyperbolic Equations, J. Comput. Phys., Vol. 48 (1982), pp. 200-222.

[32] R.F. WARMING, R.M. BEAM and H.C. YEE, Stability of Difference Approximations for Initial-Boundary Value Problems, Proceed. 3nd International Symp. Numer. Meth. Eng., 14-16 March, Paris (1983).

[33] B. GUSTAFSSON, The Choice of Numerical Boundary Conditions for Hyperbolic Systems, J. Comput. Phys. Vol. 48 (1982), pp. 270-283.

[34] H.C. YEE, Numerical approximation of boundary conditions with applications to inviscid equations of gasdynamics, NASA TM n° 81-265 (1981).

[35] D. GOTTLIEB and E. TURKEL, Boundary Conditions for multistep finite difference methods for time dependent equations, J. Comput. Phys. Vol. 26 (1978), pp. 181-196.

[36] W.M. COUGHRAN, On the approximate solution of hyperbolic initial boundary value problems, Thesis, Stanford University (1980)

[37] G.R. SHUBIN, A.B. STEPHENS and H.M. GLAZ, Steady shock tracking and Newton's method applied to one-dimensional duct flow, J. Comput. Phys. vol. 39 (1981), pp. 364-374.

[38] V. DARU, Contribution à l'étude d'une méthode numérique implicite pour résoudre les équations d'Euler, Thèse de 3ème cycle, Université Paris VI, Novembre 1983.

[39] S. ABARBANEL, D. DWOYER and D. GOTTLIEB, Improving the convergence rate of parabolic ADI Methods, AIAA Paper n° 83-1897 (1983).

ACCELERATION TO A STEADY STATE FOR THE EULER EQUATIONS

ELI TURKEL*

Abstract. A multi-stage Runge-Kutta method is analyzed for solving the Euler equations exterior to an airfoil. Highly subsonic, transonic and supersonic flows are evaluated. Various techniques for accelerating the convergence to a steady state are introduced and analyzed.

1. Introduction. In this study we shall discuss ways of accelerating the approach to a steady state for the Euler equations. We first consider the time-dependent equations. However, since we are only interested in the steady state we shall feel free to alter the equations in any way that does not affect the steady state. Since the Euler equations constitute a nonlinear system of equations, it usually can not be proven that the modified system even approaches a steady state. Instead, most of the analysis will refer to linearized versions of the equations. Assuming that the steady state is unique once the modified equations reach a steady state, it must be the steady state of the original equations. We shall concentrate on the inviscid equations, though most of the techniques are also applicable to the Navier-Stokes equations.

We assumed that a body-fitted curvilinear grid has been constructed from some package. All that we require is the (x,y) coordinates of each zone. The finite difference equations will be derived using a finite volume approach. Though the cells can be of any shape in such an approach we shall assume that all cells are quadrilaterals. Near the trailing edge some zones may degenerate and the finite volume approach still holds. The use of mapping formulas rather than finite volumes yields essentially the same finite difference formulas. The finite volume approach naturally

*Institute for Computer Applications in Science and Engineering, NASA Langley Research Center, Hampton, VA 23665 and Tel-Aviv University, Tel-Aviv, Israel under NASA Contract Nos. NAS1-16394 and NAS1-17130.

associates the value of the dependent variables with zonal averages. These averages are associated with values at the cell center. Hence, all boundaries are placed along cell faces. With a mapping formulation the natural implementation would be to place the variables at grid nodes. Furthermore, using a finite volume approach, the Jacobian of the transformation is identified with the area of a cell. This leads to a slightly different formulation for the Jacobian than would be natural for a finite difference approach. The finite volume formulation leads to a straightforward generalization for axisymmetric coordinates.

For convenience of notation we shall only consider two-dimensional flows. The Euler equations can then be written as

$$w_t + f_x + g_y = 0, \tag{1.1}$$

$$\begin{aligned} w &= (\rho, \rho u, \rho v, E)^T, \\ f &= (\rho u, \rho u^2 + p, \rho uv, \rho uh)^T, \\ g &= (\rho v, \rho uv, \rho v^2 + p, \rho vh)^T, \\ h &= \frac{E + p}{\rho}. \end{aligned} \tag{1.2}$$

Integrating (1.1) inside a cell and using the divergence theorem we get

$$\frac{\partial}{\partial t}\iint_D w\,dV + \int_{\partial D}(f\,dy - g\,dx) = 0, \tag{1.3}$$

where we have chosen D to be a quadrilateral. Let two adjacent sides of the quadrilateral be given by ξ = constant and η = constant. The component of velocity perpendicular to the curve ξ = constant, denoted by q, is given by

$$q = \frac{d\xi}{dt} = \xi_x u + \xi_y v = \frac{y_\eta u - x_\eta v}{J}, \qquad J = x_\xi y_\eta - x_\eta y_\xi, \tag{1.4a}$$

while the velocity component along the curve ξ = constant, denoted q_{11}, is given by

$$q_{11} = \frac{y_\eta v + x_\eta u}{J}. \tag{1.4b}$$

Similarly, letting r be the velocity component perpendicular to η = constant and r_{11} parallel to η = constant, we have

$$r = \eta_x u + \eta_y v = (-y_\xi u + x_\xi v)/J, \tag{1.4c}$$

$$r_{11} = (y_\xi v + x_\xi u)/J, \tag{1.4d}$$

r is proportional to q_{11}, only if the grid is orthogonal, i.e., $x_\xi y_\eta + y_\xi x_\eta = 0$.

Letting w_A denote the cell average

$$w_A = \frac{\iint w dv}{\iint dv}, \tag{1.5}$$

we replace (1.3) with

$$\frac{d}{dt}(w_A \cdot V) + \sum_{i=1}^{4} \int_{S_i} (f dy - g dx) = 0, \tag{1.6}$$

where S_i are the sides of the quadrilateral and V is its area. We evaluate the line integrals using the midpoint rule. Normalizing the mesh so that the length of each quadrilateral is one, we replace (1.6) by the approximation

$$\frac{d}{dt}(w_A \cdot V) = - \sum_{\xi=\text{const}} (f y_\eta - g x_\eta) + \sum_{\eta=\text{const}} (f y_\xi - g x_\xi). \tag{1.7}$$

With second-order accuracy we can find w at the cell face by averaging w from the cell centers. This averaging is done in FLO52ST without accounting for the difference in volumes of neighboring cells. Given w at the cell face, one can then calculate the fluxes. It is preferable to average w rather than average the fluxes to help couple the even and odd points. For efficiency the pressure is also averaged rather than computed from w. The error in this procedure is of the same order as the error in the scheme. Several numerical checks have confirmed that this pressure averaging does not introduce any additional errors.

In the following sections we discuss ways to integrate (1.8) in time and to accelerate the convergence to a steady state. To measure this acceleration we need a way of deciding when a numerical steady state has been reached. In [8] the measure used was $\Delta\rho/\Delta t$. However, once acceleration techniques are used, it is important to measure the true residual, i.e., the right-hand-side of (1.8). For exterior problems the mesh is finer near the body and is coarser in the far field. As such the maximum residual usually occurs in the far field where the volumes are large. This is true even though the true flux is small in the far field even relative to the zonal areas. An alternative is to normalize the true residual by the cell area which tends to emphasize the zones near the airfoil.

2. Runge-Kutta Methods. In 1960 Lax and Wendroff [11] introduced a one-step method which was second-order in space and time. Richtmyer [15] suggested a two-step method which is linearly equivalent to the Lax-Wendroff scheme. The advantage of the two-step scheme is that only flux evaluations are necessary rather than matrix multiplications.

A disadvantage of these two-step schemes is that the two steps are different, complicating the coding. A more important disadvantage is that the Courant number is one, even though two steps are used. The leapfrog method requires only one step and has the same stability requirement. Hence, the leapfrog scheme is twice as efficient as a two-step Lax-Wendroff method. However, the leapfrog scheme is nondissipative and hence not useful for shocked flows. Nevertheless, we are interested in multistage schemes which are more efficient than the standard two-stage schemes. Graves [7] used a three-stage Runge-Kutta scheme for the Navier-Stokes equations. Van der Houwen [21], [22] analyzed multistage schemes for both hyperbolic and parabolic problems. A fourth-order Runge-Kutta scheme for the Euler equations was popularized in [8]. For simplicity we only consider two-space dimensions although the extension to higher dimensions is straightforward. Let

$$w_t = f_x + g_y. \tag{2.1}$$

An N stage Runge-Kutta scheme is given by

$$w^{(k)} = w^{(0)} + \alpha_k \Delta t\left(D_x f^{(k-1)} + D_y g^{(k-1)}\right), \tag{2.2}$$

with $w^{(0)} = w^n$, $w^{(n)} = w^{n+1}$, $\alpha_N = 1$. We note that this method is, in general, only first-order accurate in time even when the amplification matrix agrees with that of a higher-order method. In many cases one can achieve second-order accuracy in time. When one is only interested in the steady state solution, this is not a drawback. For time-dependent problems one may wish to have higher accuracy. One can then replace (2.2) with a true higher order Runge-Kutta method. The advantage of (2.2) is that only two levels of storage are required, i.e., $w^{(0)}$ and $w^{(k)}$ at each stage.

A fundamental question is how to choose the parameters α_k, $k=1,\cdots,N-1$. To simplify the question, we only consider the linear problem with constant coefficients. We then Fourier transform (2.2) and consider the following amplification matrix

$$G = 1 + \beta_1 z + \beta_2 z^2 + \cdots + \beta_N z^n, \tag{2.3}$$

z is the Fourier transform of $\Delta t(D_x w + D_y w)$. Using central differencing in x and y, then $z = i \cdot \Delta t/\Delta x \cdot \lambda(A \sin\theta + B \sin\phi)$ where $\lambda(A)$ denotes an eigenvalue of A. θ, ϕ are the Fourier variables, $\Delta x = \Delta y$, $A = \partial f/\partial w$, $B = \partial g/\partial w$. The parameters β_i are given by

$$\begin{aligned} \beta_1 &= 1, \\ \beta_2 &= \alpha_{N-1}, \\ \beta_k &= \beta_{k-1}\,\alpha_{N-k+1}, \qquad k = 3,\cdots,N. \end{aligned} \tag{2.4}$$

One possibility is to choose the β_i so that Δt is as large as possible. If the iteration is close, in some sense, to being time accurate, this implies that we advance in time as fast as possible. This requires that $|z|$ be as large as possible while requiring that $G^* G \leq 1$. In general, this Δt is only required for some frequencies which depend on the scheme and the matrices A and B. Using central differences the values of z lie along the imaginary axis while for an upwinded scheme, z will lie on some curve in the complex plane.

Using a central difference scheme the maximum time step is given by

$$\frac{\Delta t}{\Delta x} \leq \frac{K}{\max\limits_{\theta,\phi} \rho(A \sin\theta + B \sin\phi)}, \tag{2.5}$$

where $\rho(A)$ is the spectral radius of A and K is the maximum that z can achieve in (2.3). Vichnevetsky [25] (see also [16]) has shown that when z is purely imaginary that

$$|z| \leq N-1 \quad \text{for an } N \text{ stage scheme.} \tag{2.6}$$

This maximum can be achieved and the formula is given by [16], [21]

$$G = P_N(z) = i^{N-1}\, T_{N-1}\left(\frac{-iz}{N-1}\right) + \frac{i^N}{2}\left[T_N\left(\frac{-iz}{N-1}\right) - T_{N-2}\left(\frac{-iz}{N-1}\right)\right]. \tag{2.7}$$

For N odd, this formula coincides with that given by Van der Houwen [21] and the formula is second-order accurate. For N even, the formula is only first-order accurate.

In practice, an artificial viscosity is needed to stabilize the scheme (see section 4). This can be modelled, in one dimension, by

$$w_t = w_x - \varepsilon w_{xxxx}. \tag{2.8}$$

Using central differences, z now corresponds to $z = i \sin\theta - 16\varepsilon \sin^4 \frac{\theta}{2}$. $|z|$ now lies in the left half plane. Plotting the stability region of (2.3), (2.7) we see that the stability region contains the imaginary axis up to N-1 and also includes portions of the right-half plane. The stability region also includes part of the negative real axis up to z of about two to three (depending on N). Hence, as long as ε is sufficiently small the previous results are valid. At $\theta = \pi$, $z = -16\varepsilon$ and so the stability is governed only by the artificial viscosity. A similar analysis holds when the Navier-Stokes terms are added to the differential equation.

In the previous analysis we assumed that the artificial viscosity is reevaluated at each stage. In practice the computation of the artificial viscosity is expensive and so the same artificial viscosity is used for each stage within a cycle. Hence, (2.8) is approximated by

$$w^{(k)} = w^{(0)} + \alpha_k\, \Delta t\left[D_0\, w^{(k-1)} - \varepsilon D_+^2\, D_-^2\, w^{(0)}\right]. \tag{2.9}$$

Let z be the Fourier transform of $\Delta t \cdot D_0 w$ and let y be the Fourier transform of Δt (artificial viscosity). Then the amplification factor is

$$G = 1 - (z+y)\,\beta_1 + \beta_2 z + \cdots + \beta_{N-1} z^{N-2} + \beta_N z^N + y. \tag{2.10}$$

Therefore, the previous analysis does not hold since G depends on both z and y and not on one complex number. In various computational trials the optimal parameters (2.7) worked when the artificial viscosity was reevaluated at each stage. However, for N greater than four, the optimal parameters were not stable when the viscosity was frozen. It is not clear how much of these details are dependent on the exact formulation of the artificial viscosity.

When the artificial viscosity is reevaluated at each stage, then the use of an N stage formula used twice is equivalent to some 2N stage formula. Hence, it is always more optimal to use a higher stage formula. However, when the artificial viscosity is frozen at each stage, this no longer need be true. With the second-order central difference scheme we have seen, (2.6), that the Courant number for an N stage scheme has a maximum of N-1. Thus, the efficiency per stage is (N-1)/N. Hence, for N large, we approach the efficiency of the leapfrog method. At N = 10, we already have 90% efficiency and there is not much benefit in using a higher stage method. However, since this scheme requires the reevaluation of the artificial viscosity at each stage, these higher stage schemes are not efficient. An alternative is to evaluate the viscosity M times within an N stage scheme. The relationship between M and N to achieve maximum stability is not known. Furthermore, in many cases, robustness, i.e., including sections of the left-half of the complex plane, is more important than maximal time steps.

Until now we have assumed that the stability condition is mainly governed by z near the imaginary axis, i.e., central differences with a small artificial viscosity. Equations with variable coefficients or nonlinearities require us to consider perturbations off the imaginary axis even without the presence of viscous effects, see [15]. For parabolic equations the appropriate z are along the negative real axis. The optimal parameters for this case were first considered in [17]. This is given by

$$G = T_n\left(1 + z/N^2\right), \qquad -2N^2 \le z \le 0 \tag{2.11}$$

for an N stage scheme. However, this formula is not stable for large negative z if z is slightly off the real axis. These perturbations can be caused by variable coefficients, lower order terms or boundary conditions. One type of perturbation is considered in [22]. Another perturbation of (2.11) is

$$G = (1 - \varepsilon)T_N(y) + \varepsilon\, y\, T_N(y)/N^2, \quad \text{with} \quad y = 1 + z/N^2. \tag{2.12}$$

The stability range of (2.12) is slightly smaller than but contains a large region off the real axis when $\varepsilon > 0$. In both cases the stability region for an N stage scheme is proportional to N^2. Hence, the more stages that are used the more efficient the method is. This is in contradistinction to the hyperbolic case where the efficiency per stage quickly approaches its upper bound which is given by the leapfrog scheme. Hence, when the number of stages for a parabolic problem is large, we can approach the time steps of an implicit scheme while retaining the advantages of an explicit scheme.

Using an upwinded scheme for a hyperbolic equation also results in eigenvalues that are not near the imaginary axis except for the longest of wave lengths. When one considers a pure one-sided scheme together with boundary conditions then the matrix is upper (or lower) diagonal and so all the eigenvalues lie along the negative real axis. Some experiments with this case are presented in [20].

In our analysis we have assumed that one approaches a steady state fastest by using the largest possible time step. This is intuitively obvious if one is using a time-like iteration procedure. However, when the iteration process is not time-accurate, the criterion for the optimal α_i need not be connected with the time step. This is well known for A.D.I. type schemes. In section 6 we shall show that when one uses residual smoothing, that the best strategy is not to choose the largest possible time step, even in one-space dimension. Recently, Jameson [9] has developed a multigrid code using the Runge-Kutta scheme as the smoothing operator. For this method one suspects that one would choose the α_i so as to damp the high frequencies. Hence, one choice for the parameters, in one dimension, is to choose the α_i so that

$$\min_{\alpha_i} \max_{\theta_0 \leq |\theta| \leq \pi} |G^* G(\theta)|, \tag{2.13}$$

for some θ_0, Δt and with the additional constraint that $|G^* G| \leq 1$, $-\pi \leq \theta \leq \pi$. Since there is not a well-developed theory for the multigrid method for hyperbolic equations, it is not clear that (2.13) yields the optimal parameters. Computations indicate that one wishes some combination of large time steps together with good dampling properties at high frequencies for the multigrid procedure to be efficient.

3. Time Step in Generalized Coordinates. We consider the two-dimensional equation

$$w_t + Aw_x + Bw_y = 0, \tag{3.1}$$

where A and B are constant matrices which represent the gradient of the fluxes appearing in (2.1). We shall only consider Runge-Kutta methods in time and second-order central differences in space. The effect of the artificial viscosity on the time step is ignored (see previous section for a more complete discussion). From (2.5) we see that for any multistep Runge-Kutta method that the stability criterion is of the form

$$\Delta t \cdot d \leq K, \tag{3.2}$$

where d is the maximum (over θ and ϕ) spectral radius of $D = A \sin\theta + B \sin\phi$. The constant K depends on the parameters of the Runge-Kutta scheme.

The Euler equations in general coordinates (ξ,η) are

$$(JW)_t + f_\xi + g_\eta = 0, \tag{3.3}$$

where

$$w = \begin{bmatrix} \rho \\ \rho u \\ \rho v \\ E \end{bmatrix} \qquad f = \begin{bmatrix} \rho q \\ \rho q u + y_\eta\, p \\ \rho q v - x_\eta\, p \\ q(E+p) \end{bmatrix} \qquad g = \begin{bmatrix} \rho r \\ \rho r u - y_\xi\, p \\ \rho r v + x_\xi\, p \\ r(E+p) \end{bmatrix} \tag{3.4}$$

$$J = x_\xi\, y_\eta - x_\eta\, y_\xi, \tag{3.5}$$

$$q = y_\eta\, u - x_\eta\, v, \qquad r = x_\xi\, v - y_\xi\, u. \tag{3.6}$$

The Jacobian J is equal to the volume of the cell that appears in the finite volume approach (see section 1). Let

$$w = q \sin\theta + r \sin\phi, \tag{3.7}$$

$$a = y_\eta \sin\theta - y_\xi \sin\phi, \qquad b = x_\xi \sin\phi - x_\eta \sin\theta, \tag{3.8}$$

$$h = \frac{E + p}{\rho}, \tag{3.9}$$

$$S^2 = \frac{(\gamma - 1)}{2}(u^2 + v^2), \tag{3.10}$$

then

(3.11)

$$D = \begin{bmatrix} 0 & a & b & 0 \\ aS^2 - uw & w - (\gamma-2)au & bu - (\gamma-1)av & (\gamma-1)a \\ bS^2 - vw & av - (\gamma-1)bu & w - (\gamma-2)bv & (\gamma-1)b \\ -w(h-S^2) & ah - (\gamma-1)wu & bh - (\gamma-1)wv & \gamma w \end{bmatrix}$$

Let

(3.12)

$$T = \begin{bmatrix} \frac{S^2}{\rho c} & \frac{-(\gamma-1)u}{\rho c} & \frac{-(\gamma-1)v}{\rho c} & \frac{\gamma-1}{\rho c} \\ -\frac{u}{\rho} & \frac{1}{\rho} & 0 & 0 \\ -\frac{v}{\rho} & 0 & \frac{1}{\rho} & 0 \\ S^2 - c^2 & -(\gamma-1)u & -(\gamma-1)v & \gamma-1 \end{bmatrix}$$

Then

$$D_0 = TDT^{-1} = \begin{bmatrix} w & ac & bc & 0 \\ ac & w & 0 & 0 \\ bc & 0 & w & 0 \\ 0 & 0 & 0 & w \end{bmatrix} \tag{3.13}$$

where a,b are given by (3.8) and w by (3.7). Hence,

$$d = |w| + \sqrt{a^2 + b^2}\, c. \tag{3.14}$$

Letting $\Delta\xi = \Delta\eta = 1$, then the time restriction is of the form

(3.15)

$$\Delta t \leq \frac{K\cdot J}{|w| + \sqrt{a^2+b^2}\,c} \leq \frac{K\cdot J}{|q| + |r| + \sqrt{x_\xi^2+y_\xi^2+x_\eta^2+y_\eta^2 + 2|x_\xi x_\eta+y_\xi y_\eta|}\,c},$$

where r and q are given by (3.6). In the code FLO52ST the stability criterion is implemently slightly differently as

$$\Delta t \leq \frac{K\cdot J}{|q| + |r| + \left(\sqrt{x_\xi^2 + y_\xi^2} + \sqrt{x_\eta^2 + y_\eta^2}\right)c}. \tag{3.16}$$

For most exterior problems a highly stretched mesh is used. In these cases the time step is governed by the area of the smallest cell which is much smaller than the area of the cells in the far field. To avoid this difficulty we use a different time step in each zone. This destroys the time accuracy of the solution but accelerates the convergence to the steady state. This local time step was first used in [13].

To see the effect of using a local time step we consider the radially symmetric wave equation

$$u_{tt} = \frac{1}{r}(ru_r)_r. \tag{3.17}$$

Discretizing the time variable we have

$$\lim_{\Delta t\to 0} \frac{u_j^{n+1} - 2u_j^n + u_j^{n-1}}{(\Delta t)^2} = \frac{1}{r}(ru_r)_r. \tag{3.18}$$

Let Δt be the local time step $(\Delta t)_j$ and rewrite (3.18) as

$$\lim_{\Delta t\to 0} \frac{u_j^{n+1} - 2u_j^n + u_j^{n-1}}{(\Delta t_{min})^2} = \frac{c^2(r)}{r}(ru_r)_r, \qquad c(r) = \frac{(\Delta t)_j}{(\Delta t)_{min}}. \tag{3.19}$$

Therefore, using a local time step is equivalent to introducing an artificial wave speed that increases as one goes to the far field. Hence, the further a wave goes towards infinity, the faster it goes. As an example, we consider a mesh that increases exponentially with r. Then $(\Delta t)_j$ is proportional to r and so c(r) = r. Then (3.19) becomes

$$u_{tt} = r(ru_r)_r. \tag{3.20}$$

Let s = log(r), (3.20) becomes

$$u_{tt} = u_{ss}. \tag{3.21}$$

If we only allow outgoing waves, then the solution to (3.21) is

$$u = f(s-t) = f(\log(r) - t). \tag{3.22}$$

Hence, if we begin with a wave of compact support at r = 1, t = 0, then at time t_0, the wave is centered at $r = \exp(t_0)$, i.e., the wave has moved exponentially fast.

4. Artificial Viscosity. The Runge-Kutta method with central differencing in space has two difficulties. The first difficulty is that the highest frequency is not damped, i.e., for a linear problem the neighboring points decouple. This odd-even decoupling prevents the possibility of driving the residual to machine zero. With the nonlinear equations the variables are averaged at the cell faces before forming the fluxes. This nonlinearity couples all the neighbors together. Nevertheless this coupling is weak and convergence to a steady state can be slow. In order to accelerate the convergence a fourth-order linear viscosity is added to each equation.

A second difficulty with central differences is that it does not enforce an entropy decrease across shocks. As such the central difference may converge to the wrong solution. We attempt to enforce the entropy condition by introducing an artificial viscosity. The fourth-order viscosity does not help near shocks. In fact there are theoretical indications that the fourth-order viscosity can be a destabilizing factor [14]. These observations are confirmed by numerical experiments. Hence the fourth-order viscosity is turned off in the neighborhood of shocks. In order to prevent oscillations at the shock an additional viscosity is added which behaves as a second derivative. This is in the spirit of the Navier-Stokes equations and seems to yield shocks without any overshoots. This viscosity is a nonlinear viscosity so that the formal order of accuracy is not affected by the addition of the artificial viscosity.

We wish the artificial viscosity to accomplish two contradictory purposes. We want these terms to accelerate the convergence to a steady state. This implies that we should choose the viscosity as large as possible without violating any stability restrictions. On the other hand, we wish the viscosity to be as small as feasible so as not to affect the accuracy of the solution. Formally, the viscosity is of higher order than the truncation error. However, for a finite mesh, increasing the viscosity will decrease the accuracy of the steady state.

In section 7 we will introduce an upwinded scheme that does not require the use of an artificial viscosity. The use of an artificial viscosity has both advantages and disadvantages. The basic disadvantage is its artificiality. An artificial viscosity is not aesthetically appealing. Furthermore, there are invariably constants which must be adjusted for each case. This makes the code less robust. An upwind scheme has a built-in viscosity which should automatically adjust itself to each situation. The advantage of an artificial viscosity is its flexibility. We can use the freedom of arbitrary constants or functions to tailor the code to an individual problem. This requires more work on the part of the user but it can be beneficial. The upwind schemes are more automatic. However, in cases where one does not want any viscosity, it is difficult to turn off since it is built into the algorithm.

In order to introduce the artificial viscosity into the Runge-Kutta scheme we modify (2.2) and get

$$w^{(k)} = w^{(0)} + \alpha_k \, \Delta t\left(Lw^{(k-1)} + V_2 - V_4\right), \tag{4.1}$$

where Lw represents the approximation to the Euler equations. V_4 is given by

$$V_4 = \frac{C_4}{128}\left[D_{+\xi}\, D_{-\xi}\left(\frac{J}{\Delta t}\, D_{+\xi}\, D_{-\xi}\, w\right) + D_{+\eta}\, D_{-\eta}\left(\frac{J}{\Delta t}\, D_{+\eta}\, D_{-\eta}\, w\right)\right], \tag{4.2}$$

and V_2 will be presented later. Δt is the local time step while J is the area of the zone. C_4 is an arbitrary constant (usually about 0.2) while D_+ and D_- represent forward and backward differences respectively. ξ and η are the curvilinear coordinates and we have assumed $\Delta\xi = \Delta\eta = 1$. As one approaches the boundary, V_4 can no longer be generated by (4.2). One alternative is to extrapolate the variables to an artificial position across the boundary and then to use (4.2). This is equivalent to using a one-sided approximation to (4.2). The viscosity added by V_4 is a completely dissipative mechanism. If we look at the differential level (4.2), (with V_2 = 0) leads to an approximation of

$$w_t = Lw - (Kw_{xx})_{xx}, \qquad K = \frac{c_4 J}{\Delta t}. \tag{4.3}$$

Multiplying (4.3) by w and integrating over all space we get

$$\frac{1}{2}\frac{d}{dt}\int |w|^2\, dx = \int (w,\ Lw)dx - \int w(Kw_{xx})_{xx}\, dx. \tag{4.4}$$

Integrating the last integral by parts twice and ignoring boundaries we get

$$\frac{1}{2}\frac{d}{dt}\int w^2\, dx = \int (w,\ Lw)dx - \int K(w_{xx})^2\, dx. \tag{4.5}$$

Hence, as long as K is positive the viscosity terms decrease the total energy. An alternative to (4.2) is to use

$$V_4 = \frac{c_4}{128}\left[D_{+\xi}\left(\frac{J}{\Delta t} D_{+\xi}\, D_{-\xi}^2\, w\right) + D_{+\eta}\left(\frac{J}{\Delta t} D_{+\eta}\, D_{-\eta}^2\, w\right)\right]. \tag{4.6}$$

This version contains both dissipative and dispersive characteristics. The same technique can be used on the finite difference level using summation-by-parts. One can then also include boundary terms to see their effect. Eriksson and Rizzi [4] choose boundary conditions so as to maximize the dissipation of the artificial dissipation. We shall see later that this may introduce errors into the steady state approximation.

The form of V_2 is given by

$$V_2 = \frac{c_2}{4}\left[D_{+\xi}\, \theta^x\, D_{-\xi}\, w + D_{+\eta}\, \theta^y\, D_{-\eta}\, w\right]. \tag{4.7}$$

θ is a switch which measures the gradients of the flow. In smooth regions θ should be of the order of Δ^2 while near shocks θ should be of order unity. Hence, this viscosity is of fourth-order in smooth regions and does not affect the order of the scheme. θ^x is given by

(4.8) $$\theta^{x}_{j+1/2,k} = \max \bar{\theta}_{j,k}\ \bar{\theta}_{j+1,k} .$$

Two possibilities for $\bar{\theta}$ are

(4.9) $$\bar{\theta}_{j,k} = \kappa \frac{|p_{j+1,k} - 2p_{j,k} + p_{j-1,k}|}{|p_{j+1,k} + 2p_{j,k} + p_{j-1,k}|}, \qquad \kappa \sim 1.0,$$

or

(4.9b)

$$\bar{\theta}_{j,k} = \kappa \frac{|p_{j+1,k} - 2p_{j,k} + p_{j-1,k}|^2}{|p_{j+1,k} - p_{j,k}|^2 + |p_{j,k} - p_{j-1,k}|^2 + \varepsilon}, \qquad \kappa \sim 0.05.$$

As with the fourth-order viscosity, (4.7) cannot be used directly at the points next to a boundary. Salas (see [1]) suggests extrapolating the variables to a virtual point outside the domain and then using (4.7). Eriksson and Rizzi [4] again implement the boundary terms so as to increase the dissipative effect. However, computations indicate that the viscosity V_2 introduces errors near the boundaries. This error consists of two parts. Near the leading edge false entropy is generated which then forms an entropy layer along the body. At the same time a pressure jump occurs at the trailing edge which violates the Kutta condition. This problem is especially noticeable at high angles of attack. FLO52ST was used to find the solution about a NACA 0012 at 10° angle of attack with a free stream Mach number of 0.3. The solution is completely subsonic and so the Euler and potential solutions should agree. With the standard code a noticeable pressure jump is generated at the trailing edge. Since the flow is subsonic one can set $V_2 = 0$. Without the second-order viscosity the pressure is continuous at the trailing edge. Furthermore, comparisons of the lift between the potential and Euler codes show that the Euler code underpredicts the lift by about 9% on a coarse 64×16 Omesh. On a finer mesh the lift is improved but is still much worse than the corresponding prediction of the potential code. Removing the V_2 component of the viscosity improves the lift prediction though it is still 5% too low on the 64×16 mesh. Careful checks show that the pressure jump at the trailing edge is due to the tangential component of the V_2 viscosity. Hence, this difficulty is not caused by the difficulty in evaluating (4.7) near the boundary. The cause is that the switch (4.9) becomes large near the leading and trailing edge. Hence, these regions are treated as if there were a shock and a loss of entropy is created. This entropy layer extends over the length of the airfoil and is different on the upper and lower surfaces.

Hence, the stagnation pressure on the upper and lower surfaces are different leading to a pressure jump at the trailing edge. If the tangential component of the V_2 viscosity is set equal to zero then the pressure jump disappears. However, there is no improvement in the lift prediction. Thus the lift is more sensitive to the normal component of the viscosity.

In [3] it is suggested that the viscosity V_2 be multiplied by a linear factor which is zero near the body and one in the far field. This improves the accuracy in many cases, however, it is ad hoc. An alternative is to multiply the viscosity by M^4. In stationary flow, shocks only occur in supersonic regions and so the viscosity is not changed near shocks. However, there is a stagnation point at the trailing edge and so the viscosity is turned off near the trailing edge.

5. Enthalpy Damping. With a second-order system in time one can add artificial terms that depend on the first-time derivative. Though such terms destroy the time accuracy they do not affect the steady state solution. These terms can be chosen so as to speed up the convergence to a steady state. Such applications have been used for SOR or the full potential equation. However, for a first-order system one cannot, in general, add on lower order terms since they are not zero in the steady state. For the Euler equations it is known that the total specific enthalpy is constant along each streamline in the steady state. If all the streamlines originate from a constant reservoir then the enthalpy will be constant in the entire region even in the presence of shocks. This constant enthalpy is known a priori from the inflow boundary condition. Therefore, one can add artificial terms to each equation that depend on the deviation of the local enthalpy from the steady state enthalpy. Such a forcing term is zero in the steady state and we will show that it can accelerate the approach to the steady state. For simplicity of presentation we shall assume a one-dimensional isentropic fluid. We therefore consider the following modified equations.

$$\rho_t + u\rho_x + \rho u_x = \phi \tag{5.1a}$$

$$u_t + uu_x + \frac{c^2}{\rho}\rho_x = 0 \tag{5.1b}$$

$$\phi = -\alpha\rho(h - h_0) \tag{5.1c}$$

$$h = \frac{E + p}{\rho} = \frac{c^2}{\gamma - 1} + \frac{u^2}{2}, \tag{5.1d}$$

differentiating these equations, one obtains

$$\rho_{tt} + 2u\rho_{xt} + (u^2 - c^2)\rho_{xx} + u_t\,\rho_x + \rho_t\,u_x + 2uu_x\,\rho_x$$
(5.2)
$$- \rho u_x^2 - \rho\left(\frac{c^2}{\rho}\right)_x \rho_x = \phi_t + u\phi_x,$$

also

(5.3) $$\phi_t + u\phi_x = -\,\alpha\left[(h - h_0 + c^2)\rho_t + u(h - h_0)\rho_x\right].$$

We ignore all terms involving the product of derivatives and then freeze the coefficients. We then have

(5.4)
$$\rho_{tt} + 2u\rho_{xt} - (c^2 - u^2)\rho_{xx} + \alpha(h - h_0 + c^2)\rho_t - \alpha u(h - h_0)\rho_x = 0.$$

This is a convective wave equation in which α multiplies the first time derivative of ρ. This is similar to the acceleration procedure used for the full potential equation. To reduce this to a standard wave equation we introduce new independent variables

(5.5) $$\eta = x, \qquad \tau = \frac{Mx}{c\sqrt{1 - M^2}} + \sqrt{1 - M^2}\; t, \qquad M = \frac{u}{c},$$

where we assume that $M \leq 1$. With this change, (5.4) becomes

(5.6)
$$\rho_{\tau\tau} - (c^2 - u^2)\rho_{\eta\eta} + \frac{\alpha}{\sqrt{1 - M^2}}\left[(1 - 2M^2)(h - h_0) + (1 - M^2)c^2\right]\rho_\tau$$
$$- \alpha u(h - h_0)\rho_\eta = 0$$

or

(5.7) $$\rho_{\tau\tau} - d^2\,\rho_{\eta\eta} + K\rho_\tau + L\rho_\eta = 0.$$

Using a finite difference scheme we choose K proportional to $d/\Delta\eta$ so as to reach a steady state rapidly. In terms of the original variables this implies

(5.8) $$\alpha\Delta t \text{ proportional to } \frac{1 - M}{(1 - 2M^2)(h - h_0) + c^2(1 - M^2)}.$$

Note: Since α depends on $1/\Delta t$ the enthalpy damping is not a low order term and so it affects the stability limit.

As noted above this procedure is valid only for subsonic flow. For the potential equation it is well known that adding a ϕ_t term is not advisable in supersonic regions. Using enthalpy damping for the Euler equations it has been found experimentally that using a small amount (relatively) of enthalpy damping in the supersonic regions can accelerate convergence, especially for problems which are mainly supersonic.

In the code FLO52ST, $\alpha = a_1(1 - M^2) + \alpha_2$, α_1, α_2 constant. In the above derivation we used the primitive variables. For the conservative variables, (5.1) is replaced by

$$
\begin{aligned}
&\rho_t + L_1 = -\alpha\rho(h - h_0)\\
(5.9)\qquad &(\rho u)_t + L_2 = -\alpha\rho u(h - h_0)\\
&(\rho v)_t + L_3 = -\alpha\rho v(h - h_0),
\end{aligned}
$$

where L_i represent the standard Euler terms. The finite difference approximation to (5.9) is

$$(5.10)\qquad \rho^{n+1} - \rho^n = -\Delta t L_1 - \alpha\Delta t\rho^{n+1}(h^n - h_0)$$

or

$$(5.11)\qquad \rho^{n+1} = \frac{\rho^n - \Delta t L_1}{1 + \alpha\Delta(h^n - h_0)}\,.$$

We stress then in these formulas, α is always positive independent of the sign of $(h - h_0)$. Thus, the right-hand-side of (5.9) is not the equivalent of a forcing function that gives rise to an exponential decay in an ordinary differential equation. Replacing $(h - h_0)$ by $|h - h_0|$ in the definition of ϕ destroys the enthalpy damping and can lead to divergence.

For the energy equation we append to (5.1)

$$(5.12)\qquad S_t + L_4 = 0.$$

Deriving the equation for $E - h_0\rho$ one gets a forcing term that depends on $(h - h_0)^2$ which can lead to difficulties. In [8] this was fixed in a heuristic manner. An alternative is to replace (5.12) by

$$h_t + L_5 = -\beta(h - h_0) \tag{5.13}$$

where β may depend on the Mach number. Let $\hat{E} = E - h_0 \rho$ then combining (5.13) with (5.9) yields

$$\hat{E}_t = -(h - h_0)\left(\alpha\hat{E} + \frac{\beta}{\gamma}\rho\right) = -\left[\alpha(h - h_0) + \frac{\beta}{\gamma}\hat{E}\right] + \frac{\beta}{\gamma}p. \tag{5.14}$$

We note that when β is large we force the enthalpy to be equal to its steady state value. Hence (5.14) is a generalization of the isoenergetic systems.

Computations show that the enthalpy damping is also useful in removing temporal oscillations. Using the enthalpy damping, with a Mach number dependence, yields a more monontone convergence to a steady state than in the absence of damping. This property is especially useful if one uses an acceleration procedure on top of the Runge-Kutta scheme.

6. Residual Smoothing. Residual smoothing was first introduced by Lerat [12] for use with the Lax-Wendroff scheme. Jameson [9] later introduced a similar technique in conjunction with the Runge-Kutta schemes. We shall later compare the effects on both these methods. However, we first consider a two-step Runge-Kutta method. This scheme is given by

$$\begin{aligned} u^{(1)} &= u^n + \alpha\Delta t\, Qu^{(n)} \\ u^{(2)} &= u^n + \Delta t\, Qu^{(1)} \end{aligned} \tag{6.1}$$

where Q denotes all the spatial derivatives. When an artificial viscosity is used it can be used either in both steps or else frozen at the same value for both stages. The residual smoothing is then given by

$$\prod_j \left(1 - \frac{\beta_j}{4}\delta_{xx}^{(j)}\right)\left(u^{n+1} - u^n\right) = u^{(2)} - u^n, \tag{6.2}$$

where the product is over all space dimensions. One can also replace the operator on the left-hand-side of (6.2) with a full multidimensional elliptic operator. Since this operator only involves constant coefficients on a rectangular region (in computational space), it can be inverted by a fast solver. However, we shall see that one should not use large time-steps in conjunction with (6.2) even in one-space dimension. Since the difficulties are not concerned with splitting errors there is not much of an advantage to consider multidimensional operators that are not in split form. Numerical experiments indicate that a multidimensional Laplacian is more effective in accelerating the flow to a steady state than (6.2) but not sufficiently fast to warrant the additional cost of inverting the multidimensional matrix even with a fast solver.

We now consider the constant coefficient problem in one-space dimension. Let Q be the second-order central difference and ignore the artificial viscosity. Taking the Fourier transform of (6.2) for $u_t = u_x$, we get

$$\left(1 + \beta \sin^2 \frac{\theta}{2}\right)(G-1) = i\lambda \sin \theta - \alpha\lambda^2 \sin^2 \theta \tag{6.3}$$

$$\lambda = \Delta t/\Delta x$$

or

$$G = 1 + \frac{i\lambda \sin \theta - \alpha\lambda^2 \sin^2 \theta}{1 + \beta \sin^2 \frac{\theta}{2}} . \tag{6.4}$$

The original two-step Runge-Kutta scheme ($\beta = 0$) is stable when

$$\lambda \leq \frac{\sqrt{2\alpha-1}}{\alpha} . \tag{6.5}$$

For general β, the scheme is stable if we choose

$$\alpha \geq 1/2$$

$$\beta > 2\alpha\lambda\left(\lambda - \frac{\sqrt{2\alpha-1}}{\alpha}\right) . \tag{6.6}$$

Alternatively, if we replace β by $\sigma\lambda^2$, and $\alpha \geq 1/2$, $\sigma \geq 2\alpha$, then the scheme is unconditionally stable.

In three-space dimensions, (6.4) is replaced by

(6.7) $$\left(1 + \sigma\lambda^2 \sin^2 \frac{\theta}{2}\right)\left(1 + \sigma\lambda^2 \sin^2 \frac{\phi}{2}\right)\left(1 + \sigma\lambda^2 \sin^2 \frac{\xi}{2}\right)(G-1) = i\lambda K - \alpha\lambda^2 K^2,$$

with $K = \sin\theta + \sin\theta + \sin\xi$. For stability we require that $G^* G \leq 1$. This occurs if and only if

(6.8) $$-2\alpha\left(1 + \sigma\lambda^2 \sin^2 \frac{\theta}{2}\right)\left(1 + \sigma\lambda^2 \sin^2 \frac{\phi}{2}\right)\left(1 + \sigma\lambda^2 \sin^2 \frac{\xi}{2}\right) + \alpha^2 \lambda^2 K^2 + 1 \leq 0.$$

A sufficient condition for stability is that

(6.9) $$-2\alpha\left[1 + \sigma\lambda^2\left(\sin^2 \frac{\theta}{2} + \sin^2 \frac{\phi}{2} + \sin^2 \frac{\xi}{2}\right)\right] + \alpha^2 \lambda^2 K^2 + 1 \leq 0.$$

Hence, the three-dimensional scheme is unconditionally stable if

(6.10) $$\sigma \geq \max_{\rho,\phi,\xi} \frac{\alpha[\sin\theta + \sin\phi + \sin\xi]^2}{2\left(\sin^2 \frac{\theta}{2} + \sin^2 \frac{\phi}{2} + \sin^2 \frac{\xi}{2}\right)}.$$

Therefore, we have shown that a sufficient condition for the three-dimensional scheme (6.2) to be unconditionally stable is

(6.11) $$\alpha \geq 1/2, \qquad \sigma \geq 6\alpha.$$

Hence, the three-dimensional version of (6.2) is unconditionally stable. Moreover, since the Runge-Kutta scheme gives a steady state which is independent of Δt and (6.2) is in delta form we conclude that the steady state solution to (6.2) is also independent of Δt. We point out that the version of residual smoothing proposed by Lerat does not have a steady state independent of Δt. This is because the Thommen scheme used by Lerat has a Δt dependent steady state. Since the time steps used are not very large relative to the explicit time step this dependence on Δt may not be noticeable to graphical accuracy. We also note that for the one-dimensional Lax-Wendroff two-step methods (Richtmyer, MacCormack, etc.) and for the two-dimensional Burstein scheme that the solution after the intermediate step is independent of Δt.

We wish to stress that even though (6.2) is unconditionally stable, choosing a large time step is not the best strategy. Even in one-space dimension choosing a very large time step severly retards the convergence to a steady state. This is not true, in one-space dimension, for the Lax-Wendroff scheme, if β depends on the matrix of the differential equation (see [12]). Hence, the use of residual smoothing with the Runge-Kutta scheme is fundamentally different than the backward Euler method. For the residual splitting method the inefficiency of large time steps has nothing to do with splitting errors. Let $\beta = \sigma\lambda^2$, (6.4) becomes

$$(6.12) \qquad G(\theta) = 1 + \frac{i\lambda \sin\theta - \alpha^2 \lambda^2 \sin^2\theta}{1 + \sigma\lambda^2 \sin^2\frac{\theta}{2}}, \qquad \begin{array}{l} \alpha \geq 1/2 \\ \sigma \geq 2\alpha \end{array} .$$

For large λ, (6.12) becomes

$$(6.13) \qquad G(\theta) \simeq -\frac{\alpha^2}{\sigma} \frac{\sin^2\theta}{\sin^2\theta/2} .$$

We thus see that without artificial viscosity that the highest frequency, $\theta = \pi$ is not damped.

If we add an artificial fourth-order dissipation with coefficient ν at each stage, then (6.12) is replaced by

$$(6.14) \quad G(\theta) = 1 + \frac{\lambda\left(i \sin\theta - \nu \sin^4\frac{\theta}{2}\right) + \lambda^2\left(i \sin\theta - \nu \sin^4\frac{\theta}{2}\right)^2}{1 + \sigma\lambda^2 \sin^2\frac{\theta}{2}} .$$

As λ gets larger, we get

$$(6.15) \qquad G(\pi) \sim 1 + \frac{\nu^2}{\sigma} > 1,$$

and so the scheme is not stable for large λ. We next consider the case that the same artificial viscosity is used at both stages. In that case (6.12) is replaced by

(6.16)

$$G(\theta) = 1 + \frac{i\lambda \sin\theta\left[1 + \alpha\lambda\left(i \sin\theta - \nu \sin^4\frac{\theta}{2}\right)\right] - \lambda\nu \sin^4\frac{\theta}{2}}{1 + \sigma\lambda^2 \sin^2\frac{\theta}{2}} .$$

As λ approaches infinity, the coefficient of the artificial viscosity goes to zero relative to the denominator and so we recover (6.13) in the limit. Hence, when we freeze the artificial viscosity, the scheme remains unconditionally stable. However, when λ is large, the highest frequency is damped only a small amount proportional to $1/\lambda$. Therefore, if we wish to minimize G, we do not wish to choose λ large. An alternative is to choose a finite λ so that $\int_0^\pi G^2(\theta)d\theta$ is minimized.

We also note that if one adds the viscous terms from the Navier-Stokes equations, then a similar analysis holds. Hence, if one wants the scheme to be unconditionally stable then the Navier-Stokes terms should be frozen and one evaluated once per cycle. If the Navier-Stokes terms are reevaluated at each stage then the residual smoothing should be applied at every stage and not after the second stage.

The above analysis was for a two-stage Runge-Kutta scheme. For a multi-stage scheme one should apply the residual smoothing after every even stage. If the total number of stages is odd then one should do an extra residual smoothing after the cycle is completed. One can also show that the residual smoothing will not stabilize a one-stage Runge-Kutta with central differences for a hyperbolic problem but the method is unconditionally stable for a one-stage method for parabolic problems. In fact it gives the backward Euler methods. Hence, for the Navier-Stokes equations one should use residual smoothing after every stage.

7. Highly Subsonic Flow. It is well known that for very subsonic flow that the flow can be considered as incompressible. The use of the compressible fluid equations is considered as inefficient, since the fluid velocity is much smaller than the speed of sound. The use of an explicit scheme requires Δt to be bounded by $1/c$. However, the physical properties change over time scales of order $1/u$ which is much larger. Similar arguments hold for viscous flows with a high Reynolds number. Hence, it is usually agreed that explicit schemes are inefficient for highly subsonic flows. We shall now show that if one is only interested in the steady state then a minor change to the code can greatly increase the efficiency of the explicit method. Even if one is using an implicit method the following changes should increase the efficiency of the scheme since all waves have similar speeds.

The Euler equations are expressed as

$$w_t + f_x + g_y = 0, \tag{7.1}$$

where (x,y) represent general curvilinear coordinates (see (3.3)). Since we are only interested in the steady state we replace (7.1) by the system

$$M^{-1} w_t + f_x + g_y = 0. \tag{7.2}$$

The requirements on M are that the matrix be nonsingular and that the original initial boundary value problem still be well-posed. It is straightforward to solve (7.2) with an explicit scheme. With an implicit method only the diagonal portion of the matrix to be inverted is changed. Though the code solves (7.2) we shall only analyze the constant coefficient problem

$$M^{-1} w_t + Aw_x + Bw_y = 0, \tag{7.3}$$

where the matrices M, A, B are constant and $A = \partial f/\partial w$, $B = \partial g/\partial w$. Let $w^{(0)} = Tw$, $A_0 = T A T^{-1}$, $B_0 = T B T^{-1}$, $M_0^{-1} = T M^{-1} T^{-1}$, where T is given by (3.12). Then (7.3) can be converted to

$$M_0^{-1} w_t^{(0)} + A_0 w_x^{(0)} + B_0 w_y^{(0)} = 0 \tag{7.4}$$

with

$$A_0 = \begin{bmatrix} q & c & 0 & 0 \\ c & q & 0 & 0 \\ 0 & 0 & q & 0 \\ 0 & 0 & 0 & q \end{bmatrix}$$

(7.5)

$$B_0 = \begin{bmatrix} r & 0 & c & 0 \\ 0 & r & 0 & 0 \\ c & 0 & r & 0 \\ 0 & 0 & 0 & r \end{bmatrix}$$

where q and r, defined in (3.6), are the contravariant velocity components.

If $M_0 = I$, then we have not changed the eigenvalues or the stability condition of (7.1). We now consider the case that $u^2 + v^2 \ll c^2$. We wish to choose M_0^{-1} so that the eigenvalues of M_0

A_0 and $M_0 B_0$ are independent of c. In addition, we wish M_0 to be positive definite. This will imply that (7.3) is a symmetric hyperbolic system and so is well-posed. One choice is

$$(7.6) \qquad M_0^{-1} = \begin{bmatrix} \frac{c^2}{z^2} & 0 & 0 & 0 \\ 0 & 1 & 0 & 0 \\ 0 & 0 & 1 & 0 \\ 0 & 0 & 0 & 1 \end{bmatrix},$$

where $z^2 = \max(\varepsilon, u^2 + v^2)$. ε is introduced so that the matrix M_0^{-1} is not singular at stagnation points. In particular, $\varepsilon = .01c$ seems to give reasonable results. Transforming back to the curvilinear coordinates we define

$$(7.7) \qquad Q = \begin{bmatrix} s^2 & -u & -v & 1 \\ us^2 & -u^2 & -uv & u \\ vs^2 & -uv & -v^2 & v \\ hs^2 & -uh & -vh & h \end{bmatrix} \qquad s^2 = \frac{u^2+v^2}{2}, \quad h = \frac{c^2}{\gamma-1} + s^2$$

Then

$$(7.8) \qquad M^{-1} = I + dQ, \qquad M = I + eQ,$$

where

$$(7.9) \qquad d = \frac{\gamma-1}{c^2}\left(\frac{c^2}{z^2} - 1\right), \qquad e = \frac{\gamma-1}{c^2}\left(\frac{z^2}{c^2} - 1\right).$$

We note that given the first row of Q, the following rows are derived by multiplying the first row by u,v,h, respectively. Hence, the product of Q times a vector requires only six multiplications. For use in a Runge-Kutta scheme we evaluate the fluxes in (7.2) as usual including the artificial viscosity. The vector of the changes in the variables Δw is then multiplied by the matrix M where the elements of Q are evaluated at the previous stage. The variables at the next stage can then be evaluated.

Let $\bar{M}^2 = z^2/c^2$, then the largest eigenvalue of $D = A \sin\theta + B \sin\phi$ is given by

$$\lambda = \frac{|w|(1 + \bar{M}^2) + \sqrt{w^2(1 - \bar{M}^2) + 4(a^2 + b^2)z^2}}{2}, \tag{7.10}$$

where w, a, b are given in (3.8), (3.9). We see that at a stagnation point $\bar{M} = O(\varepsilon)$ and $\lambda = O(\varepsilon)$. For $\bar{M} = 1$, $\lambda = |w| + \sqrt{a^2 + b^2}\, c$. Hence, at low Mach numbers the largest eigenvalues (and hence the time step) is independent of the sound speed c. At transonic sound speeds the largest eigenvalue is comparable to the case with M = I. Comparing with (3.16) we see that the preconditioned form (7.2) allows the use of a larger time step for all subsonic flows. Applications to the incompressible Navier-Stokes equations are presented in [19] together with extensions to supersonic flow.

8. Upwind Schemes. At each stage of the Runge-Kutta scheme we have replaced the flux derivatives by central differences. This necessitated the use of an artificial viscosity. This stabilized the scheme in both smooth regions and provided an entropy condition in shocked regions. To avoid this artificial aspect one can replace the derivatives by appropriate upwind differences.

Presently this is implemented using the flux splitting described by Van Leer [24] for the Euler equations. At each zone face ξ,η = constant, we calculate

$$\begin{aligned} w^-_{i+1/2,j} &= w_i + 1/2\, \delta_i w \\ w^+_{i-1/2,j} &= w_i - 1/2\, \delta_i w. \end{aligned} \tag{8.1a}$$

δ_i is constructed from the forward and backward differences using a switch. This switch prevents overshoots when the variables change dramatically over a zone width. The flux is then split into plus and minus contributions depending on the sign of the eigenvalues. These fluxes are calculated in a rotated system using the velocity components parallel and perpendicular to each coordinate direction. The result fluxes are then rotated back to yield the Cartesian fluxes. We then combine the plus and minus contributions from neighboring cells to obtain the flux at the face of the cell. This is done independently in each direction. Given the fluxes at all four faces of the cell we update the variables to the next stage. The coefficient of the Runge-Kutta scheme given by (2.7) were appropriate for central differences. Optimal values of the parameters for one-sided schemes are not known. Further details of the scheme are presented in [20].

Because of the extra logic involved in an upwind scheme it is more costly per stage than a central difference scheme. Though a central difference scheme requires an artificial viscosity it can be calculated once and then reused at each stage of the Runge-Kutta scheme. Since the viscosity of an upwind scheme is built in, it is more difficult to perform some operations once and then reuse them during the k stage scheme. A four-stage scheme using upwind differences requires about three times as much computer time as the corresponding four-stage central difference scheme. In addition, the enthalpy damping technique of section 5 can not be used. This occurs since the enthalpy is not a constant in the steady state when using flux splitting. The stability limit for the one-sided scheme is only about two-thirds of that allowed for the central difference scheme. Hence, the present version of the upwind scheme is about five times slower than the central difference scheme to reach a steady state with a given tolerance. The chief advantage of the upwind scheme is its robustness. There is no need to choose constants for the artificial viscosity. Preliminary test indicate that the upwind scheme works for a larger range of inflow Mach numbers than the central difference scheme.

9. Boundary Conditions. In addition to advancing the scheme in the interior, it is necessary to supply boundary conditions. At the airfoil the normal component of the velocity is zero. Using the finite volume approach, it is only necessary to know the pressure on the airfoil. This pressure is found by the normal momentum equation. It is also necessary to give boundary conditions in the far field. When the flow is subsonic at infinity one should specify three conditions at inflow and one condition at outflow.

In one-space dimension, in order to decrease the energy as fast as possible and reach a steady state rapidly, one should specify characteristic conditions. Diagonalizing the wave equation one gets

$$(9.1)\qquad \begin{array}{lll} u_t = u_x, & v_t = v_x, & 0 \le x \le 1 \\ u(1,t) = \alpha v(1,t), & v(0,t) = \beta u(0,t) & \end{array}$$

u,v are the characteristic variables. Specifying characteristic boundary conditions is equivalent to $\alpha = \beta = 0$. From (9.1) it follows that

$$\frac{d}{dt}\int_0^1 (u^2 + v^2)dx = (\alpha^2 - 1)v^2(1,t) + (\beta^2 - 1)u^2(0,t). \tag{9.2}$$

Hence, choosing $\alpha = \beta = 0$ minimizes the right-hand-side.

Therefore at inflow, one should specify three characteristic variables. However, we require that the enthalpy be constant over the entire field in the steady state. To achieve this it is necessry to specify enthalpy at inflow. This condition replaces one of the characteristic boundary conditions. For nonlinear problems it is more appropriate to specify the Riemann variables rather than characteristic variables. Hence, at inflow we specify

$$p/\rho^\gamma \tag{9.3a}$$

$$v \tag{9.3b}$$

$$h = \frac{E + p}{\rho}. \tag{9.3c}$$

u is the component of velocity perpendicular to the boundary while v is the component parallel to the boundary. The fourth boundary condition is found by extrapolation from the interior. For stability it is preferable to extrapolate characteristic variables [5]. Hence, at the inflow we extrapolate

$$u - \frac{2c}{\gamma-1}. \tag{9.3d}$$

At the outflow boundary we reverse the procedure and specify

$$u - \frac{2c}{\gamma-1}, \tag{9.4a}$$

and extrapolate

$$p/\rho^2 \tag{9.4b}$$

$$v \tag{9.4c}$$

$$h \tag{9.4d}$$

An alternative to (9.4) is to use nonreflecting boundary conditions [2].

Numerical experiments indicate that the far field boundary exerts a much greater influence on the drag and lift coefficients than does the boundary condition at the airfoil.

10. Conclusions. We have discussed many of the components of the code FLO52ST. This code gives a rapid solution to the Euler equations in both two- and three-space dimensions for a variety of geometries and a range of Mach numbers.

With central differences it is easy to increase the accuracy of the differences to fourth-order or spectral accuracy. Using an O mesh all variables are periodic around the airfoil. Therefore, high-order differences do not encounter any boundary difficulties. Using spectral methods a Fourier scheme would be appropriate. With a C mesh all boundaries are in the far field. Hence, one can simply reduce the order of accuracy near the outer boundaries where the flow is smooth enough for second-order accuracy to be adequate. The approximation normal to the boundary requires more care with regard to boundary conditions at the airfoil surface. Hence, it is reasonable to consider second-order differences normal to the airfoil while using higher-order methods parallel to the body. In this case one must be careful in approximating the metric derivatives so that the constant flow in the far field remains a solution to the finite difference equations. Work is also in progress on extending the code to the Navier-Stokes equations.

REFERENCES

[1] R. K. AGARWAL and J. E. DEESE, Transonic wing-body calculations using Euler equations, AIAA Paper 83-0501 (1983).

[2] A. BAYLISS and E. TURKEL, Far field boundary conditions for compressible flows, J. Comput. Phys., 48 (1982), pp. 182-199.

[3] H. C. CHEN, N. J. YU, P. E. RUBBERT, and A. JAMESON, Flow simulations for general Narelle configurations using Euler equations, AIAA Paper 83-0539 (1983).

[4] L. E. ERIKSSON and A. RIZZI, Analysis by computer of the convergence to steady state of discrete approximation to the Euler equations, AIAA Comput. Fluid Dynamics Conf. (1983), pp. 407-442.

[5] D. GOTTLIEB, M. GUNZBURGER, and E. TURKEL, On numerical boundary treatment of hyperbolic systems for finite difference and finite element methods, SIAM J. Numer. Anal., 19 (1982), pp. 671-682.

[6] D. GOTTLIEB and E. TURKEL, Spectral methods for time dependent partial differential equations, Third 1983 C.I.M.E. Session (1983).

[7] R. A. GRAVES and N. E. JOHNSON, Navier-Stokes solutions using Stetter's method, AIAA J., 16 (1978), pp. 1013-1015.

[8] A. JAMESON, W. SCHMIDT, and E. TURKEL, Numerical solutions of the Euler equations by finite volume methods using Runge-Kutta time-stepping schemes, AIAA Paper 81-1259 (1981).

[9] A. JAMESON, The evolution of computational methods in aerodynamics, J. Appl. Mech., 50 (1983), pp 1052-1076.

[10] A. JAMESON, Solution of the Euler equations for two-dimensional transonic flow by a multigrid method, Appl. Math. Comput., 13 (1983), pp. 327.

[11] P. D. LAX and B. WENDROFF, Difference schemes for hyperbolic equations with high-order accuracy, Comm. Pure Appl. Math., 17 (1964), pp. 381-398.

[12] A. LERAT, Une class de schémas aux différences implicites pour les systémes hyperboliques de lois de conservation, C. R. Acad. Sci. Paris, t. 288 (1979).

[13] C. P. LI, Numerical solution of the viscous reacting blunt body flows of a multicomponent mixture, AIAA Paper 73-202 (1973).

[14] M. S. MOCK, A difference scheme employing fourth-order 'viscosity' to enforce an entropy inequality, Bat Sheva Conf. Finite Elements Nonelliptic Systems (1973).

[15] R. D. RICHTMYER and K. W. MORTON, Difference Methods for Initial Value Problems, 2nd Edition, Interscience, New York, 1967.

[16] P. SONNEVELD and B. VAN LEER, The Minimax Problem Along the Imaginary Axis, to appear in Nieuw Archief voor Wiskunde, 1984.

[17] H. J. STETTER, Improved absolute stability of predictor corrector schemes, Computing, 3 (1968), pp. 286-296.

[18] E. TURKEL, Progress in computational physics, Comput. & Fluids, 11 (1983), pp. 121-144.

[19] E. TURKEL, Fast solutions to the steady state compressible and incompressible fluid equations, Proc. Ninth International Conference for Numerical Methods in Fluid Dynamics, Saclay, France, June 1984.

[20] E. Turkel and B. Van Leer, Flux vector splitting and Runge-Kutta methods for the Euler equations, Proc. Ninth International Conference for Numerical Methods in Fluid Dynamics, Saclay, France, June 1984.

[21] P. J. VAN DER HOUWEN, Construction of Integration Formulas for Initial Value Problems, North-Holland, Amsterdam, 1977.

[22] P. J. VAN DER HOUWEN and B. P. SOMMEIJER, On the internal stability of explicit m-stage Runge-Kutta methods for large m-values, ZAMM, 60 (1980), pp. 479-486.

[23] B. VAN LEER, Towards the ultimate conservative difference scheme. III: Upstream-centered finite difference schemes for ideal compressible flow, J. Comput. Phys., 23 (1977), pp. 263-275.

[24] B. VAN LEER, Flux vector splitting for the Euler equations, Springer-Verlag Lecture Notes in Physics 170 (1982), pp. 507-512.

[25] R. VICHNEVETSKY, New stability theorems concerning one-step numerical methods for ordinary differential equations, Math. Comput. Simulation, 25 (1983), pp. 199-205.

[26] N. J. YU, H. C. CHEN, S. S. SAMENT, and P. E. RUBBERT, Inviscid drag calculations for transonic flows, AIAA Comput. Fluid Dynamics Conf., 28 (1983), pp. 283-292.

RELAXATION METHODS FOR HYPERBOLIC CONSERVATION LAWS

BRAM VAN LEER* AND WILLIAM A. MULDER**

Abstract. A class of unconditionally stable integration schemes for finding steady solutions to the time-dependent Euler equations is discussed. The schemes are time-accurate for small time-steps and turn into a relaxation method for large time-steps; the spatial discretization is upwind biased. Examples of these switched evolution/relaxation (SER) schemes, and a numerical comparison, are presented. Some attention is given to the standard ADI and AF methods, which do not belong to the SER family.

1. Introduction. Numerical methods for solving problems of steady compressible flow often are based on the time-dependent flow equations. In marching toward the steady state it is useful to distinguish two phases [1].

In the first or *searching* phase the numerical solution follows a path to the steady state with reasonable time-accuracy, a safeguard against selecting an unphysical solution. In the second or *converging* phase the numerical solution feels the attraction of the steady state and converges to it rapidly.

It is possible to go through both phases with a single numerical method. The paragon is Euler's method "backward"; for the equation

$$u_t = R(u) \ , \tag{1}$$

with R(u) some nonlinear finite-difference expression, it reads

$$\Delta_t u^n = \tau^n R^{n+1} \ ; \tag{2}$$

here $\tau^n \equiv t^{n+1} - t^n$, $\Delta_t u^n \equiv u^{n+1} - u^n$. If R(u) is continuously differentiable, with

$$K(u) \equiv dR(u)/du \ , \tag{3}$$

*University of Technology, Delft, The Netherlands
**University Observatory, Leiden, The Netherlands

linearization of (2) is possible:

$$(I-\tau^n K^n)\ \Delta_t u^n = \tau^n R^n. \tag{4}$$

In the present paper we study a class of methods that includes (4), namely,

$$(I-\tau^n M^n)\ \Delta_t u^n = \tau^n R^n\ , \tag{5}$$

where M(u) is a linear operator providing numerical stability for arbitrary values of τ.

For small τ scheme (5) is time-accurate, with truncation error in $\Delta_t u^n$ of the order $O(\tau^2)$ or better; for large τ it reduces to the relaxation scheme

$$-\ M^n \Delta_t u^n = R^n\ . \tag{6}$$

This is Newton's method if M = K, in which case the right-hand side of (6), the residual, converges quadratically to zero. While this type of convergence is highly desirable, achieving it usually involves a lot of work. The matrix K may be costly to form and too costly to invert by direct methods. The classic way out, represented by scheme (5), is to replace K by a simpler matrix M, usually one with a narrower bandwidth. The utmost simplification is to replace K by a scalar.

A convenient choice of the time-step is

$$\tau_n = \varepsilon||u^n||/||R^n||\ , \tag{7}$$

where ε is a number determining the temporal accuracy. If the residual is large, as in the searching phase, (7) ensures that

$$||\Delta_t u^n||/||u^n|| \sim \varepsilon\ , \tag{8}$$

that is: the relative change of the numerical solution per time-step is of the order of ε. (In [2] the same accuracy constraint was enforced by a posteriori checking, reducing τ_n in case of violation, and recomputing $\Delta_t u^n$). Upon entering the converging phase the residual drops sharply and τ rises correspondingly, switching the scheme from (5) to (6).

Schemes of the form (5) that allow switching through (7) will be called Switched Evolution/Relaxation (SER) schemes. The aim of this paper is to indicate matrices M suitable for SER schemes approximating hyperbolic systems of conservation laws (§2), compare these in a numerical experiment (§3) and make recommendations about their use (§4).

Before getting to the main subject (§2.3 ff) we discuss two schemes that do not belong to the SER category but are very popular in aerodynamics, namely, Alternating Direction Implicit (ADI) and Approximate Factorization (AF) schemes.

2. Relaxation methods for hyperbolic equations.

2.1 General considerations. When solving flow problems in one dimension using a three-point discretization we have no excuse for not choosing Newton's method as the relaxation method. For example, the Jacobian K associated with a first-order-accurate upwind-difference approximation of the Euler equations is block-tridiagonal, and the main block-diagonal of -K is semi-dominant within a margin $O(\Delta x)$. Thus, the matrix $I-\tau K$ in (4) is block-diagonally dominant for all but the largest values of τ, and its block-LU decomposition requires no pivoting [1].

For one-dimensional upwind-biased schemes of a higher order of accuracy the formation and direct solution of the linear system (4) is more cumbersome but, possibly, worthwhile; for two-dimensional schemes of even the first order of accuracy the direct solution of (4) is out of the question. In this paper we choose to abandon (4) in favor of (5).

Nevertheless, it is fully legitimate to use a relaxation scheme of the type (4) in solving the more complete system of equations (5). The relaxation scheme may serve as a preconditioner, in conjugate-gradient methods, or as a "smoother," in multigrid methods; the latter possibility is briefly discussed in §2.5.

All classic iterative methods exploit that R is the sum of two one-dimensional expressions,

$$R(u) = R_x(u) + R_y(u) \ , \tag{10}$$

each with its own Jacobian

$$K(u) = K_x(u) + K_y(u). \tag{11}$$

This suggests relaxation schemes with

$$M(u) = M_x(u) + M_y(u) \ ; \tag{12}$$

we shall restrict ourselves to these from §2.3 onward. First we discuss two different ways to benefit from (10), not leading to SER schemes.

2.2 Alternating Direction Implicit and Approximate Factorization schemes. The ADI scheme commonly used for relaxation is a sequence of two different time-steps, each unstable by itself:

$$(I-\tau^n M_x^n) \ \Delta_t u^n = \tau^n R^n \ , \tag{13.1}$$

$$(I-\tau^{n+1}M_y^{n+1})\ \Delta_t u^{n+1} = \tau^{n+1}R^{n+1}\ , \tag{13.2}$$

$$\tau^{n+1} = \tau^n . \tag{13.3}$$

The original intention of ADI [3] was to combine unconditional stability with second-order accuracy in time through the choice $M_x = \frac{1}{2} K_x$, $M_y = \frac{1}{2} K_y$.

The related, more modern AF scheme [4] is a <u>nesting</u> of similar steps, without intermediate updating:

$$(I-\tau^n M_x^n)(I-\tau^n M_y^n)\ \Delta_t u^n = \alpha\tau^n R^n\ ; \tag{14}$$

here α is a parameter which, for the present purpose, may take the value 1 or 2. With $\alpha = 1$ and $M_x = K_x$, $M_y = K_y$, scheme (14) obviously approximates the backward Euler scheme (4) within a margin $O(\tau^3)$:

$$(I-\tau^n K^n+(\tau^n)^2 K_x^n K_y^n)\Delta_t u^n = \tau^n R^n\ . \tag{15}$$

It is equally obvious that this is not an SER scheme, since

$$\lim_{\tau^n\to\infty} \Delta_t u^n = 0\ , \tag{16}$$

regardless of the magnitude of R^n.

With $\alpha = 2$ the one-step scheme (14) advances as far in time as the two-step scheme (13). Inserting $M_x = K_x$, $M_y = K_y$ in both (13) and (14) and assuming that $R(u)$ is linear in u, e.g. $R_x = K_x u$, $R_y = K_y u$, we may rewrite these schemes as

$$\text{(ADI)}\qquad u^{n+2} = (I-\tau^n K_y)^{-1}(I+\tau^n K_x)(I-\tau^n K_x)^{-1}(I+\tau^n K_y)u^n\ , \tag{16.1}$$

$$\text{(AF},\alpha\text{=2)}\quad u^{n+1} = (I-\tau^n K_y)^{-1}(I-\tau^n K_x)^{-1}(I+\tau^n K_x)(I+\tau^n K_y)u^n\ , \tag{16.2}$$

showing their identity. Eq. (16) also reveals why ADI and AF are such effective relaxation methods for discretized <u>parabolic</u> equations like the scalar diffusion equation. In this case the eigenvalues of K_x and K_y are negative real, and the corresponding eigenvectors can be efficiently removed from the solution by cycling the value of $1/\tau$ through the spectrum of eigenvalues; see Wachspress [5].

For discretized hyperbolic equations the eigenvalues of K_x and K_y are complex or purely imaginary; therefore, time-step cycling is useless if done with real values of τ. One must resort to complex arithmetics, a fact that was well understood by the authors of [3] (J. Douglas, private communication, 1983) but seems to have fallen into oblivion. A correct implementation of time-step cycling for the Euler equations is due to Liu and Lomax [6].

The eigenvalues of K_x and K_y, whether complex or real, usually are obtained from a Fourier analysis, i.e. under the assumption that the eigenvectors are harmonics. When this assumption breaks down, e.g. for strongly variable coefficients and/or strongly nonuniform grids, time-step cycling becomes hard to implement and ADI and AF methods lose their appeal.

Abarbanel, Dwoyer and Gottlieb [7] have succeeded, for the linear diffusion equation, in reducing the sensitivity of AF to the choice of the time-step, by adding to the right-hand side of (15) a term that roughly balances the factorization error on the left-hand side:

$$(I-\tau^n K+(\tau^n)^2 K_x K_y)\ \Delta_t u^n = \tau^n (I+\gamma\tau^n K_x K_y) R^n\ . \tag{17}$$

For the parameter γ they derive an optimum value. One observes, however, that for large τ_n this scheme reduces to

$$K_x K_y \Delta_t u^n = \gamma\ K_x K_y R^n\ , \tag{18.1}$$

which is a hard way to implement the explicit scheme with time-step γ:

$$\Delta_t u^n = \gamma\ R^n . \tag{18.2}$$

An SER scheme of the form (5), with $M^n = -I/\gamma$, would achieve the same.

We conclude that ADI and AF do not have special properties that makes these schemes advantageous in finding steady solutions to hyperbolic equations.

2.3 SER schemes of first-order accuracy. It is surprising how well the classic relaxation schemes for the discretized diffusion equation, i.e. Jacobi, Gauss-Siedel and line relaxation, are suited for discretized hyperbolic equations. Block versions of these methods may be applied, in all possible combinations, to any first-order upwind discretization of the Euler equations, for an arbitrary number of space dimensions. The more powerful combinations also apply to second-order upwind discretizations.

To fix our thoughts, let us examine the possibilities in two dimensions, with flux splitting providing the upwind logic, as in [1]. The Euler equations in a Cartesian frame read

$$u_t = -(f(u))_x - (g(u))_y \; ; \tag{19}$$

here u is the state vector of conserved quantities and the vectors f(u), g(u) contain their fluxes in the x-, y-direction. The fluxes can be split in forward fluxes $f^+(u)$, $g^+(u)$ and backward fluxes $f^-(u)$, $g^-(u)$ that are continuously differentiable [8].

Now define a computational grid of adjacent rectangular volumes; the volume centered on (x_i, y_j), measuring Δx_i by Δy_j, is referred to by a subscript ij. The operators causing a shift over one volume in the forward x-, y-direction are called T_x, T_y. In this notation the first-order flux-split upwind discretization of the right-hand side of (19) becomes

$$\begin{aligned} R^n \equiv &- (f^+_{ij} - f^+_{i-1\,j} + f^-_{i+1\,j} - f^-_{ij})^n/\Delta x_i \\ &- (g^+_{ij} - g^+_{i\,j-1} + g^-_{i\,j+1} - g^-_{ij})^n/\Delta y_i \; . \end{aligned} \tag{20.1}$$

We may write the full backward-Euler scheme as

$$L^n_{ij}\Delta_t u^n_{ij} \equiv (\frac{I}{\tau^n} - K^n_{ij})\, \Delta_t u^n_{ij} = R^n_{ij} \; , \tag{20.2}$$

with

$$K^n_{ij} = -(A^x_{ij} + A^y_{ij} - B^x_{ij}T^{-1}_x - B^y_{ij}T^{-1}_y - C^x_{ij}T_x - C^y_{ij}T_y)^n \; , \tag{20.3}$$

$$A^x_{ij} = (\frac{df^+}{du} - \frac{df^-}{du})_{ij}/\Delta x_i \geq 0 \; , \tag{21.1}$$

$$A^y_{ij} = (\frac{dg^+}{du} - \frac{dg^-}{du})_{ij}/\Delta y_j \geq 0 \; , \tag{21.2}$$

$$B^x_{ij} = (\frac{df^+}{du})_{i-1\,j}/\Delta x_i \geq 0 \; , \tag{21.3}$$

$$B^y_{ij} = (\frac{dg^+}{du})_{i\,j-1}/\Delta y_j \geq 0 \; , \tag{21.4}$$

$$C^x_{ij} = -(\frac{df^-}{du})_{i+1\,j}/\Delta x_i \geq 0 \; , \tag{21.5}$$

$$C^y_{ij} = -(\frac{dg^-}{du})_{i\,j+1}/\Delta y_j \geq 0 \; . \tag{21.6}$$

Owing to the upwind differencing in (20.1), the main-diagonal blocks of $-K^n$ are comfortably large: if the matrix elements

(21.1)-(21.6) vary smoothly with i and j, $-K^n$ is semi-dominant within a margin $O(\Delta x)$. The matrix L^n will actually be block-diagonally dominant for a bias I/τ^n that is sufficiently large, but still only $O(\Delta x)$. Nevertheless, we shall not attempt to solve (20.2) directly by Gaussian elimination.

The classic relaxation schemes for the linear system (20.2) may be thought to arise from one particular way of approximating K^n: in (20.3) one or more shift operations are replaced by scalar multiplications. This is common practice in Fourier analysis, where, for a single mode with spatial frequencies ξ and η, we have

$$T_x^{\pm 1} = e^{\pm i\xi} I \ , \quad T_y^{\pm 1} = e^{\pm i\eta} I \ , \tag{22.1}$$

or

$$\Delta_t u^n_{i\pm 1\, j} = e^{\pm i\xi} \Delta_t u^n_{ij} \ , \tag{22.2}$$

$$\Delta_t u^n_{ij\pm 1} = e^{\pm i\eta} \Delta_t u^n_{ij} \ . \tag{22.3}$$

However, we wish to avoid complex arithmetics and complete dependence on Fourier analysis; see§2.2. In replacing $T_x^{\pm 1}$, $T_y^{\pm 1}$ we shall restrict ourselves to real-valued multipliers $s_x^{\pm}$, $s_y^{\pm}$:

$$T_x^{\pm 1} = s_x^{\pm} I \ , \quad T_y^{\pm 1} = s_y^{\pm} I \ , \tag{23.1}$$

or

$$\Delta_t u^n_{i\pm 1\, j} = s_x^{\pm} \Delta_t u^n_{ij} \ , \tag{23.2}$$

$$\Delta_t u^n_{ij\pm 1} = s_y^{\pm} \Delta_t u^n_{ij} \ . \tag{23.3}$$

There are three important values: $s = -1$, 0 and 1. The approximation of $T^{\pm 1}$ by $s^{\pm} = -1$ is exact for a saw-tooth component in $\Delta_t u^n$. It is seen from (20.3) and (21) that this choice makes the main-diagonal blocks more strongly dominant, leading to underrelaxation for long waves. The longest waves would be best represented by $s = 1$, but this value makes scheme (20) unstable. The stability condition on s, as shown in the Appendix, is

$$s \le 0 \ . \tag{24}$$

The value $s = 0$ will be considered the standard value. It is computationally attractive: an off-diagonal block multiplied by zero need not be computed at all, and does not influence the main diagonal.

In simplifying the right-hand side of (20.3), it is useful to distinguish the five cases listed below.

(i) All four shift operators replaced.
The block version of point/Jacobi relaxation. If the volumes with i+j even and i+j odd are updated alternatingly, checkerboard relaxation results.

(ii) Three shift operators replaced.
A combination of Jacobi and Gauss-Seidel relaxation. If, for instance, T_x^{-1} is the one operator that is not replaced, the Gauss-Seidel process requires sweeps in the forward x-direction.

(iii) Two shift operators replaced that work along different coordinate axes.
Full Gauss-Seidel relaxation. If, for instance, T_x^{-1} and T_y^{-1} are retained, one must make sweeps in the forward x-direction and the forward y-direction.

(iv) Two shift operators replaced that work along the same coordinate axis.
Line/Jacobi relaxation. If, for instance, T_y and T_y^{-1} are retained, the line relaxation is in the y direction; updating the volumes with i even and i odd alternatingly yields zebra relaxation.

(v) One shift operator replaced.
Line/Gauss-Seidel relaxation. If, for instance, T_x is the one operator that is replaced, the line relaxation is in the y-direction and the Gauss-Seidel sweep goes in the forward x-direction.

It is generally recommended in the cases (ii)-(v) to cycle through all shift operators when replacing one or more. Gauss-Seidel relaxation should not be used in the direction of a cyclic coordinate, as this causes a long-wave closure error which is hard to remove. Pattern relaxation (checkerboard, zebra) does not have this drawback.

A nonlinear variant of the Gauss-Seidel scheme results when the fluxes at t^{n+1}, needed in the original backward Euler scheme (2), are not all approximated linearly,

$$(f^{\pm})^{n+1} = (f^{\pm})^n + \left(\frac{df^{\pm}}{du}\right)^n \Delta_t u^n , \tag{25.1}$$

but are actually updated,

$$(f^{\pm})^{n+1} = f^{\pm}(u^n+\Delta_t u^n) , \tag{25.2}$$

in any volume where $\Delta_t u^n$ is already known. Implemented this way, the scheme no longer includes the off-diagonal blocks of K^n.

A very radical simplification, one that significantly reduces the relaxation power of the schemes, is achieved in replacing any of the blocks of K^n replaced by its spectral radius. Particularly well-known is the simplified point/Jacobi scheme with $s_x^+ = s_y^{\pm} = 0$, that is,

$$(I/\tau^n + \rho_{ij}^n I)\, \Delta_t u_{ij}^n = R_{ij}^n , \tag{26.1}$$

or

$$\Delta_t u_{ij}^n = \frac{\tau^n}{1+\tau^n \rho_{ij}^n} R_{ij}^n . \tag{26.2}$$

where ρ_{ij}^n is the spectral radius of $(A^x+A^y)_{ij}^n$. The factor multiplying R_{ij}^n may be interpreted as a locally adjusted time-step value, which for $\tau^n \to \infty$ approaches the local stability limit $(\rho_{ij}^n)^{-1}$.

2.4 SER schemes of second-order accuracy. The upwind spatial discretization of second-order accuracy, described and tested by Mulder and Van Leer [1] for the one-dimensional flow equations, applies in a straightforward way to the two-dimensional equations. In a Cartesian grid the two dimensions completely decouple, owing to property (10).

The relaxation schemes are essentially those of §2.3, but with the Jacobians in (20) evaluated at $(i\pm\frac{1}{2},j)$, $(i,j\pm\frac{1}{2})$, rather than (i,j), $(i\pm1,j)$, $(i,j\pm1)$; for details see [1]. Except for this adjustment, the second-order terms in the scheme are not accounted for in M_x^n or M_y^n. The second-order SER schemes therefore deviate more strongly from the full backward Euler scheme than the corresponding first-order schemes.

The stability analysis in the Appendix reveals that Jacobi relaxation is no longer stable, pattern relaxation is stable with sufficiently strong underrelaxation, whereas Gauss-Seidel relaxation is stable only when forward and backward sweeps are alternated (symmetric Gauss-Seidel). In demanding computations, e.g. on strongly non-uniform grids, with strong oblique shocks in the solution, without a preferential flow direction, it is recommended to use the best possible relaxation methods, i.e. symmetric line/Gauss-Seidel and symmetric point/Gauss-Seidel.

2.5 Multigrid relaxation. Any of the SER schemes of §2.3 or §2.4 may be used as the "smoother" in a multigrid cycle, with the symmetric Gauss-Seidel scheme as the first choice. The nonlinear version is suited for use in a "full-approximation-storage scheme" (FAS; for a review of multigrid concepts see [9]), while the linearized version is appropriate for a "correction scheme." The latter combination was successfully applied to the test problem of §3 by Mulder [10]. The FAS scheme for the Euler equations implemented by Jameson [11] does not include an SER scheme; relaxation is provided by a four-stage Runge-Kutta method with local time-step values.

3. A numerical comparison. The SER schemes of §2.3 and §2.4, and also the ADI and AF schemes of §2.2, were used to compute the steady transonic flow through a straight channel with a circular bump on the lower wall. The inflow Mach number was 0.85, the thickness of the bump was 4.2% of the chord length. The steady flow exhibits a shock almost choking the channel.

In marching toward the steady state, the isenthalpic Euler equations were used. At the walls reflection conditions were imposed with the help of mirror-image zones; the arc was described according to small-disturbance theory (thickness ignored, flow angle prescribed). Total pressure and cross-flow velocity (=0) were given at the inlet, static pressure at the outlet. The details of the equations and the boundary conditions can be found in [10], where the same problem was used for a multigrid experiment; for numerical solutions to this problem by other authors see [15].

Figure 1 shows the distributions of the pressure coefficient on both walls, obtained from solutions with first-order (a) and second-order (b) spatial accuracy, on a uniform grid of 32×16 zones.

In Table 1 and Table 2 are listed some of the many data gathered on the convergence speed achieved by the various schemes. The convergence process was monitored by the quantity RES defined by

$$\mathrm{RES}^n = \max_{k,i,j} \left(\frac{|R_k^n|}{|u_k^n| + h_k^n} \right)_{ij} , \tag{27.1}$$

where k=1,2,3 indicates the different conservation laws and h is a bias vector preventing division by zero. The time-step for SER schemes was chosen according to

$$\tau^n = \varepsilon / \mathrm{RES}^n , \tag{27.2}$$

which is similar to Eq. (7).

As a rule, the number of iterations needed to reduce the residuals by a factor of 10^{-10} is smaller when a scheme deviates less from

Experiment number	Kind of relaxation	Order of approximation	One cycle involves	Number of iterations per cycle	Number of iterations till convergence	Cpu-time spent (minutes)
1	line/GS	1	line(x),GS(+y), line(x),GS(-y); line(y),GS(+x), line(y),GS(-x)	2	123	1.2
2	line/GS	1	line(x),GS(+y), line(y),GS(+x)	1	174	1.7
3	line/GS	1	line(x),GS(+y), line(x),GS(-y)	1	204	1.7
4	line/GS	1	line(y),GS(+x), line(y),GS(-x)	1	92	0.83
5	line/GS	1	line(x),GS(+y)	½	243	2.1
6	line/GS	1	line(y),GS(+x)	½	162	1.4
7	line/GS	2	see nr. 1	2	183	2.4
8	line/GS	2	see nr. 3	1	221	2.6
9	line/GS	2	see nr. 4	1	313	3.8
10	zebra	1	line(x),pat(y)	½	243	2.0
11	zebra	1	line(y),pat(x)	½	162	1.4
12	zebra	2	line(y), pat(x;s=-0.25)	½	796	9.3
13	zebra	2	line(y), pat(x;s=-1.0)	½	575	6.8
14	line/Jacobi	1	line(x),Jac(y)	½	466	3.5
15	line/Jacobi	1	line(y),Jac(x)	½	294	2.2
16	line/Jacobi	1	line(x),Jac(y), line(y),Jac(x)	1	234	1.8
17	ADI	1	line(x),line(y)	1	486	3.7
18	AF(α=1)	1	line(x),line(y)	1	452	2.3

Table 1. Data on the convergence speed achieved by the line-relaxation schemes (SER and other) in solving the transonic flow problem of §3, on a grid of 32 x 16 zones. Problem parameters: M_∞ = 0.85, arc thickness = 4.2% of chord length. The iteration count is based on a unit including two line relaxations (regardless of their direction), making comparisons more or less meaningful. The cpu-time is given in minutes on an Amdahl V7B computer. The value of ε in the time-step formula (27.2) was 1.0 for all SER schemes. With ADI and AF the time-step, based on ε = 0.5, was frozen at the start, fixing the free-stream Courant number at a value of 4.13 for both experiments. Under-relaxation was applied only in the second-order zebra scheme (s = -0.25, -1.0), for the sake of stability.

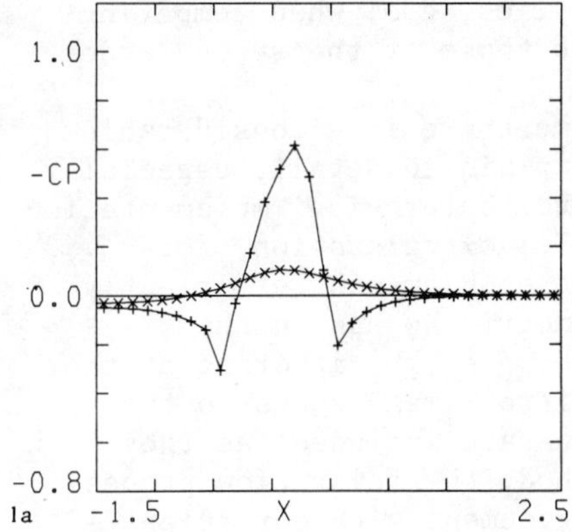

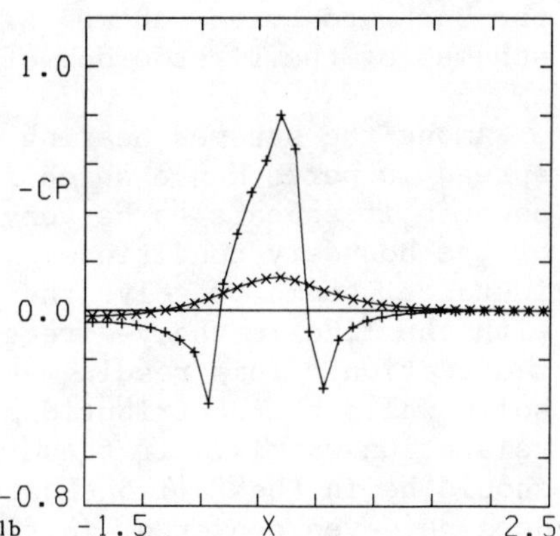

Figure 1. (a) Pressure coefficient on lower (+) and upper (x) wall for Mach 0.85 flow in a channel with a circular arc of 4.2% thickness on the lower wall, as computed from a solution on a 32×16 grid with first-order accuracy. Boundary conditions on the arc according to thin-airfoil theory. (b) As before, but computed with second-order accuracy.

Experiment number	Kind of relaxation	Order of approximation	One cycle involves	Number of iterations per cycle	Number of iterations till convergence	Cpu-time spent (minutes)
19	point/GS	1	GS(+x,+y); GS(-x,-y)	2	557	2.6
20	point/GS	1	GS(+x,+y); GS(+x,-y)	2	597	2.8
21	point/GS	1	GS(+x,+y); GS(-x,+y)	2	606	2.6
22	point/GS	1	GS(+x,+y)	1	713	3.0
23	point/GS	2	see nr. 19	2	575	4.0
24	point/Jacobi	1	Jac(x,y)	1	1367	4.4

Table 2. Convergence data for the point-relaxation schemes (SER only).

the backward Euler scheme. This rule applies, too, when comparing schemes of the first order of accuracy to those of the second order.

Among the schemes bearing the same name there is a considerable spread in performance which is hard to explain in detail, especially because it appears to be very sensitive to the precise implementation of the boundary conditions. Line/Gauss-Seidel relaxation, for instance, seems to solve the present test problem most efficiently with the line in the y-direction and symmetric sweeps in the x-direction. This result was obtained, however, by ignoring at the solid walls the contributions from the mirror-image zones to the relaxation matrix. If these contributions are included, as they should be in the full backward Euler method, the relaxation process does not even converge, in flagrant disagreement with our rule-of-thumb. Obviously, an analysis of the interference of numerical boundary conditions with relaxation schemes is due; meanwhile we chose to use the less complete linearization for all line relaxations in the y-direction.

Zebra relaxation, while second best for the first-order scheme, slows down considerably when used for the second-order scheme; this is due to the strong underrelaxation needed for stability or for convergence.

The performance of the ADI and AF schemes is found to depend critically on the value of the time-step; when regarded as relaxation methods these cannot be called robust. Attempts to optimize the choice of the time-step were not uniformly successful and therefore did not make the schemes more robust. In the present experiments a fixed time-step was used, determined by RES^0 and $\varepsilon = 0.5$. The choice $\varepsilon = 1.0$ leads to divergence for ADI and to non-convergence for AF.

When comparing the number of iterations till convergence to the cpu-time spent, one must realize that the block-elements of the matrix L^n used in the relaxation schemes were not computed at every time level. Their values, and the value of τ^n, were updated only when RES^n dropped below some control level and remained frozen until the next lower level was reached. For these levels the following sequence of fractions of RES^0 were used: 10^{-1}, 3×10^{-2}, 10^{-2}, 10^{-3}, 10^{-4}, 10^{-6}. Freezing has no significant effect on the relaxation process, except for AF, where it may change slow convergence into non-convergence (observed for $\varepsilon = 1.0$).

Along with the blocks of L^n, all blocks derived from these were frozen, i.e. the inverses of the main diagonal blocks needed for point relaxation, or the block elements of all line-wise LU-decompositions needed for line relaxation. This strategy leads to large savings on cpu-time; its weakness lies in the prerequisite that lots of storage space be available. In practice one will have to trade the upper limit to storage space for a lower limit to cpu-time; the trade-off is highly computer-dependent. Consider, for example, point/Jacobi, checkerboard and symmetric nonlinear point/Gauss-Seidel relaxation, each requiring the storage of only one block: the inverse of the main-diagonal block. While the Gauss-Seidel scheme offers the strongest relaxation per iteration, it may finish last when implemented on a supercomputer, since its update step does not vectorize.

Figures 2 through 7 show the convergence histories of some of the experiments compiled in Tables 1 and 2; RES stands for RES^n/RES^0. In most cases the residual norm does not decrease monotonically; various periodic and quasi-periodic fluctuations can be discerned. These correspond to the sequencing of sweep directions (as in Figure 6) or line directions (as in Figure 2b) and to the bouncing of disturbances from wall to wall (as in Figure 4).

4. Conclusions and recommendations. In the preceding sections it has been demonstrated that Switched Evolution/Relaxation schemes incorporating a classic relaxation method are robust means to compute steady discontinuous solutions of hyperbolic systems of conservation laws such as the Euler equations. It is crucial that the spatial discretization be upwind biased. If storage space is not restricted, the most complex relaxation methods are also the most efficient, owing to the possibility of keeping the coefficient blocks frozen during many iteration steps. Alternating-Direction Implicit and Approximate-Factorization methods are less efficient than equally complex SER methods, and not at all robust.

It is not surprising that other advocates of upwind differencing, independently or through interaction, have come to the same conclusions. Chakravarthy [12] has applied the point/Gauss-Seidel scheme to a variety of aerodynamic problems, with remarkable success; Dadone and Napolitano [13] recently turned from using AF to using SER schemes.

All SER schemes allow of underrelaxation, which improves short wave damping. It turns out that overrelaxation, a standard routine for the iterative solution of second-order elliptic equations, does not work for first-order equations (see the Appendix).

We recommend the further development of and experimentation with SER schemes requiring only one block evaluation and inversion. Neither the complexity nor the storage requirements of such schemes are extravagant, so that their application, even to three-dimensional flow problems, is within the capacity of today's computers. This view differs somewhat from Jameson's [11], who puts more emphasis on the storage aspect and therefore excludes any blocks from his

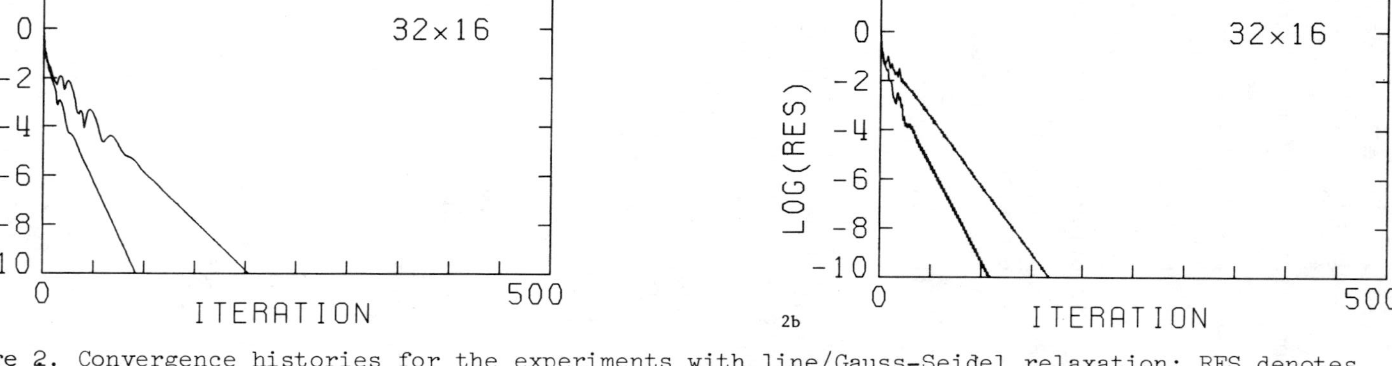

Figure 2. Convergence histories for the experiments with line/Gauss-Seidel relaxation; RES denotes RES^n/RES^0, see Eq. (27.1). (a) Experiments nr. 3 (slower convergence) and nr. 4 (faster convergence); (b) nr. 1 (1st order, fast) and nr. 7 (2nd order, slow).

2
0
-2
-4
-6
-8
-10
LOG(RES)
32×16
0
500
ITERATION

Figure 3. Convergence histories for zebra relaxation, experiments nr. 11 (1st order, fast) and nr. 13 (2nd order, slow).

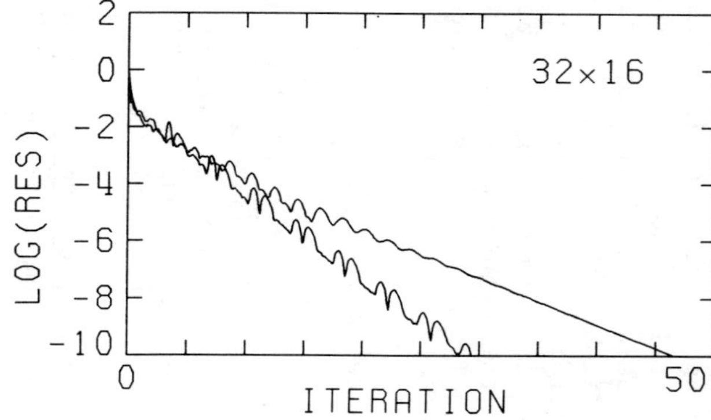

Figure 4. Convergence histories for line/Jacobi relaxation, experiments nr. 14 (slow) and nr. 15 (fast).

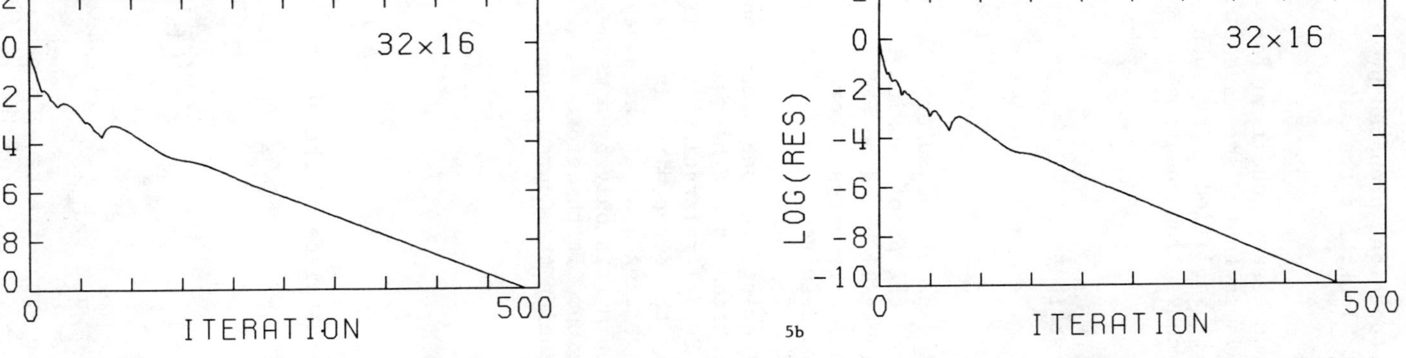

Figure 5. Convergence histories for non-SER relaxation. (a) Experiment nr. 17 (ADI); (b) nr. 18 (AF).

Figure 6. Convergence histories for point/ Gauss-Seidel relaxation, experiments nr. 19 (fast) and nr. 22 (slow).

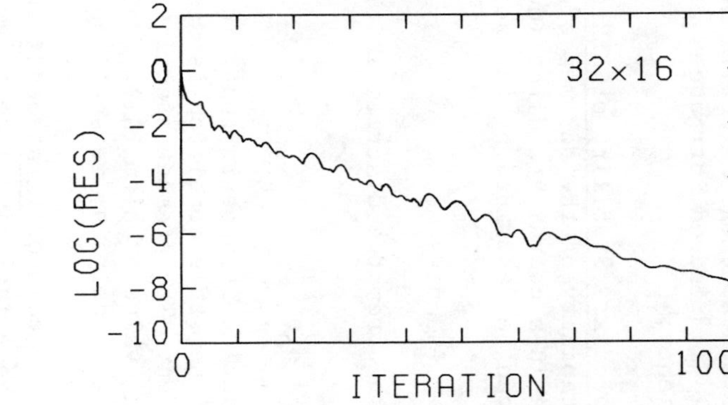

Figure 7. Convergence history for point/ Jacobi relaxation, experiment nr. 24.

relaxation schemes. A modification of his technique proposed by Turkel [14] indeed introduces a coefficient block, for the sake of efficiency.

Appendix: Stability of SER schemes. A necessary condition for the stability of the schemes of §2.3 and §2.4 is that plane waves moving in a coordinate direction shall not be amplified. Any such wave can be described by the scalar linear convection equation

$$q_t + aq_x = 0 \tag{A1.1}$$

starting from the discrete initial-value distribution

$$q_k^0 = q_0^0 e^{i\xi k}. \tag{A1.2}$$

In practice it appears that, in the special case of the Euler equations, this condition is also sufficient; the proof remains to be given. The present stability analysis is still based on the initial-value problem (A1). Pattern relaxation requires a more elaborate analysis; we shall only state some results.

In the case of first-order upwind differencing, the one-dimensional SER scheme including line relaxation is identical to the backward Euler scheme, hence is stable. With Gauss-Seidel relaxation it still is identical to the backward Euler scheme if the sweep direction is the same as the wave direction. A wave travelling against the sweep direction is accounted for in the relaxation matrix only by a main-diagonal element, just as in Jacobi relaxation. It therefore suffices to study Jacobi relaxation; introducing the Courant number

$$\sigma = \frac{a\tau}{\Delta\xi} \geq 0 , \tag{A2}$$

the SER scheme for Eq. (A1.1) reads

$$\{1+\sigma(1-s)\}\Delta_t q_k^n = - \sigma(1-T^{-1})q_k^n. \tag{A3}$$

As explained in §2.3, a scalar s has taken the place of the shift operator T^{-1}. The amplification factor $g(\xi,\sigma,s)$ of this scheme follows upon inserting (A1.2):

$$\{1+\sigma(1-s)\}(g-1) = - \sigma(1-e^{-i\xi}) , \tag{A4.1}$$

or

$$g = 1 - \frac{\sigma}{1+\sigma(1-s)} (1-e^{-i\xi}) . \tag{A4.2}$$

This corresponds to the forward Euler scheme with effective Courant number $\sigma/\{1+\sigma(1-s)\}$, which for first-order upwind differencing is stable up to a Courant-number value of 1. It follows that

$$s \le \frac{1}{\sigma} \tag{A5.1}$$

for stability; if we insist on stability for arbitrarily large σ we find

$$s \le 0 , \tag{A5.2}$$

which is Eq. (24). This condition also suffices to stabilize pattern relaxation, which is closer in performance to Gauss/Seidel than to Jacobi relaxation.

Overrelaxation, meant to damp efficiently the longest waves ($\xi \sim 0$, $T = I$), cannot be achieved through the approximation (23.1). In approximating $T^{\pm 1}$, two diagonals must be involved:

$$T^{\pm 1} = sI + (1-s)T^{\mp 1} . \tag{A6}$$

The scheme for adverse waves, e.g. with T^{-1} expressed in I and T during a backward sweep, becomes

$$\{I+\sigma(1-s)(I-T)\}\Delta_t q_k^n = -\sigma(I-T^{-1})q_k^n , \tag{A7}$$

with amplification factor

$$g = 1 - \frac{1-e^{-i\xi}}{1/\sigma+(1-s)(1-e^{i\xi})} \tag{A8}$$

For small α this reduces to

$$g = \frac{i\xi+O(\xi^2)}{1/\sigma-(1-s)i\xi+O(\xi^2)} , \tag{A9}$$

and if we choose σ much larger than the largest spatial frequency that can be represented on a finite grid, e.g. $1/\sigma = O(\xi^2)$, we finally get

$$g = 1 + \frac{1}{1-s} + O(\xi) . \tag{A10}$$

For $s = 2$, g is of the order of ξ, which is precisely the aim of overrelaxation. From (A6) we see that this choice means to replace backward differencing by forward differencing and vice versa, i.e.

$$I - T^{-1} = T - I . \tag{A11}$$

Unfortunately, $s = 2$ leads to violent growth of the shortest waves for all but the smallest values of σ; e.g., $\xi = \pi$ and $\sigma = \frac{1}{2}$ make the denominator in (A9) vanish while the numerator remains finite.

We may therefore conclude that overrelaxation with first-order upwind SER schemes is not possible.

It is tempting to investigate if underrelaxation may be improved through the use of two diagonals. If we approximate T^{-1} according to

$$T^{-1} = sI + (1+s)\,T^{+1} , \tag{A12}$$

which is satisfied by the shortest waves ($\alpha = \pi$, $T = -1$), the scheme for adverse waves becomes

$$[I+\sigma\{(1-s)I-(1+s)T\}]\Delta_t q_k^n = -\sigma(I-T^{-1})q_k^n , \tag{A13}$$

with amplification factor

$$g = 1 - \frac{1-e^{-i\xi}}{1/\sigma+\{(1-s)-(1+s)e^{i\xi}\}} \tag{A14}$$

Inserting $\xi = \pi - \phi$, with ϕ small, we get

$$g = \frac{1/\sigma - i(2+s)\phi + O(\phi^2)}{1/\sigma + 2 - i(1+s)\phi + O(\phi^2)} , \tag{A15}$$

for large enough σ, i.e. $1/\sigma = O(\phi^2)$, this leads to

$$g = -i(1+\tfrac{1}{2}s)\phi + O(\phi^2) , \tag{A16}$$

indicating extra strong damping if

$$s = -2, \tag{A17}$$

Inserting (A17) into (A12) yields

$$I + T^{-1} = -\,(I + T) , \tag{A18}$$

which, for the shortest waves, is as good an extrapolation as (A11) is for the longest waves.

It is easily verified that scheme (A13) with (A17) is unconditionally stable.

With <u>second-order</u> upwind differencing the SER schemes including line relaxation or downwind Gauss-Seidel sweeping deviate from the backward Euler scheme, as the second-order terms are not treated implicitly. For Eq. (A1.1) these schemes read

$$\{I+\sigma(I-T^{-1})\}\Delta_t q_k^n = -\sigma(I-T^{-1})\{1+\tfrac{1}{4}(T-T^{-1})\}; \tag{A19}$$

their amplification factor is

$$g = \frac{1-\sigma(1-e^{-i\alpha})\,\frac{i}{2}\sin\xi}{1+\sigma(1-e^{-i\xi})} \tag{A20.1}$$

$$= \frac{1 + 2\sigma \sin^2\frac{\xi}{2}\cos^2\frac{\xi}{2} - i\sigma\sin\xi\sin^2\frac{\xi}{2}}{1 + 2\sigma\sin^2\frac{\xi}{2} + i\sigma\sin\xi}, \tag{A20.2}$$

with modulus not exceeding 1. For $\sigma \to \infty$ we have $g = -\frac{i}{2}\sin\xi$, hence $|g| \le \frac{1}{2}$.

With Jacobi relaxation, or upwind Gauss-Seidel sweeping, the second-order SER scheme becomes unstable for any value of σ. Like scheme (A3) it is equivalent to the forward Euler scheme, used with an effective Courant number $\sigma/\{1+\sigma(1-s)\}$. The locus of the Fourier transform of the spatial-differencing operator in the complex plane now has fourth-order contact with the imaginary axis for $\alpha=0$, whereas the stability domain for the forward Euler scheme has only second-order contact. To match the stability domain with the spectral locus, a two-step technique would have to be used.

The instability is insuperable for pure Jacobi relaxation, but can be suppressed for pattern relaxation by taking negative values of s (down to $-\sqrt{2}$ if necessary), and for Gauss-Seidel relaxation by alternating between upwind and downwind sweeps. In the worst case, $\sigma \to \infty$, the amplification factors are

$$g_{\text{upwind}} = 1 - \frac{1}{1-s}(1-e^{-i\xi})(1+\frac{i}{2}\sin\xi), \tag{A21}$$

$$g_{\text{downwind}} = -\frac{i}{2}\sin\xi, \tag{A22}$$

and the modulus of their product remains safely below 1, for $s \le 0$.

Underrelaxation through the use of two diagonals works as well in the second-order case as in the first-order case. Inserting (A12) into scheme (A19) yields, for small $\phi = \pi-\xi$,

$$g = \frac{1/\sigma-i(3+s)\phi+0(\phi^2)}{1/\sigma+2-i(1+s)\phi+0(\phi^2)}, \tag{A23}$$

from which it is seen that, for sufficiently large σ, extra strong damping results when

$$s = -3. \tag{A24}$$

With this value of s the second-order scheme for the adverse waves is unconditionally stable.

It is worthwhile to mention that underrelaxation makes a scheme, whether first-order or second-order accurate, a good smoother for use in a multigrid strategy.

REFERENCES

(1) W. A. MULDER and B. VAN LEER, Experiments with implicit upwind methods for the Euler equations, J. Comp. Phys., 59 (1985), pp. 232-246.

(2) A. DADONE and M. NAPOLITANO, An implicit Lambda-Scheme, AIAA paper nr. 82-0972, AIAA-ASME 3rd Joint Thermophysics, Fluids, Plasma and Heat Transfer Conference, St. Louis, Mo., June 1982.

(3) J. DOUGLAS and J. GUNN, A general formulation of alternating direction methods, I, Numer. Math. 6 (1964), pp. 428-453.

(4) R. W. BEAM and R. F. WARMING, An implicit finite-difference algorithm for hyperbolic systems in conservation-law form, J. Comp. Phys. 22 (1976), pp. 87-110.

(5) E. WACHSPRESS, Iterative solution of elliptic systems, Prentice Hall, Englewood Cliffs, N.J., 1966.

(6) Y. LIU and H. LOMAX, Nonstationary relaxation methods for the Cauchy-Riemann and the one-dimensional Euler equations, AIAA paper 83-1901, AIAA 6th Computational Fluid Dynamics Conference, Danvers, Mass., July 1983.

(7) S. S. ABARBANEL, D.L. DWOYER and D. GOTTLIEB, Improving the convergence rates of parabolic ADI methods, ICASE Report 82-28 (1982).

(8) B. VAN LEER, Flux-vector splitting for the Euler equations, Lecture Notes in Physics 170 (1982), pp. 507-512.

(9) A. BRANDT, Guide to multigrid development, Lecture Notes in Mathematics 960 (1982), pp. 220-312.

(10) W. A. MULDER, Multigrid relaxation for the Euler equations, preprint (1984), Leiden University Observatory, submitted to J. Comp. Phys.

(11) A. JAMESON, Numerical solution of the Euler equations for compressible inviscid fluids, this volume, pp.

(12) S. R. CHAKRAVARTHY and S. OSHER, Higher-resolution applications of the Osher upwind scheme for the Euler equations, AIAA paper 83-1943, AIAA 6th Computational Fluid Dynamics Conference, Danvers, Mass., July 1983.

(13) M. NAPOLITANO and A. DADONE, Three-dimensional implicit lambda-method, NASA Contractor Report 172264 (ICASE), October 1983.

(14) E. TURKEL, Acceleration to a steady state for the Euler equations, this volume, pp.

(15) A. RIZZI and H. VIVIAND, eds., Numerical methods for the computation of inviscid transonic flows with shock waves, Notes on Numerical Fluid Mechanics, Vol. 3, Vieweg, Braunschweig, 1981.

PSEUDO-UNSTEADY SYSTEMS FOR STEADY INVISCID FLOW CALCULATION

HENRI VIVIAND*

Abstract. The pseudo-unsteady, or false transient, approach is now of classical use for the calculation of steady flows, whether viscous or inviscid. In the case of inviscid flows, with which we are concerned here, various pseudo-unsteady techniques have been implemented for transonic flow calculation, the most classical being the use of the steady Bernoulli relation, and the use of local time steps. But, to our knowledge, no systematic study of pseudo-unsteady formulations has been made in the past, and it is the purpose of this paper to present such a study.

The primary concern of this work is the calculation of steady isoenergetic flows, for which the steady Bernoulli relation holds, even when shock waves are present. However, we found that the incompressible case could be covered simultaneously in a uniform presentation, if care was taken to choose the pressure as basic variable.

Starting from the steady Euler equations for steady inviscid flows, we construct a general family of systems of first order partial differential equations of pseudo-unsteady type depending on 5 scalar parameters. These pseudo-unsteady systems (P.U.S.) are required to be in intrinsic form, that is to say to be independent of any projection axis system.

Two essential conditions must be satisfied by these systems : firstly the condition that the P.U.S. be of hyperbolic type with respect to time, and, secondly, a condition bearing on the number of eigenvalues of same sign. These conditions are discussed, and they are completely determined explicitly in terms of the parameters for a sub-family of P.U.S. which depends on 4 parameters instead of 5.

Several examples are presented and discussed in terms of the various properties which these P.U.S. can possess. In particular it is shown that there exist P.U.S. which are optimal with regard to an explicit numerical stability criterion of C.F.L. type. Other properties of interest are : conservative or non-conservative form,

+Office National d'Etudes et de Recherches Aérospatiales, 92320 Châtillon, FRANCE.

special relations with exact unsteady equations, particular forms of the eigenvalues.

List of symbols

a coefficient of pseudo-unsteady system (3.3)

c sound speed ... (Eq. (2.6))

d coefficient of pseudo-unsteady system (3.3)

d_1 coefficient of pseudo-unsteady system (3.6)

$\vec{D} = \frac{1}{\rho}$ div $[\rho\vec{U}\otimes\vec{U} + p\vec{I}]$.... (Eq. 2.15b)

e coefficient of pseudo-unsteady system (3.3)

e_1 coefficient of pseudo-unsteady system (3.6)

h specific enthalpy (Eq. (2.4))

$\vec{M} = (\vec{U}.\,\text{grad})\,\vec{U} + \frac{1}{\rho}\,\text{grad}\,p$.... (Eq. (2.14b))

p pressure

P total pressure for an incompressible flow ... (Eq (3.18))

S specific entropy ....(Eq. (2.5))

T absolute temperature .... (Eq. (2.5))

$\vec{U}$ velocity vector

U velocity modulus

Greek Letters

α coefficient of pseudo-unsteady system (3.3)

α_1 coefficient of pseudo-unsteady system (3.6)

β coefficient of pseudo-unsteady system (3.3)

β_1 coefficient of pseudo-unsteady system (3.6)

$\gamma = 1 + \left[\rho\frac{\partial h}{\partial p} - 1\right]^{-1}$ ... (Eq. (2.7))

$\delta = ad + (ed - \alpha\beta)U^2$ .... (Eq. (3.5))

$\lambda^{(k)}$ = eigenvalues .... (Eq. (4.9))

μ_0 = first component of eigenvector ν .... (Eq. (4.14)

$\vec{\mu} = (\mu_1, \mu_2, \mu_3)$, last three components of eigenvector ν .... (Eq. (4.14))

$\vec{\xi}$ arbitrary unit vector of space .... (Eq. (4.8))

ρ density

$\Psi = \text{div}\vec{U} - \frac{1}{c^2}\vec{U} \cdot \text{grad}\,\frac{U^2}{2}$ .... (Eq. (2.14a))

$\Theta = \frac{1}{\rho}\,\text{div}\,(\rho\vec{U})$........ (Eq. (2.15a)

1. Introduction. The pseudo-unsteady (or false transient) approach is the basis of several methods for the calculation of viscous or inviscid steady flows, and a review of these methods has been recently given by Peyret and Viviand [1].

We recall that, in this approach, a steady flow is determined as the asymptotic limit, for large time, of the solution of a system of unsteady equations, this system not being required to have a physical meaning as far as the transient stage is concerned.

The pseudo-unsteady approach has been developed and implemented at ONERA since many years for the calculation of steady inviscid compressible flows. References [2] to [5] give an overview of the methods used and of the problems treated. In fact, the pseudo-unsteady formulations practically used for compressible inviscid flows are few. In the case of iso-energetic rotational flows, the formulation based on the exact unsteady continuity and momentum equations in conservative form and on the steady Bernoulli relation in place of the exact energy equation is now of common use ; we shall refer to it as method "H". It is difficult to determine the precise origin of this method since it rests upon a rather obvious simplifying idea, and it seems to have been used long before the mathematical properties of the corresponding system of equations were studied [4], [6].

In the case of homentropic irrotational flows, various formulations have been discussed by Veuillot and Viviand [3], [4], and by Essers [7], [8], [9], a review of which has been presented by Viviand [2]. The practical applications of these methods up to now seem to be restricted to two-dimensional flows, and their development suffers from the existence of very efficient methods for the solution of the full potential equation.

Let us also mention the "dash-pot" methods of Essers [7], [10], which present a great theoretical interest, but which do not seem to have reached the stage of a common procedure, may be due to their complexity.

The present work is concerned with a systematic study of pseudo-unsteady systems (this expression will be hereafter abbreviated as

P.U.S.) for iso-energetic and in general rotational flows. Whereas the primary interest in this work was the calculation of compressible flows, it appeared that the case of an incompressible fluid could be treated together with that of iso-energetic compressible flows in a uniform way. The pseudo-unsteady approach does not seem to have been used for inviscid incompressible flows, but it could present a practical interest for such flows if they are rotational or if they include vortex sheets. Moreover, since such methods are used to solve the incompressible Navier-Stokes equations (e.g. see [1]), it is quite relevant to determine how they behave in regions where viscous effects become negligible, that is to say for inviscid flows.

Our study centers on the basic condition that the first order system to be solved be of hyperbolic type with respect to the time variable. This condition seems a natural one to impose in order that perturbations remain bounded in time (in linearized theory), and it gives great advantages for implementing numerical solution methods.

Indeed the numerical solution of hyperbolic systems has made important progress, and numerous explicit or implicit methods now exist. The hyperbolic nature of the system allows a systematic treatment of all boundary conditions by means of compatibility relations. This treatment can be extended to the matching of sub-domains, thus making the zonal approach an attractive and general technique for complex problems as discussed by Cambier et al. [11], [12]. All these numerical methods and techniques apply equally well to any pseudo-unsteady system, whether for compressible or for incompresible flow, as long as the system is hyperbolic with respect to time.

A more detailed version of this work exists as an ONERA report [13]. It also includes a discussion of P.U.S. for the case of an incompressible fluid, and a study of a general class of simpler P.U.S., which depend on two parameters only, for the calculation of steady compressible flows which besides being iso-energetic are also homentropic.

2. Steady equations. We consider first the continuity and the momentum equations written in conservative form for steady inviscid flows :

$$\operatorname{div}(\rho \vec{U}) = 0 \tag{2.1}$$

$$\operatorname{div}(\rho \vec{U} \otimes \vec{U} + p \vec{I}) = 0 \tag{2.2}$$

where ρ is the density, $\vec{U}$ the velocity vector, p the pressure ; the symbol $\otimes$ represents the dyadic product of vectors (a second rank tensor), and $\vec{I}$ is the unit tensor.

For an incompressible fluid, ρ is a constant and p is an independent variable. For a compressible fluid, we consider only the case of iso-energetic flows, that is to say having uniform total enthalpy, hence verifying Bernoulli's relation :

$$h + 1/2\, U^2 = H_o \ (\text{Cst}) \tag{2.3}$$

where U is the modulus of the velocity, and where h, the specific enthalpy, is a known function of p and ρ :

$$h = h\,(p, \rho) \tag{2.4}$$

Thus relation (2.3) allows to determine ρ as a function of p and U.

Still considering a compressible fluid, the specific entropy S is such that we have the differential relation :

$$TdS = dh - 1/\rho \ dp \tag{2.5}$$

where T is the absolute temperature. The sound speed c verifies $dp = c^2 \, d\rho$ if $dS = 0$; therefore, developing dh in terms of dp and $d\rho$ from (2.4), we obtain :

$$c^2 = - \rho \frac{\partial h}{\partial \rho} \left(\rho \frac{\partial h}{\partial p} - 1 \right)^{-1} \tag{2.6}$$

It is convenient to define a quantity γ , function of p and ρ , through the relation :

$$\gamma = \rho \frac{\partial h}{\partial p} \left(\rho \frac{\partial h}{\partial p} - 1 \right)^{-1} = 1 + \left(\rho \frac{\partial h}{\partial p} - 1 \right)^{-1} \tag{2.7}$$

For a perfect gas (i.e. a gas such that $h = h(T)$ and $p = \rho RT$), this definition of γ gives back the usual ratio of specific heats, but we do not need this hypothesis in all that follows. Differentiating the Bernoulli relation (2.3), and using (2.6) and (2.7), we obtain the following relation to be used later

$$c^2 \, d\rho = \gamma dp + (\gamma - 1) \, \rho \, UdU \tag{2.8}$$

We shall also consider the non-conservative forms of the continuity and momentum equations, namely :

$$\text{div} \, \vec{U} + \frac{1}{\rho} \vec{U} . \, \text{grad} \rho = 0 \tag{2.9}$$

$$(\vec{U} . \, \text{grad}) \, \vec{U} + \frac{1}{\rho} \, \text{grad} \ p = 0 \tag{2.10}$$

For a compressible fluid, we know that a consequence of these equations is the entropy equation :

$$\vec{U} . \, \text{grad} \ S = 0 \tag{2.11}$$

and that the continuity equation can be written in the form :

$$\operatorname{div} \vec{U} - \frac{1}{c^2} \vec{U} . \operatorname{grad} \frac{U^2}{2} = 0 \tag{2.12}$$

This form remains valid for an incompressible fluid if we take $1/c = 0$ in this case.

For an incompressible fluid, an equation equivalent to (2.11) can be formed with the total pressure $P = p + \frac{1}{2}\rho U^2$:

$$\vec{U} . \operatorname{grad} P = 0 \tag{2.13}$$

The system of first order partial differential equations (2.1) and (2.2), or (2.12) and (2.10), is a closed system if relations (2.3) and (2.4) are taken into account in the case of a compressible fluid.

For arbitrarily given velocity field $\vec{U}$ and pressure field p, we define quantities Ψ , $\vec{M}$, Θ , $\vec{D}$ by the identities :

$$\left.\begin{aligned} \Psi &\equiv \operatorname{div} \vec{U} - {}^1\!/_{c^2}\, \vec{U} . \operatorname{grad} \frac{U^2}{2} && \text{a)} \\ \vec{M} &\equiv (\vec{U} . \operatorname{grad})\, \vec{U} + {}^1\!/_{\rho} \operatorname{grad} p && \text{b)} \end{aligned}\right\} \tag{2.14}$$

$$\left.\begin{aligned} \Theta &\equiv {}^1\!/_{\rho} \operatorname{div} (\rho \vec{U}) && \text{a)} \\ \vec{D} &\equiv {}^1\!/_{\rho} \operatorname{div} (\rho \vec{U} \otimes \vec{U} + p\vec{I}) && \text{b)} \end{aligned}\right\} \tag{2.15}$$

In these definitions, all the terms are known functions of p and $\vec{U}$: for an incompressible fluid we have $1/c^2 = 0$ and ρ is a known constant ; for a compressible fluid ρ and c^2 are calculated by means of relations (2.3), (2.4) and (2.6) (so that in particular ρ obeys relation (2.8)). Keeping this in mind, the following relations can be established :

$$\left.\begin{aligned} \Theta &\equiv \Psi + {}^{\gamma}\!/_{c^2}\, \vec{U} . \vec{M} && \text{a)} \\ \vec{D} &\equiv \vec{M} + (\Psi + {}^{\gamma}\!/_{c^2}\, \vec{U} . \vec{M})\, \vec{U} && \text{b)} \end{aligned}\right\} \tag{2.16}$$

$$\left.\begin{aligned} \Psi &\equiv (1 + \gamma\, U^2\!/_{c^2})\, \Theta - {}^{\gamma}\!/_{c^2}\, \vec{U} . \vec{D} && \text{a)} \\ \vec{M} &\equiv \vec{D} - \vec{U}\, \Theta && \text{b)} \end{aligned}\right\} \tag{2.17}$$

These relations can be put in a convenient matrix form (with obvious multiplication conventions according to the type of the quantities : scalar, vector, or tensor) :

$$\begin{pmatrix} \Theta \\ \vec{D} \end{pmatrix} = \begin{pmatrix} 1 & \frac{\gamma}{c^2} \vec{U} \\ \vec{U} & \vec{I} + \frac{\gamma}{c^2} \vec{U} \otimes \vec{U} \end{pmatrix} \begin{pmatrix} \Psi \\ \vec{M} \end{pmatrix} \tag{2.18}$$

$$\begin{pmatrix} \Psi \\ \vec{M} \end{pmatrix} = \begin{pmatrix} 1+\gamma\frac{U^2}{c^2} & -\frac{\gamma}{c^2}\vec{U} \\ -\vec{U} & \vec{I} \end{pmatrix} \begin{pmatrix} \textcircled{H} \\ \vec{D} \end{pmatrix} \qquad (2.19)$$

3. General form of pseudo-unsteady systems

3.1 Definition

By choosing $\vec{U}$ and p (rather than $\vec{U}$ and ρ) for basic dependent variables, we can treat simultaneously both cases of incompressible or compressible fluids. Then a very general class of P.U.S. is constructed, directly in normal form with respect to $\vec{U}$ and p, by equating the derivatives of $\vec{U}$ and p with respect to the time variable t (in fact a pseudo-time) to general linear combinations of Ψ and $\vec{M}$.

We require these P.U.S. to have an intrinsic definition, that is to say independently of any projection system which might be used to determine $\vec{U}$ and $\vec{M}$. Therefore the coefficients of the linear combinations of Ψ and $\vec{M}$ must be functions of U and p only, if they are scalar, and moreover the vectorial coefficients must be proportional to $\vec{U}$. These simple considerations lead to the following general form of the P.U.S. to be considered in this study :

$$\frac{\partial p}{\partial t} + \rho\alpha\,\vec{U}.\vec{M} + \rho d\,\Psi = 0 \qquad (3.1)$$

$$\frac{\partial \vec{U}}{\partial t} + a\vec{M} + \beta\vec{U}\Psi + e\vec{U}(\vec{U}.\vec{M}) = 0 \qquad (3.2)$$

where the five scalar coefficients a, β , e, α , d are given bounded functions of p and U. Notice that a, β , α , eU^2 and d/U^2 are dimensionless. In all of what follows we shall assume $U \neq 0$.

The system (3.1), (3.2) can also be written in matrix form as follows :

$$\frac{\partial}{\partial t}\begin{pmatrix} p \\ \vec{U} \end{pmatrix} + \begin{pmatrix} \rho d & \rho\alpha\vec{U} \\ \beta\vec{U} & a\vec{I} + e\,\vec{U}\otimes\vec{U} \end{pmatrix}\begin{pmatrix} \Psi \\ \vec{M} \end{pmatrix} = 0 \qquad (3.3)$$

A first condition to be satisfied by P.U.S. of type (3.3) is that at the steady state, it admits a unique solution $\Psi = 0$ and $\vec{M} = 0$; the corresponding determinant should be nonzero, hence the conditions:

$$a \neq 0 \qquad (3.4)$$

$$\delta = ad + (ed - \alpha\beta)\, U^2 \neq 0 \tag{3.5}$$

3.2 Semi-conservative form

Expressing Ψ and $\vec{M}$ in terms of Θ and $\vec{D}$ by means of relations (2.19), we get a new form of P.U.S. (3.3) which can be said to be semi-conservative since all space derivatives appear only in divergence terms. We write this new form as follows :

$$\frac{\partial}{\partial t}\begin{pmatrix} p \\ \vec{U} \end{pmatrix} + \begin{pmatrix} \rho d_1 & \rho \alpha_1 \vec{U} \\ \beta_1 \vec{U} & a\vec{I} + e_1 \vec{U}\otimes\vec{U} \end{pmatrix}\begin{pmatrix} \Theta \\ \vec{D} \end{pmatrix} = 0 \tag{3.6}$$

where the new coefficients d_1, α_1, β_1, e_1, are related to the coefficients of (3.3) by :

$$\left.\begin{aligned} d_1 &= \left(1+\gamma\frac{U^2}{c^2}\right)d - U^2\alpha = d - U^2\alpha_1 \\ \alpha_1 &= \alpha - \gamma\frac{d}{c^2} \\ \beta_1 &= \left(1+\gamma\frac{U^2}{c^2}\right)\beta - (a + eU^2) = \beta - a - U^2 e_1 \\ e_1 &= e - \frac{\gamma}{c^2}\beta \end{aligned}\right\} \tag{3.7}$$

or, inversely :

$$\left.\begin{aligned} d &= d_1 + \alpha_1 U^2 \\ \alpha &= \alpha_1\left(1+\gamma\frac{U^2}{c^2}\right) + \frac{\gamma}{c^2}d_1 \\ \beta &= a + \beta_1 + e_1 U^2 \\ e &= e_1\left(1+\gamma\frac{U^2}{c^2}\right) + \frac{\gamma}{c^2}(\beta_1 + a) \end{aligned}\right\} \tag{3.8}$$

Let us also note that :

$$\delta = ad + (ed - \alpha\beta)U^2 = ad_1 + (e_1 d_1 - \alpha_1\beta_1)U^2 \tag{3.9}$$

3.3 Strictly conservative systems

A strictly conservative P.U.S. is a system which can be put in the form below :

$$\frac{\partial}{\partial t}\, g(p,U) + \rho\, \Theta = 0 \tag{3.10}$$

$$\frac{\partial}{\partial t}\left[\vec{U}\, F(p,U)\right] + \rho\, \vec{D} = 0 \tag{3.11}$$

where g and F are given functions of p and U such that p and U can be uniquely determined from g and $\overline{F}$ = UF.

Any system of type (3.6) cannot in general be put in the form (3.10), (3.11), but any system (3.10), (3.11) belongs to the class of systems (3.6) and we can determine the corresponding coefficients d_1, α_1, a, β_1, e_1.

By inverting (3.6) to express Θ and $\vec{D}$ linearly in terms of $\partial p/\partial t$ and $\partial\vec{U}/\partial t$, and by comparing with (3.10) and (3.11) after developing $\partial g/\partial t$ and $\partial(\vec{U}F)/\partial t$, we obtain :

$$\left.\begin{aligned}
\frac{a + e_1 U^2}{\delta} &= \frac{\partial g}{\partial p} && \text{a)}\\
\frac{\alpha_1}{\delta} &= -\frac{1}{\rho U}\frac{\partial g}{\partial U} && \text{b)}\\
\frac{\beta_1}{\delta} &= -\frac{\partial F}{\partial p} && \text{c)}\\
a &= \frac{\rho}{F} && \text{d)}\\
\frac{\alpha_1\beta_1 - e_1 d_1}{\delta} &= \frac{1}{UF}\frac{\partial F}{\partial U} && \text{e)}
\end{aligned}\right\} \tag{3.12}$$

Using (3.9), we deduce from (3.12 d) and (3.12e) :

$$\rho\,\frac{d_1}{\delta} = \frac{\partial \overline{F}}{\partial U} \quad \text{, with } \overline{F} = UF \tag{3.13}$$

To determine δ , divide (3.9) by δ^2 and express the right-hand side by means of (3.12) and (3.13). The result is :

$$\frac{1}{\delta} = \frac{1}{\rho}\left(\frac{\partial g}{\partial p}\frac{\partial \overline{F}}{\partial U} \quad \frac{\partial g}{\partial U}\frac{\partial \overline{F}}{\partial p}\right) = \frac{1}{\rho}\,\frac{\partial(g,\overline{F})}{\partial(p,U)} \tag{3.14}$$

Thus the coefficients, d_1, α_1, a, β_1, e_1 corresponding to a strictly conservative P.U.S. are completely determined in terms of g and $\overline{F}$. The conditions $\delta \neq 0$ and δ bounded ensure that p and U can be uniquely calculated from g and $\overline{F}$.

In practice, it may be convenient to explicitly give p and U as functions of g and $\overline{F}$ rather than the opposite. Then the coefficients can also be determined from the relations (considering $p = p(g,\overline{F})$, $U = U(g, \overline{F})$) :

$$\left.\begin{aligned} a + e_1 U^2 &= \rho \frac{\partial U}{\partial \overline{F}} \\ \alpha_1 &= \frac{1}{U} \frac{\partial p}{\partial \overline{F}} \\ \beta_1 &= \frac{\rho}{U} \frac{\partial U}{\partial g} \\ d_1 &= \frac{\partial p}{\partial g} \end{aligned}\right\} \qquad (3.15)$$

the coefficient a being calculated from (3.12d).

Let us also note the relation :

$$\frac{ed - \alpha\beta}{\delta} = \frac{1}{FU}\left(\frac{\partial g}{\partial U} - \frac{\partial F}{\partial U}\right) \qquad (3.16)$$

3.4 Equation for the entropy or for the total pressure

For a compressible fluid, the entropy S is related to h and p through Eq. (2.5) ; taking into account Eq. (2.3), we obtain :

$$\rho T dS = - \rho \vec{U}.d\vec{U} - dp$$

For an incompressible fluid, the total pressure $P = p + \frac{1}{2}\rho U^2$ verifies a similar relation :

$$dP = \rho \vec{U}.d\vec{U} + dp$$

We also note that $\rho \vec{U}.\vec{M}$ is equal either to $- \rho T\vec{U}.\text{grad } S$ or to $\vec{U}.\text{grad } P$. Hence either S or P obeys a pseudo-unsteady equation which is easily obtained by multiplying Eq. (3.2) by $\vec{U}$ and adding it to Eq. (3.1) :

$$\frac{\partial S}{\partial t} + (a + eU^2 + \alpha)\vec{U}.\text{grad}\, S - \frac{1}{T}(d + \beta U^2)\Psi = 0 \quad (3.17)$$

$$\frac{\partial P}{\partial t} + (a + eU^2 + \alpha)\vec{U}.\text{grad}\, P + \rho\,(d + \beta U^2)\Psi = 0 \quad (3.18)$$

It can be noticed that the condition (3.5) prevents from having $(a+eU^2+\alpha)$ and $(d+\beta U^2)$ both zero since δ can be written : $\delta = (a+eU^2+\alpha)\,d - \alpha(d+\beta U^2)$.

Of course there is not a unique form for the pseudo-unsteady entropy or total pressure equation. For example we could choose to express the material derivative of S or P in terms of Ψ and $(\vec{U}.\vec{M})$.

4. Hyperbolicity condition

4.1 Generalities

We now restrict the class of P.U.S. (3.1), (3.2), by requiring that these systems be of hyperbolic type with respect to the time variable. This condition is a natural one to impose since it ensures that perturbations remain bounded in time (at least within the framework of linearized theory) ; besides it allows a systematic treatment of boundary conditions through the compatibility relations (e.g., ref. [4], [5], [11], [12]).

Consider a cartesian coordinate system which, for the time being, can be arbitrary ; let (X_1, X_2, X_3) be the cartesian coordinates, (U_1, U_2, U_3) and (M_1, M_2, M_3) the cartesian components of $\vec{U}$ and $\vec{M}$ respectively. The quantities Ψ and $\vec{M}$ defined by (2.14) are linear functions of the space derivatives of p and $\vec{U}$, namely :

$$\left.\begin{aligned} \Psi &= \frac{\partial U_j}{\partial X_j} - \frac{1}{c^2} U_j U_k \frac{\partial U_k}{\partial X_j} \\ M_i &= U_j \frac{\partial U_i}{\partial X_j} + \frac{1}{\rho}\frac{\partial p}{\partial X_i} \qquad (i = 1, 2, 3) \end{aligned}\right\} \quad (4.2)$$

(with the summation convention for repeated indices), and we can write :

$$\begin{pmatrix} \Psi \\ M_1 \\ M_2 \\ M_3 \end{pmatrix} = B_j \frac{\partial}{\partial X_j} \begin{pmatrix} p \\ U_1 \\ U_2 \\ U_3 \end{pmatrix} \quad (4.3)$$

where B_j ($j = 1, 2, 3$) are 4 x 4 matrices functions of p and $\vec{U}$.

In the cartesian coordinate system, the P.U.S. (3.3) takes the following form :

$$\frac{\partial}{\partial t}\begin{pmatrix} p \\ U_1 \\ U_2 \\ U_3 \end{pmatrix} + C \begin{pmatrix} \Psi \\ M_1 \\ M_2 \\ M_3 \end{pmatrix} = 0 \quad (4.4)$$

where C is a 4 x 4 matrix :

$$C = \begin{pmatrix} \rho d & \rho\alpha U_1 & \rho\alpha U_2 & \rho\alpha U_3 \\ \beta U_1 & a + eU_1^2 & eU_1U_2 & eU_1U_3 \\ \beta U_2 & eU_2U_1 & a + eU_2^2 & eU_2U_3 \\ \beta U_3 & eU_3U_1 & eU_3U_2 & a + eU_3^2 \end{pmatrix} \quad (4.5)$$

Hence the developed form of system (3.1), (3.2) :

$$\frac{\partial f}{\partial t} + C B_j \frac{\partial f}{\partial x_j} = 0 \quad (4.6)$$

where f is the column matrix of the basic unknown :

$$f = (p, U_1, U_2, U_3)^T \quad (4.7)$$

4.2 Characteristic matrix

Let $\vec{\xi}$ be an arbitrary unit space vector, with components ξ_1, ξ_2, ξ_3. The characteristic matrix of system (4.6) for the direction $\vec{\xi}$ is :

$$A = C B_j \xi_j \quad (4.8)$$

and the necessary and sufficient condition for system (4.6) to be hyperbolic with respect to time is that the eigenvalues of A be all real and that a vector basis can be constructed from eigenvectors (in other words that A can be diagonalized), this being so for all unit vectors $\vec{\xi}$.

We shall note $\lambda^{(k)}$ the eigenvalues, and $\nu^{(k)} = (\mu_0^{(k)}, \mu_1^{(k)}, \mu_2^{(k)}, \mu_3^{(k)}) = (\mu_0^{(k)}, \vec{\mu}^{(k)})$ the left eigenvectors (k = 1 to 4) ; $\vec{\mu}^{(k)}$ is a space vector with components ($\mu_1^{(k)}, \mu_2^{(k)}, \mu_3^{(k)}$) in the cartesian system considered. We thus have the following relations, by definition of $\lambda^{(k)}$ and $\nu^{(k)}$:

$$\nu^{(k)} A = \lambda^{(k)} \nu^{(k)} \tag{4.9}$$

$$\det(A - \lambda^{(k)} I) = 0 \tag{4.10}$$

where I is the 4 x 4 unit matrix.

The expression of A, hence the calculation of the eigenvalues, can be much simplified by a judicious choice of the cartesian coordinate system. For such a study of the local properties of system (3.1), (3.2), it is indeed possible to use a local coordinate system, and it is an obvious choice to take one of the axis, say X_1, aligned with the local velocity $\vec{U}$ at the point and at the time considered. Then in the expression (4.5) of C we can make $U_1 = U$, $U_2 = U_3 = 0$ (but of course not in the expression (4.7) of f !). Also the expressions of the matrices B_j are much simpler. A further simplification will arise if we now choose the axis X_2 in the plane ($\vec{U}, \vec{\xi}$) when $\vec{\xi}$ is not parallel to U. In this way, we always have $\xi_3 = 0$ and the characteristic matrix A then assumes the following form :

$$A = \begin{pmatrix} \alpha U_\xi & \rho\bar{\alpha}\xi_1 & \rho d \xi_2 & 0 \\ (a + eU^2)\xi_1/\rho & (a+\bar{e})U_\xi & \beta U \xi_2 & 0 \\ a\xi_2/\rho & 0 & aU_\xi & 0 \\ 0 & 0 & 0 & aU_\xi \end{pmatrix} \tag{4.11}$$

where we have introduced some new symbols :

$$\left. \begin{aligned} U_\xi &= \vec{U}.\vec{\xi} = U\xi_1 \\ \bar{\alpha} &= \alpha U^2 + d\left(1 - \frac{U^2}{c^2}\right) \\ \bar{e} &= eU^2 + \beta\left(1 - \frac{U^2}{c^2}\right) \end{aligned} \right\} \tag{4.12}$$

Let us note the relation :

$$\bar{e}d - \bar{\alpha}\beta = (ed - \alpha\beta)U^2 \tag{4.13}$$

4.3 Eigenvectors

The eigenvectors $\nu^{(k)}$ are the solutions of the linear system

(4.9). Using the expression (4.11) of A and writing $\nu^{(k)} = (\mu_0^{(k)}, \vec{\mu}^{(k)})$, the system (4.9) can be put into an intrinsic form, independent of the coordinate system (we omit the superscript (k)) :

$$\left.\begin{aligned} &(\lambda - \alpha U_\xi)\rho\mu_0 - a\vec{\mu}.\vec{\xi} - eU_\xi\,\vec{\mu}.\vec{U} = 0 \quad &\text{a)} \\ &(\lambda - a U_\xi)\vec{\mu} - \rho\mu_0\,\vec{p} - (\vec{\mu}.\vec{U})\vec{q} = 0 \quad &\text{b)} \end{aligned}\right\} \quad (4.14)$$

where the vectors $\vec{p}$, $\vec{q}$ are linear functions of $\vec{\xi}$:

$$\left.\begin{aligned} \vec{p} &= (\alpha - \frac{d}{c^2})U_\xi\vec{U} + d\vec{\xi} \\ \vec{q} &= (e - \frac{\beta}{c^2})U_\xi\vec{U} + \beta\vec{\xi} \end{aligned}\right\} \quad (4.15)$$

4.4 Compatibility relations

The compatibility relations associated with a given unit vector $\vec{\xi}$ are obtained by multiplying system (4.6) on the left with the eigenvectors $\nu^{(k)}$.

Let $\partial f / \partial \xi$ denote the derivative of f in the direction of $\vec{\xi}$; we can decompose grad f (in fact the gradient of each of the components of f, eq. (4.7)) into two components parallel and normal to $\vec{\xi}$, which we write :

$$\left.\begin{aligned} \text{grad } f &= \vec{\xi}\,\partial f/\partial\xi + \widetilde{\text{grad}}\, f \\ &\text{with } \vec{\xi}.\widetilde{\text{grad}}\, f = 0 \end{aligned}\right\} \quad (4.16)$$

Then

$$\frac{\partial f}{\partial x_j} = \vec{e}_j \cdot \text{grad} f = \xi_j \frac{\partial f}{\partial \xi} + \vec{e}_j \cdot \widetilde{\text{grad}} f \quad (4.17)$$

where $\vec{e_1}$, $\vec{e_2}$, $\vec{e_3}$ are the unit vectors of the coordinate axes. Then using (4.17) to express $\partial f/\partial x_j$ in (4.6), and taking into account (4.9), the compatibility relations write :

$$\nu^{(k)}\left(\frac{\partial f}{\partial t} + \lambda^{(k)}\frac{\partial f}{\partial \xi}\right) = -\nu^{(k)} C\, B_j\,(\vec{e}_j \cdot \widetilde{\text{grad}}\, f) \quad (4.18)$$

This equation reveals the characteristic property of compatibility relations : they are linear combinations of the equations of the system (4.6) involving the same transport operator with speed $\lambda^{(k)}$ in the direction $\vec{\xi}$ for all the components of f. Let

$$D_{\xi}^{(k)} = \frac{\partial}{\partial t} + \lambda^{(k)} \frac{\partial}{\partial \xi} \qquad (4.19)$$

be this transport operator ; the compatibility relation (4.18) writes:

$$\nu^{(k)} D_{\xi}^{(k)} f = \mathbb{R} \qquad (4.20)$$

where the right-hand-side $\mathbb{R}$ involves only space derivatives of f in directions normal to $\vec{\xi}$.

The hyperbolicity condition can be expressed equivalently as follows : for any direction $\vec{\xi}$ there exists a system of compatibility relations which is completely equivalent to the original system of equations.

These properties of the compatibility relations form the basis for the systematic treatment of boundary conditions referred to at the beginning of section 4.1.

Going back to the original form (3.1), (3.2) of the P.U.S., the compatibility relations (4.18) can also be written in intrinsic form as follows :

$$\mu_0^{(k)} \left[\frac{\partial \rho}{\partial t} + \rho \alpha \vec{U}.\vec{M} + \rho d \Psi \right] + \vec{\mu}^{(k)} . \left[\frac{\partial \vec{U}}{\partial t} + a \vec{M} + \beta \vec{U} \Psi + e \vec{U} (\vec{U}.\vec{M}) \right] = 0 \qquad (4.21)$$

4.5 Equation for the eigenvalues

The eigenvalues are the solutions of Eq. (4.10), i.e. the equation which expresses the condition that the determinant of system (4.14) (for the unknowns μ_0 , $\vec{\mu}$) is zero :

$$(\lambda - a U_{\xi}) \left\{ (\lambda - a U_{\xi})(\lambda^2 - A_1 U_{\xi} \lambda - A_2) + A_3 \right\} = 0 \qquad (4.22)$$

where A_1, A_2, A_3 are given by :

$$\left.\begin{aligned} A_1 &= a + \bar{e} + \alpha && \text{a)} \\ A_2 &= (ed - \alpha\beta)(1 - U^2/c^2)\, U_{\xi}^2 + ad\,(1 - U_{\xi}^2/c^2) && \\ &= \delta (1 - U^2/c^2)\, \xi_1^2 + ad\, \xi_2^2 && \text{b)} \\ A_3 &= (ed - \alpha\beta)\, aU^3 \xi_1 \xi_2^2 && \text{c)} \end{aligned}\right\} \qquad (4.23)$$

where δ has been defined by (3.5).

We see that a first eigenvalue (which disappears in the case of plane flows) is

$$\lambda^{(1)} = a\, U_\xi \tag{4.24}$$

Moreover, in the important case when we have

$$ed - \alpha\beta = 0 \tag{4.25}$$

then $A_3 = 0$, and $(a\, U_\xi)$ is a double eigenvalue :

$$\lambda^{(2)} = a\, U_\xi \tag{4.26}$$

and the other two eigenvalues $\lambda^{(3)}$, $\lambda^{(4)}$ are solutions of :

$$\lambda^2 - A_1 U_\xi \lambda - A_2 = 0 \tag{4.27}$$

The discriminant of this equation is

$$\Delta = A_1^2 U_\xi^2 + 4 A_2 = (\Delta_0 - 4ad)\xi_1^2 + 4ad \tag{4.28}$$

where

$$\Delta_0 = A_1^2 U^2 + 4ad\left(1 - \frac{U^2}{c^2}\right) \tag{4.29}$$

4.6 Conditions on the signs of the eigenvalues

For a hyperbolic system, the number of boundary conditions to be imposed at a boundary is related to the signs of the eigenvalues associated with the unit vector $\vec{\xi}$ normal to this boundary. Taking this vector $\vec{\xi}$ to be directed towards the outside of the computational domain, then the number of boundary conditions to be imposed is equal to the number of negative eigenvalues. Indeed, replacing the original hyperbolic system by the equivalent system of compatibility relations (4.20), we see that those compatibility relations with $\lambda^{(k)} < 0$ correspond to a transport of information at the boundary by waves coming from outside the domain and therefore they must be discarded and replaced by boundary conditions (which are determined from the physics of the problem).

Now the number of boundary conditions to be imposed to the steady flow problem can be determined from the number of boundary conditions associated with the full system of the Euler equations. For this system, the eigenvalues (they are 5) are :

$$\lambda^{(1)} = \lambda^{(2)} = \lambda^{(3)} = U_\xi$$

$$\lambda^{(4)} = U_\xi + c \quad , \quad \lambda^{(5)} = U_\xi - c$$

and the number of boundary conditions results immediatly from the signs of these eigenvalues. However we must be careful of the fact that the energy equation has been eliminated from the steady flow problem (and replaced by Bernoulli's relation (2.3)), and since one of the 3 eigenvalues U_ξ comes from the energy equation, we must consider U_ξ only as a double eigenvalue when counting the number of negative eigenvalues.

Hence the number of boundary conditions for the steady flow problem and for the P.U.S. used, depending on the position of U_ξ with respect to -c, 0 and c, shown in table 1 below :

Table 1 : Number of boundary conditions as a function of U_ξ

U_ξ		-c		0		c	
Type of boundary	supersonic inflow	SONIC	subsonic inflow	SLIP	subsonic outflow	SONIC	supersonic outflow
number of conditions	4	3	3	1	1	0	0

Therefore, besides the hyperbolicity conditon, we also impose to the P.U.S. the condition that the number of negative eigenvalues be equal to the number of boundary conditions given by table 1. A detailed and complex discussion of these two conditions is presented in [13] ; only the results are given in the next section.

4.7 Summary of conditions on the coefficients of the P.U.S.

We recall that the P.U.S. (3.1), (3.2) are subject to the conditions (3.4), (3.5), the hyperbolicity condition, and the condition on the number of negative eigenvalues. The complete discussion of these conditions [13] leads to the following necessary and sufficient conditions on the coefficients of the P.U.S.

First we have three simple conditions :

$$a > 0 \tag{4.30}$$

$$d > 0 \tag{4.31}$$

$$\delta > 0 \tag{4.32}$$

Then we must distinguish between two cases depending on $(ed - \alpha\beta)$ being zero or not.

In the case $ed - \alpha\beta = 0$ (it is for this case that we can find many interesting examples of P.U.S.), the additional conditions are given in Table 2 below.

Table 2. Conditions (other than $a > 0$, $d > 0$) in the case $ed = \alpha\beta$

	$d+\beta U^2 \neq 0$	$d+\beta U^2 = 0$
$\bar{\alpha} \neq 0$	$\bar{\alpha}(d+\beta U^2) > 0$	$a \neq \beta\ (1 - \frac{U^2}{c^2})$
$\bar{\alpha} = 0$ (hence $\bar{e} = 0$)	$a \neq \alpha$	no condition

When $ed = \alpha\beta$, note that $\delta = ad$, hence (4.32) is a consequence of (4.30) and (4.31), and that $\bar{e}d = \bar{\alpha}\beta$ as a result of (4.13).

In the case $ed - \alpha\beta \neq 0$ (hence $\bar{e}d - \bar{\alpha}\beta \neq 0$) we have the following conditions besides (4.30), (4.31) and (4.32) :

. if $\bar{\alpha} = 0$ and $\bar{e}\ (a+\bar{e}-\alpha) > 0$, we must have $d+ \beta U^2 = 0$ (or equivalently $\alpha = \beta(1-\frac{U^2}{c^2})$)

. if $\bar{\alpha} \neq 0$ the eigenvalues $\lambda^{(2)}$, $\lambda^{(3)}$, $\lambda^{(4)}$, solutions of $(\lambda-aU_\xi)(\lambda^2-A_1U_\xi\lambda-A_2) +A_3 = 0$ must be real and distinct.

. if $U > c$, we must have

$$a\ (a + \bar{e} + \alpha\ ) + (ed - \alpha\beta\ )\ (\frac{U^2}{c^2} - 1)\ > 0$$

. if $U \geqslant c$ and $\alpha^2 + (a+eU^2)^2 \neq 0$, we must have

$$(a + \bar{e} + \alpha)\ U > 2\sqrt{\delta}\ \sqrt{\frac{U^2}{c^2} - 1}$$

5. Examples of pseudo-unsteady systems

Various examples of P.U.S. are presented and their particular properties are brought out. When a particular P.U.S. is considered, it is understood that this system satisfies all the conditions summed up in section 4.7.

All the P.U.S. considered here belong to the class characterized by the property : $ed = \alpha\beta$ which results in $(a\ U_\xi)$ being a double eigenvalue (section 4.5).

5.1 Systems involving some exact unsteady equations

5.1.1. - The exact unsteady continuity equation can be written :

$$\frac{\partial \rho}{\partial t} + \rho \, \Theta = 0 \qquad (5.1)$$

It can be checked that this equation (5.1) is a consequence of the P.U.S. (3.1), (3.2) if the coefficients verify the following two relations :

$$\left.\begin{aligned} \gamma d + (\gamma-1)\beta U^2 &= c^2 \\ \gamma \alpha + (\gamma-1)(a + e U^2) &= \gamma \end{aligned}\right\} \qquad (5.2)$$

5.1.2 - The exact unsteady momentum equation in conservative form is :

$$\frac{\partial (\rho \vec{U})}{\partial t} + \rho \vec{D} = 0 \qquad (5.3)$$

This equation results from the P.U.S. (3.1), (3.2) under the following three conditions :

$$\left.\begin{aligned} a &= 1 \\ \gamma d + [c^2 + (\gamma-1)U^2]\beta &= c^2 \\ \gamma \alpha + [c^2 + (\gamma-1)U^2] e &= 1 \end{aligned}\right\} \qquad (5.4)$$

5.1.3 - Combining the conditions (5.2) and (5.4) together, we get the well-known P.U.S. constituted by equations (5.1) and (5.3) which, to our knowledge, is the only P.U.S. commonly used for the calculation of transonic rotational flows :

$$\left.\begin{aligned} &\frac{\partial \rho}{\partial t} + \rho \, \Theta = 0 \\ &\frac{\partial (\rho \vec{U})}{\partial t} + \rho \vec{D} = 0 \end{aligned}\right\} \Longleftrightarrow \left\{\begin{aligned} &a = 1 \\ &\beta = e = 0 \\ &\alpha = \frac{1}{\gamma} \, , \quad d = \frac{c^2}{\gamma} \end{aligned}\right. \qquad (5.5)$$

This system belongs to the class of strictly conservative P.U.S. discussed in section 3.3, and it corresponds to $g = \rho$ and $F = \rho$.

5.1.4 - The exact unsteady momentum equation in non-conservative form is :

$$\frac{d\vec{U}}{dt} + \frac{1}{\rho} \operatorname{grad} p = 0 \, , \quad \text{or} \quad \frac{\partial \vec{U}}{\partial t} + \vec{M} = 0 \qquad (5.6)$$

where $d/dt = \partial/\partial t + \vec{U}.\,\text{grad}$ is the material derivative.

Equation (3.2) reduces to (5.6) when we have :

$$a = 1, \qquad \beta = e = 0 \tag{5.7}$$

The P.U.S. (5.5) above satisfies these conditions.

5.1.5 - The exact unsteady Euler equations lead to the following equation for the pressure :

$$\frac{dp}{dt} + c^2 \operatorname{div} \vec{U} = 0 \tag{5.8}$$

It can be shown that Eq. (3.1) reduces to (5.8) when we have :

$$\alpha = 1, \; d = c^2 \tag{5.9}$$

5.1.6 - Combining the conditions (5.3) and (5.9), we get the following P.U.S. :

$$\left.\begin{aligned} &\frac{dp}{dt} + \rho c^2 \operatorname{div}\vec{U} = 0 \\ &\frac{\partial(\rho\vec{U})}{\partial t} + \rho\vec{D} = 0 \end{aligned}\right\} \Longleftrightarrow \left\{\begin{aligned} &a = \alpha = 1 \\ &d = c^2 \\ &\beta = ec^2 = -\frac{(\gamma-1)c^2}{c^2+(\gamma-1)U^2} \end{aligned}\right. \tag{5.10}$$

5.1.7 - Combining the conditions (5.7) and (5.9), we get a different P.U.S. :

$$\left.\begin{aligned} &\frac{dp}{dt} + \rho c^2 \operatorname{div}\vec{U} = 0 \\ &\frac{d\vec{U}}{dt} + \frac{1}{\rho}\operatorname{grad} p = 0 \end{aligned}\right\} \Longleftrightarrow \left\{\begin{aligned} &a = \alpha = 1 \\ &d = c^2 \\ &\beta = e = 0 \end{aligned}\right. \tag{5.11}$$

5.1.8 - The exact unsteady entropy equation in non-conservative form is :

$$\frac{dS}{dt} = 0 \tag{5.12}$$

The conditions under which this equation is a consequence of the P.U.S. (3.1), (3.2) are immediately obtained by comparing (5.12) and (3.17):

$$a + eU^2 + \alpha = 1, \; d + \beta U^2 = 0 \tag{5.13}$$

From these relations, we deduce

$$(ed - \alpha\beta)\, U^2 = (1 - a)\, d$$

so that, taking into account the additional constraint $ed = \alpha\beta$, we get

$$a = 1,\ \alpha + eU^2 = 0,\ d + \beta U^2 = 0 \qquad (5.14)$$

5.1.9 - It can be shown that combining the conditions (5.4) and (5.14) together, or the conditions (5.7) and (5.14) together, leads to unacceptable P.U.S. in view of the conditions given in section 4.7.

However combining the conditions (5.2) and (5.14) leads to a valid P.U.S. characterized by the following coefficients :

$$a = \alpha = 1\,,\ eU^2 = -1\,,\ d = -\beta U^2 = c^2 \qquad (5.15)$$

Note that the relations (5.9) are satisfied, so that Equation (5.8) is satisfied by this P.U.S. which can be written :

$$\left.\begin{aligned} &\frac{dp}{dt} + \rho c^2 \operatorname{div} \vec{U} = 0 \\ &\frac{d\vec{U}}{dt} + \frac{1}{\rho} \operatorname{grad} p - \frac{\vec{U}}{U^2}\left[c^2 \operatorname{div} \vec{U} + \frac{1}{\rho} \vec{U} \cdot \operatorname{grad} p\right] = 0 \end{aligned}\right\} \qquad (5.15)$$

Transformed into semi-conservative form, this system becomes :

$$\left.\begin{aligned} &\frac{\partial \rho}{\partial t} + \rho\, \Theta = 0 \\ &\frac{\partial (\rho \vec{U})}{\partial t} + \rho \vec{D} + \frac{\vec{U}}{U^2}\left\{(\gamma-1)\vec{U} \cdot \rho \vec{D} - \left[c^2 + (\gamma-1) U^2\right] \rho\, \Theta\right\} = 0 \end{aligned}\right\} \qquad (5.17)$$

We see that the exact unsteady equations (5.1), (5.8) and (5.12) are all satisfied by this P.U.S.

Notice that none of the particular P.U.S. considered in this section 5.1 applies to an incompressible flow since some coefficients become infinite with c^2. Examples of P.U.S. valid for incompressible flows are discussed in [13].

5.2 Strictly conservative pseudo-unsteady systems

The general form of strictly conservative P.U.S. is given by Eqs. (3.10), (3.11), and the corresponding coefficients are given by relations (3.12) and (3.14).

We show that there exist P.U.S. of this kind, moreover belonging to the class with $ed = \alpha\beta$, which satisfy the conditions of section 4.7.

From relation (3.16), the condition $ed = \alpha\beta$ amounts to :

$$g - F = R(p) \tag{5.18}$$

where R(p) is an arbitrary function of p.

Then using the relations (3.12), (3.13) and (3.14) on the one hand, and the relations (3.8) on the other hand, we can determine the coefficients of the original P.U.S. (3.1), (3.2), as functions of F(p,U) and R(p). We thus find :

$$\left.\begin{aligned}
a &= \frac{\rho}{F}, \quad \frac{\beta}{\delta} = R'(p), \quad \delta = ad \\
\frac{\alpha}{\delta} &= \frac{\gamma}{\rho c^2} F - \frac{1}{\rho U}\frac{\partial F}{\partial U} \\
\frac{eU^2}{\delta} &= R'(p)\left[\gamma \frac{U^2}{c^2} - \frac{U}{F}\frac{\partial F}{\partial U}\right] \\
\frac{1}{d} &= \frac{\partial F}{\partial p} + R'(p)\left[1 + \frac{U}{F}\frac{\partial F}{\partial U}\right]
\end{aligned}\right\} \tag{5.19}$$

We also note, for the discussion of the conditions of table 2, that :

$$\left.\begin{aligned}
\frac{d + \beta U^2}{\delta} &= \frac{F}{\rho} + U^2 R'(p) \\
\bar{\alpha} &= d\left[1 + (\gamma - 1)\frac{U^2}{c^2} - \frac{U}{F}\frac{\partial F}{\partial U}\right]
\end{aligned}\right\} \tag{5.20}$$

By taking R = 0, i.e. g = F, we obtain the class of strictly conservative P.U.S. such that :

$$\beta = e = 0, \quad \frac{1}{d} = \frac{\partial F}{\partial p} \tag{5.21}$$

The conditions :

$$F > 0, \quad \frac{\partial F}{\partial p} > 0 \tag{5.22}$$

are necessary, and it is sufficient to add the condition $\bar{\alpha} > 0$. For example, taking $g = F = \rho$ gives back the classical P.U.S. (5.5).

This can be checked by calculating the coefficients from (5.19), noting that we have, from Eq. (2.8), :

$$\frac{\partial \rho}{\partial p} = \frac{\gamma}{c^2} \quad , \quad \frac{\partial \rho}{\partial U} = (\gamma - 1)\frac{\rho U}{c^2}$$

A generalization of this system is obtained by taking $g = F = \rho_o^{1-k} \rho^k$ where ρ_o is some reference density and k is a constant. The corresponding coefficients are the following :

$$g = F = \rho_o^{1-k} \rho^k \iff \begin{cases} a = (\rho/\rho_o)^{1-k} , \quad \beta = e = 0 \\ d = a c^2 / \gamma k \\ \alpha = a\left(\frac{1}{k} + \frac{1}{\gamma} - 1\right) \end{cases} \tag{5.23}$$

with the condition $k > 0$. The P.U.S. corresponding to $k = 1/2$ has been implemented by Laffont [15].

Another possibility, with $R = 0$, is to take :

$$g = F = F_o (p) \tag{5.24}$$

where Fo can be arbitrary except for the conditions (5.22). For example, choosing $F_o = p/U_o^2$ where U_o is a reference speed, gives the following P.U.S. :

$$\left.\begin{aligned} \frac{\partial}{\partial t}\left(\frac{p}{U_o^2}\right) + \rho \Theta = 0 \\ \frac{\partial}{\partial t}\left(\frac{p}{U_o^2}\vec{U}\right) + \rho \vec{D} = 0 \end{aligned}\right\} \iff \begin{cases} a = \rho U_o^2 / p , \quad d = U_o^2 \\ \beta = e = 0 \\ \alpha = \gamma U_o^2 / c^2 \end{cases} \tag{5.25}$$

A different class of P.U.S. is obtained by choosing

$$F = \rho_o = \text{const.} > 0, \tag{5.26}$$

hence, from (5.18) :

$$g = g(p) \tag{5.27}$$

The corresponding coefficients are :

$$\left.\begin{aligned} a = \frac{\rho}{\rho_o} \quad , \quad \frac{1}{d} = g'(p) \\ \beta = \frac{\rho}{\rho_o} \quad , \quad e = \gamma \frac{\rho}{\rho_o c^2} \quad , \quad \alpha = \frac{\gamma}{c^2 g'(p)} \end{aligned}\right\} \tag{5.28}$$

and the only condition for these P.U.S. is

$$g'(p) > 0 \tag{5.29}$$

A simple example of P.U.S. of this class corresponds to $g = p/U_0^2$:

$$\left.\begin{aligned} &\frac{\partial}{\partial t}\left(\frac{p}{U_0^2}\right) + \rho\,\Theta = 0 \\ &\rho_0 \frac{\partial \vec{U}}{\partial t} + \rho \vec{D} = 0 \end{aligned}\right\} \Leftrightarrow \left\{\begin{aligned} &a = \beta = \rho/\rho_0 \\ &d = U_0^2, \quad \alpha = \gamma U_0^2/c^2 \\ &e = \gamma \rho / \rho_0 c^2 \end{aligned}\right. \tag{5.30}$$

This system differs from System (5.25) only by the equation for $\vec{U}$.

Many other strictly conservative P.U.S. can also be found, for example by taking $F = F(p)$ and $g = g(p)$.

Notice that the P.U.S. (5.25) and (5.30) apply to the case of an incompressible fluid (take $1/c^2 = 0$).

5.3 Systems in semi-conservative form

The semi-conservative form has been discussed in section 3.2. From relations (3.8), we deduce $ed - \alpha\beta = e_1 d - \alpha_1 \beta$, so that a simple way to satisfy the condition $ed = \alpha\beta$ which leads to simpler P.U.S. of type (3.6) is to take : $e_1 = \alpha_1 = 0$. From relations (3.7), this is equivalent to taking :

$$e = \gamma\beta/c^2, \qquad \alpha = \gamma d/c^2 \tag{5.31}$$

and the relations (3.8) give :

$$d_1 = d\,, \quad \beta_1 = \beta - a \tag{5.32}$$

This class of P.U.S. thus depends on the coefficients a, d, β only, and can be written (Eq. (3.6)) :

$$\left.\begin{aligned} &\frac{\partial p}{\partial t} + \rho\, d\, \Theta = 0 \\ &\frac{\partial \vec{U}}{\partial t} + (\beta - a)\vec{U}\,\Theta + a\vec{D} = 0 \end{aligned}\right\} \Leftrightarrow \left\{\begin{aligned} &e = \gamma\beta/c^2 \\ &\alpha = \gamma d/c^2 \end{aligned}\right. \tag{5.33}$$

Table 2 leads to the following conditions (besides $a > 0$, $d > 0$, and noting that $\bar{\alpha} = d\,[1 + (\gamma - 1)U^2/c^2]$) :

$$\left.\begin{array}{ll} \text{either} & d + \beta U^2 > 0 \\ \text{or} & d + \beta U^2 = 0 \text{ and } a \neq \beta(1 - U^2/c^2) \end{array}\right\} \tag{5.34}$$

This class of P.U.S. applies to an incompressible fluid, being characterized in this case by $e = \alpha = 0$.

In the class of P.U.S. (5.33) we can determine a one-parameter family of systems verifying the relations (5.14), that is to say such that the equation $dS/dt = 0$, or $dP/dt = 0$ is a consequence of (5.33). This family depends on the coefficient d only and writes :

$$\left.\begin{array}{l} \dfrac{\partial p}{\partial t} + \rho d \Theta = 0 \\ \dfrac{\partial \vec{U}}{\partial t} + \vec{D} - \left(\dfrac{d}{U^2} + 1\right) \vec{U} \Theta = 0 \end{array}\right\} \quad \left\{\begin{array}{l} a = 1 \\ \beta U^2 = -d \\ \alpha = -eU^2 = \gamma d/c^2 \end{array}\right. \tag{5.35}$$

with the conditions

$$d > 0 \text{ and } d\left(\frac{U^2}{c^2} - 1\right) \neq U^2 \tag{5.36}$$

It can be noticed that for the strictly conservative P.U.S., characterized by relations (5.19), the conditions (5.31) amount to $\partial F/\partial U = 0$, i.e. g and F being functions of p only. In other words, the P.U.S. (5.33) can be put in a strictly conservative form if and only if we have :

$$a = \frac{\rho}{F(p)}, \quad d = \frac{1}{g'(p)}, \quad \beta = \frac{\rho}{F}\left(1 - \frac{F'}{g'}\right) \tag{5.37}$$

5.4 Systems with eigenvalues having simpler expressions

Here also we consider only P.U.S. which verify the condition $ed = \alpha\beta$. For such P.U.S., (aU_ξ) is a double eigenvalue :

$$\lambda^{(1)} = \lambda^{(2)} = aU_\xi \tag{5.38}$$

and the other two eigenvalues are solutions of Eq. (4.27) and can be written :

$$\lambda^{(3),(4)} = \frac{1}{2} A_1 U_\xi \pm \frac{1}{2}\sqrt{\Delta} \tag{5.39}$$

where A_1 is given by Eq. (4.23a), and Δ by Eqs. (4.28) and (4.29).

The dependency of $\lambda^{(3)}$ and $\lambda^{(4)}$ on ξ_1 is relatively complicated due to the fact that, in general, Δ itself depends upon ξ_1. But we see from Eq. (4.28) that there exists a remarkable case when Δ does not depend on ξ_1, so that $\lambda^{(3)}$ and $\lambda^{(4)}$ become linear functions of ξ_1; this is the case when the coefficients of the P.U.S. (3.1), (3.2) obey the relation :

$$\Delta_o = 4ad \tag{5.40}$$

that is to say, using Eq. (4.29) :

$$A_1^2 = 4ad/c^2 \tag{5.41}$$

In this case, the eigenvalues $\lambda^{(3)}$, $\lambda^{(4)}$ assume the simpler form :

$$\lambda^{(3),(4)} = \frac{1}{2} A_1 \, (U_\xi \pm c), \tag{5.42}$$

which, except for the factor $A_1/2$, is the form of the eigenvalues $(U_\xi \pm c)$ of the exact Euler equations.

In fact there exists an infinity of P.U.S. which have the same eigenvalues as the exact Euler equations, namely U_ξ, $U_\xi \pm c$; these systems correspond to the conditions $a = 1$, $A_1 = 2$, hence, from Eqs. (4.23a) and (5.41) :

$$a = 1, \; \alpha + \bar{e} = 1, \; d = c^2 \tag{5.43}$$

Taking into account the definitions of $\bar{\alpha}$, $\bar{e}$ given by (4.12), and using $ed = \alpha\beta$ to get e, the other coefficients can be expressed as functions of $\bar{\alpha}$:

$$\left.\begin{aligned} \alpha &= 1 + (\bar{\alpha} - c^2)/U^2 \\ \beta &= \frac{c^2}{U^2}\left(\frac{c^2}{\bar{\alpha}} - 1\right), \quad e = \alpha\beta/c^2 \end{aligned}\right\} \tag{5.44}$$

where $\bar{\alpha}$ can be given arbitrary values except zero. In fact, from (5.43) and (5.44) we deduce $\bar{\alpha}\,(d + \beta U^2) = c^4$, so that the conditions of table 2 are automatically satisfied.

An example of such a P.U.S. is system (5.11), corresponding to $\bar{\alpha} = c^2$.

Another example can be found in the class (5.33) ; it corresponds to $\alpha = \gamma$, or $\bar{\alpha} = c^2 + (\gamma - 1)\, U^2$, and it is the particular system (5.33) defined by

$$a = 1, \; d = c^2, \; \beta = -\frac{(\gamma - 1)\, c^2}{c^2 + (\gamma - 1) U^2} \tag{5.45}$$

Considering now the less restrictive condition (5.41), we can find P.U.S. of the class (5.33) which satisfy this condition, and more precisely verifying $A_1 = 2\sqrt{ad/c}$. For this, it is sufficient to note that the conditions (5.31) lead to the following expression of A_1 :

$$A_1 = a + \beta\left[1 + (\gamma - 1)\frac{U^2}{c^2}\right] + \gamma\frac{d}{c^2} \qquad (5.46)$$

We give two examples. The first one generalizes (5.45) to arbitrary (positive) values of a :

$$d = ac^2, \quad \beta = -a\,\frac{(\gamma-1)c^2}{c^2+(\gamma-1)U^2} \quad (a>0) \qquad (5.47)$$

hence $A_1 = 2a$, $\lambda^{(3),(4)} = a(U_\xi \pm c)$

The second one corresponds to $d + \beta U^2 = 0$ and is more complicated:

$$\left.\begin{aligned} d &= a\,\frac{U^2c^2}{(U+c)^2}, \quad \beta = -a\,\frac{c^2}{(U+c)^2} \\ \text{hence}\quad A_1 &= \frac{2aU}{U+c}, \quad \lambda^{(3),(4)} = \frac{aU}{U+c}(U_\xi \pm c) \end{aligned}\right\} \qquad (5.48)$$

By taking a = 1, the particular P.U.S. (5.33) defined by (5.48) turns out to also belong to the family (5.35) for which the equation $dS/dt = 0$ or $dP/dt = 0$ holds.

More general P.U.S. verifying the condition (5.41) can be determined by imposing the condition $\bar{\alpha}(d+\beta U^2) = 0$. One finds that these conditions (together with $ed = \alpha\beta$) lead to :

$$d = \frac{aU^2c^2}{(U+\varepsilon c)^2}, \quad A_1 = \frac{2aU}{U+\varepsilon c}, \quad \varepsilon = \pm 1 \qquad (5.49)$$

where the value $\varepsilon = -1$ can be accepted only if the velocity does not become sonic. Then we have two possibilities :

i) $\bar{\alpha} = 0$, hence $\bar{e} = 0$, and

$$\alpha = a\,\frac{U-\varepsilon c}{U+\varepsilon c}, \quad eU^2 = -\beta\left(1-\frac{U^2}{c^2}\right), \quad \beta \text{ arbitrary} \qquad (5.50)$$

ii) $d + \beta U^2 = 0$, hence $\alpha + eU^2 = 0$, and

$$\beta = -\frac{a c^2}{(U+\varepsilon c)^2} \quad , \; eU^2 = -\alpha \quad , \; \alpha \text{ arbitrary} \tag{5.51}$$

The P.U.S. (5.48) are a subset of (5.51) with $\varepsilon = 1$.

The P.U.S. defined by (5.49) and either (5.50) or (5.51) apply to an incompressible fluid, giving $d = aU^2$, $A_1 = 0$ (and $\lambda^{(3),(4)} = \pm\sqrt{ad} = \pm aU$) and either $\alpha = -a$, $eU^2 = -\beta$, or $\beta = -a$, $eU^2 = -\alpha$.

Finally, an example of strictly conservative P.U.S. satisfying condition (5.41) can be found in the family of P.U.S. defined by (5.23) and which depends on the constant k, at least in the case of a perfect gas with constant γ. In fact, the condition (5.41) is seen to be satisfied by the coefficients (5.23) when $k = \gamma$, which gives $A_1 = 2a/\gamma$.

6. Discussion of pseudo-unsteady systems based on numerical stability criterion

In view of the great variety of admissible P.U.S., it is natural to attempt to compare these systems from the point of view of numerical stability, assuming an explicit time integration scheme. A system which would be optimal in this sense would probably remain optimal also with an implicit scheme, since in practice implicit schemes often show some stability limitation.

We base the present discussion on a general form of the C.F.L. stability criterion which we write :

$$\Delta t \leq \Delta t_M = k \, \Delta x / V_{GM} \tag{6.1}$$

where Δt is the time step, Δx a measure of the size of the local numerical domain of dependence, K a positive constant of order unity, and where V_{GM} is the maximum with respect to $\vec{\xi}$ and with respect to l (l = 1 to 4) of the moduli of the group velocities

$\vec{V}_G^{(l)}(\vec{\xi})$.

We recall (e.g. [14]) that the group velocity associated with the eigenvalue $\lambda^{(l)}(\xi)$ (which is a phase speed) is given by :

$$\vec{V}_G^{(l)}(\vec{\xi}) = \frac{\partial \omega^{(l)}(\vec{k})}{\partial \vec{k}} \tag{6.2}$$

where $\vec{k} = k\vec{\xi}$ is the wave vector, $2\pi/k$ is the wave length and $\omega^{(l)} = k\lambda^{(l)}$ is the (angular) frequency.

The eigenvalues $\lambda^{(l)}$, solutions of Eq. (4.22), depend on $\vec{\xi}$ only through the component ξ_1 (we recall that the X_1 axis is taken along the velocity vector, i.e. $U\,\xi_1 = \vec{U}.\,\vec{\xi} = U_\xi$), $\lambda^{(l)} = \lambda^{(l)}(\xi_1)$, so that the corresponding group velocity is :

$$\vec{V}_G^{(l)} = \lambda^{(l)}\,\vec{\xi} + \frac{\vec{V_n}}{U}\,\frac{d\lambda^{(l)}}{d\xi_1} \tag{6.3}$$

where $\vec{V_n} = \vec{U} - U_\xi\,\vec{\xi}$ is the velocity component normal to $\vec{\xi}$.

Considering only the P.U.S. which satisfy ed = $\alpha\beta$, we obtain for the group velocities (the eigenvalues are given by Eqs. (5.38) and (5.39)) :

$$\vec{V}_G^{(1)} = \vec{V}_G^{(2)} = a\vec{U} \tag{6.4}$$

$$\vec{V}_G^{(3),(4)} = \frac{1}{2}A_1\vec{U} \pm \frac{1}{2\sqrt{\Delta}}\left(\Delta_0\,\xi_1\,\vec{e}_1 + 4\,ad\,\xi_2\,\vec{e}_2\right) \tag{6.5}$$

where the unit vectors $\vec{e}_1$, $\vec{e}_2$ are such that $\vec{U} = U\vec{e}_1$ and $\vec{\xi} = \xi_1\vec{e}_1 + \xi_2\vec{e}_2$.

Let $(t - t_o)$ be a fixed time interval and assume $\Delta_o \neq 0$; the extremety of vectors $(t - t_o)\,\vec{V}_G^{(3),(4)}$ describes, when $\vec{\xi}$ varies, an ellipsoïd of revolution of equation :

$$\frac{4}{\Delta_o}\left(\frac{X_1}{t-t_o} - \frac{1}{2}A_1 U\right)^2 + \frac{1}{ad}\left(\frac{Y}{t-t_o}\right)^2 = 1 \tag{6.6}$$

where Y is the radial distance from the X_1 - axis. This surface is indeed an ellipsoïd if $\Delta_o \neq 0$ because we have ad$>$0 and $\Delta_o >$ 0. It becomes a sphere for $\Delta_o = 4ad$, i.e. for these P.U.S. discussed in section (5.3). The case $\Delta_o = 0$ is rather special ; it leads to P.U.S. acceptable in supersonic flow only, and the extremity of $\vec{V}_G\,(t-t_o)$ describes a circle normal to the X_1 - axis.

Excluding the special case $\Delta_o = 0$, the maximum value $V_{GM}^{(3)}$ of the moduli of $\vec{V}_G^{(3),(4)}$ is found to be given by the following expressions :

$$\left.\begin{array}{ll} \text{. if } 4ad - \Delta_0 \leqslant \sqrt{\Delta_0}\,|A_1|\,U & \\ V_{GM}^{(3)} = 1/2\left[\sqrt{\Delta_0} + |A_1|\,U\right] & \text{a)} \\ \text{. if } 4ad - \Delta_0 > \sqrt{\Delta_0}\,|A_1|\,U & \\ V_{GM}^{(3)} = U/c \dfrac{2ad}{\sqrt{4ad - \Delta_0}} & \text{b)} \end{array}\right\} \quad (6.7)$$

In the case when $\Delta_0 = 4ad$, i.e. $A_1^2 = 4ad/c^2$, which is included in (6.7a), we have :

$$\left.\begin{array}{l} \vec{V}_G^{(3),(4)} = \dfrac{1}{2} A_1 (\vec{U} \pm c\vec{\xi}) \\ V_{GM}^{(3)} = \sqrt{ad}\,(1 + U/c) \end{array}\right\} \quad (6.8)$$

The maximum time step Δt_M can thus be expressed in the form :

$$\Delta t_M = \frac{K\ \Delta x}{a\ \operatorname{Max}\{U, V_{GM}^{(3)}/a\}} \quad (6.9)$$

where it is easily seen that $V_{GM}^{(3)}/a$ depends on the coefficients a, d, e, β , α only through the ratios d/a, e/a, β /a and α/a. This shows that 1/a can be considered as having an influence only on the local time scale, so that the optimum local stability criterion will be obtained when we have :

$$V_{GM}^{(3)}/a \leqslant U \quad (6.10)$$

For the P.U.S. considered in section 5.4, $V_{GM}^{(3)}$ is given by Eq. (6.8) and the condition (6.10) becomes :

$$\sqrt{\frac{d}{a}} \leqslant \frac{U\,c}{U + c} \quad (6.11)$$

It can be seen that the P.U.S. defined by Eqs. (5.49) and either (5.50) or (5.51) with $\varepsilon = 1$ satisfy this condition (6.11), the equality being in fact satisfied, so that the maximum time step for such P.U.S. is :

$$\Delta t_M = \frac{K\ \Delta x}{a\,U} \quad (6.12)$$

In particular the P.U.S. defined by (5.48) is of this type.

An interesting feature of optimum P.U.S. is that Δt_M can be independent of the Mach number, on condition that the coefficient a be constant or function of U only. Indeed, for the exact Euler equations for a compressible fluid, the maximum time step corresponding to the criterion (6.1) is

$$\Delta t_M^{(E)} = \frac{K \Delta x}{U + c} \tag{6.13}$$

and it is governed by the sound speed even in a very low speed region. On the contrary, the criterion (6.12) with the coefficient a chosen close to unity allows much larger time steps than (6.13) in such a region, while being also more efficient by a factor of about 2 in a transonic region. Of course, if we choose "a" such that $aU = U + c$, the criterion (6.12) becomes identical to (6.13).

The classical P.U.S. (5.5) is not optimum from the point of view of the stability criterion. For this system, we find :

$$A_1 = \frac{\gamma+1}{\gamma}, \quad ad = \frac{c^2}{\gamma}, \quad \Delta_0 = 4\frac{c^2}{\gamma} + \left(\frac{\gamma-1}{\gamma}U\right)^2 \tag{6.14}$$

so that Eq. (6.7a) applies and gives :

$$V_{G_M}^{(3)} = \frac{\gamma+1}{2\gamma} U + \sqrt{\frac{c^2}{\gamma} + \left(\frac{\gamma-1}{2\gamma}\right)^2 U^2} \tag{6.15}$$

We have $V_{G_M}^{(3)} > aU$ (here $a = 1$), and the maximum time step is :

$$\Delta t_M = \frac{K \Delta x}{V_{G_M}^{(3)}} < \frac{K \Delta x}{a U} \tag{6.16}$$

7. Shock-capturing with pseudo-unsteady systems

Shock-capturing is possible with P.U.S. written in the semi-conservative form (3.6) on condition that at convergence to a steady state the discretized equations reduce to some discretized conservative forms of the conservative steady equations (2.1), (2.2), i.e. $\rho \Theta = 0$ and $\rho \vec{D} = 0$. Moreover it is necessary that the artificial viscosity terms be directly added in conservative form to $\rho \Theta$ and to $\rho \vec{D}$ in system (3.6) (and not added to the right-hand side of (3.6)), so that the discrete steady solution approximates a weak solution of the conservative equations :

$$\begin{aligned} \rho \Theta - \operatorname{div} \vec{E} &= 0 \\ \rho \vec{D} - \operatorname{div} \vec{\tau} &= 0 \end{aligned} \tag{7.1}$$

where $\operatorname{div}\vec{E}$ and $\operatorname{div}\vec{\tau}$ are artificial viscosity terms in divergence form ($\vec{E}$ is a vector, $\vec{\tau}$ a tensor).

Practically, this amounts to replacing Θ and $\vec{D}$ in (3.6) respectively by :

$$\left.\begin{aligned} \tilde{\Theta} &= \Theta - \frac{1}{\rho}\operatorname{div}\vec{E} \\ \tilde{\vec{D}} &= \vec{D} - \frac{1}{\rho}\operatorname{div}\vec{\tau} \end{aligned}\right\} \tag{7.2}$$

For shock capturing, the artificial viscosity terms must be such as to produce an entropy rise through a shock wave. To discuss this condition, we consider the steady state entropy equation which results from the relation (already used in § 3.4) :

$$\rho T \vec{U} \cdot \operatorname{grad} S = -\rho \vec{U} \cdot \vec{M} \tag{7.3}$$

and from relations (2.17b) and (7.2) with $\rho\tilde{\Theta} = 0$, $\rho\tilde{\vec{D}} = 0$; we obtain :

$$\rho T \vec{U} \cdot \operatorname{grad} S = -\vec{U} \cdot \rho\tilde{\vec{D}} + U^2 \rho \tilde{\Theta} - \vec{U} \cdot \operatorname{div}\vec{\tau} + U^2 \operatorname{div}\vec{E} \tag{7.4}$$

But at steady state $\rho\tilde{\Theta}$ and $\rho\tilde{\vec{D}}$ vanish ; moreover the left-hand-side of (7.4) can be put in divergence form taking into account the fact that $\operatorname{div}\rho\vec{U} = \operatorname{div}\vec{E}$, hence the steady-state entropy equation :

$$\operatorname{div}(\rho\vec{U}S) = \left(S + \frac{U^2}{T}\right)\operatorname{div}\vec{E} - \frac{\vec{U}}{T} \cdot \operatorname{div}\vec{\tau} \tag{7.5}$$

where the right-hand-side groups the source terms generated by the artificial viscosity (of course these source terms should be non-negligible only in shock regions).

As with the Navier-Stokes equations, the source terms in (7.5) can be further transformed and decomposed into a reversible part in divergence form, and an irreversible part Φ/T where Φ is the artificial dissipation term :

$$\operatorname{div}(\rho\vec{U}S) = \operatorname{div}\left[\left(S + \frac{U^2}{T}\right)\vec{E} - \frac{1}{T}\vec{U}\cdot\vec{\tau}\right] + \frac{1}{T}\Phi \tag{7.6}$$

$$\frac{1}{T}\Phi = -\vec{E} \cdot \operatorname{grad}\left(S + \frac{U^2}{T}\right) + \vec{\tau}^{*} : \operatorname{grad}\left(\frac{\vec{U}}{T}\right) \tag{7.7}$$

where $\vec{\tau}^{*}$ is the transpose of $\vec{\tau}$, and the symbol : denotes the doubly contracted product of two tensors (with the Navier-Stokes equations, the entropy equation involves a heat conduction term and is in fact simpler due to $\vec{E} = 0$ and due to the cancellation of source terms coming from the momentum equation and from the energy equation).

Integrating (7.6) across the shock region $\eta_1 \leq \eta \leq \eta_2$ where η is the distance normal to the shock (and increasing in the direction of the velocity), we get the condition :

$$(\rho U_N)_1 \left(S_2 - S_1\right) = \int_{\eta_1}^{\eta_2} \frac{1}{T} \Phi \, d\eta \quad > 0 \tag{7.8}$$

where U_N is the normal velocity and the indices 1 and 2 refer respectively to the upstream and downstream sides of the shock region (we have $(\rho U_N)_2 = (\rho U_N)_1$). Of course we have made use of the hypothesis that $\vec{E}$ and $\vec{\tau}$ become negligible at $\eta = \eta_1$ and $\eta = \eta_2$.

Although this is not strictly necessary, the simplest way to ensure condition (7.8) is to have Φ always positive, or better to have Φ positive in a compression region and zero in an expansion region. These conditions can be fulfilled through appropriate choices for $\vec{E}$ and $\vec{\tau}$.

Such a choice is

$$\vec{E} = 0, \quad \vec{\tau} = \varepsilon \operatorname{grad} \vec{U} \qquad (\varepsilon \geqslant 0) \tag{7.9}$$

Indeed, we get :

$$\Phi = \varepsilon \left\{ (\operatorname{grad} \vec{U})^* : (\operatorname{grad} \vec{U}) + T \operatorname{grad} \frac{U^2}{2} \cdot \operatorname{grad} \frac{1}{T} \right\} \tag{7.10}$$

and both terms between brackets in the right-hand-side of (7.10) are always positive. The artificial viscosity coefficient ε should be chosen so as to be positive or zero and to produce non negligible dissipation only in a rapid compression region, for example :

$$\varepsilon = \varepsilon_1 \left[|\operatorname{div} \vec{U}| - \operatorname{div} \vec{U} \right], \quad \varepsilon_1 > 0 \tag{7.11}$$

If the term $\vec{E}$ is not taken equal to zero, one must be careful of the fact that it could give a negative contribution in (7.7).

8. Conclusion

A systematic study of pseudo-unsteady first-order systems for steady compressible iso-energetic flow calculation has been presented. A family of such systems, depending on 5 scalar parameters, and having an intrinsic form, has been introduced, subject to two essential conditions : hyperbolicity with respect to time and a condition concerning the number of negative eigenvalues. These conditions have been made completely explicit in terms of the parameters of the P.U.S. for the class of systems verifying the relation $ed = \alpha\beta$. Numerous examples of systems of this class are given.

This study has shown that systems having a large variety of forms and properties can be used, and that a choice of a particular system can be guided by various criteria : strictly conservative form of the system, relation of the system with some exact unsteady equation, eigenvalues having simple expressions, optimisation of a stability criterion for an explicit time-integration scheme.

The case of incompressible flows is included automatically in the present general study ; we refer to [13] for a more detailed discussion of this case as well as for a study of a class of simpler P.U.S. adapted to the case of iso-energetic and homentropic steady flows.

REFERENCES

[1] R. PEYRET and H. VIVIAND, Pseudo-Unsteady Methods for Inviscid or Viscous Flow Computation, to be published in "Recent Advances in the Aerospace Science" (edited by C. Casci).

[2] H. VIVIAND, Pseudo-Unsteady Methods for Transonic Flow Computations, Lecture Notes in Physics Vol. 141, Springer-Verlag, 1981, pp. 44-54.

[3] J.P. VEUILLOT and H. VIVIAND, Pseudo-Unsteady Methods for the Computation of Transonic Potential Flows, AIAA J., Vol. 17, n° 7, July 1979, p. 691 and ONERA T.P. n° 1978-47.

[4] H. VIVIAND and J.P. VEUILLOT, Méthodes pseudo-instationnaires pour le calcul d'écoulements transsoniques, ONERA Publication n° 1978-4, 1978.

[5] J. BROCHET, Calcul numérique d'écoulements internes tridimensionnels transsoniques, La Rech. Aérosp., n° 1980-5, 1980, p. 301-315.

[6] D. GOTTLIEB, and B. GUSTAFSSON, On the Navier-Stokes Equations with Constant Total Temperature, Studies in Appl. Math. 55, 1976, p. 167-185.

[7] J.A. ESSERS, Time-Dependent Methods for Mixed and Hybrid Steady Flows, von Karman Institute for Fluid Dynamics, Lecture Series 1978-4 on Computational Fluid Dynamics (March 13-17, 1978).

[8] J.A. ESSERS, Quasi-Natural Numerical Methods for the Computation of Inviscid Potential or Rotational Transonic Flows, Appl. Math. Modelling, Vol. 3 (Feb. 1979), p. 55-66.

[9] J.A. ESSERS and F. KAFYEKE, Application of a Fast Pseudo-Unsteady Method to Steady Transonic Flows in Turbine Cascades J. of Engineering for Power, Vol. 104 (April 1982), p. 420-428.

[10] J.A. ESSERS, New Fast Super-Dashpot Time-Dependent Techniques for the Numerical Simulation of Steady Flows - I - Numerical Formulation, Computers and Fluids, Vol. 8, n° 3 (Sept. 1980) p. 351-368.

[11] L. CAMBIER, W. GHAZZI, J.P. VEUILLOT et H. VIVIAND, Une Approche par Domaines pour le Calcul d'Ecoulements Compressibles, dans computing methods in applied sciences and engineering V" (edited by R. Glowinski and J.L. Lions), North Holland Publ. Co. (1982), p. 423-446.

[12] L. CAMBIER, W. GHAZZI, J.P. VEUILLOT and H. VIVIAND, A Multi-Domain Approach for the Computation of Viscous Transonic Flows by Unsteady Type Methods, in "Recent Advances in Numerical Methods in Fluids. Vol. III. Viscous Flow Computational Methods" (edited by W.G. Habashi), Pineridge Press (1983).

[13] H. VIVIAND, Systèmes pseudo-instationnaires pour les écoulements stationnaires de fluide parfait, Publication ONERA n° 1983-1983.

[14] G.B. WHITHAM, Linear and Non-Linear Waves, John Wiley and Sons 1974.

[15] P. LAFFONT, Etude de méthodes pseudo-instationnaires pour le calcul numérique d'écoulements transsoniques, Rapport de stage à l'ONERA, Juin 1979.

PART III:
FINITE ELEMENT METHODS

A ONE-DIMENSIONAL SHOCK CAPTURING FINITE ELEMENT METHOD AND MULTI-DIMENSIONAL GENERALIZATIONS*

THOMAS J. R. HUGHES,† MICHEL MALLET,† YOSHIHIRO TAKI,‡
TAYFUN E. TEZDUYAR§ AND ROBERTO ZANUTTA†

Abstract. Multi-dimensional generalizations of a one-dimensional finite element shock capturing scheme are proposed. A scalar model problem is used to emphasize that "preferred directions" are important in multi-dimensional applications. Schemes are developed for the two-dimensional Euler equations. One, based upon characteristics, employs the Mach lines and streamlines as preferred directions.

1. Introduction. Many finite difference schemes for the compressible Euler equations have been developed from essentially one-dimensional concepts. That is, a one-dimensional shock capturing scheme is used in a "directional splitting" format to generate the multi-dimensional scheme. Methods of this type tend to be quite efficient, but a bias may be introduced in the form of excessive numerical diffusion skew to the mesh lines. As a result, the sharp, shock-capturing behavior of the one-dimensional scheme will not be manifested when shocks are not aligned with the mesh.

In this paper we consider alternative procedures. We discuss techniques for generalizing a one-dimensional shock capturing finite element method to multi-dimensional cases. The techniques hold promise of more faithfully representing the multi-dimensional character of the compressible Euler equations. Throughout, the steady case is emphasized.

In Section 2 we review the one-dimensional theory. The method is developed from a Petrov-Galerkin superconvergence approach on a scalar singular perturbation problem involving a nonlinear flux function (Hughes [8]). The approach is aimed at high resolution of stationary shocks. The resulting numerical flux definition is very close to that proposed

*This research was sponsored by the NASA Ames Research Center under Consortium Agreement NASA-NCA2-OR745-307 and by the NASA Langley Research Center under Grant NASA-NAG-1-361.

†Division of Applied Mechanics, Durand Building, Stanford University, Stanford, California 94305.

‡Visiting Scholar from Meijo University, Nagoya, Japan.

§Department of Mechanical Engineering, University of Houston, Houston, Texas 77004

by Engquist and Osher [4]. The only difference occurs in "transonic compression". Slight alteration of the transonic compression switch yields Godunov's numerical flux. The methods generalize to one-dimensional hyperbolic systems in a straightforward manner.

As a prelude to the Euler equations, in Section 3 we consider a scalar multi-dimensional generalization of the one-dimensional theory. The methods proposed explicitly account for the preferred directional character of the governing equation and have features in common with the so-called "streamline upwind" procedure developed by Hughes and Brooks [2, 9]. The effectiveness of this procedure has been demonstrated numerically in [2, 9] and mathematically in Johnson [13] and Nävert [14]. The methodology proposed herein represents extensions and refinements of the ideas presented in [2, 9]. Several first-order "monotone" schemes are also proposed. Using concepts of anti-diffusive flux limiting, it is shown how to blend these schemes with higher-order accurate schemes to produce a so-called "high-resolution" scheme capable of producing sharp, non-oscillatory approximations to discontinuities and good accuracy in smooth flow regimes. Given a first-order diffusive scheme of a certain type, it is shown how to construct a naturally associated higher-order accurate scheme using Petrov-Galerkin/weighted-residual concepts [11]. In passing, we develop a conservative generalization of Tabata's monitone scheme on triangles which is simpler than those proposed heretofore (see, e.g., Ikeda [12]).

In Section 4 we consider the two-dimensional Euler equations. We again attempt to instill our schemes with preferred-directional character. Examination of the characteristic structure of supersonic flow suggests that there is more than one preferred direction for the Euler equations, namely, the Mach lines and streamlines. A diagonalization of the coefficient matrices is developed based upon characteristic coordinates. In the subsonic regime the diagonalized arrays become complex-valued. Nevertheless it is possible to develop a scheme applicable to all flow regimes which accounts for the characteristic information.

2. Review of One-dimensional Theory.

2.1 Singular Perturbation Problem. Consider the following singular-perturbation problem: Find $u = u(x)$, $x \in [x_L, x_R]$, such that

$$f_{,x} = \varepsilon u_{,xx} + \mathscr{f} \tag{2.1}$$

$$u(x_L) = \mathscr{g} \tag{2.2}$$

$$\varepsilon u_{,x}(x_R) = \mathscr{h} \tag{2.3}$$

where f is the flux function, ε is a positive parameter, $\mathscr{f} = \mathscr{f}(x)$ is a prescribed source term, $\mathscr{g}$ and $\mathscr{h}$ are prescribed constants, and a comma indicates differentiation. The flux function may depend upon both u and x in the most general case (see [8]).

2.2 Variational Formulation. Let $\mathcal{S}$ and $\tilde{\mathcal{V}}$ denote the set of trial solutions and weighting functions, respectively. Members of $\mathcal{S}$ are required to satisfy the Dirichlet boundary condition, namely (2.2), whereas members of $\tilde{\mathcal{V}}$ need to satisfy its homogeneous counterpart. The variational problem consists of finding $u \in \mathcal{S}$ such that for all $\tilde{w} \in \tilde{\mathcal{V}}$,

$$(2.4) \quad \int_{x_L}^{x_R} \left(\tilde{w}\, f(u, \cdot)_{,x} + \tilde{w}_{,x}\, \varepsilon\, u_{,x} \right) dx = \int_{x_L}^{x_R} \tilde{w}\, \ell\, dx + \tilde{w}(x_R)\, h$$

The Euler-Lagrange equations of (2.4) are (2.1) and (2.3).

2.3 Finite Element Formulation. In the finite element formulation we assume the trial solutions are approximated by the standard C^0 piecewise linear interpolations. This set of functions is denoted by $\mathcal{S}^h$. The discrete weighting function space, $\tilde{\mathcal{V}}^h$, is designed to optimize nodal accuracy. This necessitates solving an adjoint operator Green's function problem for each basis function. The details of this procedure are described in [8]. The finite element counterpart of (2.4) is: Find $u^h \in \mathcal{S}^h$ such that for all $\tilde{w}^h \in \tilde{\mathcal{V}}^h$,

$$(2.5) \quad \int_{x_L}^{x_R} \left(\tilde{w}^h\, f(u^h, \cdot)_{,x} + \tilde{w}^h_{,x}\, \varepsilon\, u^h_{,x} \right) dx = \int_{x_L}^{x_R} \tilde{w}^h\, \ell\, dx + \tilde{w}^h(x_R)\, h$$

Formulations of this type are referred to as Petrov-Galerkin, or generalized Galerkin, methods.

Remark 1 In the linear case in which $f(u, x) = a(x)u$, the solution is exact at the nodes. In the constant-coefficient case, in which $a = \text{const.}$, the weighting functions may be explicitly calculated [7]. This is generally not possible in the variable-coefficient case. However, in the limit $\varepsilon \to 0$, explicit formulae are possible and, in fact, are quite simple. Details are given in [8].

Remark 2 Consider the nonlinear case in which $f(u, x) = f(u)$ and f is a convex function. This ensures that $a(u^h) = D\, f(u^h)$ will have at most one zero per element. (This follows from u^h being piecewise linear.) In the limit $\varepsilon \to 0$, the weighting function associated with node number i becomes

$$(2.6) \quad \tilde{N}_i = \frac{1}{2}(1 + b \operatorname{sgn} N_{i,x}), \quad x \in [x_{i-1}, x_{i+1}], \quad i = 2, 3, \ldots, n_{np}-1$$

where

$$(2.7)\quad \begin{cases} b(x) = \displaystyle\int_{x_i}^{x_{i+1}} \operatorname{sgn} a(u^h(y))dy/h_{i+\frac{1}{2}} , & \text{if } a_i \geq 0 \geq a_{i+1} , \\ \qquad x \in [x_i, x_{i+1}] , \quad i = 1 , 2 ,\dots, n_{np}-1 & \end{cases}$$

$$(2.8)\qquad = \operatorname{sgn} a(u^h(x)) \quad , \quad \text{otherwise.}$$

The nodal coordinates are denoted by x_i , $i = 1 , 2 ,\dots, n_{np}$. The nodes are assumed to be in ascending order from left to right with $x_1 = x_L$ and $x_{n_{np}} = x_R$. N_i is the usual linear hat function above node i , $h_{i+\frac{1}{2}} = x_{i+1} - x_i$ and $a_i = a(u_i^h) = a(u^h(x_i))$. The nonzero part of $N_{n_{np}}$ is given by (2.6) with domain of definition restricted to $x \in [x_{n_{np}-1}, x_{n_{np}}]$. The degenerate case of a identically zero in an element is covered by setting $\tilde{N}_i = \frac{1}{2}$. If a has no zeros in an element then the weighting function is either zero or one in that element. If a has a zero then there are two cases: <u>transonic compression</u> (i.e., $a_i \geq 0 \geq a_{i+1}$) and <u>transonic expansion</u> (i.e., $a_i \leq 0 \leq a_{i+1}$) . In transonic compression the weighting function is constant within the element whereas in transonic expansion the weighting function has an internal discontinuity (see Figure 1). The $\tilde{w}^h$ in (2.5) is simply any linear combination of the $\tilde{N}_i$'s .

Remark 3 It is instructive to write the equations in finite difference style. Consider an internal node and assume $f = 0$. The superscript h on u^h is dropped to simplify the notation. The difference equation at node i takes on the following form:

$$(2.9)\qquad \tilde{f}_{i+\frac{1}{2}} - \tilde{f}_{i-\frac{1}{2}} = 0$$

where the numerical flux is defined by

$$(2.10)\quad \tilde{f}_{i+\frac{1}{2}} = \tilde{f}(u_{i+1}, u_i) = \frac{f_{i+1} + f_i}{2} - \frac{1}{2}\int_{u_i}^{u_{i+1}} b(u)a(u)du$$

in which

$$(2.11)\quad b(u) = \int_{u_i}^{u_{i+1}} \operatorname{sgn} a(v)dv/(u_{i+1} - u_i) , \quad a_i \geq 0 \geq a_{i+1} .$$

$$(2.12)\qquad = \operatorname{sgn} a(u) , \quad \text{otherwise} ,$$

and $f_i = f(u_i)$, $u_i = u(x_i)$. Change of variables has been used in (2.10) and (2.11). Combining (2.10)-(2.12) and noting that due to convexity $\operatorname{sgn} a = \operatorname{sgn} u$ leads to

$$(2.13) \quad \tilde{f}_{i+\frac{1}{2}} = \frac{f_{i+1} + f_i}{2} + \frac{u_{i+1} + u_i}{2} \frac{(f_{i+1} - f_i)}{(u_{i+1} - u_i)} , \quad u_i \geq 0 \geq u_{i+1} ,$$

$$(2.14) \quad = \frac{f_{i+1} + f_i}{2} - \frac{1}{2} \int_{u_i}^{u_{i+1}} |a(u)| du , \quad \text{otherwise}$$

Equation (2.14) is the <u>Engquist-Osher flux function</u> [4, 15]. Thus the present scheme is identical to Engquist-Osher with regards to flux definition except in transonic compression. A consequence is that the present method is <u>entropy satisfying</u>. Only slight modification of the transonic compression switch is required to give rise to the classical Godunov's numerical flux [5]. Specifically, replace (2.7) and (2.11) by

$$(2.15) \qquad b(u) = \operatorname{sgn} \left(\frac{f_{i+1} - f_i}{u_{i+1} - u_i} \right)$$

Then, (2.13) is replaced by

$$(2.16) \quad \tilde{f}_{i+\frac{1}{2}} = \frac{f_{i+1} + f_i}{2} + \operatorname{sgn} \left(\frac{f_{i+1} - f_i}{u_{i+1} - u_i} \right) \frac{(f_{i+1} - f_i)}{2} .$$

<u>Remark 4</u> Specializing the above definitions to Burgers' flux, $f(u) = u^2/2$, yields further insight:

$$(2.17) \quad \underline{\text{Godunov}}: \quad \tilde{f}^{G}_{i+\frac{1}{2}} = \frac{1}{2} \max \left\{ (u_i^+)^2 , (u_{i+1}^-)^2 \right\}$$

$$(2.18) \quad \underline{\text{Engquist-Osher}}: \quad \tilde{f}^{EO}_{i+\frac{1}{2}} = \frac{1}{2} \left((u_i^+)^2 + (u_{i+1}^-)^2 \right)$$

$$(2.19) \quad \underline{\text{Present method}}: \quad \tilde{f}_{i+\frac{1}{2}} = \tilde{f}^{EO}_{i+\frac{1}{2}} + \frac{1}{2} u_i^+ u_{i+1}^-$$

where $u^+ = \max\{u, 0\}$ and $u^- = \min\{u, 0\}$. The Engquist-Osher flux is C^1 whereas Godunov's and (2.19) are C^0. The explicit forward-in-time scheme,

$$u_i^{n+1} = u_i^n - \frac{\Delta t}{h_i} (\tilde{f}^n_{i+\frac{1}{2}} - \tilde{f}^n_{i-\frac{1}{2}}) ,$$

in which $h_i = (h_{i+\frac{1}{2}} + h_{i-\frac{1}{2}})/2$, Δt is the time step, and the superscripts refer to time level, is not monotonicity preserving. Both the Godunov and Engquist-Osher schemes are monotone. The effectiveness of the three methods in resolving stationary shocks is illustrated in Figure 2.

2.4 One-dimensional Systems. A generalization of the preceding ideas to one-dimensional systems is constructed as follows. Let $\underset{\sim}{F} = \underset{\sim}{F}(\underset{\sim}{U})$ be an m-dimensional flux vector. We assume that $\underset{\sim}{F}$ is a homogeneous function of degree 1 and thus we can write $\underset{\sim}{F}(\underset{\sim}{U}) = \underset{\sim}{A}(\underset{\sim}{U})\underset{\sim}{U}$ where $\underset{\sim}{A}$ is the $m \times m$ Jacobian matrix (i.e., $\underset{\sim}{A} = \partial \underset{\sim}{F}/\partial \underset{\sim}{U}$) . We recall that this is the case for the one-dimensional Euler equations. We further assume that $\underset{\sim}{F}$ defines a hyperbolic system in that

$$\text{(2.21)} \qquad \underset{\sim}{A} = \underset{\sim}{S}\, \underset{\sim}{\Lambda}\, \underset{\sim}{S}^{-1}$$

where $\underset{\sim}{\Lambda}$ is an $m \times m$ real, diagonal matrix. The weighting function matrix in the present case may be written as

$$\text{(2.22)} \qquad \underset{\sim}{\tilde{N}}_i = \frac{1}{2}(\underset{\sim}{I} + \operatorname{sgn} N_{i,x}\, \underset{\sim}{C}) \ , \quad x \in [x_{i-1}, x_{i+1}] \ , \quad i = 2\ , 3\ , \ldots, n_{np}-1$$

where $\underset{\sim}{I}$ is the $m \times m$ identity matrix and

$$\text{(2.23)} \qquad \underset{\sim}{C} = \underset{\sim}{S}\ \operatorname{diag}(b^1\ , b^2\ , \ldots, b^m)\ \underset{\sim}{S}^{-1}$$

where for $\ell = 1\ , 2\ , \ldots, m$,

$$\text{(2.22)} \qquad \begin{cases} b^\ell(x) = \displaystyle\int_{x_i}^{x_{i+1}} \operatorname{sgn} \Lambda^\ell(y)\,dy/h_{i+\frac{1}{2}} \ , & \text{if } \Lambda_i^\ell \geq 0 \geq \Lambda_{i+1}^\ell \ , \\ \qquad x \in [x_i, x_{i+1}] \ , \quad i = 1\ , 2\ , \ldots, n_{np}-1 & \end{cases}$$

$$\text{(2.25)} \qquad = \operatorname{sgn} \Lambda^\ell(x) \ , \quad \text{otherwise}$$

The corresponding numerical flux vector can be written as

$$\text{(2.26)} \qquad \underset{\sim}{\tilde{F}}_{i+\frac{1}{2}} = \underset{\sim}{\tilde{F}}(\underset{\sim}{U}_{i+1}, \underset{\sim}{U}_i) = \frac{1}{2}(\underset{\sim}{F}_{i+1} + \underset{\sim}{F}_i) - \frac{1}{2}\int_{\underset{\sim}{U}_i}^{\underset{\sim}{U}_{i+1}} \underset{\sim}{C}(\underset{\sim}{U})\underset{\sim}{A}(\underset{\sim}{U})\,d\underset{\sim}{U}$$

If exclusive use is made of (2.25), then (2.26) reduces to Osher's numerical flux vector for systems [15, 16]:

$$(2.27)\quad \tilde{F}_{i+\frac{1}{2}} = \frac{1}{2}(F_{i+1} + F_i) - \frac{1}{2}\int_{U_i}^{U_{i+1}} |A(U)|\,dU$$

A Godunov-like generalization may be obtained by replacing (2.24) with

$$(2.28)\qquad b^{\ell}(x) = \operatorname{sgn}\left(\frac{\Lambda^{\ell}_{i+1} - \Lambda^{\ell}_{i}}{\bar{U}^{\ell}_{i+1} - \bar{U}^{\ell}_{i}}\right)$$

where $\bar{U} = S^{-1} U$.

Numerical example. The explicit forward-in-time scheme was used to calculate a solution to Sod's shock-tube problem [19]. The eigenvalues of A do not change sign. Consequently, (2.24) is never actuated and thus the present method becomes identical with Engquist-Osher. Two-point Gauss quadrature was employed to calculate each element integral. (An exact procedure for calculating the integral in (2.24) has been presented in [15, 16].) Density, velocity, pressure and internal energy results are presented in Figure 3. The following parameters were used in the calculations: $h = .01$, $\Delta t = .0014$. There are no spurious oscillations. The shock is captured womewhat better than the contact discontinuity (visible only in the density and internal energy plots). The results for the first-order forward-in-time scheme are not bad, but significant improvement can be made by employing "flux-limiting" ideas such as those described in Harten [6], Roe [18] and related works. For example, an explicit finite element scheme using the flux vectors presented herein produces the excellent results shown in Figure 4 (Hughes and Mallet [10]). In the present paper our primary concern is steady phenomena, not wave propagation, so we shall omit further discussion. The interested reader should consult [10] for details.

Remark 5 It is interesting to note that the weighting function associated with Osher's methods may be succinctly written in terms of the Heaviside function. For example,

$$(2.26)\quad \tilde{N}_i = \frac{1}{2}(1 + \operatorname{sgn} a \operatorname{sgn} N_{i,x}) = H(a\, N_{i,x})$$

where $H(g) = 1$ if $g > 0$, otherwise $H(g) = 0$. Likewise, in the system case

$$(2.27)\quad \tilde{N}_i = \frac{1}{2}(I + \operatorname{sgn} N_{i,x} \operatorname{sgn} A) = H(N_{i,x}\, A)$$

where H is extended to diagonalizable matrices in the obvious way.

3. Multi-dimensional Scalar Model Problem. Shock capturing schemes, such as the ones described in the previous section, are frequently generalized to multi-dimensional situations by directional splitting [6]. When important flow phenomena, such as shock waves, are not closely aligned with the mesh a pronounced degradation of accuracy may occur [3]. Thus we will not consider directional-splitting techniques herein. To develop effective multi-dimensional concepts we will first consider a two-dimensional scalar model problem embodying predominately one-dimensional physical character, but not aligned with the mesh. Consider the following problem: Find $u = u(\underset{\sim}{x})$, $\underset{\sim}{x} \in \overline{\Omega}$ (= a closed region in $\mathbb{R}^{n_{sd}}$ where n_{sd} = the number of space dimensions), such that

$$(3.1) \qquad \operatorname{div} \underset{\sim}{f} = \varepsilon \operatorname{div} \operatorname{grad} u + \mathscr{f} \quad \text{on } \Omega$$

$$(3.2) \qquad u = \mathscr{g} \quad \text{on } \Gamma_{\mathscr{g}}$$

$$(3.3) \qquad \varepsilon \underset{\sim}{n} \cdot \operatorname{grad} u = \mathscr{h} \quad \text{on } \Gamma_{\mathscr{h}}$$

where $\underset{\sim}{f} = f \underset{\sim}{s}$, f is the flux function (as described in Section 2) and $\underset{\sim}{s}$ is a given constant vector; $\mathscr{f}: \Omega \to \mathbb{R}$, $\mathscr{g}: \Gamma_{\mathscr{g}} \to \mathbb{R}$ and $\mathscr{h}: \Gamma_{\mathscr{h}} \to \mathbb{R}$, are prescribed data; $\underset{\sim}{n}$ is the unit outward normal vector to Γ, the boundary of Ω; and

$$(3.4) \qquad \overline{\Gamma_{\mathscr{g}} \cup \Gamma_{\mathscr{h}}} = \Gamma, \quad \Gamma_{\mathscr{g}} \cap \Gamma_{\mathscr{h}} = \emptyset$$

In the inviscid limit (3.1) becomes purely one-dimensional.

A weighted residual formulation which generalizes (2.4) is given as follows: Find $u \in \mathscr{S}$ such that for all $\tilde{w} \in \tilde{\mathscr{V}}$,

$$(3.5) \qquad \int_{\Omega} \left(\tilde{w} \operatorname{div} \underset{\sim}{f} + (\operatorname{grad} \tilde{w}) \cdot (\varepsilon \operatorname{grad} u) \right) d\Omega = \int_{\Omega} \tilde{w} \mathscr{f} d\Omega + \int_{\Gamma_{\mathscr{h}}} \tilde{w} \mathscr{h} d\Gamma$$

Likewise, the generalization of (2.5) may be obtained from (3.5) by replacing u by u^h and $\tilde{w}$ by $\tilde{w}^h$. We assume simple element interpolations are employed (e.g., linear triangles or bilinear quadrilaterals in two dimensions). We wish to generalize the one-dimensional inviscid weighting function [i.e., (2.6)] to the present case. In order to simplify the developments, we shall ignore the special case of transonic compression. Thus the schemes may all be considered multi-dimensional finite element generalizations of Osher type. We have experimented with several possibilities and it is not yet clear which is the best: Two are described below:

Generalization 1. ("σ-weighting")

Let

(3.6) $\tilde{N}_i = \sigma_i/\sigma \quad , \quad \underset{\sim}{x} \in \Omega^e$

(3.7) $\sigma_i = \langle a \underset{\sim}{s} \cdot \text{grad } N_i \rangle$

(3.8) $\sigma = \sum_{i=1}^{n_{en}} \sigma_i$

Node i is assumed to be a node of element e. The domain of element e is denoted Ω^e. Element e is assumed to have n_{en} nodes. $\langle\ \rangle$ is the Macauley bracket, that is, $\langle g \rangle = g$ if g is positive, otherwise $\langle g \rangle = 0$. The idea behind this weighting function was to incorporate essential features of the one-dimensional weighting function in the streamline direction and, at the same time, produce at least continuous variation of the weighting function in the crosswind direction.

Generalization 2. ("transported weighting")

Let

(3.9) $\tilde{N}_i(\underset{\sim}{x}) = N_i(\tilde{\underset{\sim}{x}})$

(3.10) $\tilde{\underset{\sim}{x}} = \underset{\sim}{x} + \delta(\underset{\sim}{x}) \text{sgn } a(\underset{\sim}{x}) \underset{\sim}{s} \quad , \quad \delta(\underset{\sim}{x}) \geq 0$

The physical interpretation of $\tilde{\underset{\sim}{x}}$ is as follows: It represents the intersection with the boundary of the line emanating from $\underset{\sim}{x}$ in the direction $\text{sgn } a(\underset{\sim}{x})\underset{\sim}{s}$ (see Figure 5). The inspiration behind the development of this weighting function was the same as in the one-dimensional case, namely the weighting function is required to be in the kernel of the adjoint operator and take on the same downwind boundary values as N_i (see [8] for further discussion).

Remark 1 Equations (3.6) and (3.9) are reminiscent of the "streamline upwind" weighting function proposed in Hughes and Brooks [2, 9]. For example, (3.9) can be written as

(3.11) $\tilde{N}_i(\underset{\sim}{x}) = N_i(\underset{\sim}{x}) + \delta(\underset{\sim}{x}) \text{sgn } a(\underset{\sim}{x}) \underset{\sim}{s} \cdot \text{grad } N_i(\underset{\sim}{x}) + 0(\delta^2)$

The first two terms of (3.11) constitute the streamline upwind weighting function. Johnson [13] and Nävert [14] have analyzed the streamline upwind procedure for the linear case and obtained optimal error estimates for smooth solutions and have shown that discontinuities parallel to the streamline are captured in a thin layer. Numerical experiments on linear problems with the schemes defined by (3.6) and (3.9) reveal similar accuracy to that achieved by the streamline upwind method (see [2, 9] for results). A nice feature of the present formulations is that, unlike the methods described in [2, 9], no "mesh parameters" need to be defined. Observe that both generalizations reduce to the one-dimensional Osher-type weighting function under appropriate hypotheses.

Remark 2 The eth-element contribution to the flux at node i is given by the following formula:*

$$\tilde{f}_i^e = \int_{\Omega^e} \tilde{N}_i \operatorname{div} \underset{\sim}{f} \, d\Omega \tag{3.12}$$

At an internal node, assuming $\not{f} = 0$, the difference equation is written as

$$\sum \tilde{f}_i^e = 0 , \tag{3.13}$$

were the sum is taken over all elements attached to node i.

Numerical Examples. We consider numerical results for two linear test problems described in [2, 9]: advection skew to the mesh and advection in a rotating flow field ("donut problem"). We have experimented with the two weighting functions described above on bilinear quadrilaterals and linear triangles. For the triangles, one-point quadrature suffices, whereas for the quadrilaterals, two-by-two quadrature is required to suppress hourglass instabilities. The results we have obtained are about the same for both the weighting functions used in conjunction with triangles or quadrilaterals. Consequently, we shall present only one set of results which are representative of all cases. These are for (3.6) (i.e., σ-weighting) on quadrilaterals and are presented in Figures 6 and 7. Note that the profiles in Figure 6 are not monotone. On the positive side, the present technique captures the discontinuity quite well, and in the smooth solution, that is Figure 7, good accuracy is achieved. It is important in nonlinear situations that at least locally monotone behavior be attainable and therefore some generalization of the present formulation will be necessary. In the remainder of this section we shall describe some attempts to develop schemes which produce non-oscillatory approximations to discontinuities.

Tabata's scheme. Let us consider the use of σ-weighting on triangles. We assume the linear case and without loss of generality take $a = 1$. There are two possible situations: either $\underset{\sim}{s}$ points in the direction of a "vertex zone", or $\underset{\sim}{s}$ points in the direction of an "edge zone" (see Fig. 8). In the first case, the weighting function is unity for

---

*In the one-dimensional case the eth-element contribution to node i is given by [cf. (2.10)]:

$$\tilde{f}_i^e = \frac{f_{i+1} - f_i}{2} + \frac{1}{2} \operatorname{sgn} N_{i,x}(x) \int_{u_i}^{u_{i+1}} b(u)a(u)du , \quad x \in \Omega^e = [x_i, x_{i+1}]$$

The finite difference notion of numerical flux, namely (2.10), does not seem to be easily generalizable to the multi-dimensional case.

the node in the vertex zone and zero for the remaining two. A monotone difference equation is produced (i.e., the diagonal coefficient is positive and the off-diagonals are non-positive). In the second case, the weighting function is non-zero for both nodes along the edge, but zero for the remaining node. In this case the difference equations fail to be monotone. In order to achieve monotonicity one might simply ignore the edge-zone case by setting the element weighting functions to zero. Thus all non-trivial equations emanate from the triangles in which $\tilde{s}$ lies in a vertex zone. This is, in essence, Tabata's scheme (see [12, 20]). Tabata's scheme is monotone, but not conservative. In the present context conservation is guaranteed if the weighting functions can take on the value unity throughout the domain Ω . Because some elements have zero weight, conservation is violated for Tabata's scheme.

Numerical examples. Numerical results for Tabata's scheme are presented in Figures 9 and 10. As may be seen, monotone profiles are achieved for the cases of advection skew to the mesh. However, a noticeable degradation in accuracy occurs for the donut problem, a manifestation of the fact that monotone schemes are at most first-order accurate.

Characteristic difference equations. We have also experimented with a scheme in which element difference equations are developed from the method of characteristics. The essential idea is depicted in Figure 11. The difference equation for the element shown is

$$(3.14) \quad 0 = \text{meas}(\Omega^e)\,||a\,\tilde{s}||\,(u_i - u_I)/h = \text{meas}(\Omega^e)\,||a\,\tilde{s}||\,(u_i - \alpha u_j - (1-\alpha)u_k)/h \;, \quad 0 \le \alpha \le 1$$

This equation is assigned to node i and the element equations for the remaining three nodes are identically zero. The scheme is monotone and numerical results are not unlike Tabata's scheme. For advection skew to the mesh the results are identical to those in Figure 9, whereas for the donut problem they are somewhat better (see Figure 12). Conservation is also a problem with this scheme. In generalizing it to the Euler equations we have produced shocks which do not satisfy the Rankine-Hugoniot relations. Rice [17] has recently developed a refined scheme of this type.

Flux split scheme. We may easily develop a finite element analog of the split flux scheme [1]. For this purpose it is useful to transform to element natural coordinates, say ξ , η in two dimensions. In terms of these coordinates,

$$\begin{aligned} \text{div}\,\tilde{f} &= \text{div}(f\,\tilde{s}) \\ &= a\,\tilde{s}\cdot\text{grad}\,u \\ &= a(s_1 u_{,1} + s_2 u_{,2}) \\ (3.15) \qquad &= a(s_\xi u_{,\xi} + s_\eta u_{,\eta}) \end{aligned}$$

where

$$s_\xi = s_1\xi_{,x} + s_2\xi_{,y} \tag{3.16}$$

$$s_\eta = s_1\eta_{,x} + s_2\eta_{,y} \tag{3.17}$$

Ignoring source and boundary terms, the scheme is defined by

$$0 = \int_\Omega (\tilde{N}_i^\xi \, a \, s_\xi u_{,\xi} + \tilde{N}_i^\eta \, a \, s_\eta u_{,\eta}) d\Omega \tag{3.18}$$

where, corresponding to (3.6),

$$\begin{cases} \tilde{N}_i^\xi = \sigma_i^\xi / \sigma^\xi & \tilde{N}_i^\eta = \sigma^\eta / \sigma^\eta \\ \sigma_i^\xi = \langle a \, s_\xi N_{i,\xi} \rangle & \sigma_i^\eta = \langle a \, s_\eta N_{i,\eta} \rangle \\ \sigma^\xi = \sum_{i=1}^{n_{en}} \sigma_i^\xi & \sigma^\eta = \sum_{i=1}^{n_{en}} \sigma_i^\eta \end{cases} \tag{3.19}$$

and, corresponding to (3.9),

$$\begin{cases} \tilde{N}_i^\xi(\underset{\sim}{x}) = N_j(\underset{\sim}{\tilde{x}}^\xi) \, , & \tilde{N}_i^\eta(\underset{\sim}{x}) = N_i(\underset{\sim}{\tilde{x}}^\eta) \\ \tilde{\underset{\sim}{x}}_\xi = \underset{\sim}{x} + \delta_\xi(\underset{\sim}{x}) \operatorname{sgn} a(\underset{\sim}{x}) s_\xi(\underset{\sim}{x}) \underset{\sim}{e}_\xi(\underset{\sim}{x}) \, , & \tilde{\underset{\sim}{x}}_\eta = \underset{\sim}{x} + \delta_\eta(\underset{\sim}{x}) \operatorname{sgn} a(\underset{\sim}{x}) s_\eta(\underset{\sim}{x}) \underset{\sim}{e}_\eta(\underset{\sim}{x}) \end{cases} \tag{3.20}$$

in which $\underset{\sim}{e}_\xi$ and $\underset{\sim}{e}_\eta$ are covariant basis vectors (see Fig. 13). Note that this is not a weighted residual method. Rather, it corresponds to separate one-dimensional upwinding in each of the element natural directions. The scheme is not monotone, but in practice it produces smooth profiles. It is clearly conservative. Our experience with it indicates that one-point integration on quadrilaterals is sufficient.

A conservative generalization of Tabata's scheme. A conservative, monotone scheme on triangles may be constructed by combining the present ideas with Tabata's scheme. With reference to Figure 8, use Tabata's scheme if $\underset{\sim}{s}$ falls in a vertex zone (i.e., σ-weighting). If $\underset{\sim}{s}$ falls in an edge zone, resolve $\underset{\sim}{s}$ into components aligned with the element natural coordinates emanating from the opposite vertex, then use (3.18) and (3.19). The element equations are continuous functions of $\underset{\sim}{s}$ as it sweeps through the lines separating vertex and edge zones. This scheme achieves the ends strived for in [12] and seems quite a bit simpler than any of those discussed in [12].

High-resolution schemes. Monotone schemes preclude spurious oscillations about discontinuities, but are at most first-order accurate. Higher-order-accurate schemes behave well where the solution is smooth, but invariably produce oscillatory approximations to discontinuities. Clearly, to produce good results for general cases, schemes with different attributes need to be blended together in a way accounting for the local smoothness of the solution. This is the philosophy embodied in the "high-resolution"schemes developed by finite difference researchers. Many proponents of this theme were represented at the Danvers meeting on computational fluid mechanics* and the interested reader is referred there for further details and references to the literature. A one-dimensional finite element analog is presented in [10]. The key ingredients in a high-resolution scheme are: a dissipative first-order scheme exhibiting monotonicity to some degree; a second-order scheme which differs from the first-order scheme by an antidiffusive term; and a flux limiter which depends on a local smoothness monitor and acts to limit the antidiffusive flux where the solution is rough. A procedure of this type can be developed in the present context by combining scheme (3.18) with one of the preceding weighted residual formulations. Consider

$$(3.21) \quad 0 = \int_\Omega (L_i^\xi \; a \; s_\xi u_{,\xi} + L_i^\eta \; a \; s_\eta u_{,\eta}) d\Omega$$

where

$$(3.22) \quad L_i^\xi = \tilde{N}_i^\xi + \ell^\xi(\tilde{N}_i - \tilde{N}_i^\xi)$$

$$(3.23) \quad L_i^\eta = \tilde{N}_i^\eta + \ell^\eta(\tilde{N}_i - \tilde{N}_i^\eta)$$

("flux-limited weighting functions")

in which, for example, $\tilde{N}_i^\xi$ and $\tilde{N}_i^\eta$ are given by (3.19) or (3.20), and $\tilde{N}_i$ is given by (3.6) or (3.9). Note that if $\ell^\xi = \ell^\eta = 0$ we reduce to the first-order scheme (3.18), whereas if $\ell^\xi = \ell^\eta = 1$ we reduce to the higher-order scheme

$$(3.24) \quad 0 = \int_\Omega \tilde{N}_i (a \; s_\xi u_{,\xi} + a \; s_\eta u_{,\eta}) d\Omega$$

The limiters, ℓ^ξ and ℓ^η, are functions of local smoothness monitors. Note that the scheme is conservative independent of the values of ℓ^ξ and ℓ^η. Our present research is concerned with the development of schemes of this type.

---

*AIAA computational fluid dynamics conference, Danvers, Massachusetts, July 13-15, 1983, AIAA CP834, AIAA, New York, 1983.

To simplify the calculations somewhat, N_i in (3.22) and (3.23) might be taken to be

$$\tilde{N}_i = (\tilde{N}_i^{\xi} + \tilde{N}_i^{\eta})/2 \quad . \tag{3.25}$$

The philosophy embodied in this scheme is to simply add coupling terms to the first-order scheme (3.18) such that it becomes a consistent Petrov-Galerkin/weighted residual method when the limiters are set equal to one. This concept enables us to easily construct a higher-order accurate method from any given first-order method of the type (3.18). <u>The importance of the idea is that essentially first-order one-dimensional concepts automatically engender an associated higher-order accurate multi-dimensional method</u>.

4. <u>The Two-dimensional Euler Equations</u>. We adopt a "long-hand" rather than indicial notation to avoid a plethora of tensor indices. Componential definitions of arrays appearing in this section are presented in Appendix I.

Consider the steady two-dimensional Euler equations

$$\underset{\sim}{F}_{,x} + \underset{\sim}{G}_{,y} = \underset{\sim}{0} \tag{4.1}$$

where the flux vectors $\underset{\sim}{F}$ and $\underset{\sim}{G}$ are written with respect to global Cartesian coordinates (x, y). Equation (4.1) can be written in the alternative form

$$\underset{\sim}{A}\,\underset{\sim}{U}_{,x} + \underset{\sim}{B}\,\underset{\sim}{U}_{,y} = \underset{\sim}{0} \tag{4.2}$$

where $\underset{\sim}{A} = \partial\underset{\sim}{F}/\partial\underset{\sim}{U}$, $\underset{\sim}{B} = \partial\underset{\sim}{G}/\partial\underset{\sim}{U}$ and $\underset{\sim}{U}$ is the vector of conservation variables. We will also write the equations in locally defined Cartesian coordinates (x', y') :

$$\underset{\sim}{F}'_{,x'} + \underset{\sim}{G}'_{,y'} = \underset{\sim}{A}'\underset{\sim}{U}'_{,x'} + \underset{\sim}{B}'\underset{\sim}{U}'_{,y'} = \underset{\sim}{0} \tag{4.3}$$

where $\underset{\sim}{A}' = \partial\underset{\sim}{F}'/\partial\underset{\sim}{U}'$ and $\underset{\sim}{B}' = \partial\underset{\sim}{G}'/\underset{\sim}{\ }\underset{\sim}{U}'$.

The relationship between the coordinate derivatives may be written as

$$\begin{Bmatrix} \dfrac{\partial}{\partial x'} \\[2ex] \dfrac{\partial}{\partial y'} \end{Bmatrix} = \begin{bmatrix} s_1 & s_2 \\[2ex] -s_2 & s_1 \end{bmatrix} \begin{Bmatrix} \dfrac{\partial}{\partial x} \\[2ex] \dfrac{\partial}{\partial y} \end{Bmatrix} \tag{4.4}$$

where $s_1 = \cos\theta$ and $s_2 = \sin\theta$ (see Figure 14). We employ the vector $\underset{\sim}{s}$ to define a local "preferred direction". For example, $\underset{\sim}{s}$ might be defined to be the unit normal to a discontinuity in the flow. Davis [3] has adopted this definition in his rotationally-biased shock-capturing scheme. Another possibility is to align $\underset{\sim}{s}$ with the flow (i.e., $\underset{\sim}{s} = \underset{\sim}{u}/||\underset{\sim}{u}||$ where $||\underset{\sim}{u}|| = (\underset{\sim}{u} \cdot \underset{\sim}{u})^{1/2}$ is the length of the velocity vector), etc. Note that (4.2) and (4.3) are related by

$$\text{(4.5)} \quad \underset{\sim}{A}\,\underset{\sim}{U}_{,x} + \underset{\sim}{B}\,\underset{\sim}{U}_{,y} = \underset{\sim}{Q}(\underset{\sim}{A}'\underset{\sim}{U}'_{,x'} + \underset{\sim}{B}'\underset{\sim}{U}'_{,y'})$$

where $\underset{\sim}{Q}$ is a rotation matrix given by (I10) in Appendix I.

Flux vector splitting. This scheme is the system analog of (3.18). The eth-element contribution to the flux at node i is given by

$$\text{(4.6)} \quad \tilde{\underset{\sim}{F}}{}_i^e = \int_{\Omega^e} \underset{\sim}{Q}(\tilde{\underset{\sim}{N}}{}_i^{\xi}\,\underset{\sim}{A}'_{\xi}\underset{\sim}{U}'_{,\xi} + \tilde{\underset{\sim}{N}}{}_i^{\eta}\,\underset{\sim}{B}'_{\eta}\underset{\sim}{U}'_{,\eta})\,d\Omega$$

where

$$\text{(4.7)} \quad \underset{\sim}{A}'_{\xi} = \xi_{,x'}\underset{\sim}{A}' + \xi_{,y'}\underset{\sim}{B}' = \underset{\sim}{S}'_{\xi}\underset{\sim}{\Lambda}_{\xi}(\underset{\sim}{S}'_{\xi})^{-1}$$

$$\text{(4.8)} \quad \underset{\sim}{B}'_{\eta} = \eta_{,x'}\underset{\sim}{A}' + \eta_{,y'}\underset{\sim}{B}' = \underset{\sim}{S}'_{\eta}\underset{\sim}{\Lambda}_{\eta}(\underset{\sim}{S}'_{\eta})^{-1}$$

$$\text{(4.9)} \quad \begin{cases} \tilde{\underset{\sim}{N}}{}_i^{\xi} = \underset{\sim}{S}'_{\xi}\underset{\sim}{\sigma}{}_i^{\xi}(\underset{\sim}{\sigma}^{\xi})^{-1}(\underset{\sim}{S}'_{\xi})^{-1}, & \tilde{\underset{\sim}{N}}{}_i^{\eta} = \underset{\sim}{S}'_{\eta}\underset{\sim}{\sigma}{}_i^{\eta}(\underset{\sim}{\sigma}^{\eta})^{-1}(\underset{\sim}{S}'_{\eta})^{-1} \\ \underset{\sim}{\sigma}{}_i^{\xi} = \langle N_{i,\xi}\underset{\sim}{\Lambda}_{\xi}\rangle, & \underset{\sim}{\sigma}{}_i^{\eta} = \langle N_{i,\eta}\underset{\sim}{\Lambda}_{\eta}\rangle \\ \underset{\sim}{\sigma}^{\xi} = \sum_{i=1}^{n_{en}} \underset{\sim}{\sigma}{}_i^{\xi}, & \underset{\sim}{\sigma}^{\eta} = \sum_{i=1}^{n_{en}} \underset{\sim}{\sigma}{}_i^{\eta} \end{cases}$$

or

$$\text{(4.10)} \quad \begin{cases} \tilde{\underset{\sim}{N}}{}_i^{\xi} = \underset{\sim}{S}'_{\xi}\underset{\sim}{N}_i(\tilde{\underset{\sim}{x}}_{\xi})(\underset{\sim}{S}'_{\xi})^{-1} \\ \underset{\sim}{N}_i(\tilde{\underset{\sim}{x}}_{\xi}) = \operatorname{diag}\left(N_i(\tilde{\underset{\sim}{x}}{}_{\xi}^{1}),\ N_i(\tilde{\underset{\sim}{x}}{}_{\xi}^{2}),\ \ldots,\ N_i(\tilde{\underset{\sim}{x}}{}_{\xi}^{m})\right) \\ \tilde{\underset{\sim}{x}}{}_{\xi}^{\ell} = \delta_{\xi}^{\ell}(\underset{\sim}{x})\,\operatorname{sgn}\Lambda_{\xi}^{\ell}(\underset{\sim}{x})\,\underset{\sim}{e}_{\xi}(\underset{\sim}{x}), \quad \ell = 1, 2, \ldots, m \end{cases}$$

(con'd.)

$$\begin{cases} \tilde{N}_i^{\eta} = \underset{\sim}{S}'_{\eta} \underset{\sim}{N}_i(\underset{\sim}{\tilde{x}}_{\eta})(\underset{\sim}{S}'_{\eta})^{-1} \\ \underset{\sim}{N}_i(\underset{\sim}{\tilde{x}}_{\eta}) = \operatorname{diag}\left(N_i(\underset{\sim}{\tilde{x}}^1_{\eta}), N_i(\underset{\sim}{\tilde{x}}^2_{\eta}), \ldots, N_i(\underset{\sim}{\tilde{x}}^m_{\eta})\right) \\ \underset{\sim}{\tilde{x}}^{\ell}_{\eta} = \delta^{\ell}_{\eta}(\underset{\sim}{x}) \operatorname{sgn} \Lambda^{\ell}_{\eta}(\underset{\sim}{x}) \underset{\sim}{e}_{\eta}(\underset{\sim}{x}), \quad \ell = 1, 2, \ldots, m \end{cases}$$

that the computations indicated in (4.6) may be simplified as follows:

$$(4.11) \quad \underset{\sim}{\tilde{N}}{}^{\xi}_i \underset{\sim}{A}'_{\xi} \underset{\sim}{U}'_{,\xi} = \underset{\sim}{S}'_{\xi} \underbrace{\underset{\sim}{\sigma}{}^{\xi}_i (\underset{\sim}{\sigma}{}^{\xi})^{-1} \underset{\sim}{\Lambda}_{\xi}}_{\text{diagonal}} (\underset{\sim}{S}'_{\xi})^{-1} \underset{\sim}{U}'_{,\xi}, \text{ etc.}$$

<u>Numerical example</u>. Our numerical experiments with this scheme have produced the expected results, namely, smooth, over-diffuse solutions. Figure 15 presents the solution of a supersonic oblique-shock calculation (see Davis [3] for details). As may be seen the profile is monotone, but the shock is spread over too many elements.

<u>High-resolution scheme employing scalar flux limiters</u>. To reduce diffusion, we may employ the concepts developed at the end of Section 3. The proposed high-resolution scheme takes the form

$$(4.12) \quad \underset{\sim}{\tilde{F}}{}^e_i = \int_{\Omega^e} \underset{\sim}{Q}(\underset{\sim}{L}{}^{\xi}_i \underset{\sim}{A}'_{\xi} \underset{\sim}{U}'_{,\xi} + \underset{\sim}{L}{}^{\eta}_i \underset{\sim}{B}'_{\eta} \underset{\sim}{U}'_{,\eta}) d\Omega$$

where

$$(4.13) \quad \underset{\sim}{L}{}^{\xi}_i = \underset{\sim}{\tilde{N}}{}^{\xi}_i + \ell^{\xi}(\underset{\sim}{\tilde{N}}_i - \underset{\sim}{\tilde{N}}{}^{\xi}_i)$$

$$(4.14) \quad \underset{\sim}{L}{}^{\eta}_i = \underset{\sim}{\tilde{N}}{}^{\eta}_i + \ell^{\eta}(\underset{\sim}{\tilde{N}}_i - \underset{\sim}{\tilde{N}}{}^{\eta}_i)$$

$$(4.15) \quad \underset{\sim}{\tilde{N}}_i = \frac{1}{2}(\underset{\sim}{\tilde{N}}{}^{\xi}_i + \underset{\sim}{\tilde{N}}{}^{\eta}_i)$$

The antidiffusive flux limiters, ℓ^{ξ} and ℓ^{η}, play the same role as before and the remarks made after (3.25) are relevant here too. We have not yet experimented with (4.12)-(4.15). A perceived potential shortcoming is that the scalar limiters do not afford sufficient control over the various solution components. It is not apparent how to introduce additional control given the fact that $\underset{\sim}{\tilde{N}}{}^{\xi}_i$ and $\underset{\sim}{\tilde{N}}{}^{\eta}_i$ are diagonalized by different matrices. We shall return to this issue shortly.

A method based upon diagonalization. The directional bias which needs to be introduced in two-dimensions can only be unambiguously defined if the system is diagonalizable.* It does not seem to be widely known that the coefficient matrices for the two-dimensional Euler equations are in fact simultaneously diagonalizable. This will be shown to be the case in what follows. In the supersonic regime the diagonalization corresponds to transformation to characteristic coordinates. The characteristic directions are the streamline and Mach lines. For large Mach number, the Mach lines approach the streamline. However, as a sonic point is approached the Mach lines become perpendicular to the streamline (see Figure 16). This indicates that a single preferred direction is not enough if optimal behavior is sought. The diagonalization of the coefficient matrices also holds in the subsonic regime, but the transformations and Mach-like directions become complex-valued. Nevertheless, it appears that essentially similar techniques, exploiting diagonal structure, can be used to generate the discrete equations in both the subsonic and supersonic regimes.

Diagonalization. To describe the diagonalization we shall begin with the Euler equations in local coordinates in which $\underset{\sim}{s} = \underset{\sim}{u}/||\underset{\sim}{u}||$:

$$\underset{\sim}{A}'\underset{\sim}{U}'_{,x'} + \underset{\sim}{B}'\underset{\sim}{U}'_{,y'} = \underset{\sim}{0} \tag{4.16}$$

Let $\underset{\sim}{A}'$ be written in the usual way as

$$\underset{\sim}{A}' = \underset{\sim}{S}'\underset{\sim}{\Lambda}'(\underset{\sim}{S}')^{-1} \tag{4.17}$$

Substituting (4.17) into (4.16) leads to

$$\underset{\sim}{S}'\left(\underset{\sim}{\Lambda}'(\underset{\sim}{S}')^{-1}\underset{\sim}{U}'_{,x'} + (\underset{\sim}{S}')^{-1}\underset{\sim}{B}'\underset{\sim}{S}'(\underset{\sim}{S}')^{-1}\underset{\sim}{U}'_{,y'}\right) = \underset{\sim}{0} \tag{4.18}$$

There exists a matrix $\underset{\sim}{R}'$ such that

$$(\underset{\sim}{R}')^{T}\underset{\sim}{\Lambda}'\underset{\sim}{R}' = \underset{\sim}{D}' \tag{4.19}$$

$$(\underset{\sim}{R}')^{T}(\underset{\sim}{S}')^{-1}\underset{\sim}{B}'\underset{\sim}{S}'\underset{\sim}{R}' = \underset{\sim}{E}' \tag{4.20}$$

where $\underset{\sim}{D}'$ and $\underset{\sim}{E}'$ are diagonal. Substituting (4.19) and (4.20) into (4.18) results in

$$\underset{\sim}{S}'(\underset{\sim}{R}')^{-T}\left(\underset{\sim}{D}'(\underset{\sim}{S}'\underset{\sim}{R}')^{-1}\underset{\sim}{U}'_{,x'} + \underset{\sim}{E}'(\underset{\sim}{S}'\underset{\sim}{R}')^{-1}\underset{\sim}{U}'_{,y'}\right) = \underset{\sim}{0} \tag{4.21}$$

Remark 1 In the case in which $\underset{\sim}{A}'$ and $\underset{\sim}{B}'$ are assumed constant, (4.21) is equivalent to the diagonal system

*For diagonalizable systems the ideas of Section 3 are relevant.

(4.22) $$\tilde{D}'\tilde{\tilde{U}}'_{,x} + \tilde{E}'\tilde{\tilde{U}}'_{,y} = \tilde{0}$$

where $\tilde{\tilde{U}} = (\tilde{S}'\tilde{R}')^{-1}\tilde{U}$.

Remark 2 In the supersonic regime all arrays are real, whereas in the subsonic regime $\tilde{R}'$ and $\tilde{E}'$ contain imaginary entries. (See Appendix I.)

A high-resolution scheme with flux limiters applied to each characteristic component. A scheme which generalizes (3.21)-(3.23) to each characteristic component of the system may now be derived. The following scheme employs a generalization of σ-weighting. Let

(4.23) $$\tilde{L}_i^{\xi} = (\tilde{T}'\tilde{Y}_i^{\xi} + \tilde{\bar{T}}'\tilde{Z}_i^{\xi})(\tilde{T}')^{-1}$$

(4.24) $$\tilde{L}_i^{\eta} = (\tilde{T}'\tilde{Y}_i^{\eta} + \tilde{\bar{T}}'\tilde{Z}_i^{\eta})(\tilde{T}')^{-1}$$

where

(4.25) $$\tilde{T}' = \tilde{S}'(\tilde{R}')^{-T}$$

(4.26) $$\tilde{Y}_i^{\xi} = |\tilde{X}_i^{\xi}| + \tilde{\ell}^{\xi}(|\tilde{X}_i| - |\tilde{X}_i^{\xi}|)$$

(4.27) $$\tilde{Y}_i^{\eta} = |\tilde{X}_i^{\eta}| + \tilde{\ell}^{\eta}(|\tilde{X}_i| - |\tilde{X}_i^{\eta}|)$$

(4.28) $$\tilde{Z}_i^{\xi} = \tilde{\bar{X}}_i^{\xi} + \tilde{\ell}^{\xi}(\tilde{\bar{X}}_i - \tilde{\bar{X}}_i^{\xi})$$

(4.29) $$\tilde{Z}_i^{\eta} = \tilde{\bar{X}}_i^{\eta} + \tilde{\ell}^{\eta}(\tilde{\bar{X}}_i - \tilde{\bar{X}}_i^{\eta})$$

(4.30) $$\tilde{X}_i = \tilde{W}_i\tilde{W}^{-1}$$

(4.31) $$\tilde{X}_i^{\xi} = \tilde{W}_i^{\xi}(\tilde{W}^{\xi})^{-1}$$

(4.32) $$\tilde{X}_i^{\eta} = \tilde{W}_i^{\eta}(\tilde{W}^{\eta})^{-1}$$

(4.33) $$\tilde{W} = \sum_{i=1}^{n_{en}} |\tilde{W}_i|$$

$$(4.34)\qquad \underset{\sim}{W}^{\xi} = \sum_{i=1}^{n_{en}} |\underset{\sim i}{W}^{\xi}|$$

$$(4.35)\qquad \underset{\sim}{W}^{\eta} = \sum_{i=1}^{n_{en}} |\underset{\sim i}{W}^{\eta}|$$

$$(4.36)\qquad \underset{\sim i}{W} = \underset{\sim i}{W}^{\xi} + \underset{\sim i}{W}^{\eta}$$

$$(4.37)\qquad \underset{\sim i}{W}^{\xi} = N_{i,\xi}(\xi_{,x}\underset{\sim}{D}' + \xi_{,y}\underset{\sim}{E}')$$

$$(4.38)\qquad \underset{\sim i}{W}^{\eta} = N_{i,\eta}(\eta_{,x}\underset{\sim}{D}' + \eta_{,y}\underset{\sim}{E}')$$

and ℓ^{ξ} and ℓ^{η} are real diagonal matrices of flux limiters which act independently on characteristic components. All of the calculations indicated in (4.26)-(4.38) involve complex-valued diagonal matrices, and the superposed bars in (4.23), (4.24), (4.28) and (4.29) denote complex conjugate. In the supersonic case all matrices are real. It may be verified that $\underset{\sim i}{L}^{\xi}$ and $\underset{\sim i}{L}^{\eta}$ are matrices with real coefficients even in the subsonic case. A simplification like that discussed previously, namely

$$(4.39)\qquad \underset{\sim i}{X} = \frac{1}{2}(\underset{\sim i}{X}^{\xi} + \underset{\sim i}{X}^{\eta})$$

may also be used as an economical alternative to (4.30), (4.33) and (4.36). Equations (4.23) and (4.24) should be used in (4.11) to define the nodal flux.

The following facts may be verified: The method is conservative in that

$$(4.40)\qquad \sum_{i=1}^{n_{en}} \underset{\sim i}{L}^{\xi} = \sum_{i=1}^{n_{en}} \underset{\sim i}{L}^{\eta} = \underset{\sim}{I}$$

If $\underset{\sim}{\ell}^{\xi} = \underset{\sim}{\ell}^{\eta} = \underset{\sim}{I}$, then

$$(4.41)\qquad \underset{\sim i}{L}^{\xi} = \underset{\sim i}{L}^{\eta} = (\underset{\sim}{T}'|\underset{\sim i}{X}| + \overline{\underset{\sim}{T}}'\underset{\sim i}{X})(\underset{\sim}{T}')^{-1}$$

The modulus term in (4.41) engenders central-difference like character, whereas the complex-conjugate term introduces compensating numerical diffusion. At a supersonic point, (4.41) becomes

$$\underset{\sim}{L}_i^{\xi} = \underset{\sim}{L}_i^{\eta} = \underset{\sim}{T}'(|\underset{\sim}{X}_i| + \underset{\sim}{X}_i)(\underset{\sim}{T}')^{-1} \tag{4.42}$$

which is the system analog of the Petrov-Galerkin/weighted-residual scheme for scalar equations presented in Section 3 (see (3.6)-(3.8) and (3.24)).

The method defined by (4.42) was tested on the oblique shock problem. Triangles were used for the discretization and results are presented in Figure 17. As may be seen the shock is very sharp, but due to the higher-order accuracy of the scheme oscillations appear. The Rankine-Hugoniot conditions are accurately satisfied. By modifying the weighting function to achieve Tabata's scheme (as described in Section 3) the monotone profile of Figure 18 is obtained. Due to the lack of conservation of Tabata's scheme, a slight violation of the Rankine-Hugoniot relations is noted (the density should attain a value of 1.460 on the plateau, but Tabata's scheme produces a value of 1.475).

Using the characteristic difference method on each characteristic separately produced the results of Figure 19. Lack of conservation caused the density to be 1.520, a somewhat larger violation of the Rankine-Hugoniot relations than that registered by Tabata's scheme.

Time-dependent schemes. The generalization to time-dependent phenomena is a subject that we will not dwell upon here. We will content ourselves with making two observations. The first concerns the structure of the explicit, forward-in-time scheme. In the present case we have

$$\underset{\sim}{U}_i^{n+1} = \underset{\sim}{U}_i^{n} - \frac{\Delta t}{h_i}\underset{\sim}{\tilde{F}}_i^{n} \tag{4.38}$$

where $\underset{\sim}{\tilde{F}}_i^{n}$ is the sum of all element flux contributions to node i and

$$h_i = \sum_e \left(\frac{1}{n_{en}} \text{meas } (\Omega^e)\right) \tag{4.39}$$

where the summation is over all elements attached to node i. We have used this scheme to calculate the steady Euler flows presented in this section.

The second observation concern the development of high-resolution time-dependent schemes. We will just sketch our basic ideas. Let us first develop a semi-discrete time-deptendent scheme corresponding to the scalar flux-limited steady scheme defined by (4.12)-(4.15). The diffusive first-order semi-discrete scheme will be taken to be

$$(4.40)\qquad \underset{\sim}{0} = \int_\Omega \underset{\sim}{Q}(\psi_i \underset{\sim}{I}\, \underset{\sim}{U}'_{,t} + \tilde{\underset{\sim}{N}}{}^\xi_i\, \underset{\sim}{A}'_\xi \underset{\sim}{U}'_{,\xi} + \tilde{\underset{\sim}{N}}{}^\eta_i\, \underset{\sim}{B}'_\eta \underset{\sim}{U}'_{,\eta})\,d\Omega$$

where $\psi_i = 1/n_{en}$ in each element connected to node i, and is zero otherwise. Note $\sum \psi_i = 1$. The higher-order accurate consistent Petrov-Galerkin semi-discrete scheme will be taken to be

$$(4.41)\qquad \underset{\sim}{0} = \int_\Omega \underset{\sim}{Q}\, \tilde{\underset{\sim}{N}}_i(\underset{\sim}{U}'_{,t} + \underset{\sim}{A}'_\xi \underset{\sim}{U}'_{,\xi} + \underset{\sim}{B}'_\eta \underset{\sim}{U}'_{,\eta})\,d\Omega$$

Combining these in the usual way leads to

$$(4.42)\qquad 0 = \int_\Omega \underset{\sim}{Q}(\underset{\sim}{L}{}^t_i\, \underset{\sim}{U}'_{,t} + \underset{\sim}{L}{}^\xi_i\, \underset{\sim}{A}'_\xi \underset{\sim}{U}'_{,\xi} + \underset{\sim}{L}{}^\eta_i\, \underset{\sim}{B}'_\eta \underset{\sim}{U}'_{,\eta})\,d\Omega$$

where $\underset{\sim}{L}{}^\xi_i$ and $\underset{\sim}{L}{}^\eta_i$ are defined by (4.13) and (4.14), and

$$(4.43)\qquad \underset{\sim}{L}{}^t_i = \psi_i \underset{\sim}{I} + \ell^t(\tilde{\underset{\sim}{N}}_i - \psi_i \underset{\sim}{I})$$

in which ℓ^t is an additional scalar, antidiffusive flux limiter.

A semi-discrete generalization of the "characteristic scheme" also can be put in the form (4.42). However, in this case, $\underset{\sim}{L}{}^\xi_i$ and $\underset{\sim}{L}{}^\eta_i$ are defined by (4.23) and (4.24), respectively, and

$$(4.44)\qquad \underset{\sim}{L}{}^t_i = \psi_i \underset{\sim}{I} + \underset{\sim}{\ell}{}^t \left((\underset{\sim}{T}'|\underset{\sim}{X}_i| + \overline{\underset{\sim}{T}}{}'\underset{\sim}{X}'_i)(\underset{\sim}{T}')^{-1} - \psi_i \underset{\sim}{I}\right)$$

where $\underset{\approx}{\ell}{}^t$ is a diagonal matrix of antidiffusive flux limiters.

This formalism has been successfully used to develop a one-dimensional high-precision finite element scheme in Hughes and Mallet [10].

5. Conclusions. In this paper we have proposed ideas for generalizing a one-dimensional finite element shock capturing scheme to multidimensional cases. Particular attention is paid to the two-dimensional steady Euler equations. Based on earlier studies it is felt that "preferred directions" are the key to good behavior in the multi-dimensional case. This point of view is developed in the schemes advocated herein. For the two-dimensional Euler equations the characteristics distinguish themselves as preferred directions and a scheme based upon characteristics is proposed. Further research needs to concern itself with the definitions of flux limiters appearing in the high-resolution formalism.

Appendix I Definitions of Matrices Appearing in Section 4.

$$\text{(I1)} \qquad \underset{\sim}{F} = \begin{Bmatrix} \rho u \\ \rho u^2 + p \\ \rho uv \\ \rho u(e + p) \end{Bmatrix}$$

$$\text{(I2)} \qquad \underset{\sim}{G} = \begin{Bmatrix} \rho v \\ \rho vu \\ \rho v^2 + p \\ \rho v(e + p) \end{Bmatrix}$$

$$\text{(I3)} \qquad \underset{\sim}{F}' = \begin{Bmatrix} \rho u' \\ \rho (u')^2 + p \\ \rho u'v' \\ \rho u'(e + p) \end{Bmatrix}$$

$$\text{(I4)} \qquad \underset{\sim}{G} = \begin{Bmatrix} \rho v' \\ \rho v'u' \\ \rho (v')^2 + p \\ \rho v'(e + p) \end{Bmatrix}$$

(I5)
$$\underset{\sim}{A} = \begin{bmatrix} 0 & 1 & 0 & 0 \\ \bar{\gamma}||\underset{\sim}{u}||^2/2 - u^2 & u(2-\bar{\gamma}) & -\bar{\gamma}v & \bar{\gamma} \\ -uv & v & u & 0 \\ (\bar{\gamma}||\underset{\sim}{u}||^2 - \gamma e)u & \varepsilon - \bar{\gamma}u^2 & -\bar{\gamma}uv & \gamma u \end{bmatrix}$$

(I6)
$$\underset{\sim}{B} = \begin{bmatrix} 0 & 0 & 1 & 0 \\ -vu & v & u & 0 \\ \bar{\gamma}||\underset{\sim}{u}||^2/2 - v^2 & -\bar{\gamma}u & v(2-\bar{\gamma}) & \bar{\gamma} \\ (\bar{\gamma}||\underset{\sim}{u}||^2 - \gamma e)v & -\bar{\gamma}vu & \varepsilon - \bar{\gamma}v^2 & \gamma v \end{bmatrix}$$

(I7)
$$\underset{\sim}{A}' = \begin{bmatrix} 0 & 1 & 0 & 0 \\ \bar{\gamma}||\underset{\sim}{u}||^2/2 - (u')^2 & u'(2-\bar{\gamma}) & -\bar{\gamma}v' & \bar{\gamma} \\ -u'v' & v' & u' & 0 \\ (\bar{\gamma}||\underset{\sim}{u}||^2 - \gamma e)u' & \varepsilon - \bar{\gamma}(u')^2 & -\bar{\gamma}u'v' & \gamma u' \end{bmatrix}$$

(I8)
$$\underset{\sim}{B}' = \begin{bmatrix} 0 & 0 & 1 & 0 \\ -v'u' & v' & u' & 0 \\ \bar{\gamma}||\underset{\sim}{u}||^2/2 - (v')^2 & -\bar{\gamma}u' & v'(2-\bar{\gamma}) & \bar{\gamma} \\ (\bar{\gamma}||\underset{\sim}{u}||^2 - \gamma e)v' & -\bar{\gamma}v'u' & \varepsilon - \bar{\gamma}(v')^2 & \gamma v' \end{bmatrix}$$

$$
\text{(I9)} \qquad \underset{\sim}{U} = \begin{Bmatrix} \rho \\ \rho u \\ \rho v \\ \rho e \end{Bmatrix} , \qquad \underset{\sim}{U}' = \begin{Bmatrix} \rho \\ \rho u' \\ \rho v' \\ \rho e \end{Bmatrix}
$$

$$
\text{(I10)} \qquad \underset{\sim}{Q} = \begin{bmatrix} 1 & 0 & 0 & 0 \\ 0 & s_1 & -s_2 & 0 \\ 0 & s_2 & s_1 & 0 \\ 0 & 0 & 0 & 1 \end{bmatrix}
$$

In the following definitions, $\underset{\sim}{s} = \underset{\sim}{u}/||\underset{\sim}{u}||$. Thus $||\underset{\sim}{u}|| = u'$ and $v' = 0$.

$$
\text{(I11)} \qquad \underset{\sim}{A}' = \begin{bmatrix} 0 & 1 & 0 & 0 \\ (\bar{\gamma}/2-1)(u')^2 & u'(2-\bar{\gamma}) & 0 & \bar{\gamma} \\ 0 & 0 & u' & 0 \\ (\bar{\gamma}(u')^2 - \gamma e)u' & \varepsilon - \bar{\gamma}(u')^2 & 0 & \gamma u' \end{bmatrix} .
$$

$$
\text{(I12)} \qquad \underset{\sim}{B}' = \begin{bmatrix} 0 & 0 & 1 & 0 \\ 0 & 0 & u' & 0 \\ \bar{\gamma}(u')^2/2 & -\bar{\gamma}u' & 0 & \bar{\gamma} \\ 0 & 0 & \varepsilon & 0 \end{bmatrix}
$$

(I13) $$\underset{\sim}{U}' = \begin{Bmatrix} \rho \\ \rho u' \\ 0 \\ \rho e \end{Bmatrix}$$

(I14) $$\underset{\sim}{F}' = \begin{Bmatrix} \rho u' \\ \rho (u')^2 + p \\ 0 \\ \rho u'(e + p) \end{Bmatrix} , \qquad \underset{\sim}{G}' = \begin{Bmatrix} 0 \\ 0 \\ p \\ 0 \end{Bmatrix}$$

(I15) $$\underset{\sim}{\Lambda}' = \text{diag}(u' , u' , u' + c , u' - c)$$

(I16) $$\underset{\sim}{S}' = \frac{\rho}{\sqrt{2}} \begin{bmatrix} \sqrt{2}/\rho & 0 & 1/c & 1/c \\ u' \sqrt{2}/\rho & 0 & u'/c + 1 & u'/c - 1 \\ 0 & -\sqrt{2} & 0 & 0 \\ (u')^2/(\rho \sqrt{2}) & 0 & (u')^2/(2c) + u' + c/\bar{\gamma} & (u')^2/(2c) - u' + c/\bar{\gamma} \end{bmatrix}$$

(I17) $$\underset{\sim}{R}' = \begin{bmatrix} 1 & 0 & 0 & 0 \\ 0 & 0 & -\sqrt{2(u' - c)/(u' + c)} & \sqrt{2(u' - c)/(u + c)} \\ 0 & -1 & (u' - c)/(u' + c) & (u' - c)/(u' + c) \\ 0 & 1 & 1 & 1 \end{bmatrix}$$

(I18) $\underset{\sim}{D}' = \text{diag}(u', 2u', z_1, z_1)$

(I19) $z_1 = 4u'(u' - c)/(u' + c)$

(I20) $\underset{\sim}{E}' = \text{diag}(0, 0, z_2, -z_2)$

(I21) $z_2 = 4u'c\sqrt{(u' - c)/(u' + c)^3}$

The notation is as follows:

ρ = density;

u = x-component of velocity;

v = y-component of velocity;

u' = $s_1 u + s_2 v$ = x'-component of velocity;

v' = $-s_2 u + s_1 v$ = y'-component of velocity;

p = $(\gamma - 1)\rho\left(e - ||\underset{\sim}{u}||^2/2\right)$ = pressure;

γ = ratio of specific heats;

$\bar{\gamma}$ = $\gamma - 1$;

ε = $\gamma e - \bar{\gamma}||\underset{\sim}{u}||^2/2$; and

c = $(\gamma p/\rho)^{\frac{1}{2}}$ = acoustic speed.

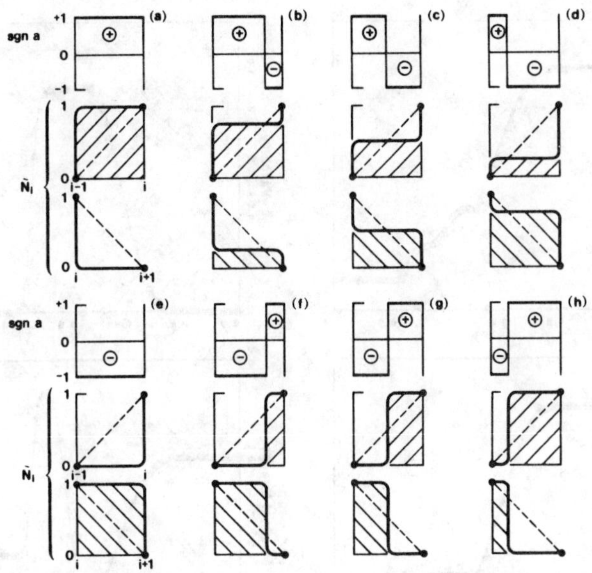

Figure 1. Weighting functions for the case $\varepsilon \to 0$. Case (a) is the supersonic case; cases (b), (c) and (d) are transonic compressions; case (e) is the subsonic case; and cases (f), (g) and (h) are transonic expansions. If cases (f), (g) and (h) are used in place of cases (b), (c) and (d), respectively, the weighting functions generate Engquist-Osher numerical flux vectors.

Initial data: $u_i^0 = +1, \quad 1 \le i \le k$
$= \omega_i, \quad k < i < l, \quad -1 \le \omega_i \le +1$
$= -1, \quad l \le i \le n_{np}$

Scheme: Explicit, forward-in-time, uniform meshing

n_{np}	$\sum_{i=1}^{n_{np}} u_i^0$	G	EO	Present Method
{even, odd}	{even, odd}	+1, 0, −1	+1, +α, 0, −α, −1; $\alpha = 1/\sqrt{2}$	+1, 0, −1
{odd, even}	{even, odd}	+1, 0, −1	+1, 0, −1	+1, 0, −1
{odd, even} (G); {even, odd} (EO and present)	{even, odd} + γ, $-1 \le \gamma \le +1$	+1, γ, 0, −1	+1, α, 0, β, −1; $\alpha^2+\beta^2=1, \ \alpha+\beta=\gamma$	α, +1, 0, β, −1; $\alpha^2+\beta^2+\alpha\beta=1, \ \alpha+\beta=\gamma$

Figure 2. Resolution of stationary shocks for Burgers' equation.

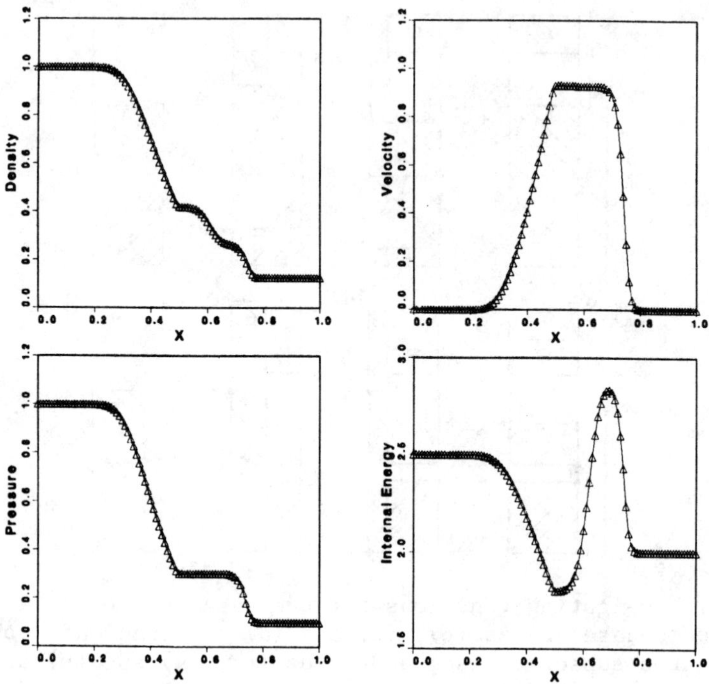

Figure 3. First-order, explicit, forward-in-time solution of Sod's problem.

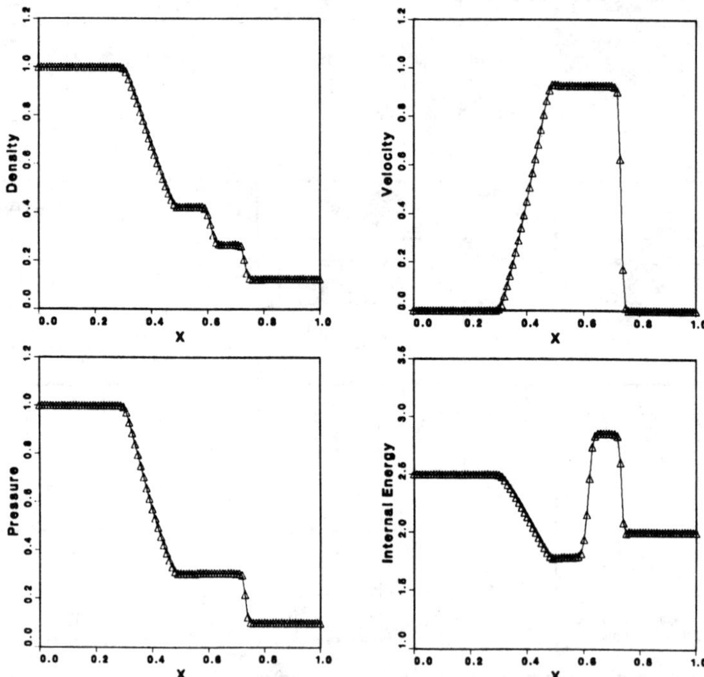

Figure 4. High-resolution explicit solution of Sod's problem [10].

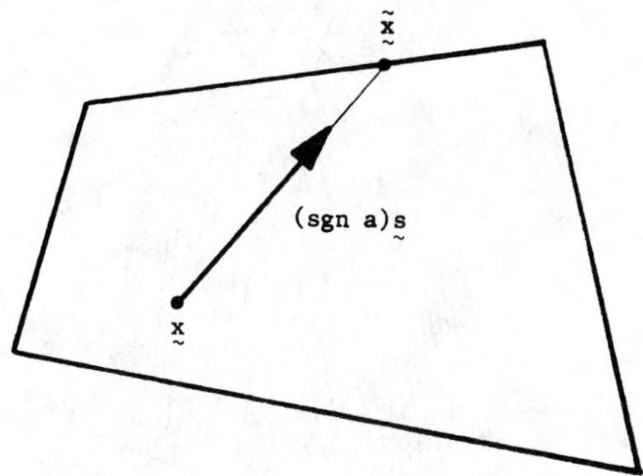

Figure 5. Illustration of $\tilde{x}$ for the transported weighting function on a quadrilateral.

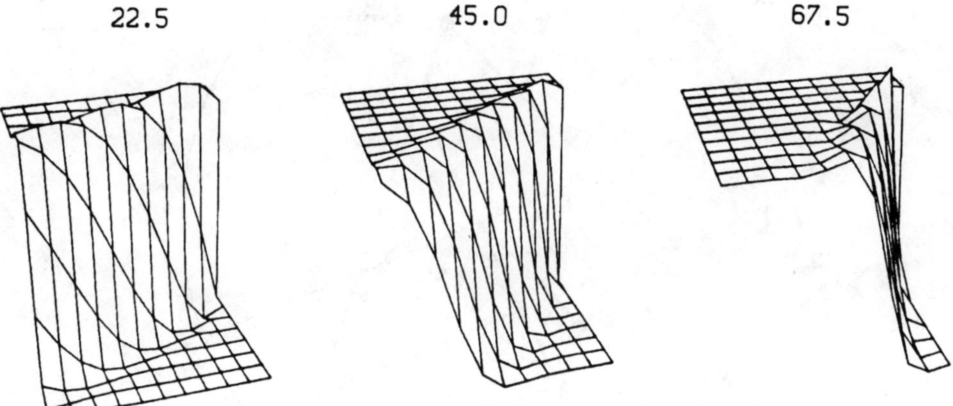

Figure 6. Advection skew to the mesh; σ-weighting on bilinear quadrilaterals.

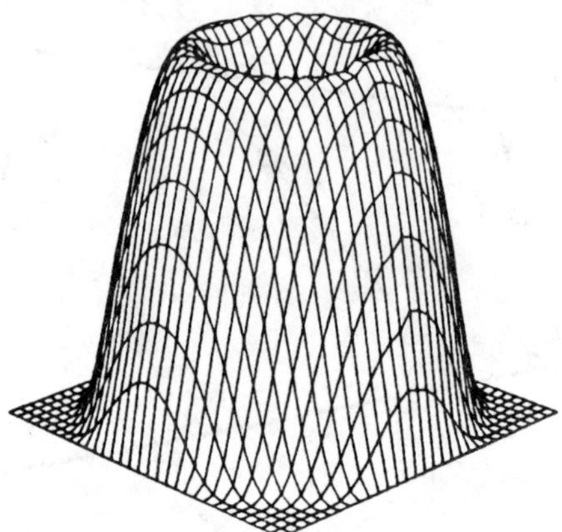

Figure 7. Advection in a rotating flow field; σ-weighting on bilinear quadrilaterals.

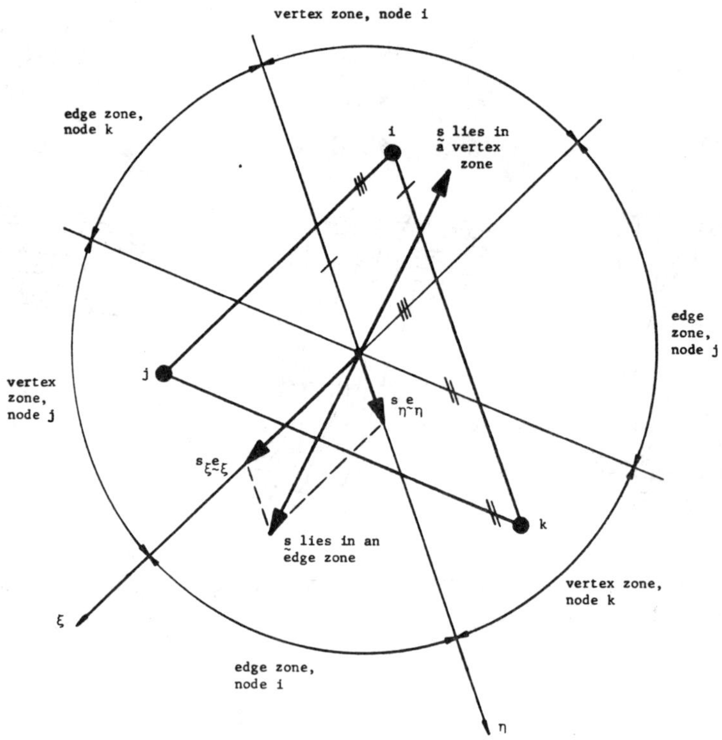

Figure 8. When $\underset{\sim}{s}$ lies in a vertex zone on a triangle, σ-weighting produces monotone difference equations. When $\underset{\sim}{s}$ lies in an edge zone, σ-weighting produces non-monotone difference equations. If the edge case is simply omitted, we have Tabata's non-conservative, monotone scheme. The edge-case treatment can be modified by resolving $\underset{\sim}{s}$ into ξ and η components and using flux-splitting ideas to generate a simple, monotone, conservative generalization of Tabata's scheme.

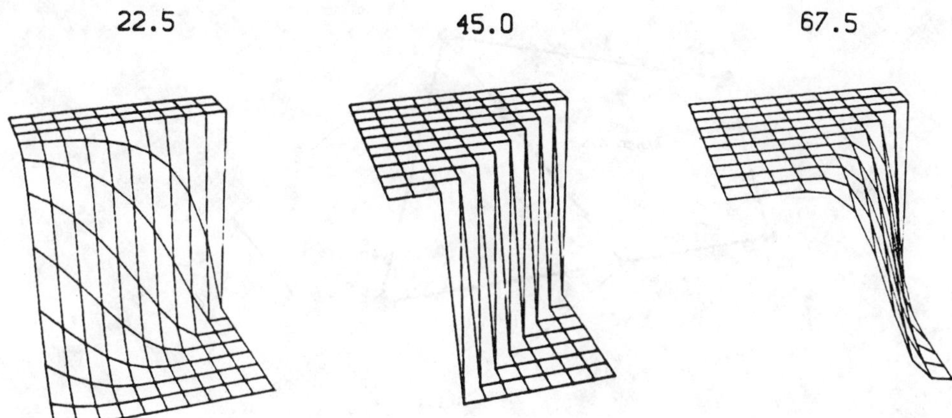

Figure 9. Advection skew to the mesh with Tabata's scheme. Each quadrilateral is subdivided into two triangles. The diagonals run in the northeast-southwest direction.

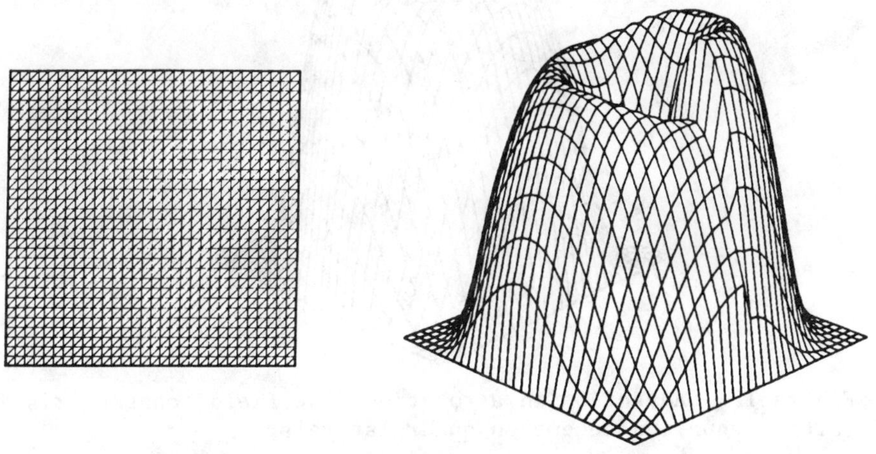

Figure 10. Advection in a rotating flow field; finite element mesh and results for Tabata's scheme.

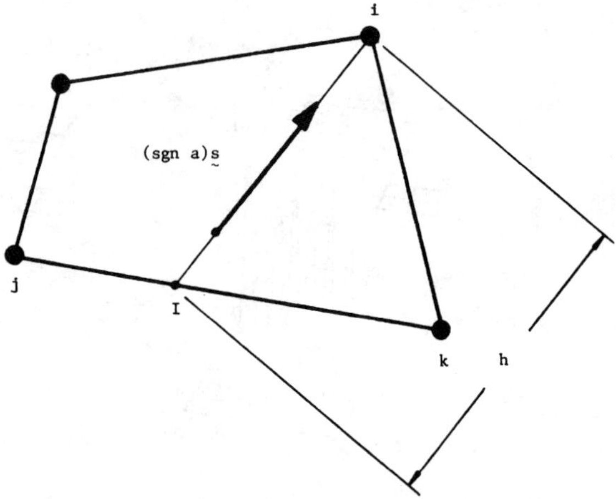

Figure 11. Geometrical aspects of characteristic difference equation scheme.

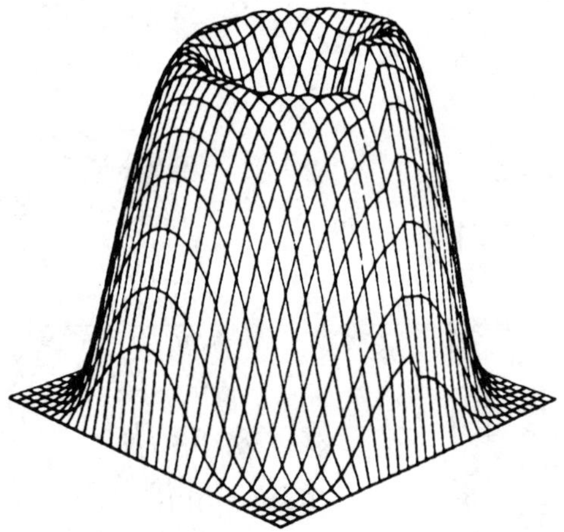

Figure 12. Advection in a rotating flow field; characteristic difference equation scheme on quadrilaterals.

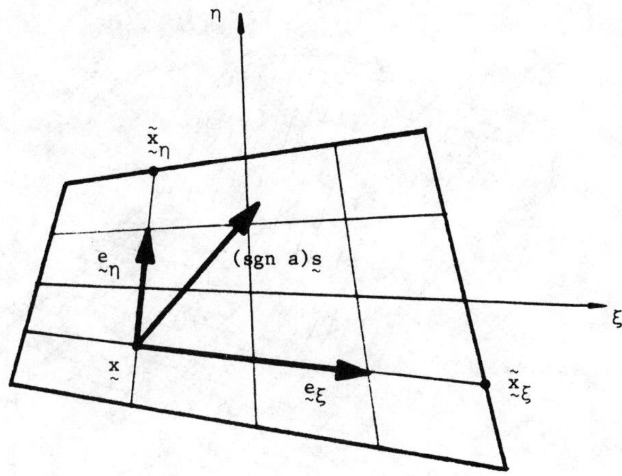

Figure 13. Illustration of $\tilde{x}_\xi$ and $\tilde{x}_\eta$ for transported weighting functions in flux-splitting scheme.

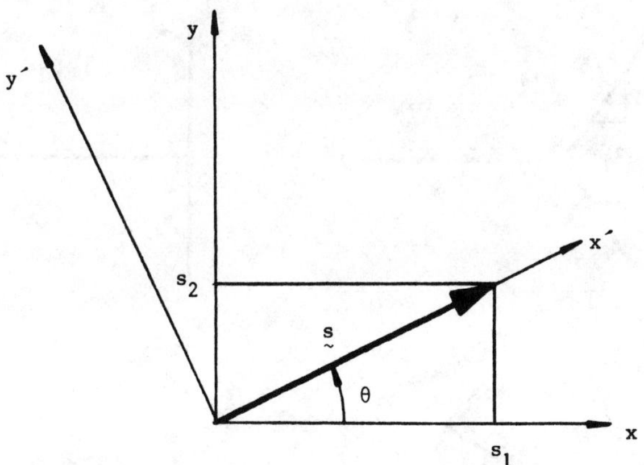

Figure 14. Local preferred direction.

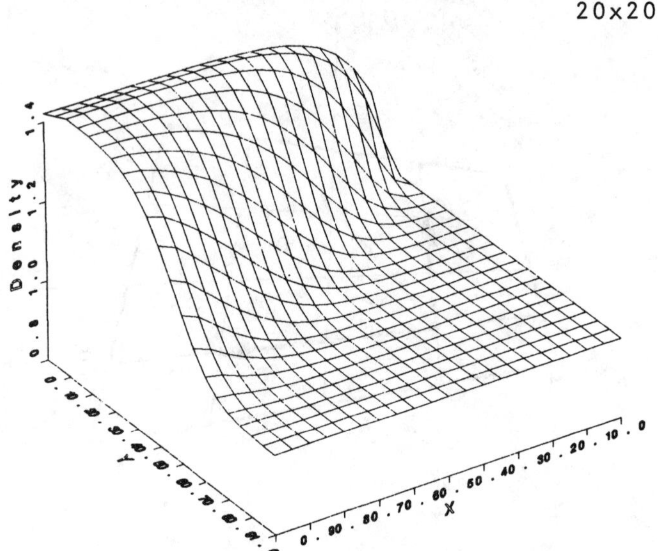

Figure 15. Supersonic oblique shock. The first-order flux-vector splitting scheme with σ-weighting on quadrilaterals and one-point Gaussian quadrature was used. This scheme is conservative.

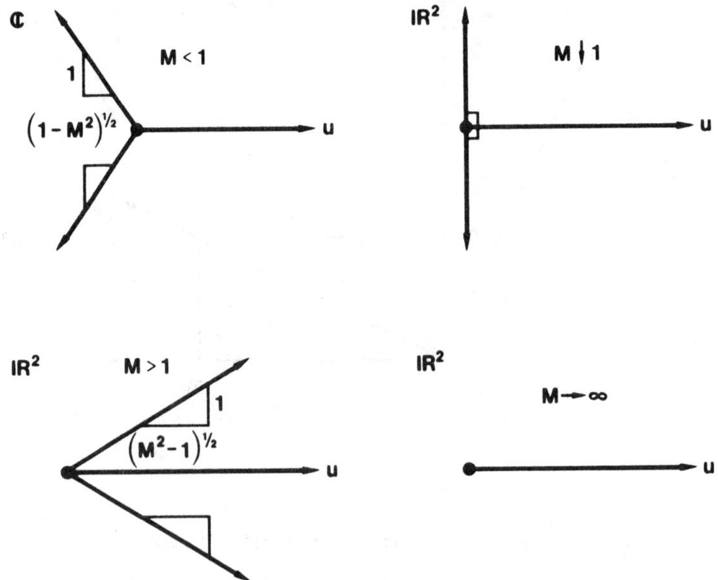

Figure 16. Geometry of the characteristics for the two-dimensional steady Euler equations.

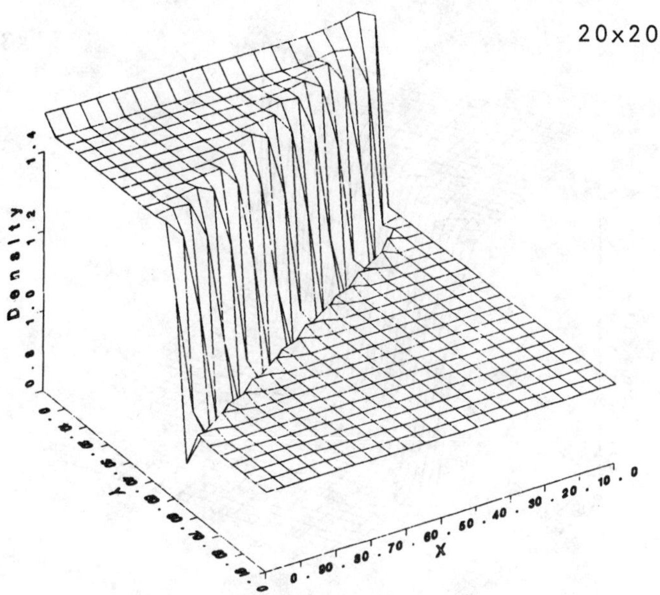

Figure 17. Supersonic oblique shock. The characteristic Petrov-Galerkin formulation with σ-weighting on triangles and one-point Gaussian quadrature was used. This scheme is conservative.

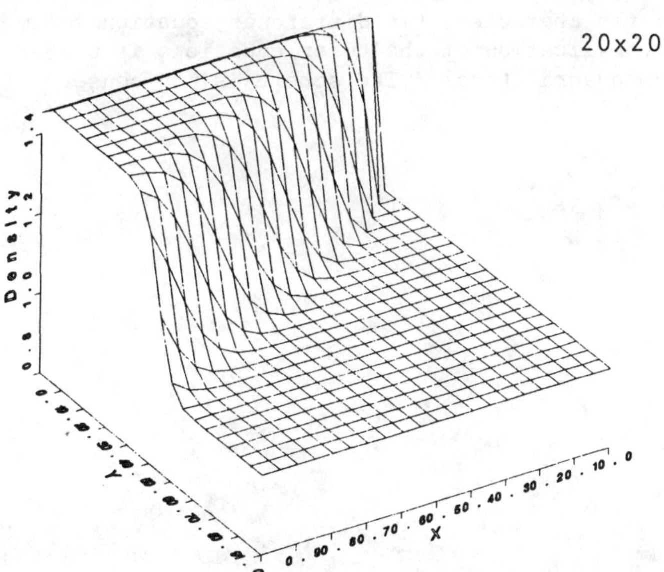

Figure 18. This is the same as in Figure 17 except the "edge zone" cases were omitted. This amounts to a system version of Tabata's non-conservative, monotone scheme.

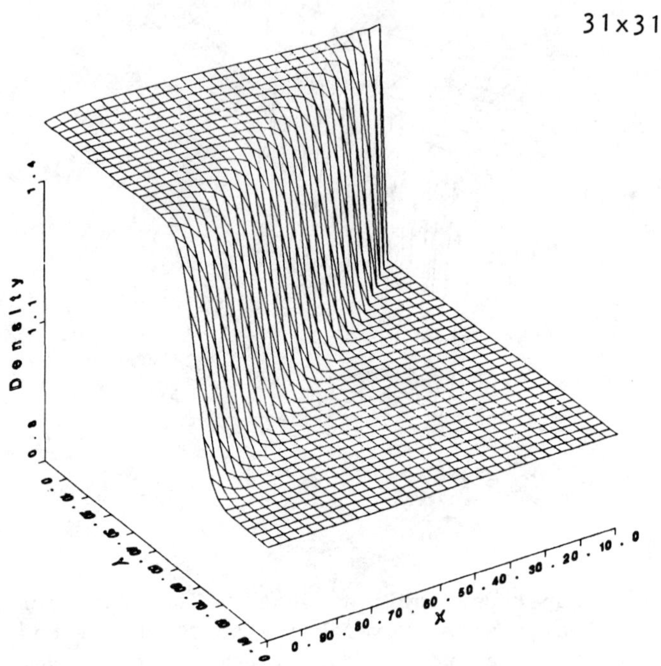

Figure 19. Supersonic oblique shock. Each characteristic component is treated by the characteristic difference equation scheme (see Figure 11). The diagonalization of the Euler equations is performed at the center of each quadrilateral. The scheme is non-conservative.

REFERENCES

[1] N. N. ANUCINA, Some difference schemes for hyperbolic systems, pp. 1-13 in Difference methods for solutions of problems of mathematical physics I, ed. N. N. Yanenko, American Mathematical Society, Providence, RI, 1967.

[2] A. N. BROOKS and T.J.R. HUGHES, Streamline-upwind/Petrov-Galerkin formulations for convection dominated flows with particular emphasis on the incompressible Navier-Stokes equations, Computer Methods in Applied Mechanics and Engineering, 32(1982), pp. 199-259.

[3] S. F. DAVIS, A Rotationally biased upwind difference scheme for the Euler equations, NASA Contractor Report, Contract No. NAS1-17070, ICASE, NASA Langley Research Center, Hampton, Virginia, July 1983.

[4] B. ENGQUIST and S. OSHER, Stable and entropy satisfying approximations for transonic flow calculations, Mathematics of Computation, 34(1980), pp. 45-75.

[5] S. K. GODUNOV, A Finite difference method for the numerical calculation of discontinuous solutions of the equations of fluid dynamics, Mat. Sb., 47(1959), pp. 271-290.

[6] A. HARTEN, High resolution schemes for hyperbolic conservation laws, Journal of Computational Physics, 49(1983), pp. 357-393.

[7] P. W. HEMKER, A numerical study of stiff two-point boundary value problems, Thesis, Mathematisch Centrum, Amsterdam, 1977.

[8] T.J.R. HUGHES, A shock capturing finite element method, Sixth International Conference on Computing Methods in Applied Sciences and Engineering, INRIA, Versailles, December 12-16, 1983 (Proceedings to be published by North-Holland).

[9] T.J.R. HUGHES and A. N. BROOKS, A theoretical framework for Petrov-Galerkin methods with discontinuous weighting functions: Application to the streamline upwind procedure, pp. 47-65 in Finite Elements in Fluids, Vol. 4, eds. R. H. Gallagher et al., John Wiley and Sons, London, 1982.

[10] T.J.R. HUGHES and M. MALLET, A high-precision finite element method for shock-tube calculations, preprint.

[11] T.J.R. HUGHES, T. E. TEZDUYAR and A. N. BROOKS, A Petrov-Galerkin finite element formulation for systems of conservation laws with special reference to the compressible Euler equations, pp. 97-125 in Numerical Methods for Fluid Dynamics, eds. K. W. Morton and M. J. Baines, Academic Press 1982.

[12] T. IKEDA, Maximum principle in finite element models for convectional-diffusion phenomena, North-Holland, Amsterdam, 1983.

[13] C. JOHNSON, Finite element methods for convection-diffusion problems, Fifth International Symposium on Computing Methods in Engineering and Applied Sciences, INRIA, Versailles, December 1981.

[14] U. NÄVERT, A finite element method for convection-diffusion problems, Ph.D. Thesis, Department of Computer Sciences, Chalmers University of Technology, Göteborg, Sweden, 1982.

[15] S. OSHER, Shock modelling in aeronautics, pp. 179-217 in Numerical Methods for Fluid Dynamics, eds. K. W. Morton and M. J. Baines, Academic Press, London, 1982.

[16] S. OSHER and F. SOLOMON, Upwind difference schemes for hyperbolic systems of conservation laws, Mathematics of Computation, 38(1982), pp. 339-374.

[17] J. RICE and R. J. SCHNIPKE, A new streamline upwind finite element method for convection dominated flows, preprint.

[18] P. L. ROE, Some contributions to the modelling of discontinuous flows, Presented at the AMS/SIAM Summer Seminar on Large Scale Computations in Fluid Mechanics, Scripps Institution of Oceanography, University of California, San Diego, June 26th to July 8th, 1983.

[19] G. A. SOD, A survey of several finite difference methods for systems of nonlinear hyperbolic conservation laws, Journal of Computational Physics, 27(1978), pp. 1-31.

[20] M. TABATA, A finite element approximation corresponding to the upwind finite differencing, Mem. Numer. Math., 4(1977), pp. 47-63.

IMPLICIT FINITE ELEMENT METHODS FOR THE EULER EQUATIONS

BRUNO STOUFFLET

Abstract. We develop here implicit Finite Element Methods for the calculation of steady states of the full Euler equations for compressible inviscid fluids. Firstly, three linearly unconditionally stable methods are constructed and studied for a class of one dimensional first-order schemes : a totally linearized implicit scheme, an approximate factored Δ-scheme and an inconsistent scheme for which a result of convergence is shown in the scalar case. Extension to the solution of the 2-D Euler equations is formulated via a control volume type approximation. Secondly, we utilize the so-constructed solvers to derive an implicit version of a centered second-order Finite Element scheme. Numerical experiments on 2-D flows in a channel are presented in order to compare the efficiency of the different proposed implicit methods with a first-order Godunov-type upwind scheme. Accurate solutions of flows around bodies are computed with the implicit second-order scheme with very large CFL numbers and are compared with other available results.

1. Introduction. In recent years, the appearance of implicit methods has considerably contributed to the improvement of the efficiency of numerical methods for solving hyperbolic systems such as the Euler equations. Limitations on the time-step can be very restrictive if explicit methods are used. Implicit formulations permit greater time-steps and may reduce the computing time in a significant way.
Several difference implicit methods have been developed in the last ten years.
One of the earliest and important contributions to the study of implicit difference schemes is due to Beam and Warming 5 ; their scheme leads to the solution of block-tridiagonal linear systems. It is extended to the case of several space variables with an ADI technique.
Lerat [17],[18] builds a class of centered second-order accurate implicit difference schemes, containing that of Beam and Warming, and leading to the solution of block-tridiagonal linear systems. More recently Yee, Harten and Warming [23] describe two implicit conservative and non-conservative versions of Harten's scheme [8], extended in two di-

*INRIA, Domaine de Voluceau, Rocquencourt, 78153 LE CHESNAY CEDEX, France and Ecole Polytechnique.

mensions by an ADI technique. The linear systems are also block-tridiagonal.
Specifically for the solution of the Euler equations, Steger and Warming [24] present an upwind flux-splitting implicit scheme requiring the solution of a triangular linear system.
Mac Cormack also has proposed an implicit scheme [20]. His is centered and the linear systems to be solved are bidiagonal. Casier, Deconinck and Hirsch [6] recently studied a class of centered bidiagonal schemes of second-order accuracy extending Mac Cormack's scheme, and allowing the use of very large CFL numbers. Another approach is that of Jameson [10] which consists in smoothing the residual with an elliptic-type operator giving a linearly unconditionally stable scheme.

This paper deals with implicit Finite Element approximations of steady Euler flows. The choice of Finite Element Methods allows the use of non-structured triangulations ; the mesh generation becomes more flexible and adapted to complex geometries. Conversely, the Finite Element approach precludes the use of x-y spatial ADI algorithms as used in the Finite Difference approach and appropriate linear solvers are required.

Two general observations can be noted in order to choose 1) the spatial approximation and 2) the implicit solver.
Upwinding is a good principle to construct robust implicit schemes with nice matrix properties such as diagonally dominance.
Centered Galerkin P1 (i.e. continuous, linear by triangle) approximations are accurate and suitable in the physical step of an implicit algorithm. These two observations will guide us to construct Finite Element implicit methods to solve Euler Equations.
Section 2 reviews some properties of the hyperbolic equations and of their approximations. Section 3 presents an analysis of a totally linearized implicit version of a class of explicit first-order accurate (mainly upwind) schemes in one-space variable requiring the solution of a complete linear system ; a triangular factored version is constructed. In Section 4, an efficient inconsistent version is derived from the same class of first-order schemes with a block-diagonal matrix inversion ; a non-linear analysis is performed in a scalar case and a result of convergence is shown with no CFL number restriction. In Section 5, the schemes are extended to the case of several space variables. Then, in Section 6, an implicit second-order accurate P1-Galerkin centered scheme is constructed with the totally linearized implicit solver as a mathematical step. Finally, in Section 7 numerical results are given.

2. Notations and recalls

ⓐ Weak solutions of a one dimensional hyperbolic system and Kruzkov's theory

We consider the Cauchy problem

$$(2.1) \qquad \frac{\partial W}{\partial t} + \frac{\partial F(W)}{\partial x} = 0 \ , \quad \{x,t\} \in \mathbb{R} \times]0,T[\quad ,$$

(2.2) $W(x,0) = W_o(x), \quad \forall x \in \mathbb{R}$,

where $F \in C^1(\mathbb{R}^m)$, $W_o \in (L^\infty(\mathbb{R}))^m$, $W \in \mathbb{R}^m$, $T > 0$.

The Jacobian matrix $A(W) = F'(W)$ is diagonalizable at any point W and has real eigenvalues. There exists a transformation T such that

(2.3) $\forall W \in \mathbb{R}^m$, $A(W) = T^{-1}(W)\, \Lambda(W)\, T(W)$,

where $\Lambda(W) = [\lambda_i(W)\, \delta_{ij}]_{1\le i,j\le m}$ is diagonal.

For every real function g, we will note g(A) the matrix similar to $g(\Lambda)$ given by $g(A) = T^{-1}(W)g(\Lambda)T(W)$ where $g(\Lambda) = [g(\lambda_i)\delta_{ij}]_{1\le i,j\le m}$.

Problem (2.1), (2.2) does not admit in general a classical solution defined for all t>0. We must introduce the notion of weak solution to admit singular solutions (see Lax [13] and Hopf [9]). The uniqueness of a weak solution of (2.1)-(2.2) is not generally ensured and one must add an entropy condition to select the good physical solution of the problem, as in Lax [14].

In the case m=1, the problem has been extensively studied. Kruzkov [12] and Oleinik [28] have shown the existence and uniqueness of the entropic solution. We recall here the theory developped by Kruzkov for the scalar equation.

In the scalar case, we denote the functions by lowercase letters. Thus, the Cauchy problem is written as

(2.4) $\dfrac{\partial u}{\partial t} + \dfrac{\partial f(u)}{\partial x} = 0$, $\{x,t\} \in \mathbb{R} \times]0,T[$,

(2.5) $u(x,0) = u_o(x)$, $\forall x \in \mathbb{R}$,

where $f \in C^1(\mathbb{R})$, $u_o \in L^\infty(\mathbb{R})$.

<u>Definition 2.1.</u>
A solution in Kruzkov's sense of (2.4)-(2.5) is a function $u \in L^\infty(\mathbb{R} \times]0,T[\,)$ such that :

(i) $\forall k \in \mathbb{R}$, $\forall \phi \in C_o^2(\mathbb{R} \times]0,T[)$ one has

(2.6) $$\iint_{\mathbb{R}\times]0,T[} [\,|u-k|\,\frac{\partial \phi}{\partial t} + sg(u-k)(f(u)-f(k))\frac{\partial \phi}{\partial x}\,]dx\, dt \ge 0,$$

(ii) $\forall R > 0$, for a negligible set $\mathcal{E} \in]0,T[$,

(2.7) $$\lim_{\substack{t\to 0\\ t\notin \mathcal{E}}} \int_{|x|<R} |u(x,t)-u_o(x)|\,dx = 0 \ .$$

The existence and uniqueness of such a solution is proved in Kruzkov [11] (see also Leroux [14]).

(b) Approximation of the system

We introduce here the notion of conservative difference schemes (see, among others, Lerat [17] for this concept).
Let $h > 0$ be the mesh size in space and $\Delta t > 0$ be the time-discretization step ; $\Delta t = \sigma h$ where σ is a constant non negative coefficient. We intend to let the parameter h tend to zero.

The line $\mathbb{R}$ is divided into an infinity of intervals $I_j=[(j-\frac{1}{2})h,(j+\frac{1}{2})h]$ for $j \in \mathbb{Z}$, and $[0,T]$ into N+2 ($N =[\frac{T}{\Delta t}]$) intervals $J_o= [0, \frac{\Delta t}{2}[$ ... $J_n = [(n-\frac{1}{2})\Delta t,(n+\frac{1}{2})\Delta t[$, ...

Let χ_j and χ^n be the characteristic functions of I_j and J_n respectively.
We now consider the two spaces :

$$\mathcal{V}_h = \{W | W = \sum_{i\in\mathbb{Z}} W_i \chi_i\} ,$$

$$\mathcal{V}^{\Delta t} = \{W | W = \sum_{n=0}^{N} W^n \chi^n\} .$$

For all function $W(x,t) \in (L^\infty(\mathbb{R} \times]0,T[))^m$ we construct two functions :

$$W_h \in \mathcal{V}_h , \quad W_h(x,t) = \sum_{n=0}^{N} W_i^n(\chi_i \otimes \chi^n)(x,t) ,$$

where $W_i^n = \frac{1}{h}\int_{I_i} W(x,n\Delta t)dx$; and

$$W^{\Delta t} \in \mathcal{V}^{\Delta t} , \quad W^{\Delta t}(x,t) = \sum_{n=0}^{N} W^n(x)\chi^n(t),$$

where $W^n(x) = W(x,n\Delta t)$.

We define now two approximate problems :

Definition 2.2

The time-discretized problem with two time levels

$$\frac{W^{n+1} - W^n}{\Delta t} + G(W^n, W^{n+1})_x = 0 \tag{2.8}$$

where $W^{\Delta t} = \Sigma\, W^n \chi^n \in \mathcal{V}^{\Delta t}$, is consistent with (2.1) if $G(W,W)=F(W)$ for all $W \in \mathbb{R}^m$.

Definition 2.3

The conservative scheme with two time levels and three points in space given by

$$(2.9)\qquad \frac{W_i^{n+1} - W_i^n}{\Delta t} + \frac{1}{\Delta x}(\phi_{i+1/2}^n - \phi_{i-1/2}^n) = 0,$$

where $W_h = \Sigma W_i^n \chi_i \otimes \chi^n \in \mathcal{V}_h \otimes \mathcal{V}^{\Delta t}$ and $\phi_{i+1/2}^n = \phi(W_i^n, W_{i+1}^n, W_i^{n+1}, W_{i+1}^{n+1})$, is consistent with (2.8) if $\phi(U,U,V,V) = G(U,V)$ for all $U,V \in \mathbb{R}^m$.

(c) The specific case of the Euler equations

The Euler equations of gasdynamics constitute a nonlinear hyperbolic system, where the flux functions are homogeneous. In 1-D, these equations are written as (2.1), (2.2), where F satisfies the property

$$(2.10)\qquad F(W) = A(W)W \quad \text{for all} \quad W \in \mathbb{R}^m \quad (m=3) .$$

In the following, we are interested in constructing implicit versions of a given first-order explicit scheme solving the 1-D Euler equations. The scheme will be characterized by its numerical flux function ϕ and will take into account property (2.10) in the consistency condition of Definition 2.3.

The class of explicit schemes is given by

$$(2.11.a)\qquad \frac{W_i^{n+1} - W_i^n}{\Delta t} + \frac{1}{\Delta x}(\phi_{i+1/2}^n - \phi_{i-1/2}^n) = 0,$$

$$(2.11.b)\qquad \phi_{i+1/2}^n = \phi(W_i^n, W_{i+1}^n),$$

$$(2.11.c)\qquad \phi(U,V) = H_1(U,V)U + H_2(U,V)V, \ \forall U, V \in \mathbb{R}^m ,$$

$$(2.11.d)\qquad H_1(U,U) + H_2(U,U) = A(U), \quad \forall U \in \mathbb{R}^m \quad \text{(new consistency condition)}.$$

Here are some examples of such schemes :

(α) The Godunov-type scheme of Vijayasundaram [27] corresponding to

$$\phi^V(U,V) = A^+\left(\frac{U+V}{2}\right)U + A^-\left(\frac{U+V}{2}\right)V ;$$

(β) The Godunov-type Q-schemes family of Van Leer [26] corresponding to

$$\phi^{VL}(U,V) = \frac{1}{2}[A(U)+Q(A)\left(\frac{U+V}{2}\right)]U + \frac{1}{2}[A(V)-Q(A)\left(\frac{U+V}{2}\right)]V$$

where Q is a positive real function ;

(γ) The scheme of Steger and Warming [24] corresponding to

$$\phi^{SW}(U,V) = A^+(U)U + A^-(V)V .$$

In the following, we consider that the computational domain is bounded, the physical quantities being constant in the farfield ; the domain is chosen such that no perturbation reaches the boundary during the computation. Then, we only manipulate finite dimensional linear systems. We do not take boundary conditions into consideration in the discussion of the stability of the schemes.

3. Construction of a totally linearized implicit scheme and of a Δ-scheme

We now look for conservative implicit schemes which are linearly solvable at each time step and linearly unconditionally stable in ℓ_2. Following Beam and Warming [5], we can design an implicit scheme from (2.11). The totally implicit scheme corresponding to (2.11) is

$$\frac{W_i^{n+1} - W_i^n}{\Delta t} + \frac{1}{\Delta x}(\phi_{i+1/2}^{n+1} - \phi_{i-1/2}^{n+1}) = 0 . \tag{3.1}$$

The nonlinear flux functions are evaluated at t^{n+1} which exclude a direct calculation of W^{n+1}. In order to overcome this difficulty, a scheme, linear in W^{n+1}, can be obtained by a linearization using a Taylor expansion of ϕ^{n+1} about W^n. Then $\phi_{i+1/2}^{n+1}$ is replaced by

$$\tilde{\phi}_{i+1/2}^{n} = \phi_{i+1/2}^{n} + \frac{\partial \phi_{i+1/2}^{n}}{\partial U}(W_i^{n+1} - W_i^n) + \frac{\partial \phi_{i+1/2}^{n}}{\partial V}(W_{i+1}^{n+1} - W_{i+1}^{n}) . \tag{3.2}$$

The resulting scheme is consistent with (2.8) where $G(W^n, W^{n+1}) = F'(W^n)W^{n+1}$. When $\Delta t \to +\infty$, the scheme reduces to Newton's method for the search of the zeros of the function $\psi(W^n) = \{\phi_{i+1/2}^{n} - \phi_{i-1/2}^{n}\}_{i \in \mathbb{Z}}$, i.e. the stationary solution of (2.11) if it exists. If ψ have some required properties, we can hope that the implicit scheme will converge to a steady-state solution in a small number of iterations.
But the evaluation of (3.2) is not trivial because of the presence of derivative terms of $\phi_{i+1/2}^{n}$. Using the following result, a simpler scheme is derived which preserves the interesting properties.
Let us define the numerical flux function ϕ^I by

$$\phi^I(U,V,W,Z) = H_1(U,V)W + H_2(U,V)Z . \tag{3.3}$$

Proposition 3.1.
The scheme defined by (3.2) and the following scheme

$$\frac{W_i^{n+1} - W_i^n}{\Delta t} + \frac{1}{\Delta x}(\phi_{i+1/2}^{I,n} - \phi_{i-1/2}^{I,n}) = 0 \tag{3.4}$$

have the same second-order equivalent system. □

The proof can be found in Stoufflet [25].
Scheme (3.4) has a very simple expression and is equivalent to Newton's method up to second-order terms when $\Delta t \to + \infty$. All the implicit sche-

mes of (3.4) type derived from the above explicit schemes (α), (β), (γ) are ℓ_2 unconditionally linearly stable (an easy proof is obtained by performing a Fourier analysis).

Proposition 3.2.
The second-order equivalent systems of the explicit scheme (2.11) and of the derived implicit scheme (3.4) are respectively

$$W_t + F(W)_x = \Delta t\,[D^e(W)W_x]_x \quad \text{where } D^e(W) = \tfrac{1}{2}[H(W) - \sigma A^2],$$

$$W_t + F(W)_x = \Delta t\,[D^I(W)W_x]_x \quad \text{where } D^I(W) = \tfrac{1}{2}[H(W) + \sigma A^2],$$

where $H(W) = (H_1 - H_2)(W,W)$.

The implicit scheme is more diffusive than the initial explicit scheme. One time-step of the scheme (3.4) leads to the solution of a linear system of the form $M^n W^{n+1} = W^n$. The following result gives a sufficient condition for the invertibility of M^n.

Proposition 3.3.
The matrix M^n appearing in (3.4) is invertible in the scalar case ($m=1$) if $H_1 \geq 0$ and $H_2 \leq 0$. This condition is satisfied by schemes (α) and (γ) and also by scheme (β) if the function Q satisfies $Q(x) \geq |x|$, $\forall x \in \mathbb{R}$.

Moreover, the matrix is diagonally dominant in the scalar linear case for schemes (α) and (γ) and for scheme (β) if Q is such that $Q(x) \geq x$. □

As mentionned before, (3.4) requires the computation of the solution of a complete linear system (in fact, in 1-D, this system is block-tridiagonal). In order to obtain simpler systems to solve, one can follow Steger and Warming [24] and derive a method in which triangular systems are solved. We rewrite (3.4) as a Δ-scheme

$$\text{(3.5a)}\quad \sigma H^n_{2,i+1/2}\,\delta W^{n+1}_{i+1} + [I + \sigma H^n_{1,i+1/2} - \sigma H^n_{2,i-1/2}]\,\delta W^{n+1}_i - \sigma H^n_{1,i-1/2}\,\delta W^{n+1}_{i-1} = -\sigma[\phi^n_{i+1/2} - \phi^n_{i-1/2}],$$

$$\text{(3.5b)}\quad W^{n+1}_i = W^n_i + \delta W^{n+1}_i .$$

The form (3.5) suggests an approximate factorization of the left hand-side term.
The solution procedure becomes :

$$\text{(3.6a)}\quad [I + \sigma H^n_{1,i+1/2}]\,\delta W^{n+1/2}_i - \sigma H^n_{1,i-1/2}\,\delta W^{n+1/2}_{i-1} = -\sigma[\phi^n_{i+1/2} - \phi^n_{i-1/2}],$$

$$\text{(3.6b)}\quad [I - \sigma H^n_{2,i-1/2}]\,\delta W^{n+1}_i + \sigma H^n_{2,i+1/2}\,\delta W^{n+1}_{i+1} = \delta W^{n+1/2}_i ,$$

(3.6c) $W_i^{n+1} = W_i^n + \delta W_i^{n+1}$.

The $m \times m$ block matrices are invertible for all Δt for schemes (α) and (γ), and also scheme (β) if the function Q is such that $Q(x) \geq |x|$.

The schemes (3.6) are linearly unconditionally stable for all schemes (α)-(γ).

For schemes (α)-(γ), we note that $H_1(U,U).H_2(U,U) = 0$ and therefore an analysis of (3.6) based on constant coefficients would conclude that no error is introduced by the factorization of the matrix.

4. Construction of an inconsistent scheme

We now construct a scheme inconsistent with the time-dependent equation but consistent with the stationary one $F(W)_x = 0$ at the steady-state. We will show that the resulting non-conservative schemes are unconditionally stable for a large class of spatial approximations. Let the numerical flux function ϕ be as in Section 3 $\phi(U,V) = H_1(U,V)U + H_2(U,V)V$ and let (2.11) define the associated scheme.

The idea is to view (3.1) as a Jacobi relaxation procedure and, then to construct the corresponding Gauss-Seidel relaxation procedure of (3.1). We can find the application of this idea in Peyret-Viviand [23] for the Navier-Stokes equations.

The non-conservative scheme that we present is the following

$$(4.1) \quad \left\{ \begin{aligned} W_i^{n+1} = W_i^n - \sigma[& H_1(W_i^n, W_{i+1}^n) W_i^{n+1} + H_2(W_i^n, W_{i+1}^n) W_{i+1}^n \\ & - H_1(W_{i-1}^{n+1}, W_i^n) W_{i-1}^{n+1} - H_2(W_{i-1}^{n+1}, W_i^n) W_i^{n+1}]. \end{aligned} \right.$$

We can easily calculate the modified equation equivalent to (4.1).

<u>Proposition 4.1.</u>
<u>Scheme (4.1) is consistent with the PDE</u>

$$(4.2) \quad [I - \sigma H_2(W,W) - \sigma \frac{\partial (H_1+H_2)}{\partial U}(W,W)] W_t + F(W)_x = 0,$$

<u>up to first-order terms.</u> □

The next proposition gives us a characterization of the linear stability of (4.1).

<u>Proposition 4.2.</u>
<u>Scheme (4.1) is linearly ℓ_2-stable if and only if</u> $(\sigma h_1+1)(h_1+h_2) \geq 0$, <u>where</u> h_1 <u>and</u> h_2 <u>are the eigenvalues of</u> H_1 <u>and</u> H_2, <u>respectively</u>. □

The numerical flux functions of schemes (α) and (γ) satisfy this condition for all $\Delta t > 0$.

Remark 1.
Generalization of (4.1) type schemes can be given for non-homogeneous systems on Q-schemes family of Van Leer.
This inconsistent scheme will be defined by :

$$(4.3)\quad \begin{cases} W_i^{n+1} = W_i^n - \frac{\sigma}{2}(F(W_{i+1}^n)-F(W_{i-1}^{n+1})) \\ \qquad + \tilde{Q}_{i+1/2}^n \left(\frac{W_{i+1}^n - W_i^{n+1}}{2}\right) - \tilde{Q}_{i-1/2}^{n+1/2}\left(\frac{W_i^{n+1} - W_{i-1}^{n+1}}{2}\right), \end{cases}$$

where the coefficients $\tilde{Q}_{i+1/2}^n$ and $\tilde{Q}_{i-1/2}^{n+1/2}$ remain to be defined. □

In the remainder of this section, we show a convergence result for the inconsistent scheme (4.3) in the scalar case. Le Roux [15],[16] has proved the convergence of the approximate solution corresponding to the scalar explicit Q-schemes family

$$(4.4)\quad \begin{cases} u_i^{n+1} = u_i^n - \frac{\sigma}{2}(f(u_{i+1}^n)-f(u_{i-1}^n)) \\ \qquad + q_{i+1/2}^n\left(\frac{u_{i+1}^n - u_i^n}{2}\right) - q_{i-1/2}^n\left(\frac{u_i^n - u_{i-1}^n}{2}\right), \end{cases}$$

in which it is assumed that

$\forall i \in \mathbb{Z}$, $q_{i+1/2}^n$ depends only on u_i^n and u_{i+1}^n,

to the entropic solution of (2.4), (2.5) under a restriction on the CFL number and provided a good choice of the artificial viscosity parameters $q_{i+1/2}^n$ is made.

Let u_o be given in $L^\infty(\mathbb{R})$ and $M = \|u_o\|_{L^\infty(\mathbb{R})}$; we have $u_i^o = \frac{1}{h}\int_{I_i} u_o(x)\,dx$.

We suppose that u_o has a compact support and $u_o \in BV_{loc}(\mathbb{R})$ (space of functions of locally bounded variation).

Let R_o be a real number such that $|x| \geq R_o \Rightarrow u_o(x) = 0$;
there exists $i_o \in Z$ such that $R_o \in I_{i_o}$.

We construct an operator T_h by a restriction of (4.3) to the scalar case ; it is obvious that the approximation u^1 obtained from u_o by (4.3) has a compact support, so does u^n for $n \in \{0,\ldots,N\}$. T_h will be constructed if we impose that for $n \in \{0,\ldots,N\}$, there exists some real number $R_n \in I_{\ell_n}$ such that $u_i^n = 0$ for $i < l_n$; from the definition of the scheme (4.3), we can take $l_n = -i_o - n$.

The operator T_h is defined by $u^{n+1} = T_h u^n$ where $u^n(x) = u_i^n$ if $x \in I_i$ and

$$
(4.5)\quad \begin{cases} \text{for } i < i_{n+1},\ u_i^{n+1} = 0, \\ \text{for } i \geq i_{n+1},\ u_i^{n+1} = u_i^n - \frac{\sigma}{2}(f(u_{i+1}^n) - f(u_{i-1}^{n+1})) \\ \qquad + \tilde{q}_{i+1/2}^n \left(\frac{u_{i+1}^n - u_i^{n+1}}{2}\right) - \tilde{q}_{i-1/2}^{n+1/2}\left(\frac{u_i^{n+1} - u_{i-1}^{n+1}}{2}\right) . \end{cases}
$$

We define $u_h(x,t) = u_i^n$ if $(x,t) \in I_i \times J_n$.

Using the local bounded variation of u_o and a result of compactness as in Le Roux [15], Conway and Smoller [7], we can prove the convergence of a subsequence of $\{u_h\}_h$ in $L^1_{loc}(\mathbb{R} \times]0,T[)$.

Theorem 4.1.
Under the assumptions

$$(4.6)\quad u_o \in L^\infty(\mathbb{R}) \cap BVloc(\mathbb{R}),\ u_o \text{ has compact support,}$$

$$(4.7)\quad \sup_{|\xi| \leq M} |f'(\xi)| < \infty ,$$

$$(4.8)\quad \forall i \in \mathbb{Z},\ \forall n \in \{1,\ldots,N\},\ \mathrm{Inf}(\tilde{q}_{i+1/2}^n, \tilde{q}_{i-1/2}^{n+1/2}) \geq \sigma |f'(\xi_{i+1/2}^{n+1/2})| ,$$

where $\xi_{i+1/2}^{n+1/2}$ is a real number such that

$$f(u_{i+1}^n) - f(u_{i-1}^{n+1}) = f'(\xi_{i+1/2}^{n+1/2})(u_{i+1}^n - u_{i-1}^{n+1}) ,$$

the family $\{u_h\}_h$ contains a subsequence in $L^1_{loc}(\mathbb{R} \times]0,T[)$ that converges to a function $u \in L^\infty(\mathbb{R} \times]0,T[)$. □

Remark 4.2.
Condition (4.8) indicates how to choose the viscosity parameters $\tilde{q}_{i+1/2}^n$ and $\tilde{q}_{i-1/2}^{n+1/2}$ on which depend the stability and convergence of the scheme. Le Roux gave in [15] a similar condition for the scalar Q-schemes family of Van Leer (4.4) where $q_{i+1/2}^n$ depended only on u_i^n and u_{i+1}^n. The non-conservative character of the scheme (4.5) allows us to choose $\tilde{q}_{i+1/2}^n$ and $\tilde{q}_{i-1/2}^{n+1/2}$ verifying (4.8) and then depending on u_{i+1}^n and u_{i-1}^{n+1}.

Condition (4.8) implies that T_h conserves the L^∞-norm, the BV-norm in space and the BV-norm in time in a certain sense. The operator T_h does not satisfy exactly the required properties needed to apply Le Roux's theorem [15, p.32] ; the BV-estimates in space and time are in this case infinite series, but the proof of Le Roux can easily be adapted because the support of u_o is compact. □

In order to show the convergence of the whole sequence to a function u verifying a certain equation, we have to make a more restrictive hypothesis and suppose that

$$(4.9)\quad \begin{cases} \forall i \in \mathbb{Z},\ \forall n \in \{1,\dots,N\}, \\ \tilde{q}_{i-1/2}^{n+1/2} = \tilde{q}_{i+1/2}^{n} = q = cste \geq \sigma \sup_{|\xi| \leq M} |f'(\xi)|, \end{cases}$$

(we can take $q = \sigma \sup_{|\xi| \leq M} |f'(\xi)|$).

We define a real function $g \in C^1(\mathbb{R})$ coïnciding with f on $[-M, M]$ and satisfying

$$(4.10)\quad q \geq \sigma \sup_{\xi \in \mathbb{R}} |g'(\xi)| ;$$

g is defined by

$$(4.11)\quad \begin{cases} g(y) = f'(-M)(y+M) & \text{for } y < -M, \\ g(y) = f(y) & \text{for } |y| \leq M, \\ g(y) = f'(M)(y-M) & \text{for } y > M. \end{cases}$$

<u>Theorem 4.2.</u>
<u>Under assumption</u> (4.9), <u>the family</u> $\{u_h\}_h$ <u>converges in</u> $L^1_{loc}(\mathbb{R} \times]0,T[)$ <u>to a function</u> $u \in L^\infty(\mathbb{R} \times]0,T[)$ <u>such that</u> $s(u)$ <u>is the solution in Kruzkov's sense of the Cauchy problem</u>

$$(4.12a)\quad \frac{\partial v}{\partial t} + \frac{\partial\, g(s^{-1}(v))}{\partial x} = 0,$$

$$(4.12b)\quad v(x,0) = s(u_o(x)),$$

<u>where</u> $s(y) = (1 + \frac{q}{2})y - \frac{\sigma}{2} g(y)$. □

By virtue of (4.10), s is invertible on $\mathbb{R}$ and (4.12a) has a sense. Theorem 4.2 proves the convergence of $\{u_h\}_h$ without any CFL restriction. Proofs of theorems 4.1 and 4.2 can be found in Stoufflet [25].

5. <u>Two dimensional extensions</u>

The two-dimensional Euler equations in conservation form are

$$(5.1)\quad \frac{\partial W}{\partial t} + \frac{\partial F(W)}{\partial x} + \frac{\partial G(W)}{\partial y} = 0 \quad + \text{Boundary Conditions},$$

with

$$W = \begin{bmatrix} \rho \\ \rho u \\ \rho v \\ e \end{bmatrix}, F(W) = \begin{bmatrix} \rho u \\ \rho u^2 + p \\ \rho uv \\ (e+p)u \end{bmatrix}, G(W) = \begin{bmatrix} \rho v \\ \rho uv \\ \rho v^2 + p \\ (e+p)v \end{bmatrix},$$

and $p = .4\,[e - \frac{1}{2}\rho(u^2+v^2)]$,

where ρ is density, p is pressure, e is the total energy and u and v are the components of the velocity.
(5.1) is a hyperbolic system, i.e. $\lambda F'(W)+\mu G'(W)$ is diagonalizable with real eigenvalues for all $\{W,\lambda,\mu\} \in \mathbb{R}^m \times \mathbb{R} \times \mathbb{R}$ (here m=4).

We first describe the extension of a conservative scheme to solve (5.1) via a control volume type integration. For each function $H : \mathbb{R}^m \to \mathbb{R}^m$, we note $\phi_H(U,V,W,Z)$ the numerical flux function characterizing the one dimensional scheme of (2.9) type.

We have

$$(5.2) \qquad \phi_H(U,U,U,U) = H(U) ,$$

$$(5.3) \qquad \tilde{H}(U,W) = \phi_H(U,U,W,W) .$$

Using (5.3) a two-level time-discretization of (5.1) is written as follows :

$$(5.4) \qquad \frac{W^{n+1}-W^n}{\Delta t} + \tilde{F}(W^n,W^{n+1})_x + \tilde{G}(W^n,W^{n+1})_y = 0.$$

We consider a triangulation $\mathcal{T}_h$ of a polygon Ω_h which approximates the domain Ω ; K(i) denotes the set of the indices of the vertices adjacent to the vertex a_i :

$$K(i) = \{i_1,\ldots,i_{q_i}\} .$$

Let G_{ij} be the centroïd of the triangle whose vertices are a_i, a_{i_j}, a_{i_j-1} assuming that the convention $i_o = i_{q_i}$ is made and that I_{i_j} is the mid-point of the segment $[a_i, a_{i_j}]$.
We define the cell around a_i, noted supp(a_i), as the bounded region limited by the segments $[G_{i1},I_{i1}]$, ..., $[I_{iq_i} G_{i_1}]$; ∂C_{ij} is the side $G_{ij}\ I_{ij}\ G_{ij+1}$ (see Fig. 1).

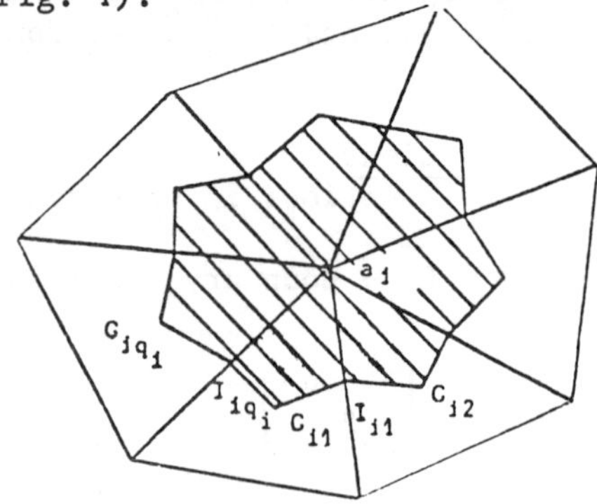

Figure 1 : The cell around a_i (represented by the shaded region).

Let $\mathcal{V}_h$ be the functional space ,

(5.5) $\mathcal{V}_h = \{v \in L^\infty(\Omega_h),\ v|_{supp(a_i)} = cste = v_i,\ \forall a_i \text{ vertex of } \mathcal{T}_h\}$.

A basis for the space $\mathcal{V}_h$ is given by the set of characteristic functions $\{\chi_i,\ \forall a_i$ vertex of $\mathcal{T}_h\}$ defined by :

$$\chi_i(x) = \begin{cases} 1 & \text{if } x \in supp(a_i), \\ 0 & \text{otherwise} \end{cases}$$

By analogy with a control volume formulation, we use an integration by parts

(5.6) $$\iint_{supp(a_i)} \frac{W^{n+1}-W^n}{\Delta t} dx\, dy + \int_{\partial supp(a_i)} (\tilde{F}(W^n,W^{n+1})\nu_x + \tilde{G}(W^n,W^{n+1})\nu_y) d\sigma = 0,$$

where $\vec{\nu} = (\nu_x,\nu_y)$ denotes the outward vector normal to $\partial\, supp(a_i)$.
Then (5.6) is rewritten as

(5.7) $$\frac{W_i^{n+1} - W_i^n}{\Delta t} a(i) + \sum_{j\in K(i)} \underbrace{\int_{\partial C_{ij}} (\tilde{F}(W^n,W^{n+1})\nu_x + \tilde{G}(W^n,W^{n+1})\nu_y) d\sigma}_{H_{ij}} = 0,$$

where $a(i)$ is the area of $supp(a_i)$.

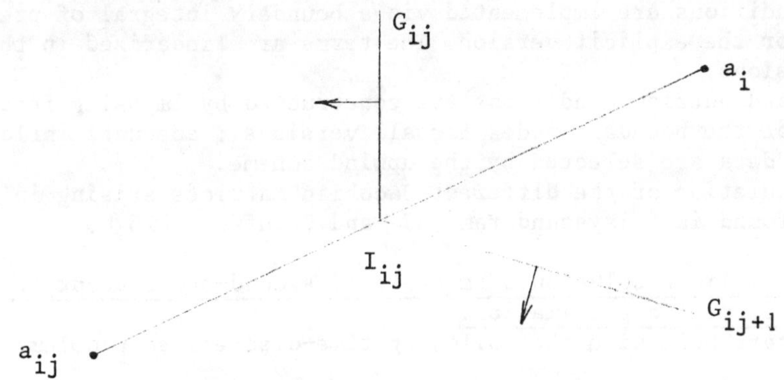

Figure 2 : Outward vector normal to ∂C_{ij}.

To each approximation of the integral in (5.7) corresponds a scheme for the problem (5.1). We now specify the approximation in order to extend the above one dimensional scheme.

We take

(5.8) $$H_{ij} = \phi_{\eta_{x_{ij}} F + \eta_{y_{ij}} G}(W_i^n, W_j^n, W_i^{n+1}, W_j^{n+1}),$$

where $\eta_{x_{ij}} = \int_{\partial C_{ij}} \nu_x \, d\sigma$ and $\eta_{y_{ij}} = \int_{\partial C_{ij}} \nu_y \, d\sigma$.

Actually, the evaluation of (5.8) corresponds to the one dimensional calculation of the flux through ∂C_{ij} along the direction $a_i a_j$ with the numerical flux function evaluated at I_{ij}.

Example : The implicit linearized version of Vijayasundaram's scheme becomes

$$\frac{W_i^{n+1} - W_i^n}{\Delta t} + \sum_{j \in K(i)} [P_{ij}^+ (\frac{W_i^n + W_j^n}{2}) W_i^{n+1} + P_{ij}^- (\frac{W_i^n + W_j^n}{2}) W_j^{n+1}] = 0 ,$$

where $P_{ij}(.) = (\eta_{x_{ij}} F' + \eta_{y_{ij}} G')\,(.)$.

No particular problems arise to extend the inconsistent schemes constructed in Section 4. If the mesh nodes are ordered in some specific way, we are able to construct the 2-D extension of these schemes since in fact, it is equivalent to a Gauss-Seidel sweep.

The scheme used to solve equation (5.1) will condition the approach to deal with boundary conditions. We present here the treatment of boundary conditions for the different implicit versions of the two explicit schemes we have tested numerically : Vijayasundaram's scheme (α) and a Q-scheme of Van Leer (β) with $Q(x) = |x|$.
These two schemes are upwind schemes and for both of them we have used the same treatment of the boundary conditions. Wall (or profile) boundary conditions are implemented via a boundary integral of pressure terms for the explicit version. The terms are linearized in the implicit versions.
Inflow and outflow conditions are constructed by imposing freestream values at the boundary nodes for all versions ; adequate inflow and outflow data are selected by the upwind scheme.
The calculation of the different Jacobian matrices arising in (5.8) can be found in Vijayasundaram [27] and Stoufflet [25] .

6. Implicit solution of a centered second-order accurate Finite Element approximation.

We start here with the following time-discretized problem

$$\text{(6.1)} \quad \frac{W^{n+1} - W^n}{\Delta t} + F(W^{n+1})_x + G(W^{n+1})_y = 0 .$$

A Taylor series expansion of the flux functions F and G about W^n gives the following Δ-problem

$$\text{(6.2a)} \quad \frac{1}{\Delta t} \delta W^n + (A(W^n)\delta W^n)_x + (B(W^n)\delta W^n)_y = -F(W^n)_x - G(W^n)_y ,$$

$$\text{(6.2b)} \quad W^{n+1} = W^n + \delta W^n .$$

The left-hand side of (6.2a) is called the mathematical phase and the right hand-side the physical phase.

From the previous triangulation, we define the classical P_1 approximation space $\mathcal{W}_h$ and the P_1-interpolation operator r_h as follows :

$$\mathcal{W}_h = \{v_h \in L^\infty(\Omega_h),\ v_h \text{ is continuous, } v_h \text{ is linear in each triangle } T \text{ of } \mathcal{T}_h\},$$

$$\forall v \in L^\infty(\Omega_h),\ r_h(v) \in \mathcal{W}_h, \text{ and } r_h(v)(a_i) = v(a_i),\ \forall a_i \text{ vertex of } \mathcal{T}_h.$$

We define the projection operator $\mathcal{S}$ from $\mathcal{W}_h$ to $\mathcal{V}_h$ as follows :

$$\forall v \in \mathcal{W}_h,\ \mathcal{S}v|_{\mathrm{supp}(a_i)} = v(a_i)\ .$$

is a mass-lumping operator as in Baba and Tabata [4] .

The variational formulation of the approximate problem will be

$$(6.3)\quad \begin{cases} W^n \in \mathcal{W}_h \ ;\ \text{find } \delta W^n \in \mathcal{V}_h \text{ such that } \forall v \in \mathcal{W}_h\ , \\ \iint_{\Omega_h} \{ \frac{1}{\Delta t}\, \delta W^n + (A(W^n)\delta W^n)_x + (B(W^n)\delta W^n)_y\} \mathcal{S}v \, dx\, dy = \\ \iint_{\Omega_h} \{-F(W^n)_x - G(W^n)_y\}\, v\, dx\, dy, \end{cases}$$

$$(6.4)\quad W^{n+1} = W^n + r_h\, \delta W^n.$$

Let $\{N_i\}_i$ be the set of basis functions of $\mathcal{W}_h$ defined as follows :

$$\forall a_j \text{ vertex of } \mathcal{T}_h,\ N_i(a_j) = \delta_{ij}, \text{ and } N_i \in \mathcal{W}_h\ .$$

For $v = N_i$, (6.3) is rewritten as

$$(6.5)\quad \begin{cases} \delta W_i^n \frac{a(i)}{\Delta t} + \int_{\partial \mathrm{supp}(a_i)} ((A(W^n)\delta W^n)\nu_x + (B(W^n)\delta W^n)\nu_y)\, d\sigma = \\ \sum_{K \in \mathcal{T}(i)} \{ \frac{\partial N_i}{\partial x} \iint_K F(W^n)\, dx\, dy + \frac{\partial N_i}{\partial y} \iint_K G(W^n)\, dx\, dy\} \\ - \int_{\partial \Omega_h} N_i [F(W^n)\nu_x + G(W^n)\nu_y]\, d\sigma\ , \end{cases}$$

with $\mathcal{T}(i)$ the set of the triangles of $\mathcal{T}_h$ having a_i as a vertex.

The <u>right-hand side</u> is calculated as in Angrand and al. [3] , Part 4.

The boundary conditions are inserted via the boundary integrals $\int_{\partial\Omega_h}$ for the wall and profile conditions as well as the conditions at infinity. Good shock resolution is reached if we introduce a Von Neumann-Richtmyer type numerical viscosity term, which is a discretization of

$$(6.6) \qquad \hat{\chi} \left\{ \frac{\Delta x^2}{2} \frac{\partial}{\partial x} \left[\left|\frac{\partial W_h}{\partial x}\right| \frac{\partial W_h}{\partial x} \right] + \frac{\Delta y^2}{2} \frac{\partial}{\partial y} \left[\left| \frac{\partial W_h}{\partial y} \right| \frac{\partial W_h}{\partial y} \right] \right\} .$$

The resulting scheme is a second-order accurate approximation.

The mathematical part (left-hand side) of (6.5) is discretized according to the ideas discussed in Section 5 and using a first order upwind scheme

$$(6.7) \qquad \int_{\partial \mathrm{supp}(a_i)} (.) \; d\sigma \rightarrow \phi_{\nu_{x_{ij}} F + \nu_{y_{ij}} G} (W_i^n, W_j^n, \delta W_i^n, \delta W_j^n) .$$

We have used the totally implicit linearized scheme of Vijayasundaram for which $\phi_{\nu_{x_{ij}} F + \nu_{y_{ij}} G} = P_{ij}^{+} \left(\frac{W_i^n + W_j^n}{2}\right) \delta W_i^n + P_{ij}^{-} \left(\frac{W_i^n + W_j^n}{2}\right) \delta W_j^n$,

and the corresponding approximate factored Δ-scheme with triangular matrices.

7. Numerical experiments

(a) Steady flow in a channel

The calculation of the steady flow in a channel past a bump was a test case proposed at the GAMM Workshop in 1979 [28] . The bump is a 4,2 % thick circular arc with length 1. in a 2.073 high channel. Freestream values correspond to a Mach number of .85, for which the flow is transonic. For consistency with the GAMM test, we use a triangulation with 72 × 21 vertices, which gives 2840 triangles.

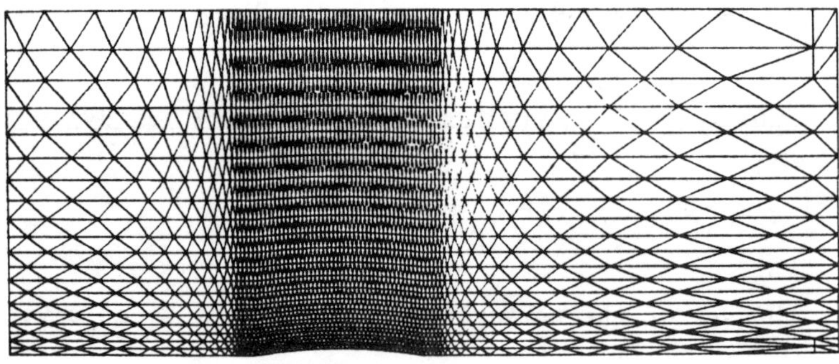

Figure 3 : Triangulation of the GAMM channel (1512 nodes). The inflow is on the left.

We have solved this problem with 2-D extensions of the two first-order upwind schemes - Vijayasundaram scheme (α) and the Q-scheme of Van Leer (β) with $Q(x) = |x|$. The number of unknowns is 1512 × 4.
The C_p distributions are given in Fig. 4.a and 4.b. With such a refined mesh, the diffusion of the first order upwind approximation is not apparent. The shock is not at all smeared with the scheme (α) but we observe a slight overshoot also present on Mach distribution corresponding to an undershoot on entropy distribution in Fig. 5.a (see Angrand and al. [1]). Scheme (β) gives a shock smeared on two points without overshoot (Fig. 4.b) ; the solution of scheme (β) does not violate the entropy condition (Fig. 5.b). Both methods give a solution without oscillations either before and after the shock. The convergence to the steady solution of the two discretization schemes is very similar.

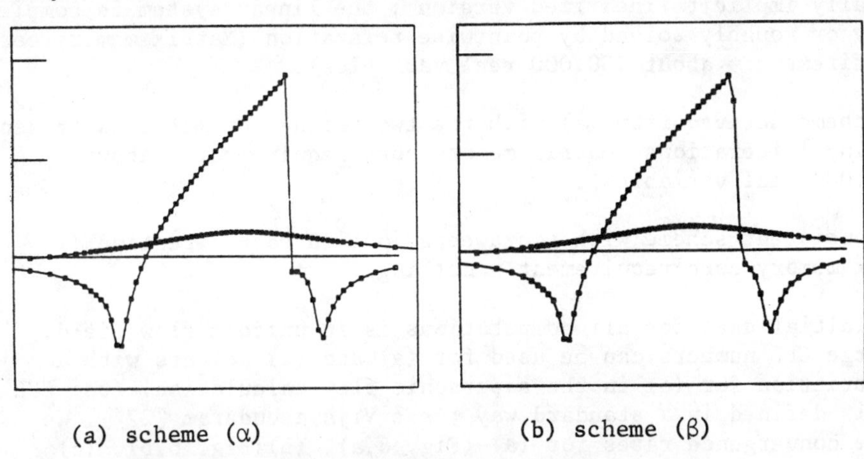

(a) scheme (α) (b) scheme (β)

Figure 4 : Transonic flow in a channel at M_∞ = .85. C_p distribution.

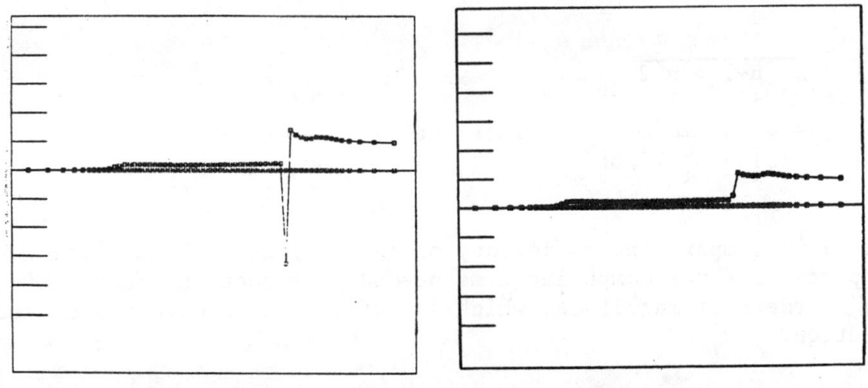

(a) scheme (α) (b) scheme (β)

Figure 5 : Transonic flow in a channel at M_∞ = .85. Entropy distribution.

The results given by the two schemes are comparable to the second-order accurate simulations of Angrand and al [1] and Lerat and Sides [19].

In order to compare the behaviour of the different implicit solvers constructed in the previous section, three test cases differing by the values of the freestream Mach number were chosen : $M_\infty = .50$ (completely supersonic flow), $M_\infty = .85$ (previous case ; transonic flow), $M_\infty = 1.50$ (completely supersonic flow). Only the results produced by scheme (α) are presented here. The experiments conducted with the various solvers and scheme (β) indicate a very similar behaviour, for details see Stoufflet [25].

The implicit solvers have been implemented in the following way :

(a) Totally implicit linearized version : the linear system is completely or roughly solved by pointwise relaxation (Matrix memory core requirement = about 150.000 real variables).

(b) Δ-scheme derived from (a) with the two triangular matrices frozen during λ iterations (Matrix memory core requirement = about 150.000 real variables).

(c) Inconsistent scheme with two sweeps, one in each direction (matrix memory core requirement = nothing).

The initial data for all computations is an uniform flow field. Very large CFL numbers can be used for (a) and (c) solvers with a CFL restriction for (c) in the supersonic flow calculation. (the CFL number is defined in a standard way ; see Vijayasundaram [27]). We give the convergence rates for (a) (Fig. 6.a), (b) (Fig. 6.b), (c) (Fig. 6.c) on the three test cases. The slowest convergence of the flow-field occurs in the transonic flow. The mean residual considered here is defined by

$$\text{RES} = \frac{\sqrt{\sum_i (\rho_i^{n+1} - \rho_i^{n})^2}/\Delta t^n}{\sqrt{\sum_i (\rho_i^{1} - \rho_i^{o})^2}/\Delta t^o} \quad \text{representing} \quad \frac{\left\| \partial\rho/\partial t \right\|_{L^2}}{\left\| \partial\rho/\partial t\,(o) \right\|_{L^2}} .$$

In Table 1 we compare the efficiency of (a), (b) and (c) versions and in particular the computing time needed to reduce the mean residual by 3 orders of magnitude, which is sufficient to obtain a converged solution.

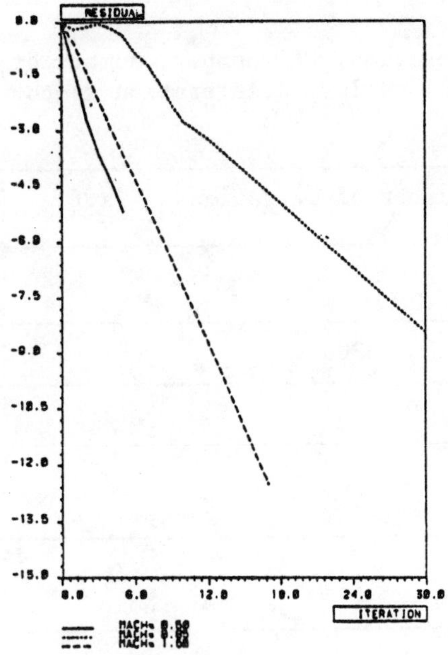

(a) Implicit linearized scheme with about 50 relaxation sweeps on the linear system per time step.

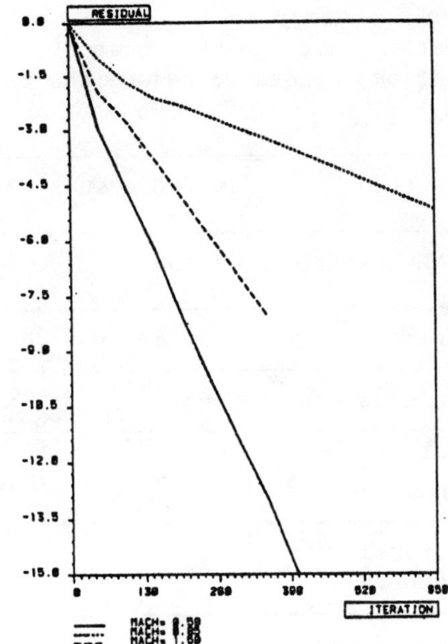

(b) Δ-factored scheme with $\lambda = 10$.

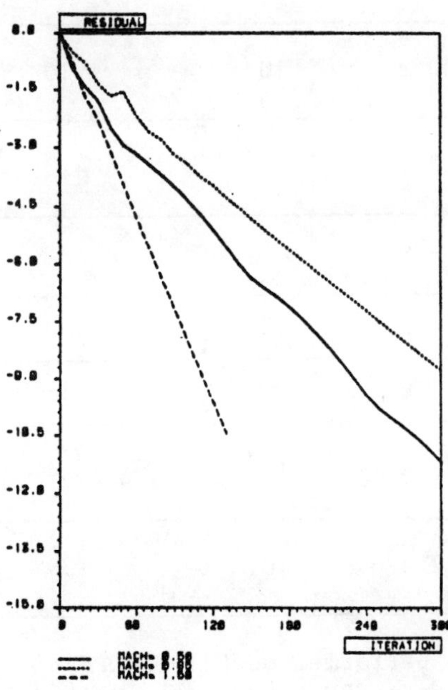

(c) Inconsistent scheme.

Figure 6 : Steady flow in a channel. Convergence histories.

TABLE 1

Method, average CPU* cost of one iteration, CFL number, number of iterations needed to reduce the residual by 10^3, acceleration factor**.

Method	av.CPU cost	CFL	Number of iterations	a.f.
GAMM CHANNEL $M_\infty = .50$				
Explicit	5.75 s	.8	3200	
Inconsistent	65 s	10^6	81	3.5
Implicit linearized	720 s	10^3	3	8.6
Factored Δ-scheme	22 s	15.	123	6.8
GAMM CHANNEL $M_\infty = .85$				
Explicit	5.75 s	.8	5100	
Inconsistent	65 s	10^6	100	4.2
Implicit linearized	400 s	10^3	10	7.5
Factored Δ-scheme	22 s	15	420	3.2
GAMM CHANNEL $M_\infty = 1.50$				
Explicit	5.75 s	.8	1500	
Inconsistent	65 s	50.	36	3.7
Implicit linearized	228 s	10^3	4	9
Factored Δ-scheme	22 s	10.	135	3

* Computations were performed on CII-HB DPS 68

** Acceleration factor = CPU time of the method/CPU time of explicit method.

We observe that solver (a) with an incompletely converged solution of the linear system gives the best efficiency with an acceleration ratio of approximately 8/1 on the three cases but with a maximum memory core requirement.
The inconsistent scheme represents a good trade-off between cost expense and memory requirements.
For the transonic flow problem (M_∞ = .85), Fig. 8 shows the curve representing the build up of the number of points in the supersonic zone, which is a good convergence estimator. Results given in Table 1 are confirmed.

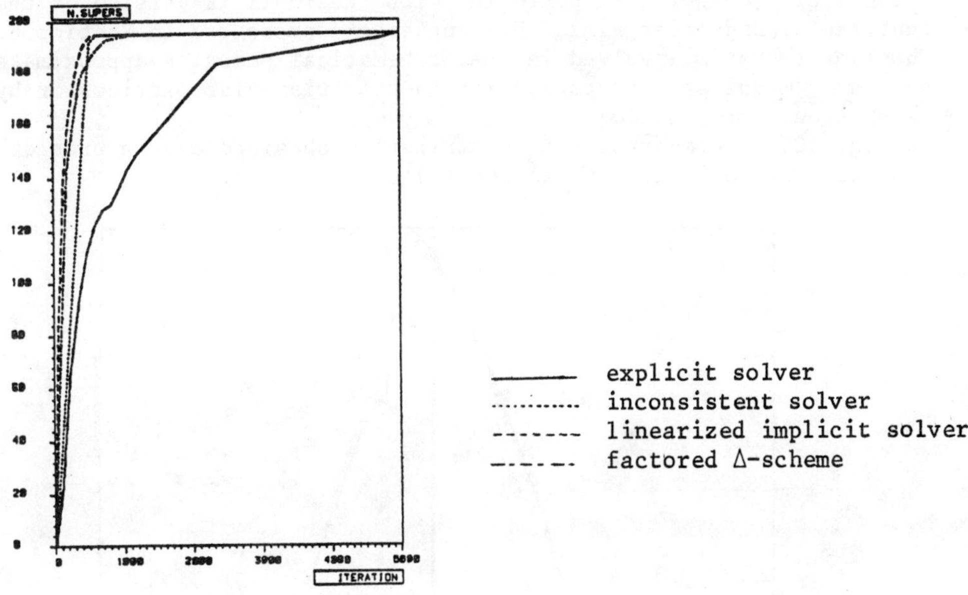

Figure 8 : Build-up of the number of supersonic points (time unity = CPU time of one explicit iteration).

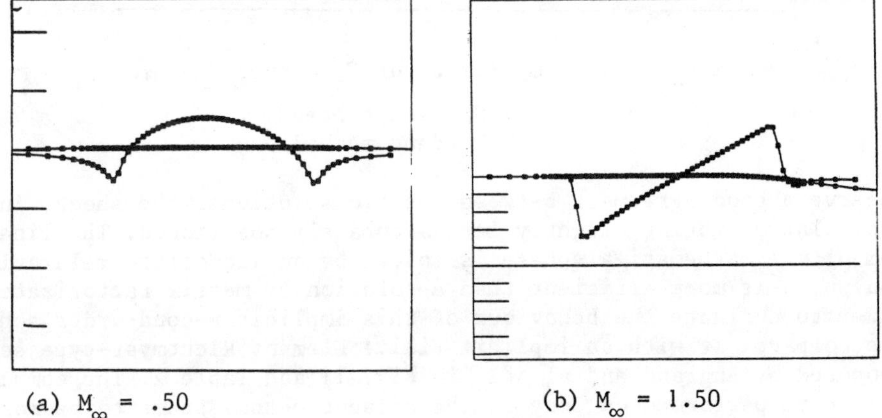

(a) M_∞ = .50 (b) M_∞ = 1.50

Figure 9 : Mach number distribution.

We present the steady-state Mach number distributions obtained using the discretization of scheme (α) in Figs. 9.a and 9.b for the subsonic case (M_∞ = .50) and the supersonic case (M_∞ = 1.50). Comparable results can be found for similar simulations in Ni [21] .

(b) Steady flow past a cylinder

This problem has been solved among others by Jameson [10],[11]. The computational domain outside the cylinder is bounded "at infinity" at about 25 Chords. Only a half cylinder is considered because of the symmetry of the problem and the triangulation has 845 nodes.
The calculation has been performed with the implicit version of the centered second-order Finite Element scheme presented in Section 6. The linear system involved in the mathematical phase is approximately solved by means of a factorization in two triangular matrices or by a relaxation-type procedure.
On Fig. 10, we compare the C_p distribution obtained by the present method to that obtained by Jameson [11].

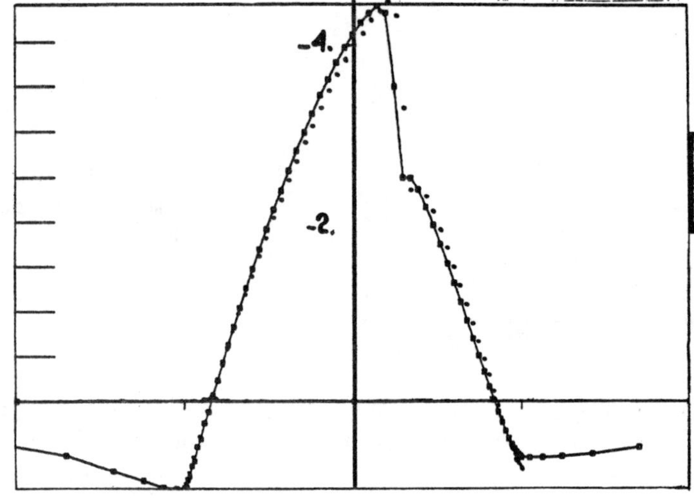

Figure 10 : Comparaison of C_p curves
• Jameson scheme
▪ Present method

We observe a good agreement between the two solutions, the shock, in our calculation, being slightly less accurately positioned. The linear system of the mathematical phase is solved by an incomplete relaxation method which is more efficient than a solution by matrix factorization. In order to evaluate the behaviour of this implicit second-order scheme, we compare it with an implicit Finite Element Richtmyer-type scheme proposed by Angrand and al [2] in Fig. 11 and Table 2. The computations were performed on Cray 1. The present method seems to be more efficient in CPU time cost but the matrix to be stored requires a

memory core about 5 times greater. On the other hand, the present method converges to a steady-state independent of the time-step and this constrasts with the behaviour of the implicit Richtmyer scheme.
The simulation converges with a CFL number greater than 5 if we let it increase continuously from 1 to the desired value in about 50 iterations.

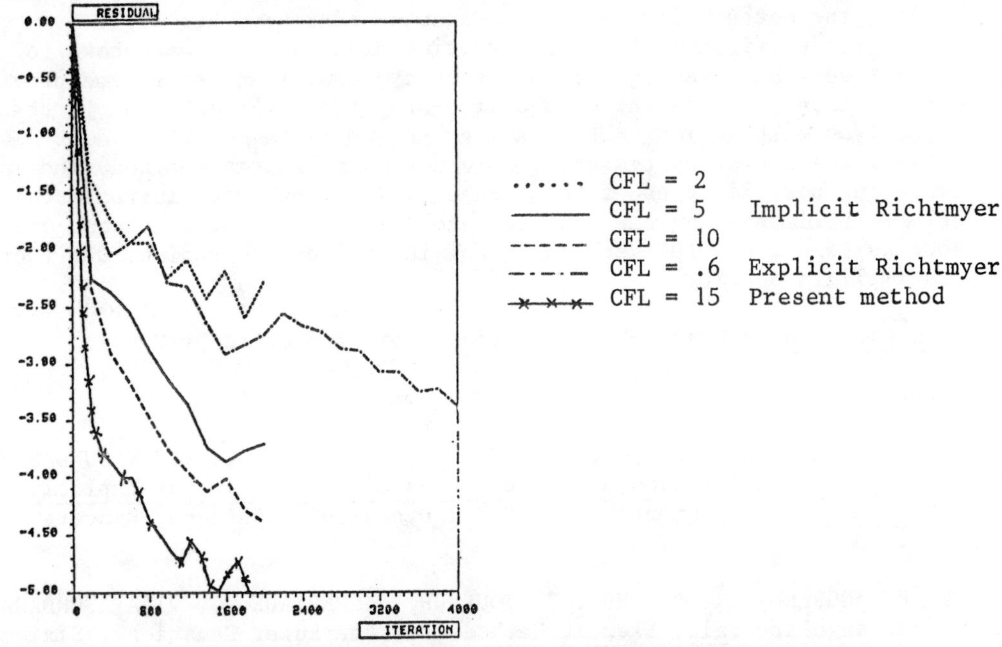

Figure 11 : Comparison between the explicit Richtmyer scheme, the implicit Richtmyer scheme with different CFL numbers and the present method.

TABLE 2

Method, average CPU cost of one iteration, CFL number, number of iterations needed to reduce the residual by 10^3, needed CPU time.

METHOD	av. CPU cost	CFL	Number of iterations	CPU time
Explicit Richtmyer type	0.14	.6	3200	7'52"
Implicit Richymyer type	0.31	15	500	2'35"
Present Method	0.35	20	300	1'45"

8. Conclusion

We have presented and compared various implicit methods derived from a first-order upwind scheme. We have emphasized that upwind schemes have nice matrix properties. Moreover, no parameter adjustment is needed and these schemes have proved their robustness. The implicit linearized upwind scheme and the scheme factored with triangular matrices were shown to have the desirable properties for adequately being used in the mathematical part of a second-order accurate implicit Δ-scheme. The first of the two resulting order schemes was shown to present several advantages ; for accuracy, the presented scheme provides solutions independent of the time-step ; for CPU efficiency, the comparison with an implicit Richtmyer scheme is favourable to our scheme, the linear system being solved by a relaxation method. The interesting possibilities of this scheme are currently evaluated precisely (experiments on new problems, comparison with other second-order schemes) and a new ADI-version is now in progress to replace the factored implicit scheme.

Acknowledgment : Cray 1-S computations were supported by GCCVR.

References

[1] F.ANGRAND, A. DERVIEUX, V. BOULARD, J. PERIAUX, G. VIJAYASUNDARAM, Transonic Euler Simulations by Means of Finite Element Explicit Schemes, AIAA Computational Fluid Dynamics Conference, Danvers, 1983.

[2] F. ANGRAND, A. DERVIEUX, V. BOULARD, J. PERIAUX, G. VIJAYASUNDARAM, Triangular Finite Element Methods for the Euler Equations, Sixième Colloque International sur les méthodes de calcul scientifique et technique, Versailles, 1983.

[3] F. ANGRAND, A. DERVIEUX, L. LOTH, G. VIJAYASUNDARAM, Simulations of Euler Transonic Flows by Means of Finite Element-type Schemes, INRIA Report, no 250, 1983.

[4] K. BABA, M. TABATA, On a Conservative Upwind Finite Element Scheme for Convective Diffusion Equations, RAIRO Numerical Analysis, 15 (1981), pp. 3-25.

[5] R.M. BEAM, R.F. WARMING, An Implicit Finite-Difference Algorithm for Hyperbolic Systems in Conservation-Law Form, J. Comp. Phys., 22 (1976), pp. 87-110.

[6] F. CASIER, H. DECONINCK and C. HIRSCH, A Class of Central Bidiagonal Schemes With Implicit Boundary Conditions for the Solution of Euler's Equations, AIAA 83-0126 paper, 1983.

[7] E. CONWAY and A. SMOLLER, Global Solutions of the Cauchy Problem for Quasi-linear First-Order Equations in Several Space Variables, Comm. Pure and Appl. Math., 19 (1966), pp. 95-105.

[8] A. HARTEN, On a Class of High Resolution Total-Variation-Stable Finite Difference Schemes, NYU Report, 1982.

[9] E. HOPF, On the Right Weak Solution of the Cauchy Problem for a Quasi-linear Equation of First Order, J. Math. Mec., 19, (1969/70), pp. 483-487.

[10] A. JAMESON, The Evolution of Computational Methods in Aerodynamics, MAE Report 1608, 1983.

[11] A. JAMESON, W. SCHMIDT and E. TURKEL, Steady State Solution of the Euler Equations for Transonic Flow, AIAA paper 81-1259, 1981.

[12] S.N. KRUZKOV, First Order Quasi-linear Equations in Several Independent Variables, Math. USSR Sbornik, 10 (1970), pp. 217-243.

[13] P.D. LAX, Weak solutions of Nonlinear Hyperbolic Equations and their Numerical Computation, Comm. on Pure and Appl. Math., 7 (1954), pp. 159-193.

[14] P.D. LAX, Hyperbolic Systems of Conservation Laws II, Comm. on Pure and Appl. Math., 10 (1957), pp. 537-566.

[15] A.Y. LEROUX, RESOLUTION NUMERIQUE DU PROBLEME DE CAUCHY POUR UNE EQUATION HYPERBOLIQUE QUASI-LINEAIRE DU PREMIER ORDRE A UNE OU PLUSIEURS VARIABLES D'ESPACE, Thèse de 3e cycle, Université de Rennes, 1974.

[16] A.Y. LEROUX, A Numerical Conception of Entropy of Quasilinear Equations, Math. Comput., 31 (1977), pp. 848-872.

[17] A. LERAT, SUR LE CALCUL DES SOLUTIONS FAIBLES DES SYSTEMES HYPERBOLIQUES DE LOIS DE CONSERVATION A L'AIDE DE SCHEMAS AUX DIFFERENCES, Thèse d'Etat, Université Paris VI, 1981.

[18] A. LERAT, Implicit Methods of Second-order Accuracy for the Euler Equations, AIAA Computational Fluid Dynamics Conference, Danvers, 1983.

[19] A. LERAT and J.SIDES, Finite Volume Method for the Solutions of Euler Equations, Notes on Numerical Fluid Mech., 3 (1981), pp. 142-152.

[20] R.W. MAC CORMACK, A Numerical Method for Solving the Equation of Compressible Viscous Flows, AIAA Journal, 20 (1982), pp. 1275-1281.

[21] R.H. NI, A Multiple Grid Scheme for Solving the Euler Equations, AIAA Journal, 20 (1982), pp. 1565-1571.

[22] O.A. OLEINIK, Uniqueness and Stability of the Generalized Solution of the Cauchy Problem for a Quasilinear Equation, Amer. Math. Soc. Transl. ser. 2, 33 (1963), pp. 285-290.

[23] R. PEYRET, H. VIVIAND, Computation of Viscous Compressible Flows Based on Navier-Stokes Equations, AGARD n° 212, 1975.

[24] J.L. STEGER, R.F. WARMING, Flux Vector Splitting of the Inviscid Gasdynamic Equations with Application to Finite Difference Methods, J. Comp. Phys., 40 (1981), pp. 263-293.

[25] B. STOUFFLET, RESOLUTION NUMERIQUE DES EQUATIONS D'EULER DES FLUIDES PARFAITS COMPRESSIBLES PAR DES SCHEMAS IMPLICITES EN ELEMENTS FINIS, Thèse de Docteur-Ingénieur, Université de Paris VI, 1984.

[26] B. VAN LEER, Towards the Ultimate Conservative Difference Schemes (III) Upstream Centered Finite Difference Schemes for Ideal Compressible Flow, J. Comp. Phys., 23 (1977), pp. 263-275.

[27] G. VIJAYASUNDARAM, RESOLUTION NUMERIQUE DES EQUATIONS D'EULER POUR DES ECOULEMENTS TRANSSONIQUES AVEC UN SCHEMA DE GODUNOV EN ELEMENTS FINIS, Université de Paris VI, 1982.

[28] H. VIVIAND, Note on Numerical Fluid Mechanics, Vol.5, Proceedings of the fourth GAMM Conference on Numerical Methods in Fluid Mechanics, 1981.

[29] H.C. YEE, R.F. WARMING, A. HARTEN, Implicit Total Variation Diminushing (TVD) Schemes for Steady-State Calculations, NASA Technical Memorendum SU342, 1983.

PART IV:
INCOMPRESSIBLE FLOW CALCULATIONS AND SPECIAL NUMERICAL TECHNIQUES

VORTEX-SHEET CAPTURING IN NUMERICAL SOLUTIONS OF THE INCOMPRESSIBLE EULER EQUATIONS

ARTHUR RIZZI AND LARS-ERIK ERIKSSON*

Abstract. Numerical solutions to the Euler equations that contain vorticity appear qualitatively correct, but how the vorticity is generated is still an open question. The simpler case of incompressible flow where the only allowable discontinuity is a vortex sheet is a good model to use for studying this question. Our approach is the artificial compressibility method which leads to a hyperbolic system of equations that we solve by finite-volume differences centered in space and explicit multistage time stepping. The stability of this novel system is analyzed, its allowable discontinuities are described, and appropriate farfield and solid wall boundary conditions are introduced. Results are presented for both two and three dimensional flows. Whether vorticity is produced or not depends very strongly on the body geometry and the transient discontinuities that evolve in the flowfield. The results are analyzed for the entropy produced in the flowfield, and for the diffusion of the vortex sheets.

1. Introduction. The recent discovery that a vortex sheet can be generated and captured in the solution of a flowfield with separation from a leading edge of a delta wing at high angle of attack has stirred up a lot of excitement about numerical methods to solve the Euler equations. The initial euphoria surrounding this result stems from the qualitatively realistic character of the computed flowfields. What lies ahead of us is to look quantitatively and critically at these solutions in order to arrive at a better understanding of the correctness of the computed flow. One troubling question is why do all the solutions so far display a loss in total pressure, or equivalently a production of entropy, when no shock waves are present? Another feature is that vortex sheets captured in the solution tend to spread over a number of grid points, and the extent of this diffusion needs to be calibrated.

---

* FFA The Aeronautical Research Institute of Sweden,
S-161 11 BROMMA, Sweden

The production of entropy as well as the generation of vorticity and shedding of vortex sheets is usually associated with an irreversible flow process. In the context of the Euler equations a fluid particle can undergo such a process only across a discontinuity. The numerical method itself, however, may introduce another loss mechanism into the flow particularly if its truncation error or artificial viscosity model is crude. So for those flows which we know to be isentropic, the degree of total pressure loss in the computed solution is one measure of that solution's accuracy. When the flowfield does contain vorticity the debate on how this vorticity is initially created continues to rage on. Some argue for a discontinuity in a transient state as its source while others point to the discretization error and numerical viscosity. In this paper we strive for insight into these questions by computing and analyzing several solutions to the incompressible Euler equations. Apart from the purely practical interest in incompressible flow we believe this approach is well motivated because more analytic solutions to these equations are known, mathematically they are better understood, and only one type of discontinuity is allowed, but perhaps most important for detailed comparison there exist computed potential solutions in which the free vortex sheet is fitted to the surrounding irrotational flow and allowed to roll up into a vortex.

We take the artificial compressibility approach to solve the incompressible Euler equations, show that it leads to a hyperbolic system, carry out a numerical study of its condition, set forth the CFL stability limit for time integration, and examine the types of discontinuities that it admits. Appropriate numerical farfield and solid wall boundary conditions are formulated also. We present several computed solutions of 2D and 3D flows and discuss them in terms of vorticity generation, entropy production, and diffusion of vortex sheets. When compared with relevant results from existing potential methods known to be accurate, we find good overall quantitative agreement. For the delta wing we believe it is vital to resolve the conical flow singularity at the apex.

2. Artificial compressibility method. If the interest is only in steady incompressible flow, Chorin's artificial compressibility concept[1,2] is an attractive approach to solve the problem in either two or equally well three dimensions because it defines a special iteration path that is hyperbolic and therefore allows full use of the standard time-marching procedures developed for compressible flow. Surprisingly enough this concept has attracted little attention until now for application to the strictly inviscid incompressible problem[3].

Hyperbolicity

In two-dimensional Cartesian coordinates the equation to solve is

$$q_t + A\, q_x + B\, q_y = 0 \tag{1}$$

where

$$q = \begin{bmatrix} p/\rho_o \\ u \\ v \end{bmatrix} \qquad A = \begin{bmatrix} 0 & c^2 & 0 \\ 1 & 2u & 0 \\ 0 & v & u \end{bmatrix} \qquad B = \begin{bmatrix} 0 & 0 & c^2 \\ 0 & v & u \\ 1 & 0 & 2v \end{bmatrix}$$

ρ_o is the constant density of the flow and c is a free real parameter that can be chosen to accelerate the time decay. The quasilinear system (1) is called hyperbolic at the point (x,y,t,q) if there exists a nonsingular matrix $T(\alpha,\beta)$ that diagonalizes the linear combination $C = \alpha A + \beta B$

$$T^{-1}CT = \begin{bmatrix} \lambda^{(1)} & & 0 \\ & \lambda^{(2)} & \\ 0 & & \lambda^{(3)} \end{bmatrix}$$

where the eigenvalues λ of C are real and the norms of T and T^{-1} are uniformly bounded for arbitrary real α and β. The eigenvalues of C are found to be, with the definitions $U = \alpha u + \beta v$ and $a^2 = U^2 + c^2(\alpha^2+\beta^2)$, $\lambda^{(1)} = U - a$, $\lambda^{(2)} = U$, $\lambda^{(3)} = U + a$ and are always real. Using matrix C's complete set of linearly independent right eigenvectors as the columns of matrix T we find assuming $\alpha^2+\beta^2$

$$T = \begin{bmatrix} 0 & c^2 a & c^2 a \\ -\beta c^2 & u(U+a)+\alpha c^2 & -u(U-a)-\alpha c^2 \\ \alpha c^2 & v(U+a)+\beta c^2 & -v(U-a)-\beta c^2 \end{bmatrix} \tag{2}$$

Its inverse is formed from the left eigenvectors of C

$$T^{-1} = \begin{bmatrix} \dfrac{\beta u - \alpha v}{a^2 c^2} & -\dfrac{vU+\beta c^2}{a^2 c^2} & \dfrac{uU+\alpha c^2}{a^2 c^2} \\ -\dfrac{U-a}{2a^2 c^2} & \dfrac{\alpha}{2a^2} & \dfrac{\beta}{2a^2} \\ \dfrac{U+a}{2a^2 c^2} & \dfrac{\alpha}{2a^2} & \dfrac{\beta}{2a^2} \end{bmatrix} \tag{3}$$

Choosing the maximum norm one can then go on to complete the demonstration of hyperbolicity by making estimates that show that these last two matrices are uniformly bounded if $|u|$ and $|v|$ are bounded.

The equations that we actually solve, however, are not the Cartesian set (1) but the general finite-volume form[4]

$$\frac{\partial}{\partial t} \int q \; dvol + \iint \mathbf{H} \cdot \mathbf{n} \; ds = 0 \tag{4}$$

where $\mathbf{H}\cdot\mathbf{n} = [c^2\mathbf{V}\cdot\mathbf{n},\; u\mathbf{V}\cdot\mathbf{n} + p/\rho_o\, \mathbf{n}\cdot\mathbf{e}_x,\; v\mathbf{V}\cdot\mathbf{n} + p/\rho_o\, \mathbf{n}\cdot\mathbf{e}_y]$ is the vector flux of q across the surrounding quadrilateral cells with volume VOL and vector areas are $\mathbf{S}_I = SIX\; \mathbf{e}_x + SIY\; \mathbf{e}_y$ and $\mathbf{S}_J = SJX\; \mathbf{e}_x + SJY\; \mathbf{e}_y$. We can analyze eq.(4) locally after semi-discretization using centered space differences by holding the metrics of the cell constant to obtain

$$\frac{d}{dt}(q \; vol)_{ij} + (\tilde{A}\; \delta_I + \tilde{B}\; \delta_J) q_{ij} = 0 \tag{5}$$

where

$$\tilde{A} = \frac{\partial(\mathbf{H}\cdot\mathbf{S}_I)}{\partial q} = \begin{bmatrix} 0 & c^2SIX & c^2SIY \\ SIX & U+uSIX & uSIY \\ SIY & vSIX & U+vSIY \end{bmatrix} \quad \tilde{B} = \frac{\partial(\mathbf{H}\cdot\mathbf{S}_J)}{\partial q} = \begin{bmatrix} 0 & c^2SJX & c^2SJY \\ SJX & V+uSJX & uSJY \\ SJY & vSJX & V+vSJY \end{bmatrix}$$

with U= uSIX+vSIY and V= uSJX+vSJY. Notice that the two matrices $\tilde{A}$ and $\tilde{B}$ in eq.(5) are linear combinations of A and B, eq.(5) then is hyperbolic, and its eigenvalues and diagonalizing matrices follow from (2) and (3) by absorbing the metrics into α and β.

Value for Parameter c

Although the finite-volume transformation does not destroy the hyperbolicity of the problem, specifying an inappropriate value for the parameter c may degrade its condition if the different wave speeds in the system become too disparate. In order to guide us in choosing a value for c we need to look at a measure more quantitative than the boundedness of the transformation matrices. In fact what we want to know is how the bound varies with c. This can be determined numerically for given values of $\mathbf{V}$ and c by computing the eigenvalues σ of T^*T since the L_2 norms are $\|T\| = (\sigma_{max})^{1/2}$ and $\|T^{-1}\| = (\sigma_{min})^{-1/2}$. For the specified values of $\mathbf{V}$ and c a good measure of the condition of the system is then the number $K = \|T\|\; \|T^{-1}\| = (\sigma_{max}/\sigma_{min})^{1/2}$. One can surmise, and we have verified it in an actual computation, that this condition number K depends only on the ratio $c^2/(u^2+v^2)$. Computed numerically and plotted in Fig. 1 as the radial coordinate of the polar diagram (K, θ) where the wave angle θ defines $\alpha = \sin\theta$ and $\beta = \cos\theta$ for $0 \leq \theta \leq 2\Pi$, the condition number K is displayed in Fig. 1 for three different values of $c^2/(u^2+v^2)$. When the ratio $r = c^2/(u^2+v^2)$ is greater than unity the pressure waves dominate over the convection waves and the system is less directionally dependent and better conditioned. This analysis applies to the constant coefficient matrix, but for actually solving the flowfield the parameter c need not be a global constant. In the spirit of local-time-step scaling we set it proportional to the local velocity squared $c^2=r(u^2+v^2)$ where, based on this study, r is a constant in the range $1<r<5$.

CFL Condition

In order to solve the hyperbolic system (4), assuming that under appropriate boundary conditions it does converge to a steady state, we straight-forwardly apply our time-marching finite-volume procedure developed for the compressible Euler equations which uses an explicit three-stage Runge-Kutta type time integration scheme. In the absence of boundaries the usual linearized Fourier analysis of eq.(5) specifies the limit on the time step for which the integration locally is stable, i.e. the CFL condition is $\Delta t \leqslant (CFL)/(|\lambda|max)$ where the constant CFL depends on the particular multistage method that is used. A conservative estimate for the maximum eigenvalue $|\lambda|max$ of the 2D spatial difference operator leads to the local-time-step condition

$$\Delta t_{ij} \leqslant CFL \left[\frac{VOL}{\tilde{U} + (\tilde{U}^2+c^2S^2)^{1/2}}\right]_{ij} \tag{6}$$

where $\tilde{U} = (|SIX| + |SJX|)|u| + (|SIY| + |SJY|)|v|$

$S^2 = (|SIX| + |SJX|)^2 + (|SIY| + |SJY|)^2$

And the 3D condition follows in direct analogy to this. The computed results we present here have all been achieved using this step size in a local-time-step integration scheme.

Discontinuities

In steady flow of course the incompressible equations admit only the tangential discontinuity with jump conditions $[p]= 0$ and $\mathbf{V}_1 \cdot \mathbf{n} = \mathbf{V}_2 \cdot \mathbf{n} = 0$. The artificial compressibility method, however, approaches steady flow only asymptotically in time so we must investigate what transient discontinuities are allowed in this pseudo-system of equations. Following the standard analysis of discontinuities for conservation laws, we shrink the integration region in Eq.(4) around the discontinuity surface to obtain in the limit

$$\begin{aligned} s[p] &= c^2[V_n] \\ s[V_n] &= [V^2_n+p] = 2V^*_n[V_n] + [p] \\ s[V_t] &= [V_nV_t] = V^*_n[V_t] + V^*_t[V_n] \end{aligned} \tag{7}$$

where s is the speed of the discontinuity in the direction of its normal $\mathbf{n}$, parallel to the surface is the tangential direction $\mathbf{t}$, the asterisk indicates the average of the value on each side of the discontinuity e.g. $V_n^* = (V_{n_1} + V_{n_2})/2$, and the square brackets their difference $[V_n] = V_{n_2} - V_{n_1}$. System (7) is linear homogeneous for given s and average velocities

$$\begin{pmatrix} -s & c^2 & 0 \\ 1 & 2V_n^* - s & 0 \\ 0 & V_t^* & V_n^* - s \end{pmatrix} \begin{pmatrix} [p] \\ [V_n] \\ [V_t] \end{pmatrix} = 0 \tag{8}$$

and a nontrivial solution exists if

$$(s - V_n^*)\,(s^2 - 2sV_n^* - c^2) = 0$$

The discontinuity therefore may move with any of three different speeds $s_1 = V_n^*$, $s_2 = V_n^* + (V_n^{*2} + c^2)^{1/2}$, or $s_3 = V_n^* - (V_n^{*2} + c^2)^{1/2}$. The jump relations for the first speed then are eigenvectors of (8) and work out to be

$$\begin{pmatrix} [p] \\ [V_n] \\ [V_t] \end{pmatrix} = k \begin{pmatrix} 0 \\ 0 \\ 1 \end{pmatrix} \qquad \text{with } s = V_{n_1} = V_{n_2} = V_n^* \tag{9}$$

and k is an arbitrary constant. We immediately recognize this solution as the unsteady tangential discontinuity that corresponds directly to the steady one. The second and third eigenvectors associated with the speeds $s = s_2$ and $s = s_3$ respectively

$$\begin{pmatrix} [p] \\ [V_n] \\ [V_t] \end{pmatrix} = k \begin{pmatrix} c^2 \\ s \\ sV_t^*(V_n^{*2} + c^2)^{-1/2} \end{pmatrix} \tag{10}$$

are more unexpected since they allow jumps in both pressure and velocity across a discontinuity. With $s = s_2$ the shock travels downstream, and the pressure and normal velocity of a fluid particle passing through it rises or falls together depending on the sign of k. With $s = s_3$ the shock moves upstream and the jumps in pressure and normal velocity take opposite signs. One of these solutions may be spurious in the sense that it violates some sort of entropy condition that a real flow should fulfill, but we shall not pursue this matter further here. Of the two the second seems more physical, and specifies vanishing jumps in the limit as c^2 goes to zero. In any case we presume that a fluid particle traveling through either one undergoes an irreversible process that produces a rise (or fall) in its entropy function and a corresponding loss (or gain) in its total pressure.

Boundary Conditions

Boundary conditions of course specify the particular problem and two different types are of concern here: flow conditions on a solid wall and at the farfield boundary of the mesh. Since our mesh is aligned to the wall, for the first type we set the velocity flux through the wall to zero and determine the pressure p on it from the incompressible normal momentum equation $\underset{\sim}{V}\cdot(\underset{\sim}{V}\cdot\text{grad})\,\underset{\sim}{n} = \underset{\sim}{n}\cdot\text{grad}\,p/\rho_o$ which is exactly analogous to the conditions for compressible flow[5]. When it is differenced to formally first-order accuracy the pressure on the surface is deduced from the interior values.

Our farfield conditions are a form of Engquist's first approximation[6] which uses the characteristic variables for the one-dimensional version of Eq.(5) that is normal to the boundary in order to focus on a particular set of characteristic planes, those whose normals point toward the boundary and whose slopes in time are the eigenvalues λ of $\tilde{A}$ and are given above. The transformation matrices T^{-1} and T diagonalize this one- dimensional equation so that

$$\phi_t+\Lambda\phi_x = 0 \text{ where } \phi= T^{-1}q \quad\text{and}\quad \Lambda= T^{-1}\tilde{A}\,T= \text{diag}\{\lambda_1\,\lambda_2\,\lambda_3\}$$

The combination of boundary conditions determined from outside the domain and auxilliary conditions set from inside follows in the now-standard way according to whether the associated characteristic directions enter or leave the domain. When $|U| < a$ our implementation is to set the two ingoing characteristic variables $\phi^{(1)}$ and $\phi^{(2)}$ to their freestream values, linearily extrapolate the third $\phi^{(3)}$ from the computational field, and then solve for the original unknowns $q= T\phi$. At outflow it is $\phi^{(2)}$ that is given the values of undisturbed flow, and $\phi^{(1)}$ and $\phi^{(2)}$ are extrapolated.

Artificial Viscosity Model

Our rationale for introducing an artificial viscosity operator into the numerical procedure has been put forward before[7]. To counteract nonlinear aliasing associated with centered differences, to set the entropy condition across shock waves, and perhaps even to insure the existence of a steady state are among the reasons that bespeak such a model. The dissipative operator Dq that we add to the convective differencing of the compressible Euler equations contains a nonlinear second difference term and a linear fourth-difference term. Both of these are implemented with boundary conditions so that the complete operator has the negative semi-definite property $(q^T Dq) \leqslant 0$ which has been found important for minimizing the injection of spurious entropy layers at solid boundaries. Since shock waves are not present in incompressible flow, at least in steady state, there is not much need for the second difference term to fulfill an entropy condition. So the dissipative operator that we use for the incompressible results in this paper is comprised only of the fourth difference term.

3. Computed results

Potential or Nonpotential Solution

To focus the discussion on when and under what conditions an irrotational solution to the incompressible Euler equations can be expected, consider the physical steady system of the continuity equation and the curl of the momentum equation

$$\operatorname{div} \underset{\sim}{V} = 0 \quad ; \quad \frac{\partial}{\partial t} \operatorname{curl} \underset{\sim}{V} = \operatorname{curl} (\underset{\sim}{V} \times \operatorname{curl} \underset{\sim}{V}) \tag{11}$$

Clearly the irrotational solution curl $\underset{\sim}{V} = 0$ satisfies this nonlinear system as well as the Laplace equation, but rotational solutions are equally admissible. Why or in what way during the course of the numerical solution of this nonlinear system does the computation method arrive at either the Laplace solution or the rotational one we cannot say much about right now. But what we do attempt to do with the computed results presented here is to show examples of both types of solutions and grope for plausible explanations.

2D Irrotational Solution

Our first example is inviscid incompressible flow past a circle for which the analytical solution is known. Its comparison in Fig. 2 with the C_p pressure distribution computed by this method on the upper half circle shows that indeed we are obtaining the irrotational solution. And the vector diagram of velocity directions (Fig. 3) indicates the fore and aft symmetry of the overall flowfield and the loss-free character of the solution.

2D Solution with Circulation

The next example is also two dimensional but the entire plane is included now and the geometry is the NACA 0012 airfoil at 5 deg. incidence. Measured by the decay of the average and maximum time difference of pressure in the entire field and the evolution of lift and drag, the convergence of the solution computed upon a mesh with 128 cells around the airfoil and 28 outward is given in Fig. 4. Although the comparison of computed C_p on the airfoil with that of an accurate boundary integral (singularity) method[8] is generally good (Fig. 5), there are small discrepancies at the leading edge pressure peak and the trailing edge, the latter due undoubtedly in part to the mesh being unable to resolve completely the flow singularity there. But perhaps the best check on accuracy is the degree to which the Bernoulli relation along a streamline in steady flow $p/\rho_o + 1/2\ V^2 = p_t$ (constant) is satisfied. In this flow all streamlines originate from a constant freestream so the total pressure p_t takes the same value on every streamline. Figure 6 confirms that there are errors (loss in total pressure $1-p_t/p_{t_\infty}$) in the vicinity of the leading and trailing edges, but they are small in magnitude and confined locally to these two regions. We conclude that an almost irrotational solution is obtained with very nearly the correct circulation even without a bound-vortex-correction technique in the farfield boundary conditions. Furthermore contours of constant flow angle (Fig. 7) demonstrate that even without our invoking a Kutta condition the flow leaves the trailing edge smoothly which is an essential feature of potential flow. The central question before us therefore is how does the Euler-equation method arrive at the correct circulatory flow? Although no definitive answer is yet at hand, some clues point to the creation of discontinuities in the transient stages of the flowfield's evolution. Figure 8 gives the values of lift C_L and circulation $\Gamma = \int \underset{\sim}{V} \cdot \underset{\sim}{d\ell}$ around a circuit close to the airfoil obtained in the solution on each of its first 25 time steps after starting from initial conditions of freestream flow which do not satisfy the boundary conditions on the airfoil. We see that a nonzero value for lift is reached even after the first iteration. The circulation, however, tends to lag behind the lift and in fact is even negative. When we look in Fig. 9 at the contour maps of pressure and vorticity $\Omega = \underset{\sim}{e}_z \cdot \text{curl}\ \underset{\sim}{V}$ of these solutions we see what can be interpreted as a blast wave moving out from the nose of the airfoil leaving behind it a high pressure region of rotational flow. The vorticity distribution on the airfoil surface has peaks at the leading and trailing edges which achieve a maximum on the fifth iteration but then decay to the small residual level in the converged state by being swept out of the flowfield. Presumably the vorticity created in the field is counterbalanced by a vortex bound in the airfoil which accounts for the circulation. We hypothesize, therefore, that the genesis of the circulation lies in irreversible processes that occur across transient shock waves. Once generated, the part that this vorticity plays, no matter how unphysical it may be, is to usher the iterative solution toward the rotational root of Eq.(11). The trailing edge evidently is the controlling factor, namely the shocks subside only when the flow leaves the trailing edge smoothly.

3D Rotational Solution

In this section we present the computed solution of incompressible flow around a 70 deg. swept delta wing of zero thickness and unit length at 20 deg. angle of attack. The flow separates from the leading edge in a vortex sheet which then rolls up to form a vortex over the wing. The mesh we use is an O-O type[9] constructed with a polar singular line at the apex and a parabolic singular line at the tip of the trailing edge and has 64 cells around the half span, 28 each on the upper and lower chord, and 20 outwards. Figure 10 presents the convergence history of the solution in the first (coarse) of two grids. Barring the effect of the trailing edge, the solution should be very nearly conical near the apex of the wing. Among others Hoeijmakers[10] has developed a potential boundary integral method for this sort of conical flow which fits a vortex sheet to the surrounding irrotational flowfield and allows it to rollup under its own influence, and his results for this case are available for comparison[11]. In Fig. 11 3D isobar maps and contour lines of constant vorticity magnitude drawn in plane projection survey the computed Euler-equation solution in three (nonplanar) spanwise mesh surfaces and reveal the vorticity and low pressure trough in the field over the wing. Plotted in frontal view and to larger scale are the vorticity contours in the 30% chord section, where the flow should be nearly conical, together with the rolled-up fitted vortex-sheet discontinuity of Hoeijmakers' computation. The position of his vortex core and the maximum of our vorticity field agree remarkably well. This is the first quantitative demonstration so far that a vortex sheet separating from a leading edge can be captured in a vorticity field with a reasonable degree of realism. The presence of the vortex of course can be detected in measured quantities too. The most common is surface pressure on the wing. Compared in Fig. 12 are spanwise C_p distributions of our results, Hoeijmakers', and measured[12] values at 3 chord stations. Except for the secondary vortex separation near the leading edge in the laminar flow experiment, our results are in fair correspondence with the measurements in those sections farthest from the apex. Nearest the apex the overall pressure level is about right but the suction peak is too broad which may indicate that our captured vortex sheet has diffused more than the measured one has. Contours of total pressure (Fig. 13) also support this opinion since in Fig. 13 we find substantial losses spread over the entire vortex core region. And this observation is difficult to interpret at present. Qualitatively similar patterns have been obtained before in wind tunnels, but if we reason strictly about inviscid flow, the streamlines in the regions between the coils of the vortex sheet all start in the freestream, and hence steady flow should be loss free. Of course the streamlines forming the vortex sheet itself also begin in the freestream, but on route some or all of them pass through the conical flow singularity at the apex, integration becomes meaningless, and the value of p_t is indeterminate. The question at hand then is, how well do we resolve the conical flow at the apex? The surface contours of total pressure near the apex show some conical behaviour, but closer inspection makes us conclude that this solution has missed much detail there. Whether we can improve upon it by constructing a mesh with properties locally at the apex that better match the conical flow singularity remains to be answered in a future paper.

4. Concluding remarks. The artificial compressibility method is an interesting one for solving the incompressible Euler equations because its solutions provide a lot of insight into the types of flows that can occur. A free parameter can be chosen to alter the wave-propagation speeds, and unexpected transient shocks have been uncovered which may play a role as the vorticity-creation mechanism. A steady 3D flowfield with a free-shear layer has also been computed and its features have been studied. Comparison with an accepted solution was fair, and points to the need for improved accuracy at the apex.

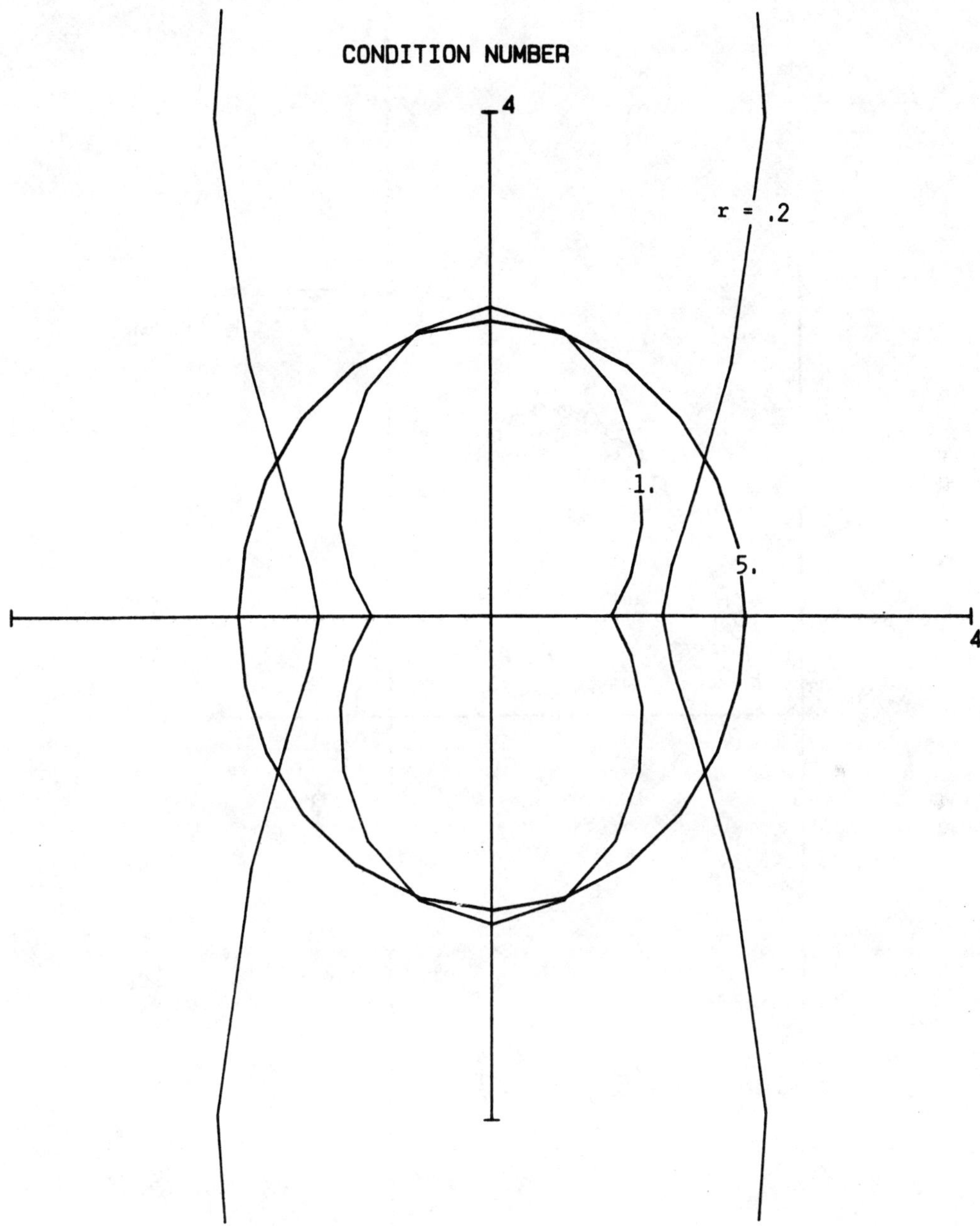

Fig. 1 Polar diagram of condition number $K = \|T\|_{L_2} \|T^{-1}\|_{L_2}$ of the hyperbolic system (1) as a function of the plane wave angle θ for three values of the ratio $r = c^2/(u^2+v^2) = 0.2$, 1, and 5. Poorly conditioned when $r < 1$.

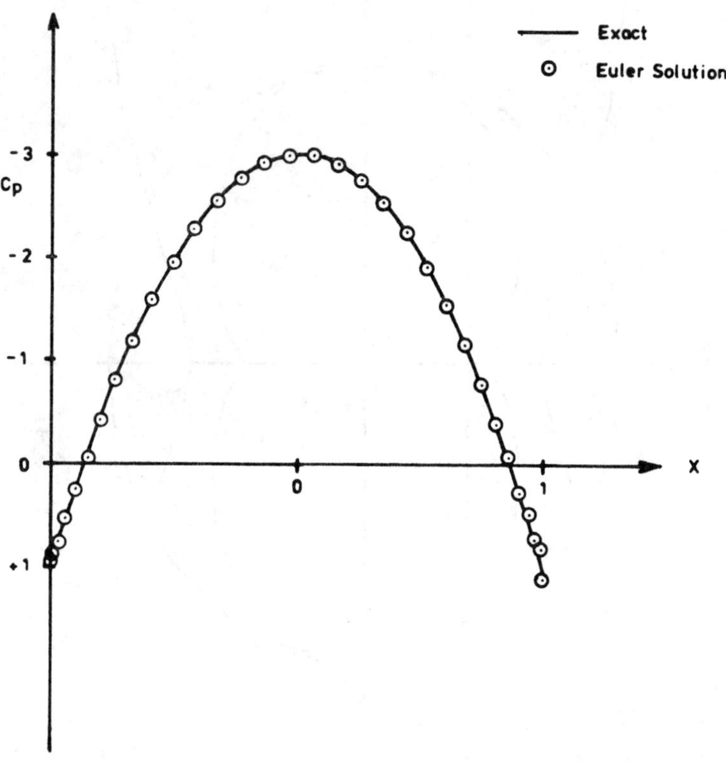

Fig. 2 Comparison of the exact solution to the Laplace equation and the numerical solution to the Euler equations for incompressible flow around a circle.

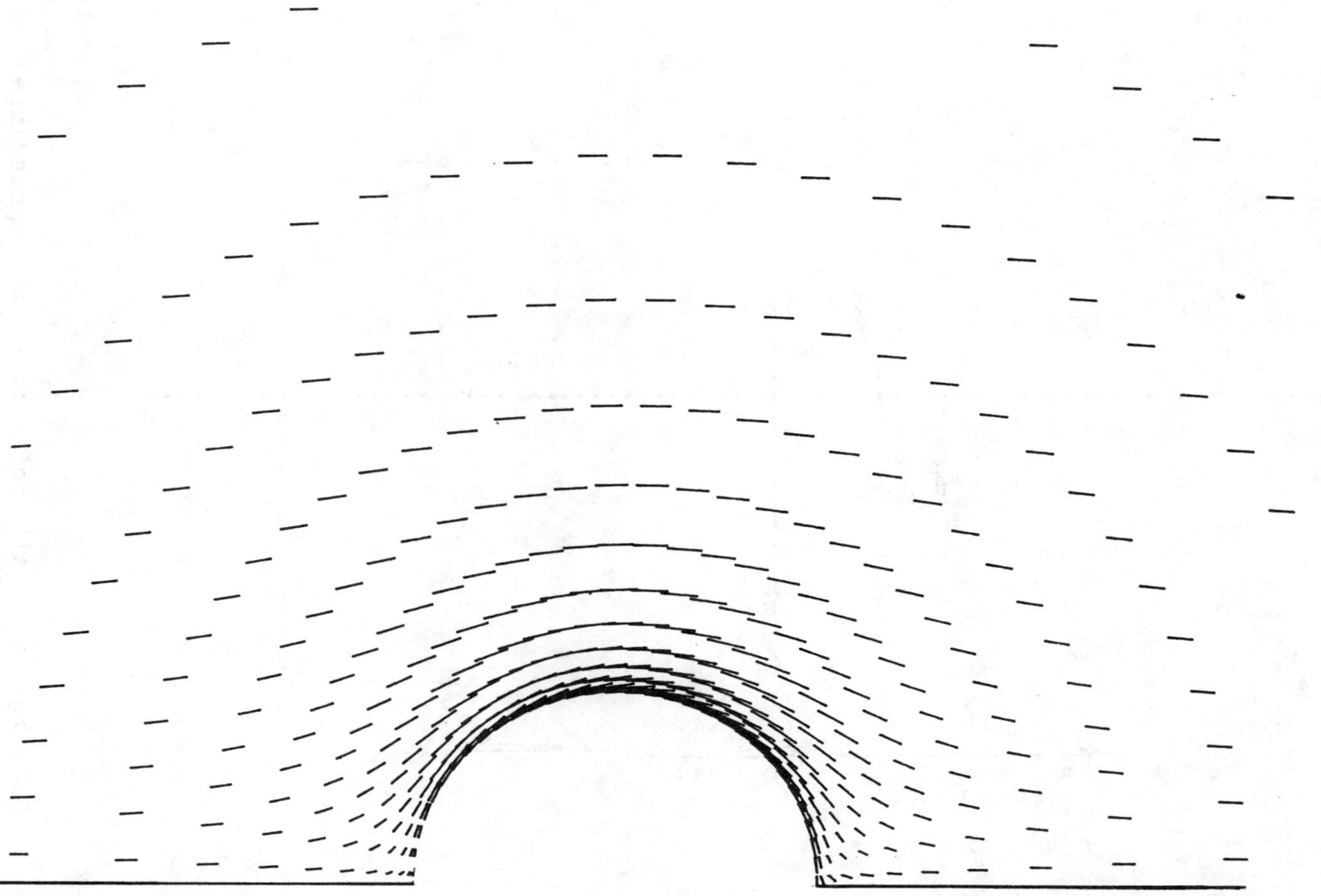

Fig. 3 Vector plot of the velocity field of the computed Euler-equation solution to flow around the circle.

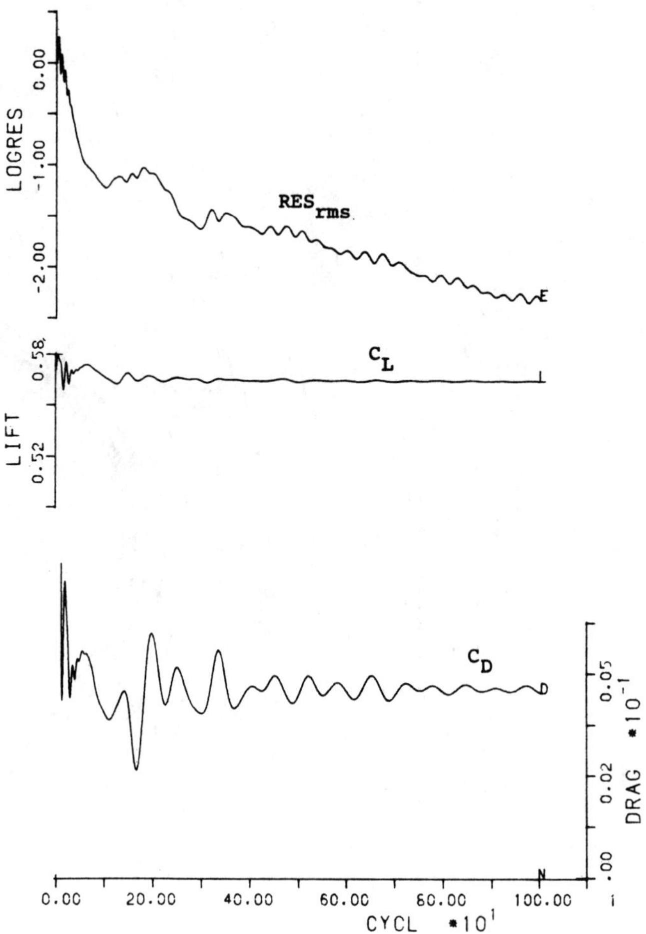

Fig. 4 Convergence of the solution of incompressible flow past the NACA 0012 airfoil indicated by the decay of the average time difference of pressure and the evolution of lift and drag. $M_\infty = 0.$ $\alpha = 5$ deg.

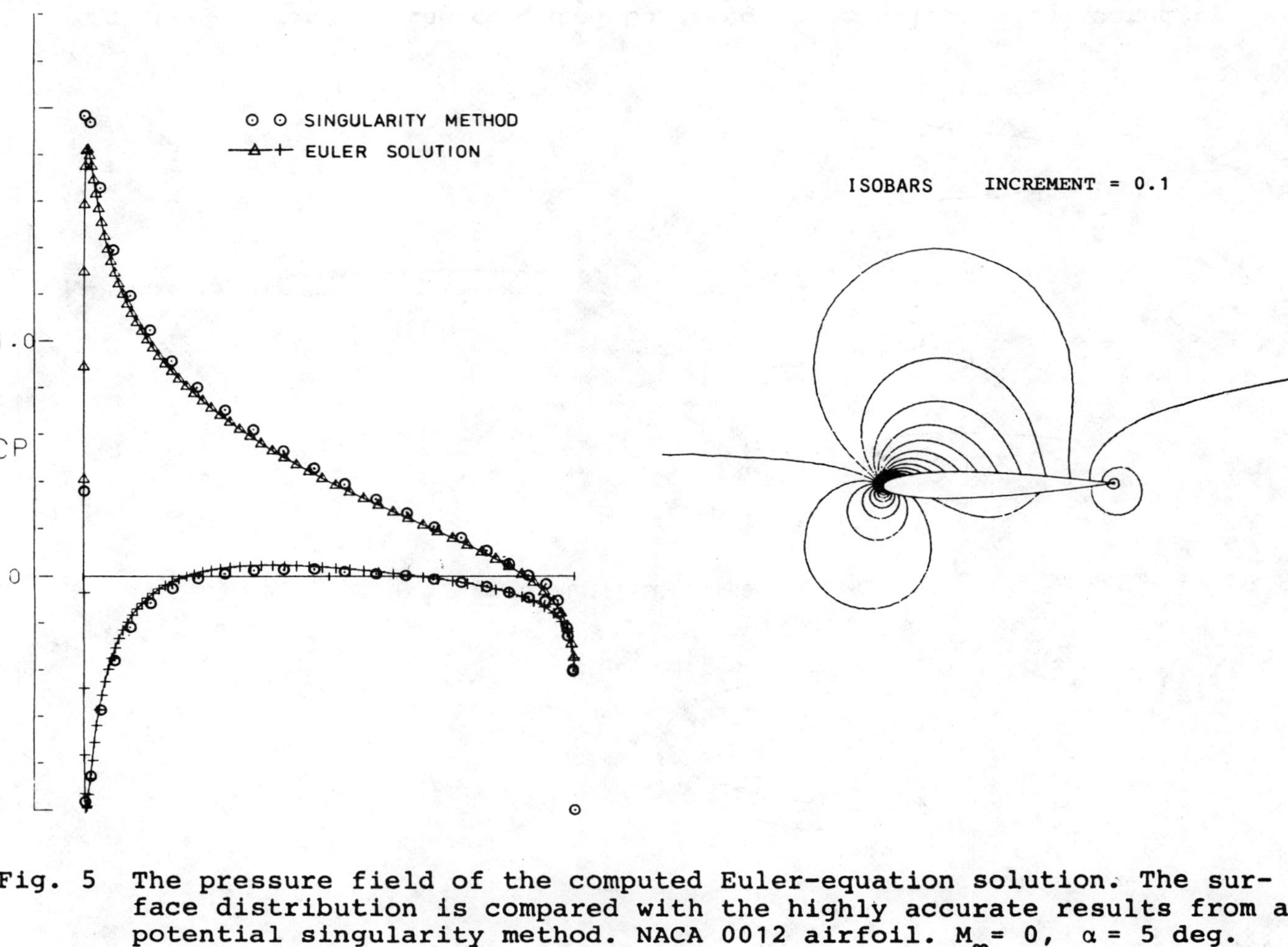

Fig. 5 The pressure field of the computed Euler-equation solution. The surface distribution is compared with the highly accurate results from a potential singularity method. NACA 0012 airfoil. $M_\infty = 0$, $\alpha = 5$ deg.

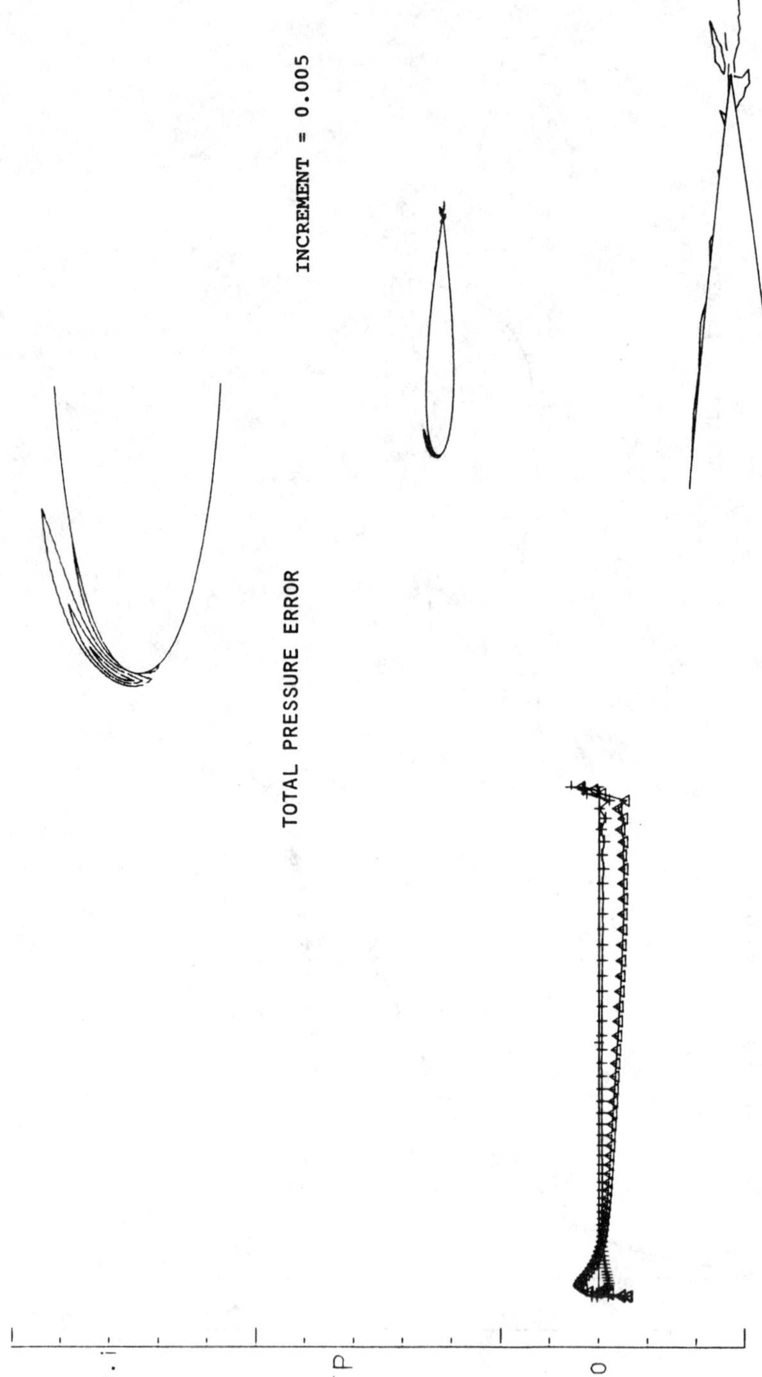

Fig. 6 Accuracy of the computed Euler-equation solution indicated by total pressure error $1-p_t/p_{t_\infty}$. NACA 0012 airfoil, $M_\infty = 0.$ $\alpha = 5$ deg.

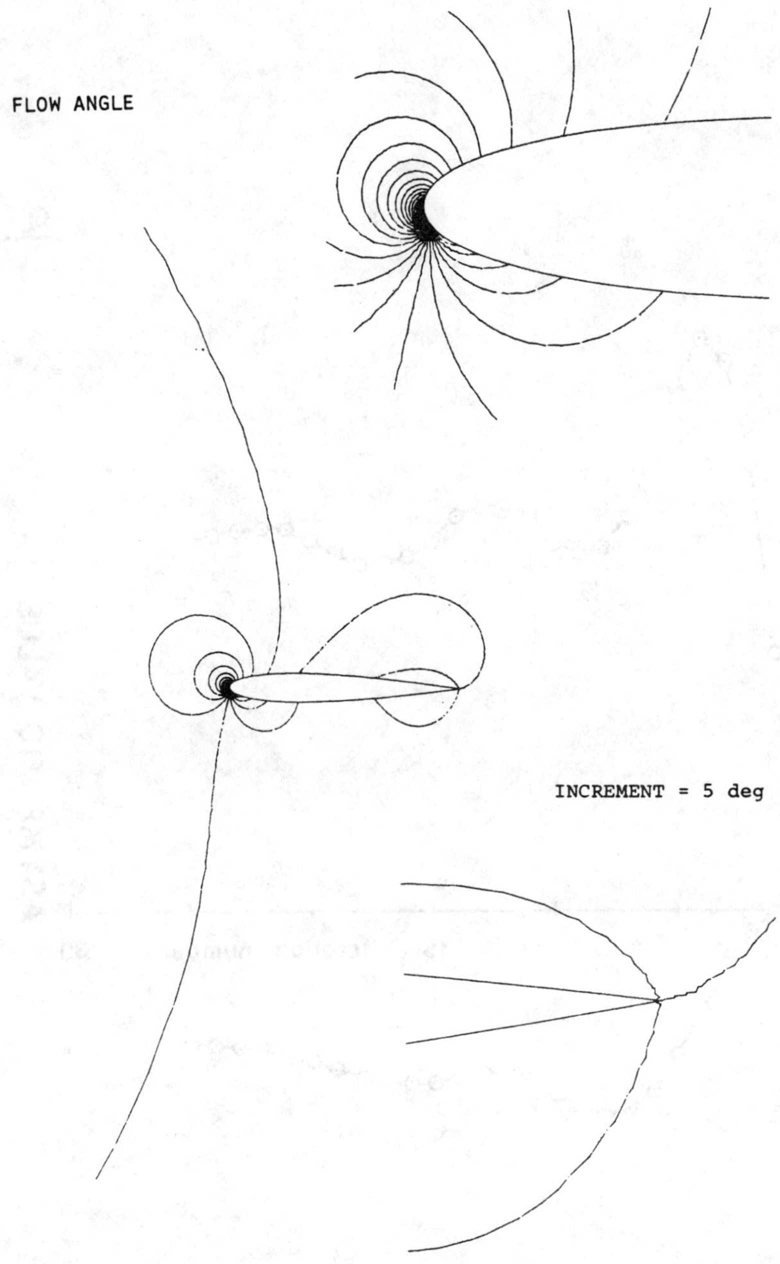

Fig. 7 Contours of constant flow angle in the computed solution demonstrate that the flow leaves the trailing edge smoothly. NACA 0012, $M_\infty = 0$, $\alpha=5$ deg.

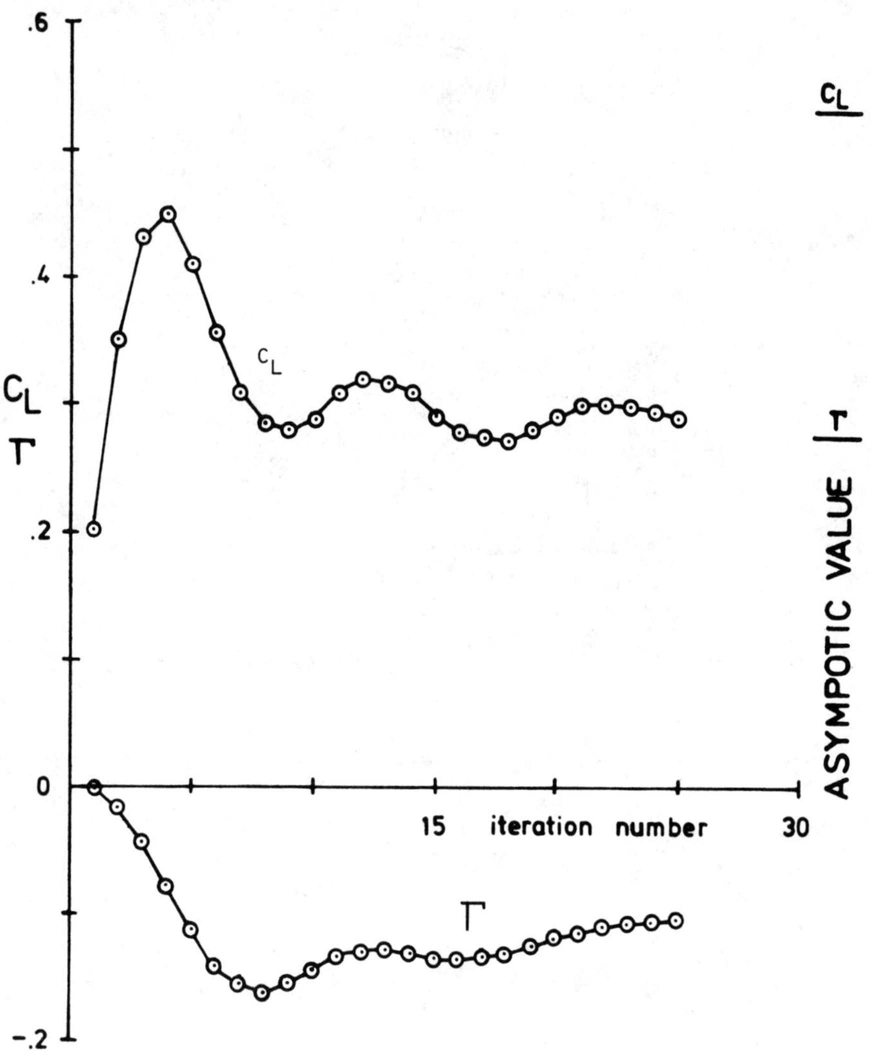

Fig. 8 Evolution of lift and circulation during the first 25 iterations after an impulsive start from free-stream initial conditions. NACA 0012, $M_\infty = 0$, $\alpha = 5$ deg.

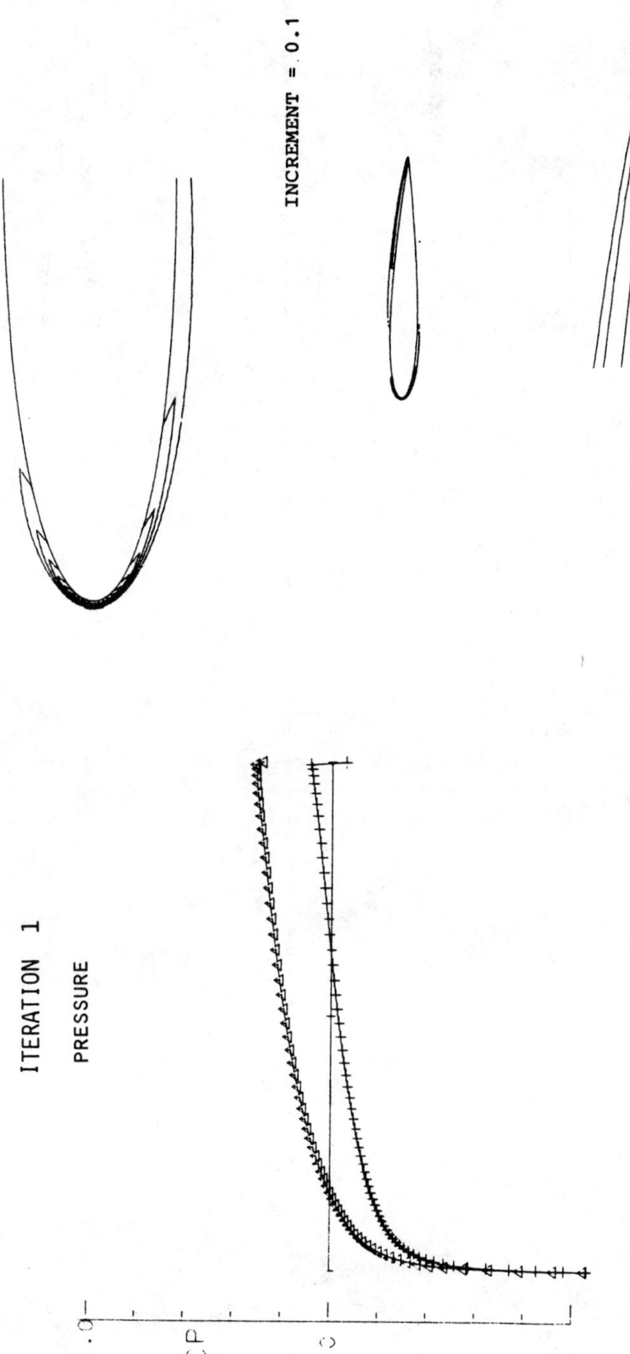

Fig. 9 Contour maps of the pressure and vorticity fields during those early iterations. NACA 0012, $M_\infty = 0$, $\alpha = 5$ deg.

a) pressure field iteration 1.

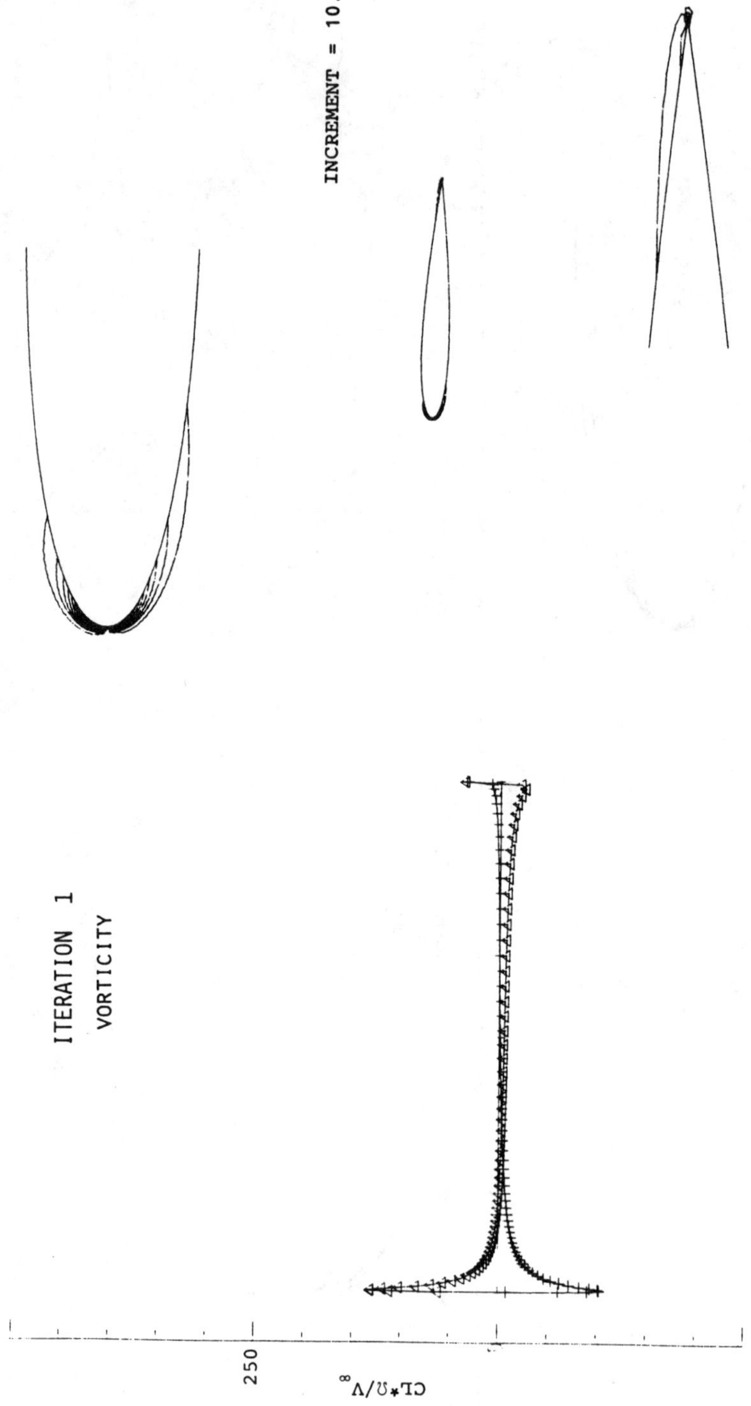

Fig. 9 b) vorticity field iteration 1.

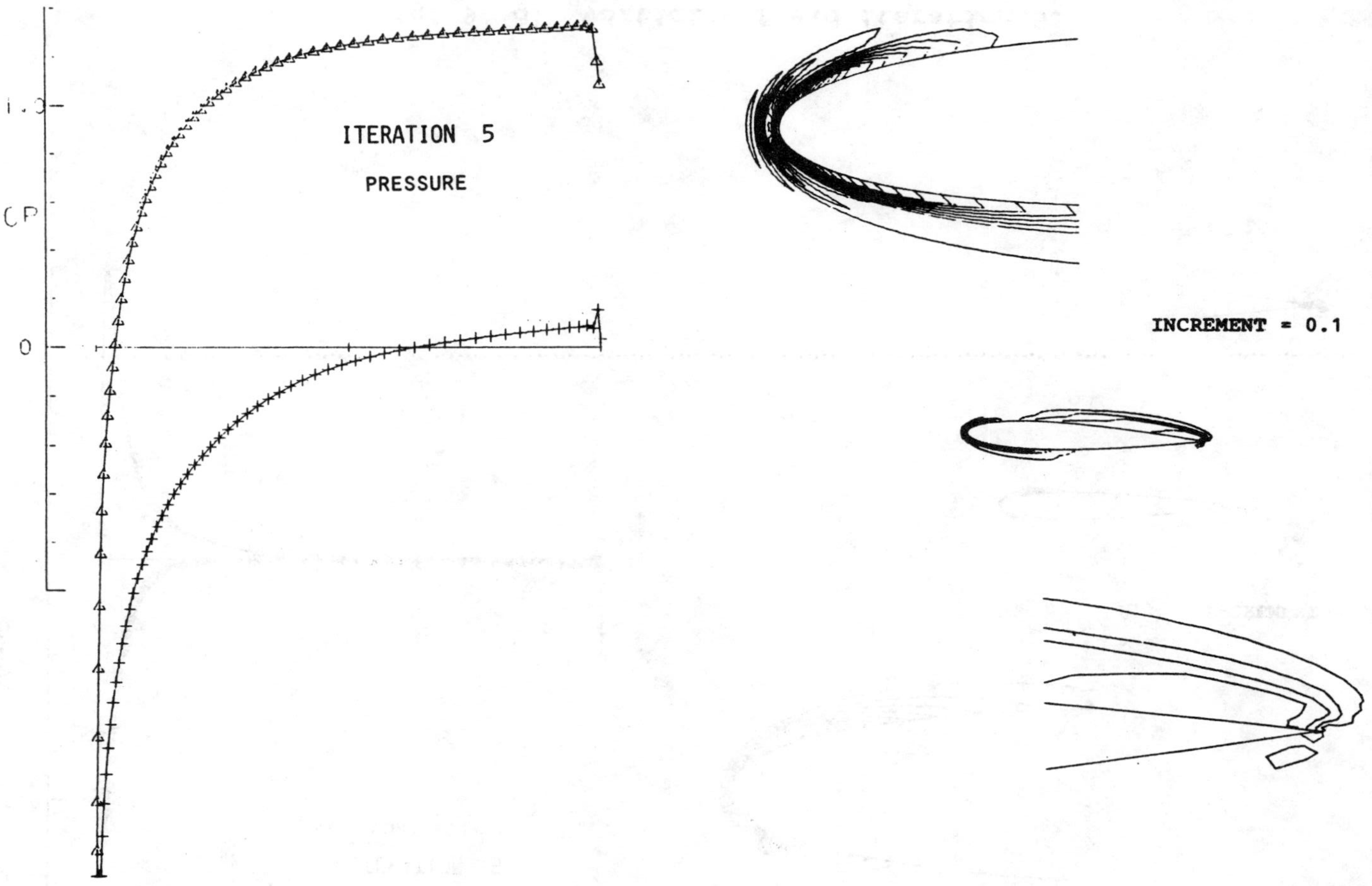

Fig. 9 c) pressure field iteration 5.

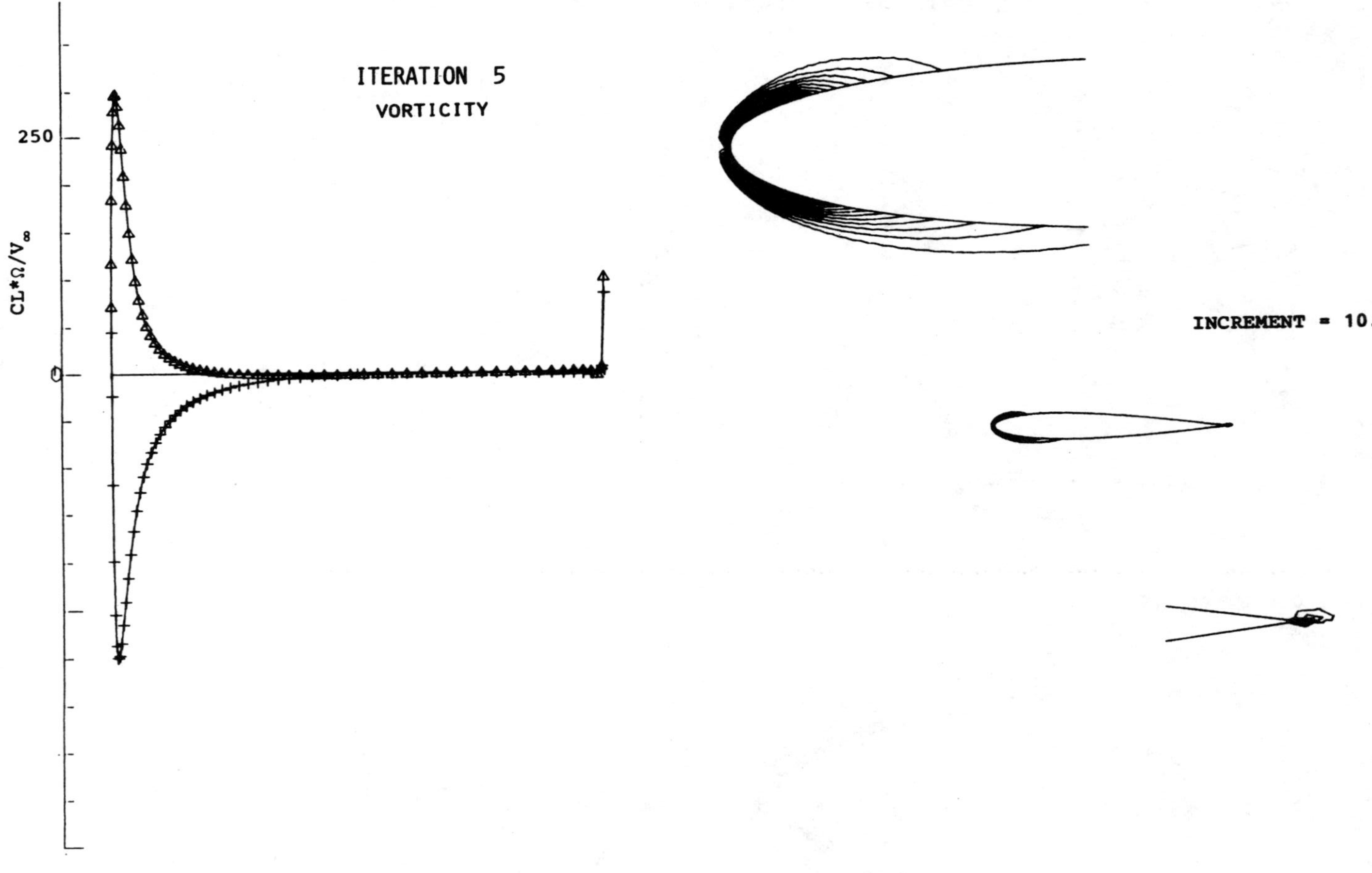

Fig. 9 d) vorticity field iteration 5.

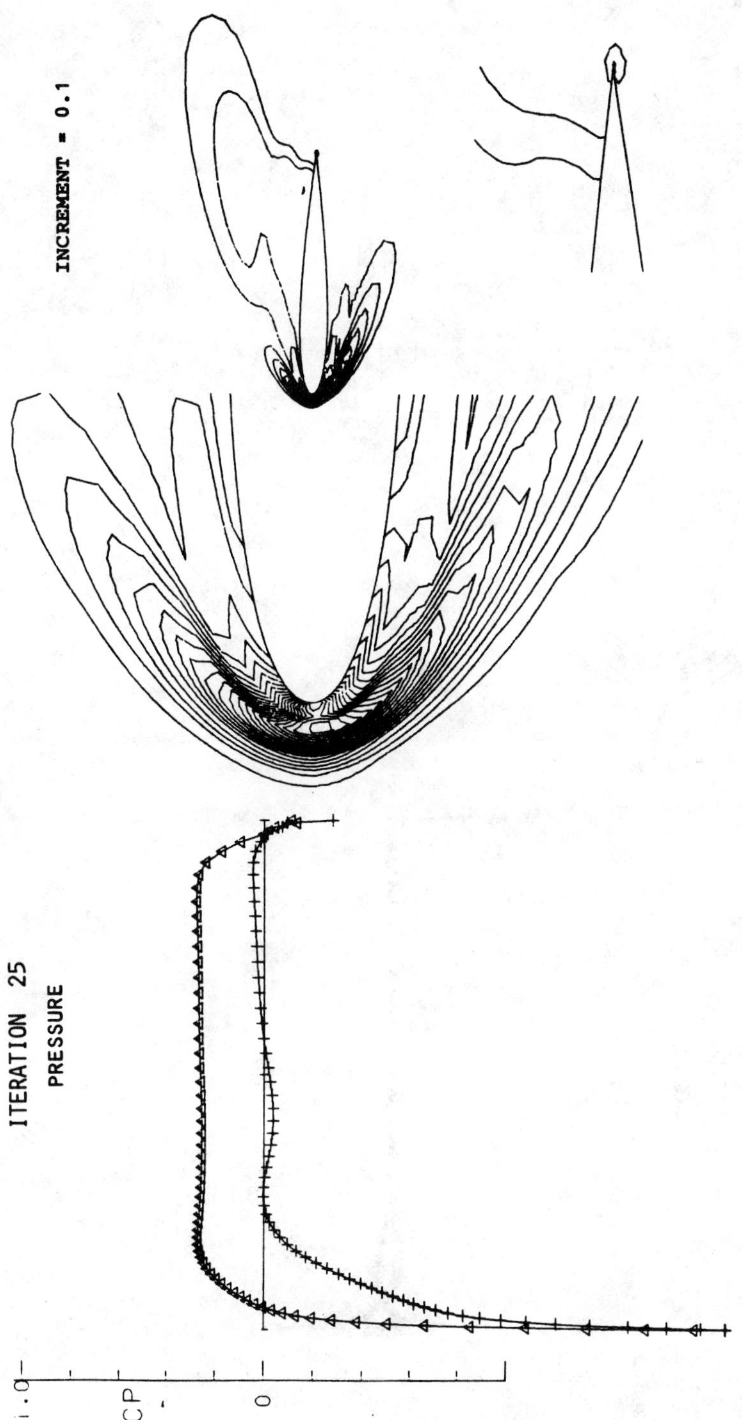

Fig. 9 e) pressure field iteration 25.

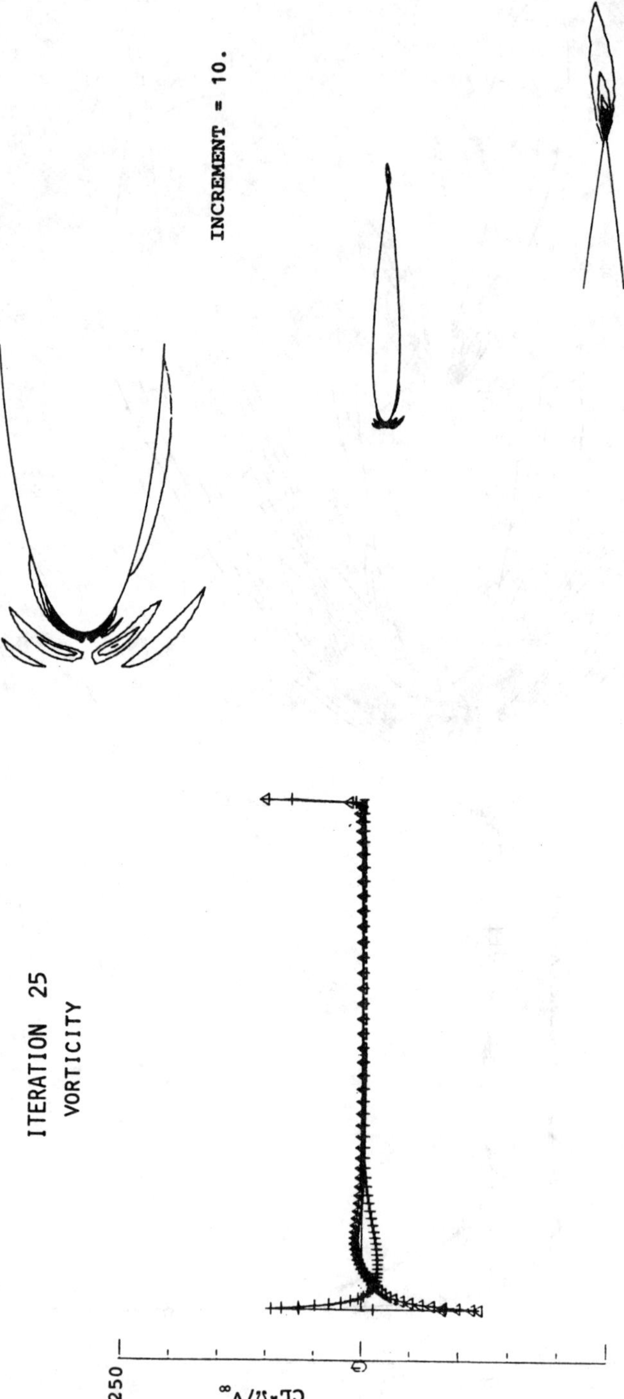

Fig. 9. f) vorticity field iteation 25.

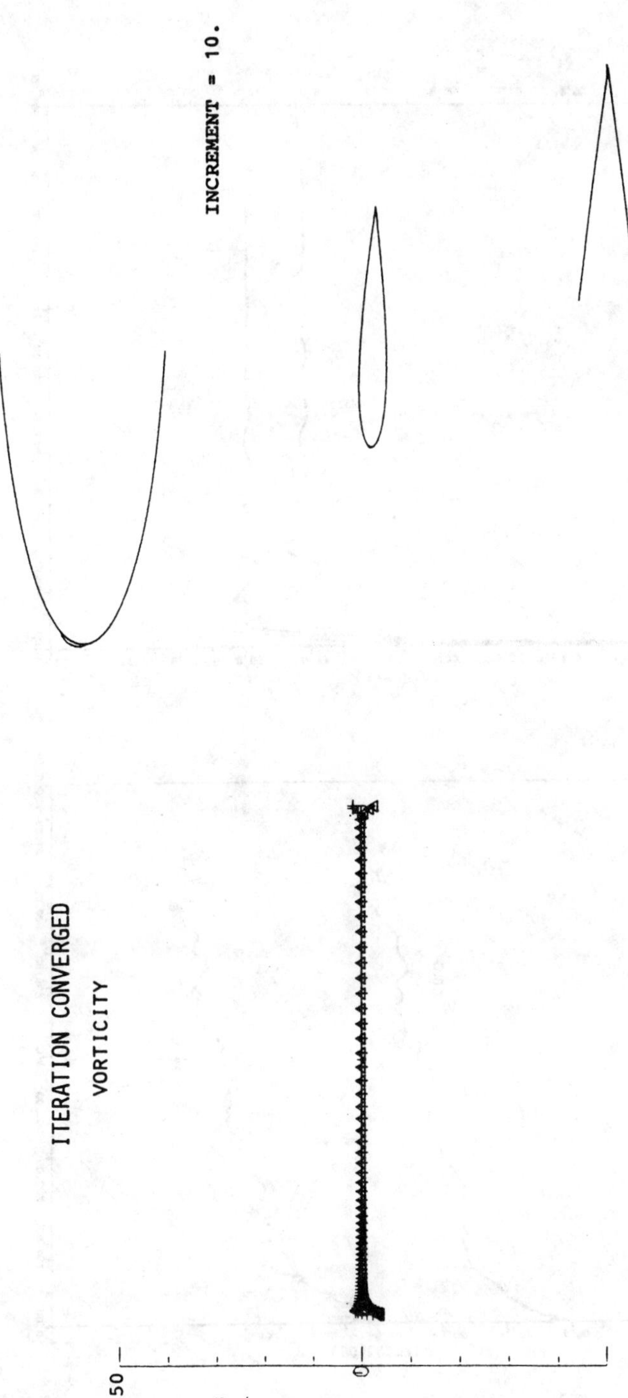

Fig. 9 g) converged vorticity field.

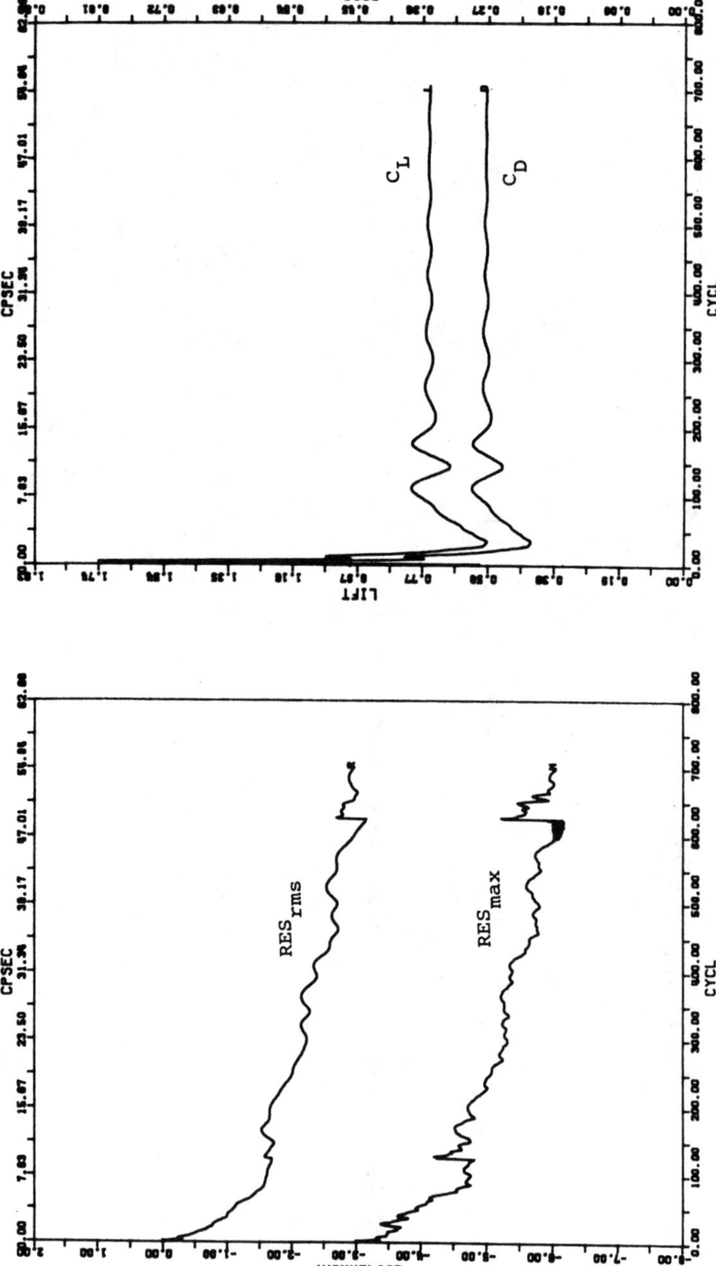

Fig. 10 Convergence of the 3D Euler-equation soluton for flow separating from the leading edge of a swept flat plate delta wing indicated by the residual decay and evolution of lift and drag. $M_\infty = 0$, $\alpha = 20$ deg. CYBER 205.

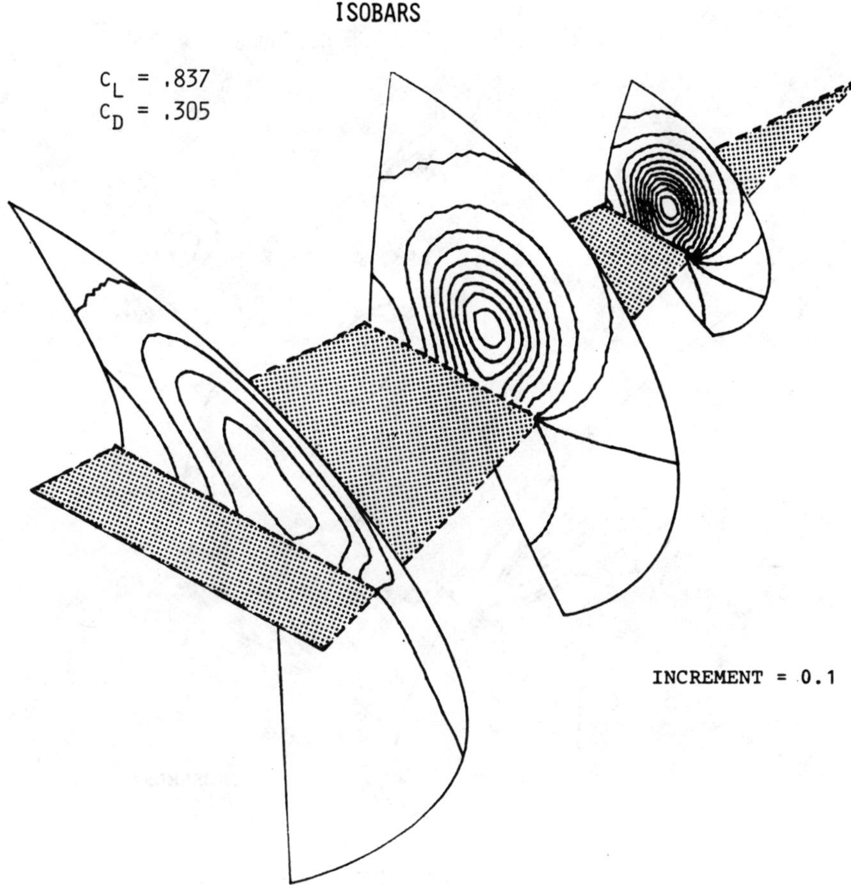

Fig. 11 Contour maps of the computed Euler-equation solution for flow past a 70 deg. swept flat plate delta wing reveal the presence of a vortex over the wing. $M_\infty = 0$, $\alpha = 20$ deg.

a) isobars in three nonplanar mesh surfaces.

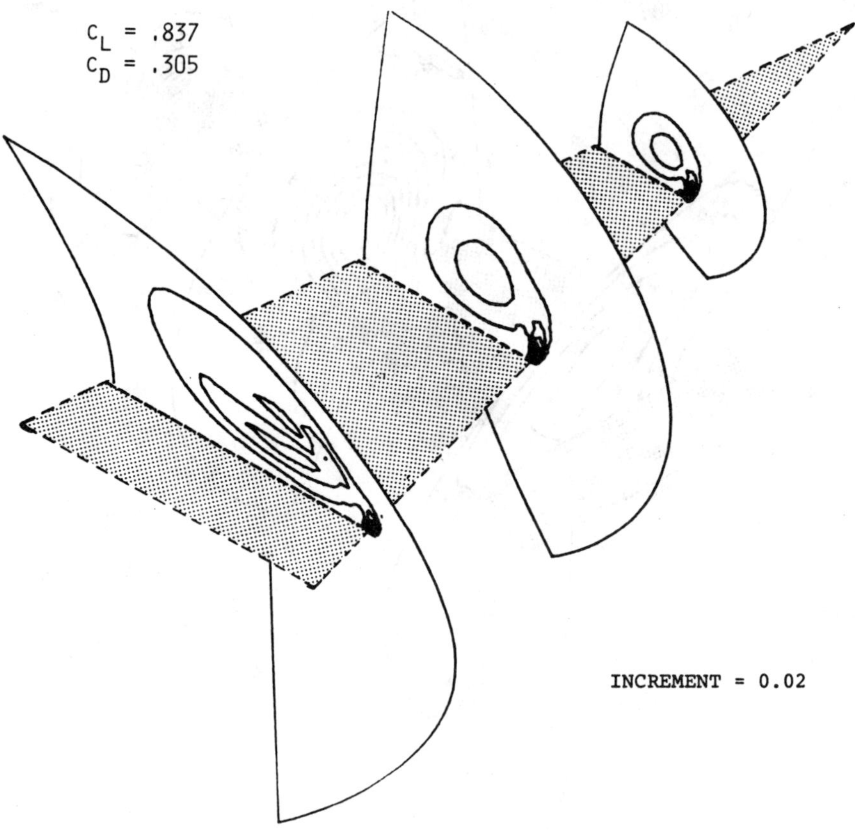

Fig. 11 b) contours of constant vorticity magnitude in three nonplanar mesh surfaces.

COMPARISON OF CAPTURED VORTICITY FIELD AND
FITTED VORTEX SHEET

31% x-station

—— VORTICITY MAGNITUDE CONTOURS

- - - FITTED VORTEX SHEET[11]

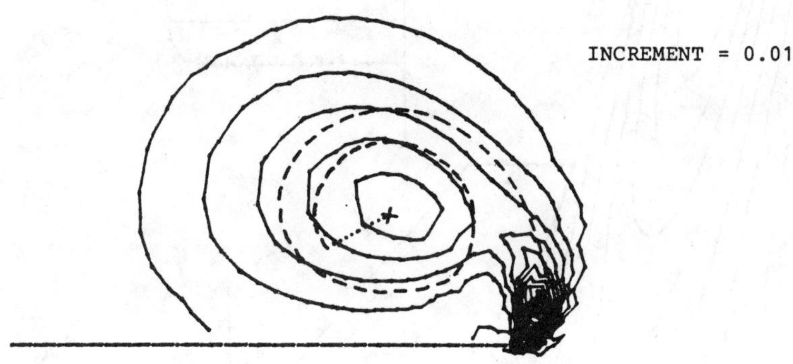

Fig. 11 c) comparison in the x= 0.31 mesh surface between contours of the vorticiyt field computed from the Euler equations and the shed vortex sheet that is fitted as a discontinuity to a surrounding potential solution potential solution which assumes conical flow.

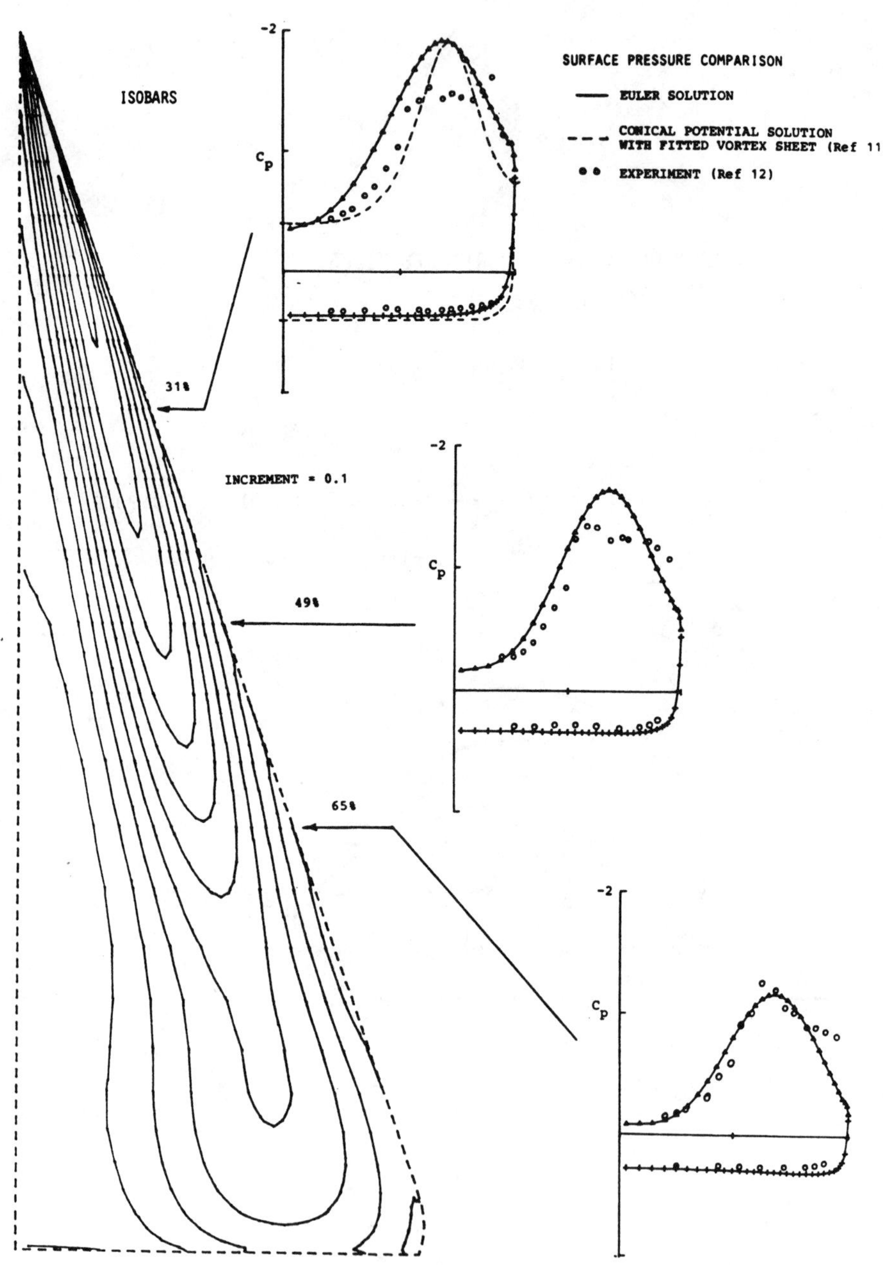

Fig. 12 Computed pressure field on the upper surface of the wing compared measured values in three stations and the potential flow results in the first station where the flow is nearly conical. $M_\infty = 0.$ $\alpha = 20$ deg.

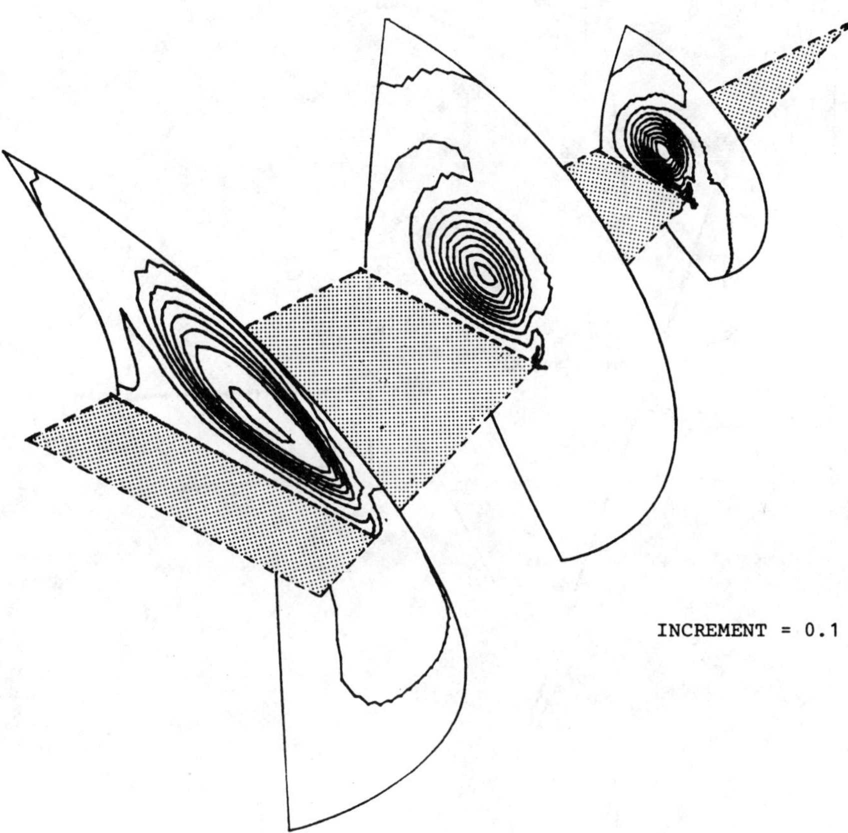

Fig. 13

Contours of total pressure error $1-p_t/p_{t_\infty}$ in the computed Euler-equation solution for flow $M_\infty = 0$, $\alpha = 20$ deg. in

a) three nonplanar mesh surface around the 70-deg. swept flat plate wing.

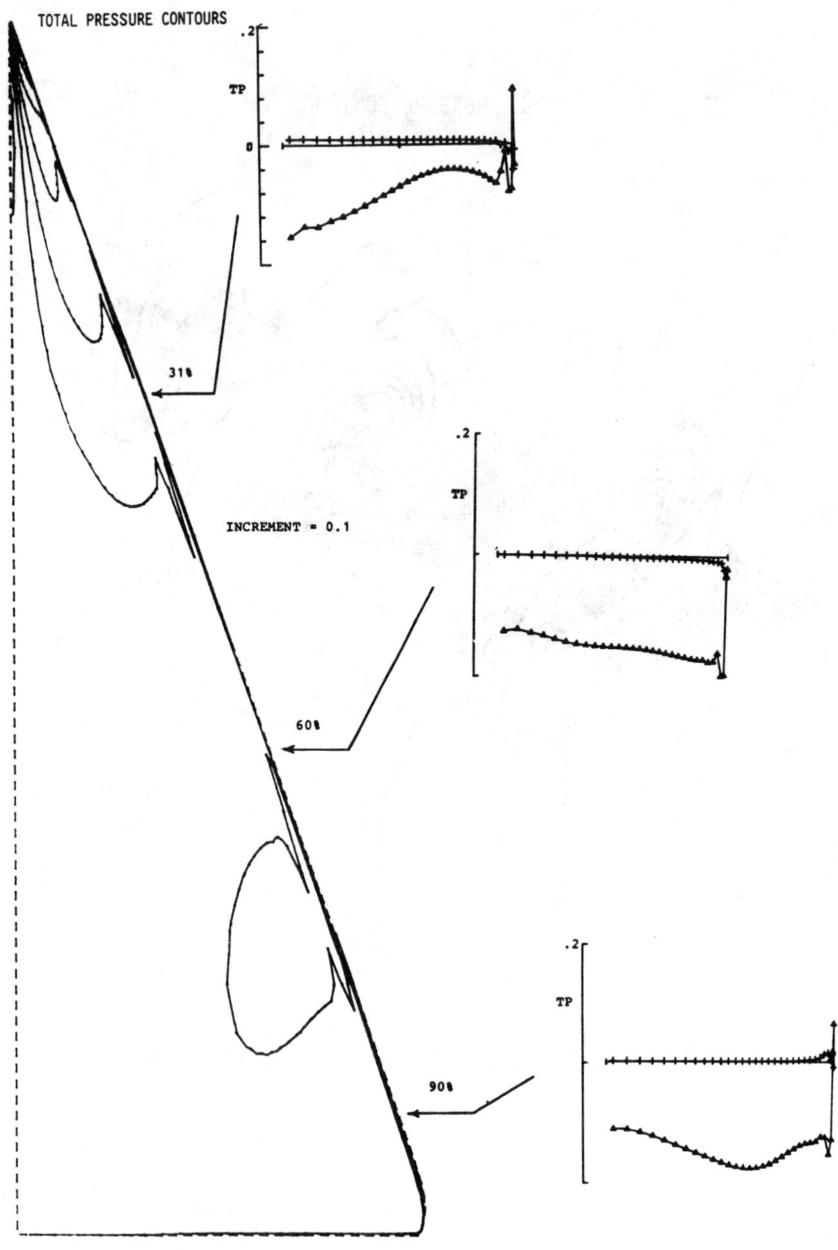

Fig. 13 b) on the upper surface of the wing.

REFERENCES

[1] A.J. CHORIN, A Numerical Method for Solving Incompressible Viscous Flow Problems, J. Comp. Phys., 2, pp. 12-26, 1967.

[2] R. PEYRET and T. TAYLER, Computational Methods for Fluid Flow, Springer, New York, 1982.

[3] L.-E., ERIKSSON, A. RIZZI and J.P. THERRE, Solving the Steady Incompressible Euler Equations to Model Inviscid Rotational Flow, AIAA J., in preparation.

[4] A. RIZZI, Damped Euler-Equation Method to Compute Transonic Flow Around Wing-Body Combinations, AIAA J., Vol. 20, pp. 1321-1328, Oct. 1982.

[5] A.W. RIZZI, Numerical Implementation of Solid-Body Boundary Conditions for the Euler Equations, ZAMM, Vol. 58, pp. T301-T304, 1978.

[6] B. ENGQVIST and A. MAJDA, Absorbing Boundary Conditions for the Numerical Simulation of Waves, Math, Comp., Vol. 31, pp. 629-651, July 1977.

[7] L.-E. ERIKSSON and A. RIZZI, Analysis by Computer of the Convergence to Steady State of a Discrete Approximation to the Euler Equations, AIAA Paper No. 83-1951 presented at the 6th CFD Conf., Danvers, MA, 1983.

[8] L.-E. ERIKSSON, Calculation of Two-Dimensional Potential Flow Wall Interference for Multicomponent Airfoils in Closed Low Speed Wind Tunnels, FFA TN AU-1116, Part 1, Stockholm 1975.

[9] L.-E. ERIKSSON, Generation of Boundary-Conforming Grids Around Wing-Body Configurations Using Transfinite Interpolation, AIAA J., Vol. 20, pp. 1313-1320, Oct. 1982.

[10] H.W.M. HOEIJMAKERS, W. VAATSTRA and N.G. VERGAAGEB, Vortex Flow Over Delta and Double-Delta Wings, J. Aircraft, Vol. 21, No. 9, Sept. 1983.

[11] H.W.M. HOEIJMAKERS, Private communication.

[12] D.J.,MARSDEN, R.W.SIMPSON and W.J. RAINBIRD, The Flow Over Delta Wings at Low Speeds with Leading Edge Separation, CoA Report, No. 114, Cranfield, 1958.

A SUB-DOMAIN APPROACH FOR THE COMPUTATION OF COMPRESSIBLE INVISCID FLOWS

JEAN-PIERRE VEUILLOT AND LAURENT CAMBIER*

Abstract. A sub-domain approach for the computation of inviscid flows is proposed within the framework of pseudo-unsteady methods. A general matching technique of two perfect fluid sub-domains is presented in the case where the systems of equations solved in each sub-domain are hyperbolic with respect to time. This technique is based on the interpretation of selected compatibility relations as transport equations. The proposed method provides a solution to the problem of fitting discontinuities (shocks and wakes). Numerical applications are presented for realistic 2D problems.

1. Introduction. An attractive way to cope with the more and more complex problems considered in Computational Fluid Dynamics seems to be the sub-domain (or multi-domain) approach. In such an approach, the computation of the flow in a domain $\mathcal{D}$ is split up into several simpler coupled problems defined in sub-domains $\mathcal{D}_1, \mathcal{D}_2, \ldots\ldots\ldots\ldots$, the union of which is the domain $\mathcal{D}$. This decomposition may be motivated by one or several of the following points :

- The mesh generation can be made easier either for representing complex geometries or for refining the mesh only in limited regions of the field where important flow gradients require such a refinement. Thus for a given accuracy, a better adaptation of the mesh allows one to save on computation cost.

- In each sub-domain the mathematical model can be simplified by some physical assumptions valid in the corresponding flow region (e.g. negligible viscous effects, irrotational flow, ...). The numerical method can also differ from one sub-domain to an other one.

- The sub-domain approach offers a natural way to treat the problem of discontinuity fitting.

*Office National d'Etudes et de Recherches Aérospatiales, BP 72, 92322 Châtillon Cedex, France.

- Finally, the decomposition into sub-domains can be motivated by the computer architecture, for instance, by the use of multi-processor computers.

2. Matching technique based on compatibility relations

We describe here a technique which allows the matching of the solutions defined in two neighbouring sub-domains $\mathcal{D}_1$ and $\mathcal{D}_2$ divided by an internal boundary Σ. This technique is based on the properties of t-hyperbolic systems and is an extension of the treatment of external boundaries proposed in [1]. The case of a general t-hyperbolic system is considered in [2]. In the following, the matching technique is detailed in 2D for the classical pseudo-unsteady system which is derived from the Euler equations by replacing the unsteady energy equation by the steady Bernoulli relation. Henceforth this system will be referred to as the H-system denoting that the total enthalpy H is constant. The interpretation of compatibility relations as transport equations is the basis of our method, and we first recall these compatibility relations for the H-system.

2.1. H-system compatibility relations

The H-system is made up of the two unsteady equations for the density ρ and the momentum $\vec{Q} = \rho\vec{V}$, and can be written in the conservative form :

$$(1)\qquad \begin{cases} \dfrac{\partial \rho}{\partial t} + \operatorname{div} \rho\vec{V} = 0 \\[2ex] \dfrac{\partial \vec{Q}}{\partial t} + \operatorname{div}\left(\vec{Q}\otimes\vec{V} + p\,\bar{\bar{1}}\right) = 0 . \end{cases}$$

The static pressure p is derived from the steady Bernoulli relation. For an ideal gas with constant specific heat ratio γ , p is given by :

$$(2)\qquad p = \frac{\gamma-1}{\gamma}\,\rho\left(H - \frac{V^2}{2}\right) .$$

The H-system has been proved to be t-hyperbolic (see for instance [1]). Consequently, considering plane or axisymmetric flows, for any unit vector ξ of the plane there exist three independent linear combinations of system (1) which can be written in the following remarkable form :

$$(3)\qquad \alpha_0^{(i)}\left(\frac{\partial \rho}{\partial t} + \lambda^{(i)}\frac{\partial \rho}{\partial \xi}\right) + \vec{\alpha}^{(i)}\cdot\left(\frac{\partial \vec{Q}}{\partial t} + \lambda^{(i)}\frac{\partial \vec{Q}}{\partial \xi}\right) = RHS^{(i)} .$$

Each of these combinations, called compatibility relations, is obtained by adding the continuity equation multiplied by the scalar $\alpha_0^{(i)}$ to the momentum equation multiplied by the vector $\vec{\alpha}^{(i)}$. The eigenvalues $\lambda^{(i)}$ arranged in increasing order, are given by :

(4) $$\lambda^{(1)} = \vec{V}.\vec{\xi} - C^+ , \quad \lambda^{(2)} = \vec{V}.\vec{\xi} , \quad \lambda^{(3)} = \vec{V}.\vec{\xi} - C^- ,$$

and the coefficients of the compatibility relations (3) are given by :

(5) $$\begin{cases} \alpha_0^{(1)} = \dfrac{a^2}{\gamma} - \vec{V}.\vec{\alpha}^{(1)} & \vec{\alpha}^{(1)} = C^-\vec{\xi} - \dfrac{\gamma-1}{\gamma}\vec{V} \\ \alpha_0^{(2)} = \vec{V}.\vec{\eta} & \vec{\alpha}^{(2)} = -\vec{\eta} \\ \alpha_0^{(3)} = \dfrac{a^2}{\gamma} - \vec{V}.\vec{\alpha}^{(3)} & \vec{\alpha}^{(3)} = C^+\vec{\xi} - \dfrac{\gamma-1}{\gamma}\vec{V} , \end{cases}$$

$\vec{\eta}$ being a unit vector normal to $\vec{\xi}$ ($\vec{\eta}.\vec{\xi} = 0$).

In these expressions C^+ and C^- are two characteristic speeds which play the same role as the algebraïc speeds of sound $+a$ and $-a$ in the original unsteady Euler equations. C^+ and C^- can be written :

(6) $$C^{\pm} = \frac{\gamma-1}{2\gamma}\vec{V}.\vec{\xi} \pm \sqrt{\frac{a^2}{\gamma} + \left(\frac{\gamma-1}{2\gamma}\vec{V}.\vec{\xi}\right)^2} .$$

Note that C^+ and C^- tend to $+a$ and $-a$ when γ tends to unity.

In the compatibility relations (3) $\partial/\partial\xi = \xi_i \partial/\partial x_i$ represents the derivation along the direction defined by the unit vector $\vec{\xi}$, and the right-hand sides ($RHS^{(i)}$) only involve derivatives along the vector $\vec{\eta}$ normal to $\vec{\xi}$. If these derivatives are known at time t_0, the compatibility relations may then be considered as transport equations along the characteristic lines of slope $1/\lambda^{(i)}$ in the plane (ξ , t) and enable us to determine the solution at time $t_1 > t_0$. This interpretation is the basis of the general technique of boundary condition treatment described in [1] and the matching technique detailed in 2.2 is an extension of this technique.

If the non conservative variables p and $\vec{V}$ are chosen instead of the conservative variables ρ and $\vec{Q} = \rho\vec{V}$, then the compatibility relations (3) may be written :

(7) $$\beta_0 \left(\frac{\partial p}{\partial t} + \lambda^{(i)} \frac{\partial p}{\partial \xi} \right) + \vec{\beta}^{(i)} . \left(\frac{\partial \vec{V}}{\partial t} + \lambda^{(i)} \frac{\partial \vec{V}}{\partial \xi} \right) = RHS^{(i)} ,$$

where the coefficients $\beta_0^{(i)}$ and $\vec{\beta}^{(i)}$ (i = 1 to 3) have a much simpler form than the coefficients $\alpha_0^{(i)}$ and $\vec{\alpha}^{(i)}$ given by (5).

$$(8) \quad \begin{cases} \beta_0^{(1)} = 1 & \vec{\beta}^{(1)} = \rho C^- \vec{\xi} \\ \beta_0^{(2)} = 0 & \vec{\beta}^{(2)} = \vec{\eta} \\ \beta_0^{(3)} = 1 & \vec{\beta}^{(3)} = \rho C^+ \vec{\xi} . \end{cases}$$

The previous results obtained in 2D can easily be extended to 3D by considering that the eigenvalue $\lambda^{(2)} = \vec{V}.\vec{\xi}$ is double and by replacing in (5) and (8) the vector $\vec{\eta}$ by two independent unit vectors, $\vec{\eta}_1$ and $\vec{\eta}_2$, normal to $\vec{\xi}$.

2.2. Matching of two adjacent sub-domains

Let P be a point of an interface Σ between two sub-domains $\mathcal{D}_1$ and $\mathcal{D}_2$ of the computational domain $\mathcal{D}$. The solutions respectively defined in $\mathcal{D}_1$ and $\mathcal{D}_2$ may be either continuous or discontinuous across the interface Σ which is possibly moving. The normal velocity of Σ is denoted by $W = \vec{W}.\vec{\nu}$ where $\vec{\nu}$ is the unit normal to Σ at P, pointing for instance from $\mathcal{D}_1$ into $\mathcal{D}_2$.

At the point P the variables ρ and $\vec{Q}$ are double-valued : ρ_1 and $\vec{Q}_1$ when P is considered as belonging to $\mathcal{D}_1$, ρ_2 and $\vec{Q}_2$ when P is considered as belonging to $\mathcal{D}_2$. These two sets of unknowns and possibly the normal velocity W are derived from a system consisting of selected compatibility relations and additional relations which will be called matching relations.

The choice of the compatibility relations which must be used is made in the following way :

- we consider the compatibility relations associated with the unit normal $\vec{\nu}$, which implies that the unit vector $\vec{\xi}$ must be replaced by $\vec{\nu}$ in (3) to (8) ;

- in $\mathcal{D}_1$, we retain only the compatibility relations which transport information from $\mathcal{D}_1$ into $\mathcal{D}_2$, that is to say, the relations (3) corresponding to the eigenvalues $\lambda_1^{(i)}$ which satisfy the inequality,

$$(9) \quad \lambda_1^{(i)} \geq W \; ;$$

- in the same way we retain in $\mathcal{D}_2$ only the compatibility relations which transport information from $\mathcal{D}_2$ into $\mathcal{D}_1$. These relations correspond to the eigenvalues $\lambda_2^{(i)}$ such that,

$$(10) \quad \lambda_2^{(i)} \leq W .$$

The configuration of the characteristic lines at the point P is sketched in Figure 1 : the path of P in the plane (η, t) is represented by a dashed line the slope of which is equal to $1/W$, and the compatibility relations retained in $\mathcal{D}_1$ (resp. in $\mathcal{D}_2$) correspond to the characteristic lines of type (a) (resp. type (b)).

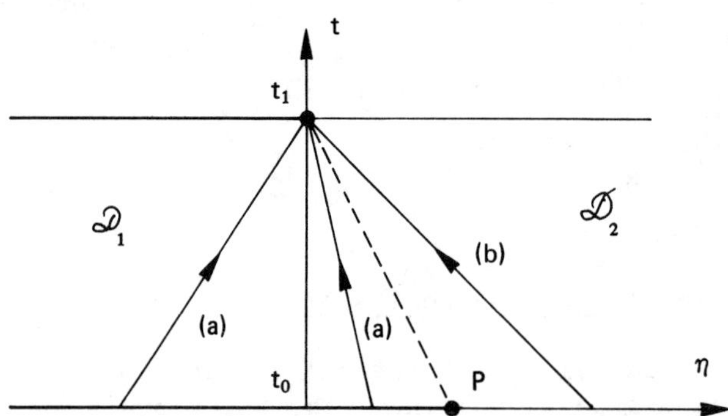

Figure 1. Characteristic lines configuration.

Note that in the inequalities (9) and (10) the border-line case ($\lambda^{(i)} = W$) corresponds to a transport along the interface Σ.

When the interface Σ is fixed ($W = 0$), the position of the eigenvalues with respect to W (i.e. the determination of their sign) depends only on the position of the algebraïc normal Mach number $M_\nu = \vec{V}.\vec{\nu}/a$ with respect to -1, 0 and +1. This result is similar to the one obtained with the full unsteady Euler equations.

In the case of the full Euler equations, the position of the eigenvalues with respect to W for a moving interface ($W \neq 0$), is determined in the same way by considering the position of the relative normal Mach number $M'_\nu = (\vec{V}.\vec{\nu} - W)/a$ with respect to -1, 0 and + 1. On the other hand, in the case of the H-system, the discussion is more complicated and M'_ν must be compared with C^-/a, 0 and C^+/a, $C^\pm/a$ depending on the absolute normal Mach number :

$$\frac{C^\pm}{a} = \frac{\gamma-1}{2\gamma} M_\nu \pm \sqrt{\frac{1}{\gamma} + \left(\frac{\gamma-1}{2\gamma} M_\nu\right)^2} \; . \tag{11}$$

In the sub-domain $\mathcal{D}_1$ the position of the relative Mach number M'_ν with respect to C^-/a, 0 and C^+/a is illustrated in Figure 2 : the two curves $M'_\nu = C^\pm/a$ have been plotted in a diagram (M_ν, M'_ν) and, with the axis $M'_\nu = 0$, they bound four areas (① to ④) in which the number NC_1 of the compatibility relations to be retained in $\mathcal{D}_1$ is given.

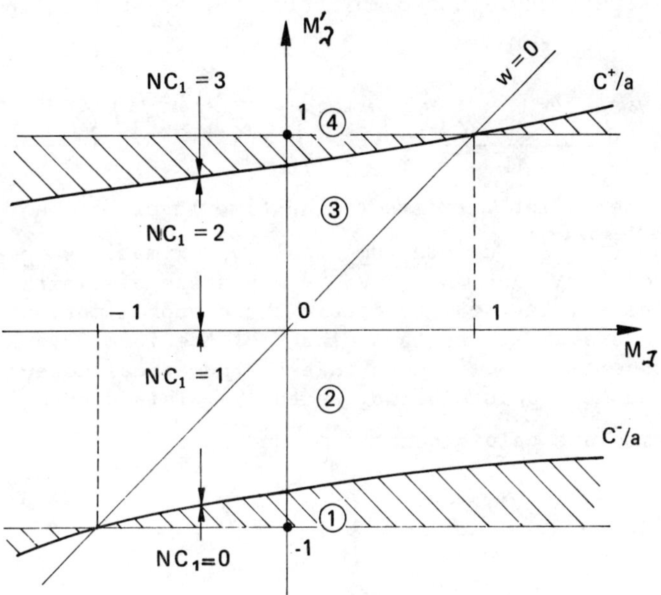

Figure 2. Number NC_1 of the compatibility relations to be retained in $\mathcal{D}_1$.

Due to the choice of the normal orientation ($\vec{\nu}$ pointed from $\mathcal{D}_1$ into $\mathcal{D}_2$), the number NC_2 of the compatibility relations to be retained in $\mathcal{D}_2$ varies from $NC_2 = 3$ in area ① to $NC_2 = 0$ in area ④ . Note that for an interface Σ across which the solution is discontinuous, the two points ($M_{\nu 1}$, $M'_{\nu 1}$) and ($M_{\nu 2}$, $M'_{\nu 2}$) do not necessarily belong to the same area.

In order to make clear the foregoing, different types of interfaces will be examined in sections 2.4. to 2.6.. But, first, a description of the discretization of a compatibility relation will be given.

2.3. Discretization of a compatibility relation

In practice the discretization of a compatibility relation, for instance in the sub-domain $\mathcal{D}_1$, is easily achieved in the computational mesh starting from the linear combination of the conservative system (1) equivalent to (3) :

$$\alpha_0^{(i)}\left(\frac{\partial \rho}{\partial t} + \operatorname{div} \rho\vec{V}\right) + \vec{\alpha}^{(i)} \cdot \left(\frac{\partial \vec{Q}}{\partial t} + \operatorname{div}\left[\vec{Q}\otimes\vec{V} + p\,\bar{\bar{1}}\right]\right) = 0 . \tag{12}$$

The sub-domain subscript 1 will be omitted in this section for conciseness sake.

After discretizing, this combination may be written in the following compact form :

$$(13)\quad \alpha_0^{(i)\,n+1/2}\left(\frac{\rho^{n+1}-\rho^{n}}{\Delta t}+A^{n+1/2}\right)+\vec{\alpha}^{(i)\,n+1/2}\cdot\left(\frac{\vec{Q}^{n+1}-\vec{Q}^{n}}{\Delta t}+\vec{B}^{n+1/2}\right)=0,$$

where the superscript n refers to the time level $t^n = n\Delta t$ and where $A^{n+1/2}$ and $\vec{B}^{n+1/2}$ denote numerical approximations of the space derivatives $\operatorname{div}(\rho\vec{V})$ and $\operatorname{div}(\vec{Q}\otimes\vec{V}+p\,\bar{\bar{1}})$ by means of finite difference formulae. The simplest way to obtain these approximations consists in discretizing system (1) at the point P of the interface Σ (the basic numerical scheme is modified so that the numerical dependence domain only contains mesh points belonging to $\mathcal{D}_1$). This discretized system gives intermediate values denoted ρ^* and $\vec{Q}^*$:

$$(14)\quad \left\{\begin{array}{l} \dfrac{\rho^*-\rho^n}{\Delta t}+\left\{\operatorname{div}\rho\vec{V}\right\}^{n+1/2}=0 \\[2ex] \dfrac{\vec{Q}^*-\vec{Q}^n}{\Delta t}+\left\{\operatorname{div}\left(\vec{Q}\otimes\vec{V}+p\,\bar{\bar{1}}\right)\right\}^{n+1/2}=0, \end{array}\right.$$

and the approximations $A^{n+1/2}$ and $\vec{B}^{n+1/2}$ in (13) are defined by the discretized terms $\left\{\operatorname{div}\rho\vec{V}\right\}^{n+1/2}$ and $\left\{\operatorname{div}\left(\vec{Q}\otimes\vec{V}+p\,\bar{\bar{1}}\right)\right\}^{n+1/2}$.

Then we obtain the very simple relation :

$$(15)\quad \alpha_0^{(i)\,n}\left(\rho^{n+1}-\rho^*\right)+\vec{\alpha}^{(i)\,n}\cdot\left(\vec{Q}^{n+1}-\vec{Q}^*\right)=0.$$

This relation has been linearized by replacing $\alpha_0^{(i)\,n+1/2}$ and $\vec{\alpha}^{(i)\,n+1/2}$ by $\alpha_0^{(i)\,n}$ and $\vec{\alpha}^{(i)\,n}$. The final values ρ^{n+1} and $\vec{Q}^{n+1}$ are different from the intermediate values ρ^* and $\vec{Q}^*$ except in the special case where the three compatibility relations defined in $\mathcal{D}_1$ must be retained.

The relation (15) can be transformed in terms of the non conservative variables p and $\vec{V}$:

$$(16)\quad \beta_0^{(i)\,n}\left(p^{n+1}-p^*\right)+\vec{\beta}^{(i)\,n}\cdot\left(\vec{V}^{n+1}-\vec{V}^*\right)=0.$$

In the following we will use these last relations (16) which are simpler than the relations (15).

2.4. Artificial interface

In this case, the solution $(\rho,\vec{Q})$ is continuous across the interface Σ and the matching relations simply express the continuity of ρ and $\vec{Q}$. Then it stands to reason that the three compatibility relations (3) written in $\mathcal{D}_1$ are identical with those written in $\mathcal{D}_2$. Considering one of these three relations the following cases can occur :

- the relation expresses a transport from $\mathcal{D}_1$ into $\mathcal{D}_2$ and must be discretized in $\mathcal{D}_1$; therefore the terms p^* and $\vec{V}^*$ in (16) are evaluated in $\mathcal{D}_1$;

- the relation expresses a transport in the opposite direction and must be discretized in $\mathcal{D}_2$ (p^* and $\vec{V}^*$ evaluated in $\mathcal{D}_2$) ;

- the relation expresses a transport along Σ (border-line case) and the choice of the sub-domain in which it must be discretized does not matter.

As an example, suppose that the fluid goes through a fixed interface Σ from $\mathcal{D}_1$ to $\mathcal{D}_2$ with a subsonic normal velocity ($0 < M_\nu < 1$). On this assumption the two compatibility relations corresponding to the positive eigenvalues $\lambda^{(2)}$ and $\lambda^{(3)}$ must be retained in $\mathcal{D}_1$ and the compatibility relation corresponding to the negative eigenvalue $\lambda^{(1)}$ must be retained in $\mathcal{D}_2$. Taking into account the matching relations which express the continuity across Σ, it is not necessary to specify the sub-domain subscript in the left-hand sides of the discretized compatibility relations (16) which are then written :

$$
(17) \qquad \begin{cases} p^{n+1} + (\rho c^-)^n V_\nu^{n+1} = \left[p^* + (\rho c^-)^n V_\nu^*\right]_2 \\ V_\eta^{n+1} = \left[V_\eta^*\right]_1 \\ p^{n+1} + (\rho c^+)^n V_\nu^{n+1} = \left[p^* + (\rho c^+)^n V_\nu^*\right]_1 . \end{cases}
$$

We recall that the subscript 1 or 2 specifies the sub-domain ($\mathcal{D}_1$ or $\mathcal{D}_2$) in which the right-hand sides are evaluated. Solving system (17) is always possible without any difficulty, and the values of the conservative variables (ρ^{n+1} and $\vec{Q}^{n+1}$) are computed from p^{n+1}, V_ν^{n+1} and V_η^{n+1}.

To conclude, the case of a point located on an artificial interface differs from the case of an ordinary interior point by the fact that the discretization of the compatibility relations leads to a numerical scheme upwinded along $\vec{\nu}$ according to the speeds of wave propagation. In the 1D case, the extension of this treatment to every mesh point may be related to the scheme proposed by Courant, Isaacson and Rees [3].

2.5. Shock

We assume now that Σ is a shock, that is to say a discontinuity line associated with a weak solution of the H-system (1) and fulfilling the "entropy condition" of Lax [4]. Suppose for instance that the fluid goes across Σ from $\mathcal{D}_1$ to $\mathcal{D}_2$ ($V_{\nu 1} - W > 0$ and $V_{\nu 2} - W > 0$), then this "entropy condition" may be expressed by the following inequalities :

$$
(18) \qquad \left\{ \begin{array}{c} W < \lambda_1^{(1)} \\ \lambda_2^{(1)} < W < \lambda_2^{(2)} \end{array} \right. .
$$

The first inequality implies that all the three compatibility relations must be retained in the usptream sub-domain $\mathcal{D}_1$, whereas the second inequality implies that only the compatibility relation associated with $\lambda_2^{(1)}$ must be retained in the downstream sub-domain $\mathcal{D}_2$. In the diagram (M_ν, M'_ν) (Figure 2), the points representing the upstream and the downstream conditions respectively belong to the areas (4) and (3) (the areas (1) and (2) would correspond to a shock crossed by the flow from $\mathcal{D}_2$ to $\mathcal{D}_1$). In the hatched areas the H-system exhibits a peculiar behaviour in comparison with the Euler system : shocks corresponding to these areas have either an upstream relative normal Mach number $M'_{\nu 1}$ less than unity, or a downstream relative Mach number $M'_{\nu 2}$ greater than unity. In order to determine the seven scalar unknowns ($p_1, V_{\nu 1}, V_{\eta 1}, p_2, V_{\nu 2}, V_{\eta 2}$ and W) the four compatibility relations retained in $\mathcal{D}_1$ and in $\mathcal{D}_2$ are completed by three matching relations which are the jump relations associated with the H-system. These matching relations may be written :

$$
(19) \qquad \left\{ \begin{array}{c} \rho_1 (V_{\nu 1} - W) = \rho_2 (V_{\nu 2} - W) \\ p_1 + \rho_1 V_{\nu 1} (V_{\nu 1} - W) = p_2 + \rho_2 V_{\nu 2} (V_{\nu 2} - W) \\ V_{\eta 1} = V_{\eta 2} \end{array} \right. .
$$

Due to the conservation of the absolute total enthalpy H, these jump relations differ from the Rankine-Hugoniot relations associated with the Euler system and which express the conservation of the relative total enthalpy $H' = \frac{\gamma - 1}{\gamma} \frac{p}{\rho} + \frac{(\vec{V} - \vec{W})^2}{2}$. This discrepancy disappears at steady state.

It can be proved that the "entropy condition" (18) added to the jump relations (19) implies that the density and the static pressure increase and that the normal velocity decreases when the flow goes across the shock Σ. For these quantities the H-system behaves as the Euler system. However, for unsteady non physical H-shocks, it may happen that the entropy jump be negative.

Numerically the system composed of the four selected compatibility relations and of the three jump relations (19) is solved in the following way. Since all the compatibility relations are discretized in $\mathcal{D}_1$, the values p_1^*, $V_{\nu 1}^*$ and $V_{\eta 1}^*$ given by the numerical scheme are kept without any change ($p_1^{n+1} = p_1^*$, $V_{\nu 1}^{n+1} = V_{\nu 1}^*$ and $V_{\eta 1}^{n+1} = V_{\eta 1}^*$) according to the discretized compatibility relations (16). The remaining unknowns p_2^{n+1}, $V_{\nu 2}^{n+1}$, $V_{\eta 2}^{n+1}$ and W are derived from a non-linear system consisting of the discretized compatibility relation retained in $\mathcal{D}_2$ and of the jump relations (19).

2.6. Contact discontinuity

The weak solutions of the H-system also include contact discontinuities. Then Σ is a discontinuity line which the flow does not cross. The normal relative mass flow is zero and we have :

$$V_{\nu 1} = V_{\nu 2} = W \,. \tag{20}$$

In this case, the two compatibility relations associated with the eigenvalues $\lambda_1^{(2)}$ and $\lambda_1^{(3)}$ (resp. $\lambda_2^{(1)}$ and $\lambda_2^{(2)}$) which satisfy the inequality (9) (resp. (10)) must be retained in $\mathcal{D}_1$ (resp. $\mathcal{D}_2$). The seven unknowns ($p_1, V_{\nu 1}, V_{\eta 1}, p_2, V_{\nu 2}, V_{\eta 2}$ and W) are derived from a system consisting of these four compatibility relations, of the two relations (20) and of one matching relation which expresses the continuity of the static pressure. The tangential components of the velocity ($V_{\eta 1}$ and $V_{\eta 2}$) are given by the numerical scheme according to the discretized compatibility relations (16) associated with the eigenvalues $\lambda_1^{(2)} = V_{\nu 1}$ and $\lambda_2^{(2)} = V_{\nu 2}$. The common values $p = p_1 = p_2$ and $V_\nu = V_{\nu 1} = V_{\nu 2}$ of the static pressure and of the normal velocity are derived from the two other discretized compatibility relations :

$$\begin{cases} p^{n+1} + (\rho c^+)_1^n V_\nu^{n+1} = p_1^* + (\rho c^+)_1^n V_{\nu 1}^* \\ p^{n+1} + (\rho c^-)_2^n V_\nu^{n+1} = p_2^* + (\rho c^-)_2^n V_{\nu 2}^* \,. \end{cases} \tag{21}$$

Finally the normal velocity of the discontinuity Σ is directly given by (20).

Although the H-system is based on the assumption that the total enthalpy H is constant, it can be noticed that the value H_1 of H in $\mathcal{D}_1$

may differ from the value H_2 of H in $\mathcal{D}_2$. Consequently, the proposed method allows for instance the determination of the slip line between two flows with different stagnation conditions (pressure and temperature).

3. Numerical method

In this section, we briefly present the main features of the numerical method which has been used in the various applications described in section 4., focusing on the points which are directly related to the sub-domain approach. The pseudo-unsteady system is solved by means of an explicit predictor-corrector scheme. The equations are discretized in conservative form using MacCormack's finite difference scheme directly in the physical plane on an arbitrary curvilinear mesh. This discretization method and the numerical treatment of the external boundaries by compatibility relations are described in [1]. In order to save on computation cost the well-known local time step technique has been implemented.

In general the meshes in two neighbouring sub-domains $\mathcal{D}_1$ and $\mathcal{D}_2$ do not coïncide along their interface Σ. This can lead us to discretize a compatibility relation at a point P of Σ which is a mesh point of only one of the two sub-domains. To be more explicit we return to the example of section 2.4. : P is a point of an artificial interface Σ between $\mathcal{D}_1$ and $\mathcal{D}_2$, and we suppose now that P is a mesh point of $\mathcal{D}_1$ which does not coïncide with a mesh point of $\mathcal{D}_2$ (see Figure 3).

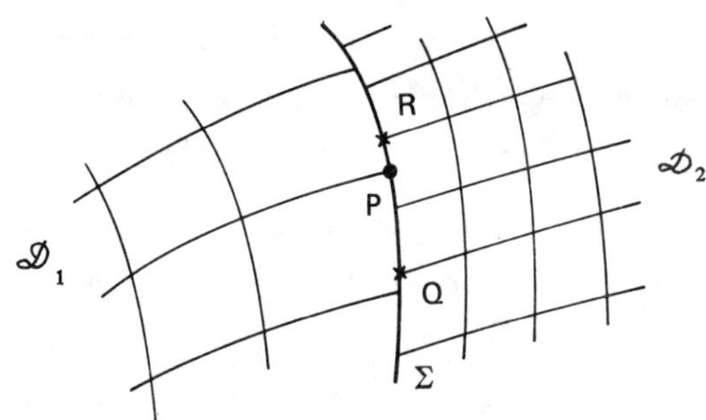

Figure 3. Mesh points along the interface Σ.

When the right-hand sides of the discretized compatibility relations must be evaluated in $\mathcal{D}_2$ - for instance, the first equation of (17) - the intermediate values (p^*, V_ν^* and V_η^*) at the point P_* are derived from a linear interpolation of the values (p^*, V_ν^* and V_η^*) at the neighbouring mesh points Q and R of $\mathcal{D}_2$.

Contrary to some authors who fit discontinuities in a fixed mesh, we consider discontinuities as moving mesh lines. The displacement velocity $\dot{P}$ of a point P of a moving discontinuity Σ is related to the normal velocity W through :

$$\dot{\vec{P}} \cdot \vec{\nu} = W, \tag{22}$$

but the component of $\dot{\vec{P}}$ tangent to Σ can be given arbitrarily. In general it is convenient to assume that P moves along a curve which may be either a mesh line or a coordinate line. For instance, when fitting a slip line Σ with $\dot{\vec{P}}$ assumed to be parallel to the y-axis, the ordinate $y(x,t)$ of the point P of Σ is given by integrating the equation :

$$\frac{\partial y}{\partial t} + u \frac{\partial y}{\partial x} - v = 0, \tag{23}$$

where (u, v) are the cartesian components of the fluid velocity. This partial derivative equation is directly derived from the relation (20) giving W and must be properly discretized in order to insure stability.

Because of the displacement of the discontinuities, the discretization of the pseudo-unsteady system must be carried out in a moving mesh. To this end, the numerical scheme could be applied in a moving mesh, like in [1], by modifying the time derivatives in system (1) : $\partial\rho/\partial t$, for instance is replaced by $\partial\rho/\partial\tau - \text{grad}\rho \cdot \dot{\vec{M}}$, where $\partial/\partial\tau$ represents the time derivative for a given mesh point M of velocity $\dot{\vec{M}}$. We use here another technique. The quantities (ρ^{n+1}, $\vec{Q}^{n+1}$) are computed in a fixed mesh defined at time t^n , and are projected upon a new mesh defined at time t^{n+1}. If M^n and M^{n+1} denote the position of a mesh point respectively at time t^n and at time t^{n+1} , then the value of a quantity ϕ computed in the t^n-mesh is obtained from a first order Taylor's expansion :

$$\phi(M^{n+1}) = \phi(M^n) + \Delta t \, \text{grad}\,\phi \cdot \dot{\vec{M}} \; . \tag{24}$$

4. Applications

The present sub-domain approach has been implemented for various numerical applications. In this paper the presentation of applications is limited to some demonstrative numerical results obtained by solving the H-system for 2D inviscid compressible flows around airfoils. But we first recall some other applications which have already been presented. Calculations of transonic flows in channels are presented in [2] and in [5], showing some examples of the shock fitting technique. The numerical simulation of a shock wave-turbulent boundary layer interaction is detailed in [2]. In this high Reynolds number calculation the computational domain is divided into "inviscid" sub-domains where the Euler equations are solved and into "viscous" sub-domains where the averaged Navier-Stokes equations are solved. The interfaces between these sub-domains are located in inviscid regions in such a way that the matching technique for t-hyperbolic systems can be applied. Figure 4 shows the sub-domain decomposition and the computed iso-Mach number contours. Note that the shock is fitted in the "inviscid" part of the domain and captured in the "viscous" one. The shock fitting technique has also been applied to the problem of the interaction of an axisymmetric vortex with a shock normal to the streamwise axis of the vortex [6].

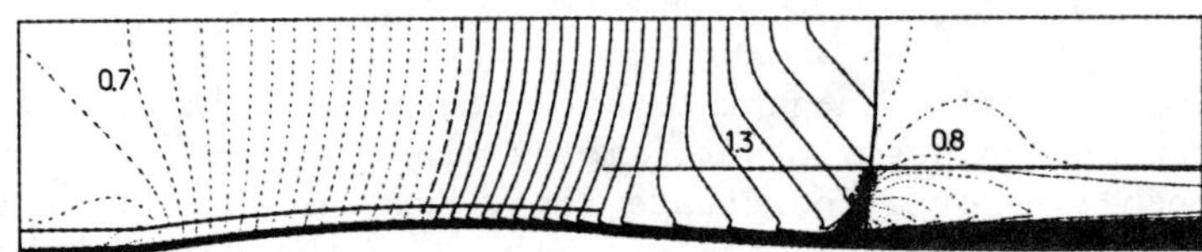

Figure 4. Shock-boundary-layer interaction. Iso-Mach number contours ($\Delta M = 0.02$).

The fitting technique of contact discontinuities is well suited to inviscid calculations of aircraft afterbodies. Due to the different stagnation conditions between the external flow, the primary jet and the secondary jet if it exists, the flow field exhibits strong discontinuities across the slip lines bounding the jets. References [7] and [9] contain examples of such calculations.

Calculations around airfoils

The sub-domain approach has been tested for the computation of the inviscid transonic flow around the NLR 7301 profile at the following free-stream conditions : $M_\infty = 0.721$ and $\alpha_\infty = -0.194°$. This application emphasizes two advantages of the decomposition of the computational domain into sub-domains : on the one hand, the use of meshes adapted to the flow gradients in order to obtain a better accuracy without increasing the computation cost, on the other hand, the fitting of the slip line issued from the trailing edge. This last feature allows an accurate implementation of the Kutta-Joukowski condition.

The computational domain is partitioned into six sub-domains the meshes of which are represented on Figure 5. First the C-type domain is divided into a "lower" region (sub-domains ①, ② and ⑤) and an "upper" region (sub-domains ③, ④ and ⑥) by means of a horizontal cut line issued from the leading edge and the fitted slip-line issued from the trailing edge. This decomposition allows one to reduce the discretization errors due to the use of a non-symmetrical numerical scheme (in this way, symmetry of the computed solution is exactly satisfied in the case of a symmetrical profile at zero angle of attack). Because of the displacement of the slip line which is an interface between the sub-domains ① and ④, the mesh points of these sub-domains move during the iterative process. In the sub-domains ② and ③ close to the leading edge, the meshes are strongly refined as can be seen on the enlargement shown on Figure 6. This local refinement allows a reduction of the entropy production due to the discretization errors around the leading edge where the flow exhibits a strong expansion. So the relative error on the wall stagnation pressure does not exceed 10^{-3}.

Figure 7 shows the computed iso-Mach number contours. In Figure 8 it can be seen that the strong discontinuity between the meshes does not damage the smoothness of the curves across the artificial inter-

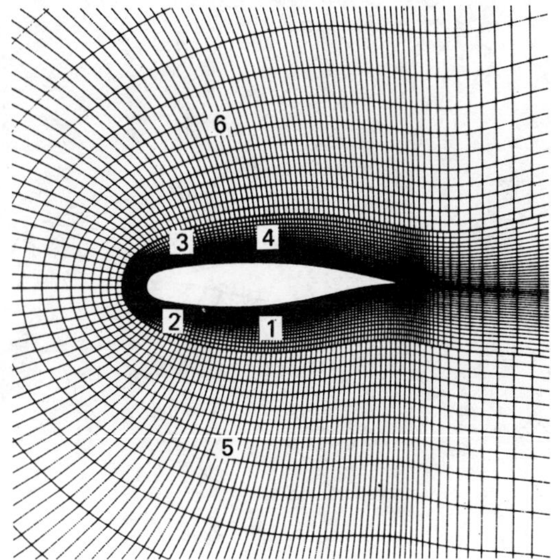

Figure 5. NLR 7301. Meshes of the six sub-domains.

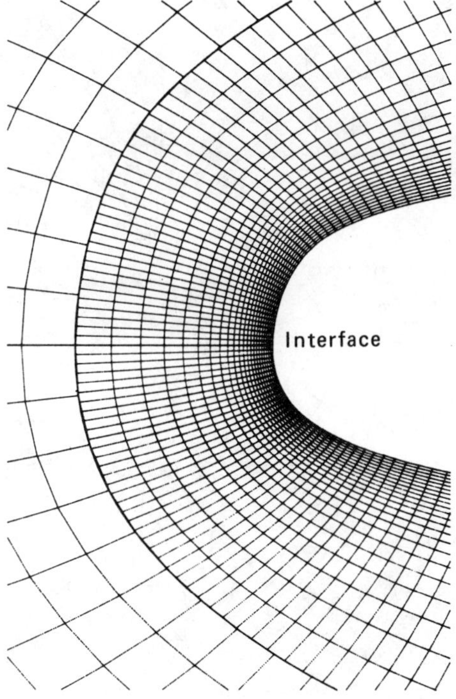

Figure 6. NLR 7301. Enlargement near the leading edge.

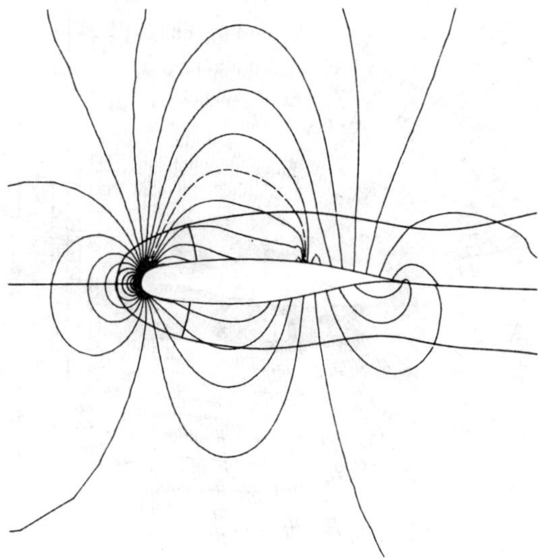

Figure 7. NLR 7301. Iso-Mach number contours, $M_\infty = 0.721, \alpha_\infty = -0.194°$.

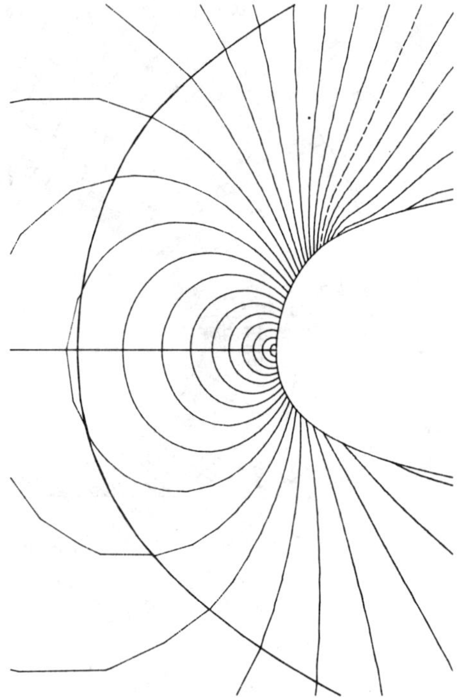

Figure 8. NLR 7301. Iso-Mach number contours near the leading edge.

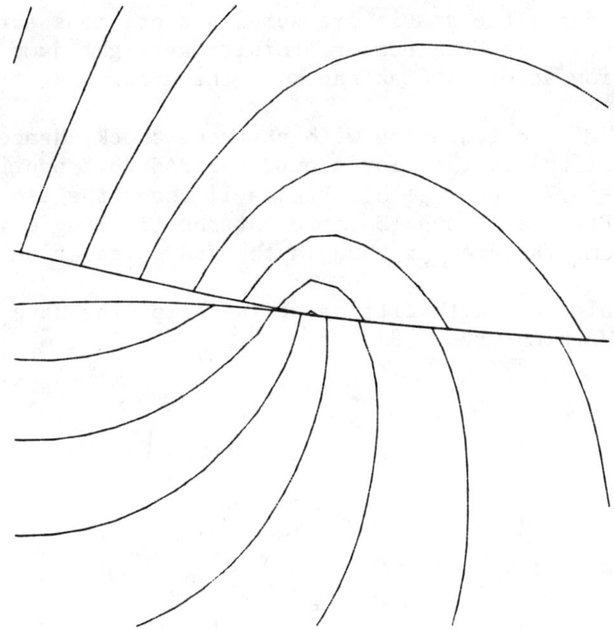

Figure 9. NLR 7301. Iso-Mach number contours near the trailing edge.

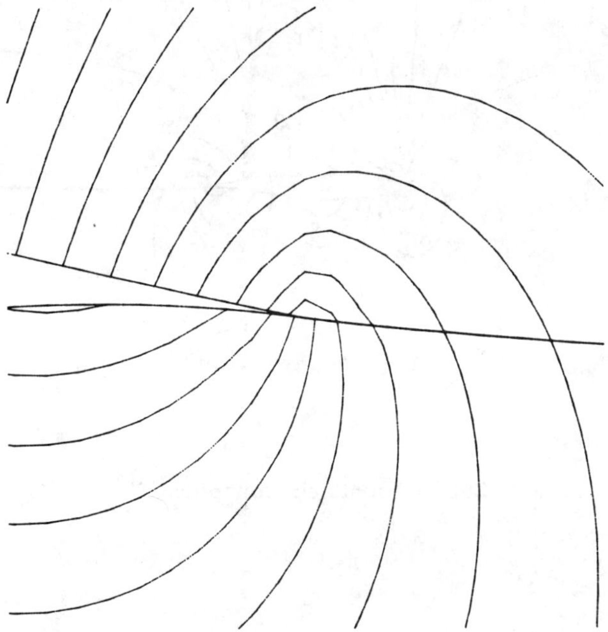

Figure 10. NLR 7301. Isobar contours near the trailing edge.

faces. Figures 9 and 10 represent the behaviour of the solution near the trailing edge : the static pressure is continuous across the slip line, in contrast the Mach number exhibits a slight jump due to the entropy production across a weak shock on the upper side of the profile.

An example of a calculation with a strong shock (hence with a great entropy production) is the flow around the RAE 2822 profile for the conditions $M_\infty = 0.75$ and $\alpha_\infty = 3°$. Figure 11 shows the iso-Mach number contours and Figures 12 and 13 represent the trailing edge flow with a Mach number jump far greater than in the NLR calculation.

Other calculations with fitting of the slip-line have been provided for an AGARD Working Group [8].

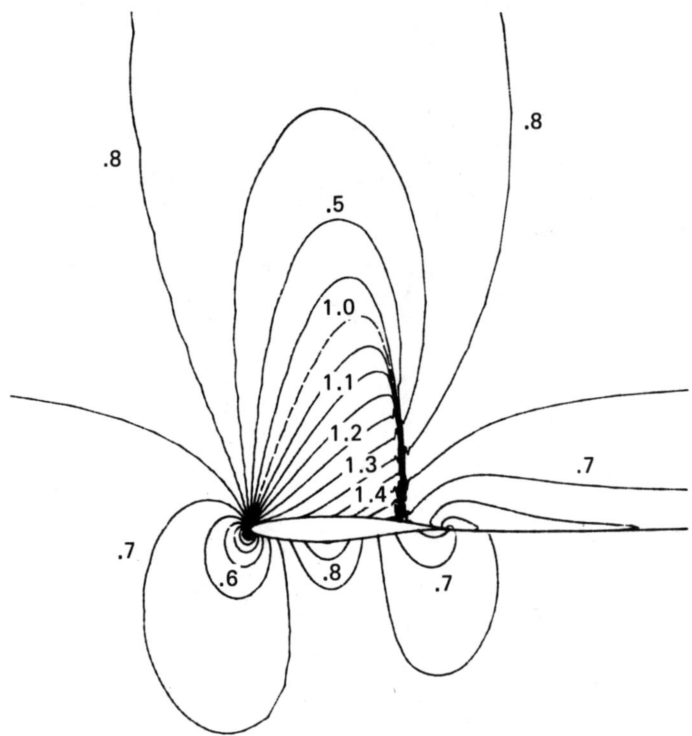

Figure 11. RAE 2822. Iso-Mach contours, $M_\infty = 0.75$, $\alpha_\infty = 3°$.

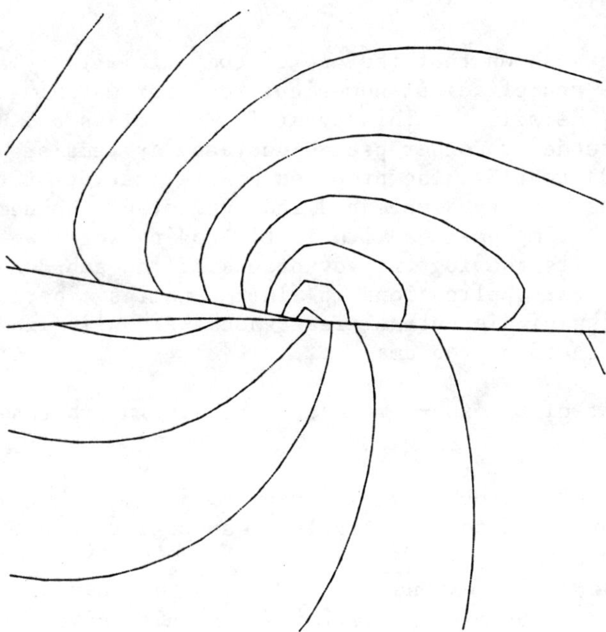

Figure 12. RAE 2822. Iso-Mach contours near the trailing edge.

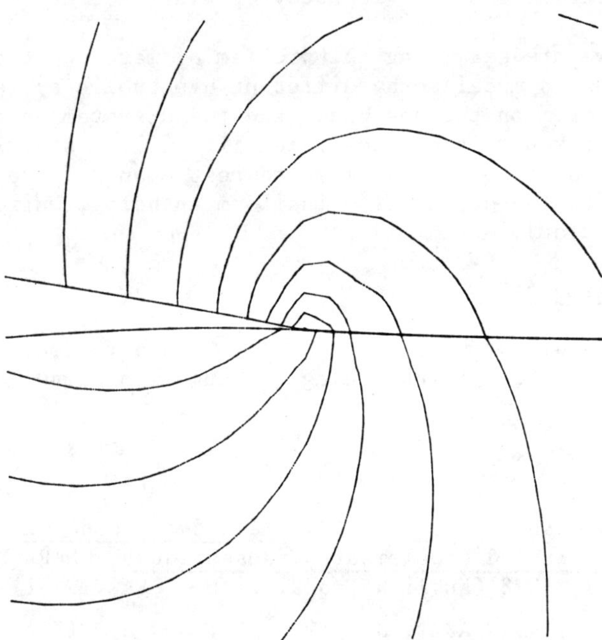

Figure 13. RAE 2822. Isobar contours near the trailing edge.

5. Conclusion

We have thus shown that the use of compatibility relations provides an attractive and efficient numerical tool for matching sub-domains. The technique detailed in this paper for the classical H-system can easily be extended to other pseudo-unsteady systems such as those presented in [10] to [12]. The proposed discretization of the compatibility relations is very simple and independent of the numerical scheme, and thus can be implemented with little coding work. We have obtained numerical results showing the advantages of the sub-domain approach for various practical applications in fluid dynamics : better mesh adaptation, fitting of discontinuities (shocks and slip-lines), viscous-inviscid interaction problems, ...

Development of the sub-domain approach is in progress in the following areas :

- the fitting of a shock wave in transonic flow on a profile should be achieved in spite of some difficulties encountered in the region where the shock vanishes ;
- work is underway to extend the method to the case of overlapping sub-domains, and some promising preliminary results have been obtained ;
- a 3D extension of the sub-domain approach does not raise theoretical problems and should allow one to tackle the computation of flows around complex geometries such as wing/body or wing/nacelle configurations.

Finally, we are examining the problem of matching two sub-domains in which the flow is modelled by different hyperbolic systems, such as the full Euler system on the one hand, and the H-system on the other hand. Care must be taken to avoid possible difficulties stemming from the fact that the hyperbolic systems used rest upon different simplifying assumptions for the steady flow (uniform enthalpy, uniform entropy, irrotational flow).

Acknowledgements

The authors wish to thank A.M. Vuillot and V. Couaillier for their efficient cooperation in obtaining the numerical results.

[1] H. VIVIAND and J. P. VEUILLOT, Méthodes Pseudo-Instationnaires pour le calcul d'Ecoulements Transsoniques, ONERA Publication N° 1978-4, 1978 (English Translation : ESA TT 561).

[2] L. CAMBIER, W. GHAZZI, J.P. VEUILLOT and H. VIVIAND, A multi-domain approach for the computation of viscous transonic flows by unsteady methods, Recent Advances in Numerical Methods in Fluids, Vol. 3, Pineridge Press, Swansea, 1984.

[3] R. COURANT, E. ISAACSON and M. REES, On the solution of non-linear hyperbolic differential equations by finite differences, Comm. Pure Appl. Math. 5 (1952), pp. 243-255.

[4] P.D. LAX, Hyperbolic systems of conservation laws II, Comm. Pure Appl. Math. 10 (1957), pp. 537-566.

[5] L. CAMBIER, W. GHAZZI, J.P. VEUILLOT and H. VIVIAND, Une Approche par Domaines pour le Calcul d'Ecoulements Compressibles, Computing Methods in Applied Sciences and Engineering V, North-Holland, Amsterdam, 1982, pp. 423-446.

[6] E. HOROWITZ, Contribution à l'Etude des Eclatements Tourbillonnaires : Interaction Choc-Tourbillon, Thèse de docteur-ingénieur Université PARIS VI, 1984.

[7] Ph. MORICE, L. CAMBIER and J.P. VEUILLOT, Numerical computation of transonic flow past an axisymmetric nacelle, 14th ICAS Congress, Toulouse, 1984.

[8] AGARD FDP WG-07, Test cases to steady inviscid transonic or supersonic flows, Report (to appear).

[9] AGARD FDP WG-08, Test cases for aerodynamics of aircraft afterbodies, Report (to appear).

[10] H. VIVIAND, Systèmes Pseudo-Instationnaires pour les Ecoulements Stationnaires de Fluide Parfait, ONERA Publication N° 1983-4, 1983.

[11] H. VIVIAND, Pseudo-unsteady systems for steady inviscid flow calculations, Contributed to the present volume.

[12] A. RIZZI, Entropy production and vortex-sheet capturing in numerical solutions of the Euler equations, Contributed to the present volume.

PART V:
PHYSICAL APPLICATIONS

SIMULATION OF THE KELVIN-HELMHOLTZ INSTABILITY OF A SUPERSONIC SLIP SURFACE WITH THE PIECEWISE-PARABOLIC METHOD (PPM)

PAUL R. WOODWARD*

Abstract

The Piecewise-Parabolic Method (PPM) has been used to study the nonlinear development of the Kelvin-Helmholtz instability of a Mach 2 slip surface in both a gamma-law gas and in an isothermal gas. A simplified version of PPM appropriate to this and other problems with only weak shocks is described. The instability calculations demonstrate the usefulness of discontinuity steepening in PPM and they illustrate the complexity in a flow problem which this method can treat accurately on Cray-I-class computers. The simulations also bring to light characteristic combinations of nonlinear waves which arise from finite-amplitude perturbations of the slip surface and which exhibit an approximately self-similar growth. After passing through a fairly chaotic phase of development, the mixing layer generated by the instability achieves a relatively ordered state which does not appear to depend greatly upon the nature of the initial perturbation, but which does depend upon the length scale over which strict periodicity is enforced in the simulation.

I. INTRODUCTION

The simulations of unstable supersonic slip surfaces presented here are motivated by earlier simulations of the unstable oscillations of supersonic jets emanating from the nuclei of active galaxies [1-3]. However, these simulations of slip surfaces are also interesting from a purely fluid dynamical viewpoint, and they are mathematically interesting as a demonstration of the power of the piecewise-parabolic method, or PPM, in simulating complex nonlinear flow phenomena. The simulations presented here will be followed by a more exhaustive study using PPM, to be undertaken in collaboration with Drs. M. L. Norman and K.-H. Winkler.

II. THE PIECEWISE-PARABOLIC METHOD (PPM)

The Piecewise-Parabolic Method, or PPM, has been designed to solve time-dependent gas dynamical problems involving strong shocks. Its general approach is described and compared to several others in [4], the details of the scheme are described in [5], and a simplified dissipation mechanism for one formulation of the scheme is described in [2]. PPM has grown out of the earlier MUSCL scheme [6-9], which in turn derives ideas from the scheme of Godunov [10]. Early results of PPM formulated differently than the scheme in [4], [5], and [2] are given in [11].

The PPM scheme is a second-order-accurate scheme of the Godunov type. It updates conserved quantities -- mass, momentum, and total energy -- in computational cells or zones by differencing time-averaged fluxes of these quantities computed at the zone interfaces. These fluxes are obtained by solving Riemann problems (shock tube problems) representing the nonlinear interaction of two constant states of the gas. These states are chosen to represent spatial averages over the domains of dependence of the two families of sound waves impinging on the interface during the timestep (see Fig. 1). The averages over these domains are computed using interpolated parabolic structures within the zones which are constrained to be monotone increasing or decreasing (see Fig. 2). The scheme is therefore upstream-centered in all characteristic families and it preserves the monotone character of features in the flow where and when it is appropriate to do so (obviously, it is not always appropriate to preserve monotonicity). The formulation of PPM used here splits the calculation of two-dimensional flow into a sequence of one-dimensional passes, symmetrized to give second-order accuracy. Each of these passes is in turn split into a Lagrangian calculation followed by a remapping to the fixed Eulerian grid (see Fig. 3).

The version of PPM used here differs from the detailed description given in [5] by using the simplified dissipation mechanism described in [2] and by using a simplified Riemann problem solver appropriate to the relatively weak shocks encountered in these simulations. Also, an isothermal equation of state is used in some of the simulations. This allows the energy equation to be eliminated and exact solutions to

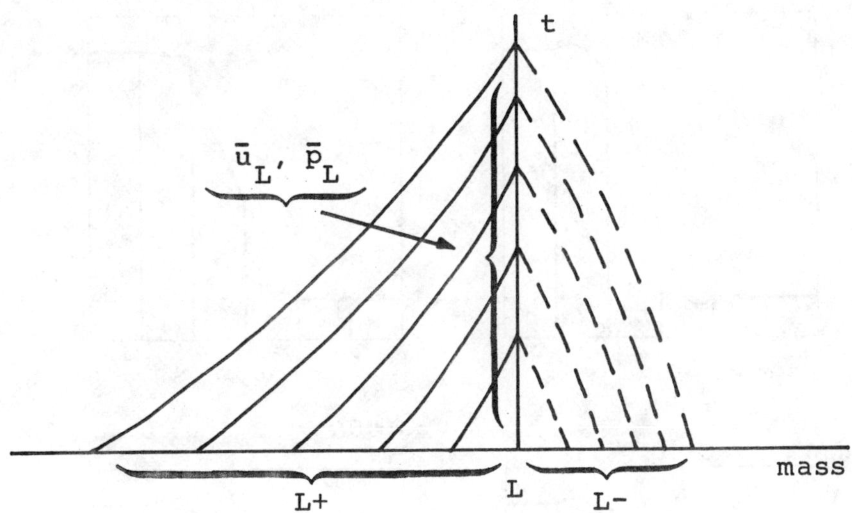

Fig. 1 A space-time diagram showing the domains of dependence L+ and L- for the zone interface L. Sound waves travel to the right (or left) toward the interface from the domain L+ (or L-) following the characteristic paths indicated by the solid (or dashed) lines. Constant states for Riemann's problem are obtained by averaging over the domains L±. The solution of Reimann's problem gives a constant state at the interface which is used to approximate the true time-averaged state there. This state gives the time-averaged velocity and pressure at the interface, $\bar{u}_L$ and $\bar{p}_L$, which are the fluxes of specific volume and momentum needed to update the Lagrangian zone-averaged variables.

Riemann's problem to be obtained without iteration. For the isothermal calculations, no additional dissipation mechanisms as described in either [2] or [5] were used. It should also be noted that the contact discontinuity steepening described in [5] and illustrated in Fig. 2 is inappropriate in isothermal calculations. In all calculations a passive variable, the original vertical position of the fluid element y_o, was advected along with the fluid flow using the PPM advection scheme with discontinuity steepening. This procedure proved essential in unravelling the very complex sequence of events following the initial excitation of vortical motions at the slip surface.

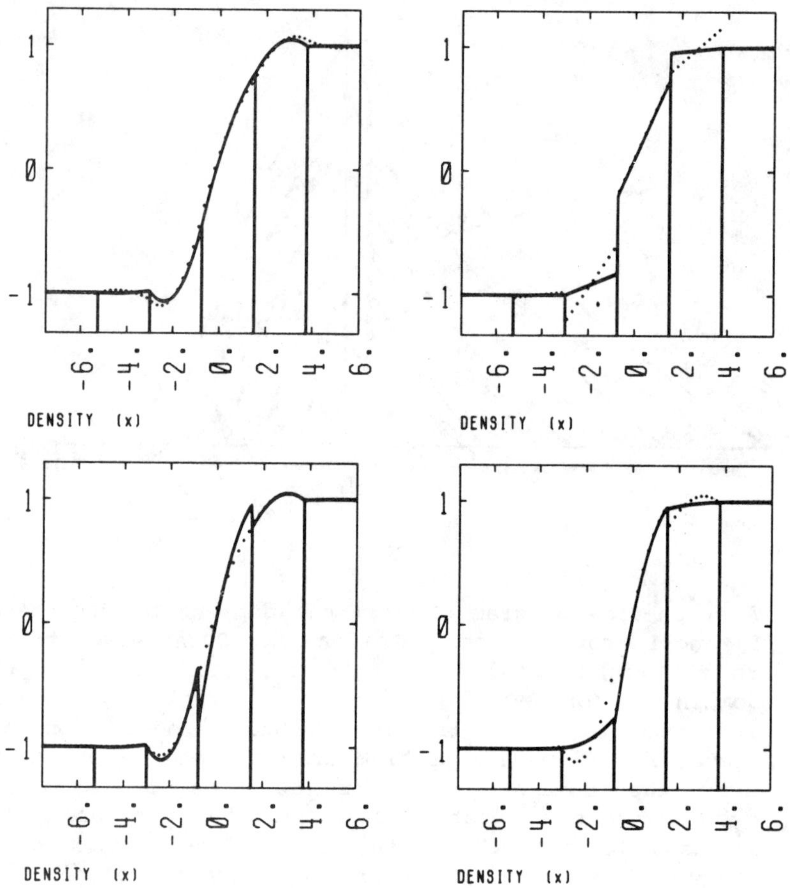

Fig. 2 The PPM representation of tanh (x) is constructed in several stages. First values are interpolated at zone interfaces using a formula based upon a cubic curve which has the proper average values in four zones nearest to a zone interface. The fomula is cast in terms of average values and slopes in the two nearest zones. These interface values together with the zone averages define parabolae within each zone. The dotted lines in (a) show parabolae based on the interpolation formula using the slopes indicated by the dotted lines in (b). When these slopes are constrained to give $\rho(x)$ monotone as shown by the solid lines in (b), the interpolation formula gives the parabolae shown by the solid lines in (a). These parabolae are steepened in the two middle zones in (c) by an algorithm that detects and steepens discontinuities. Finally, the parabolae are constrained to be monotone by disregarding one interface value, if necessary, and forcing a zero slope at the opposite interface. The final representation is shown as the solid lines in (d).

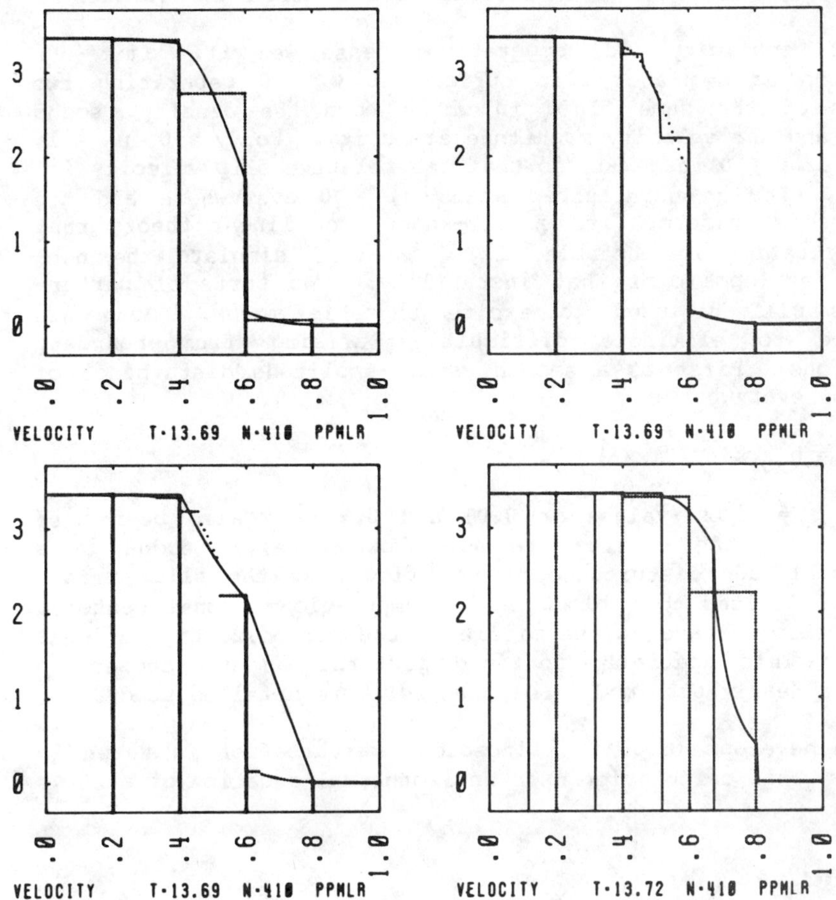

Fig. 3 The propagation of an isolated strong shock on a uniform grid by the PPM scheme. (a) Original zone-averaged data (dotted lines) is used to generate monotone interpolation parabolae (solid lines) within the five zones shown. (b) By averaging over the interpolation parabolae (dotted lines) within the domains of dependence of the zone interfaces, constant states (solid lines) are obtained for Riemann problems. (c) Solution of the Riemann problems (dotted lines) yields interface velocities which imply compressional velocity gradients (solid lines) for the Lagrangian zones. (d) Interpolation parabolae (solid lines) are generated for the compressed Lagrangian zones which yield upon intergation the new average volocities in the original Eulerian zones (dotted lines). The shock is again one zone wide after moving 3/4 of a zone width to the right (from Ref. 5).

III. PPM SIMULATIONS OF UNSTABLE SUPERSONIC SLIP SURFACES

For simplicity and computational ease we will limit our attention to a planar slip surface at $y = 0$ separating two regions of the same fluid in which both the density, sound speed, and the velocity magnitude are unity. For $y > 0$, $u_x = 1$, while for $y < 0$, $u_x = -1$, so that the relative slip velcoity is Mach 2. In the unperturbed state $u_y = 0$ everywhere and the pressure is uniform. It is well-known from linear theory that this situation is unstable [12]. We will simulate the non-linear development of that instability. Two sorts of perturbations will be used to excite unstable modes. Both are periodic to eliminate difficulties arising from boundary conditions. First is a smooth, small-amplitude disturbance of the flow everywhere:

$$u_y = u_{yo} \sin (2\pi x/3) \quad ,$$

with $u_{yo} = 0.02$ (values of 0.01 and 0.2 have also been used for u_{yo} but the results are not shown here). Second is a large-amplitude disturbance of the flow near the slip surface which is limited to a block of 19 computational zones centered on the slip surface at the middle of the periodic computational domain. This spatially confined perturbation was chosen to excite a nearly pure modal response for more detailed study.

The development of the sinusoidal perturbation is shown in Fig. 4. This calculation uses an isothermal equation of state

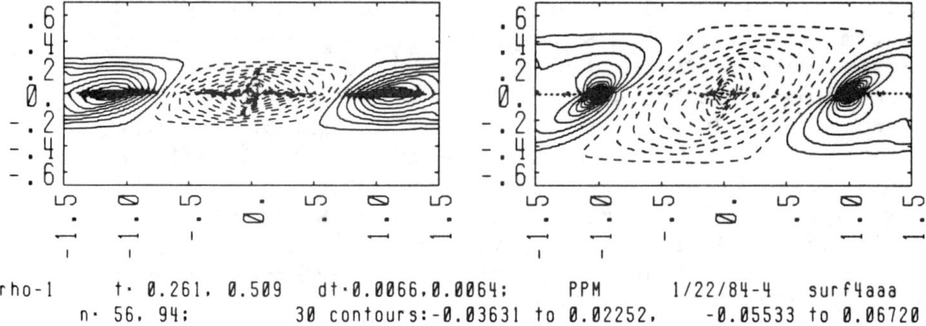

Fig. 4 An unstable, Mach 2 slip surface in an isothermal gas is perturbed at time 0 and shown here at times 0.26 and 0.51. Thirty equally spaced contours of $(\rho - 1)$ are shown, with negative contours dotted.

in which the pressure and density are always equal (unit sound speed). The evolution for a gamma-law gas with $\gamma = 1.4$ is qualitatively similar but is marred by numerical viscous heating and subsequent expansion of the gas near the slip surface. To avoid this artificial dissipation of kinetic energy into heat when opposing flows are blended within a zone the isothermal equation of state was introduced. For all the calculations shown here a fine grid of 360x360 zones was used in the upper half plane, and the solution in the lower half plane was obtained via symmetry conditions. In the region $-1.5 < x < 1.5$ and $0 < y < 2$ the grid is uniform with $\Delta x = \Delta y = 1/120$. The y grid lines steadily increase in separation above $y = 2$, reaching $y = 20$ in 120 intervals. At $y = 20$ all y-derivatives are forced to vanish, although no signals reach this boundary during most of the calculation. The performance of PPM on such a nonuniform grid is excellent, and the calculation in the region shown should be quite free from contamination by waves reflected by the nonuniform grid.

In Figs. 4a and 4b thirty equally spaced contours of the quantity $\rho - 1$ are displayed, with negative contours dotted.

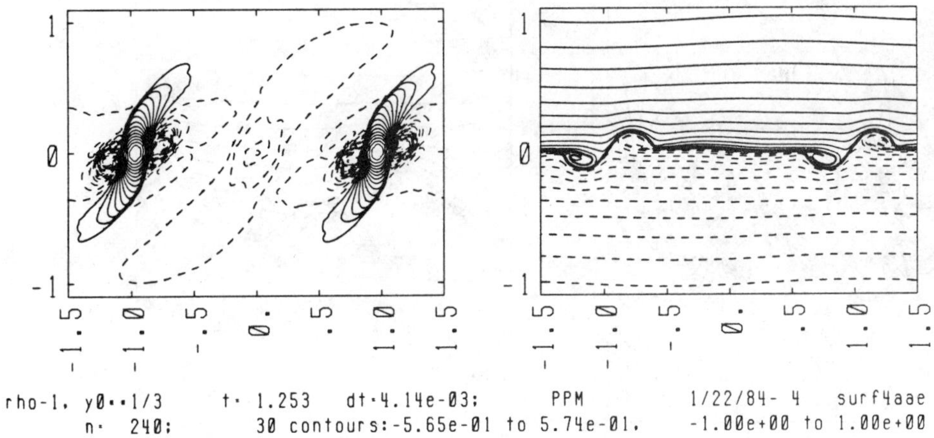

Fig. 5 The flow in Fig. 4 is here shown at time 1.25. Thirty equally spaced contours of $(\rho - 1)$ and $y_0^{1/3}$, the cube root of the original vertical position are shown with negative contours dotted. The compression in Fig. 4 has steepened and excited two nonlinear wave systems associated with localized kinks in the slip surface. These grow nearly self-similarly until they interact with each other by a collision of their component vortices.

The flow is shown at times 0.26 and 0.51, and an expansion centered at x = 0 and a compression at x = ±1.5 are evident. Despite the small size of u_{yo}, the perturbation amplitude, this compression steepens rapidly near x = ±1 to form two distinct systems of nonlinear waves by time 0.75. These systems of nonlinear waves are characteristic features of perturbed supersonic slip surfaces in these and many other PPM simulations. They tend to grow steadily at the expense of the kinetic energy of the shear. By time 1.25 they have grown large enough to be well resolved on the grid, and they are shown in Fig. 5.

In Fig. 5a we display density contours as before, while in Fig. 5b we show 30 equally spaced contours of the quantity $y_0^{1/3}$, the cube root of the original vertical position of the fluid elements. Again negative contours are dotted, and these therefore refer to the fluid which was originally moving to the left. The original slip surface is traced out by the transition from dotted to solid contours in Fig. 5b. Each nonlinear wave system is associated with a fairly distinct and isolated kink in the slip surface. These kinks are associated

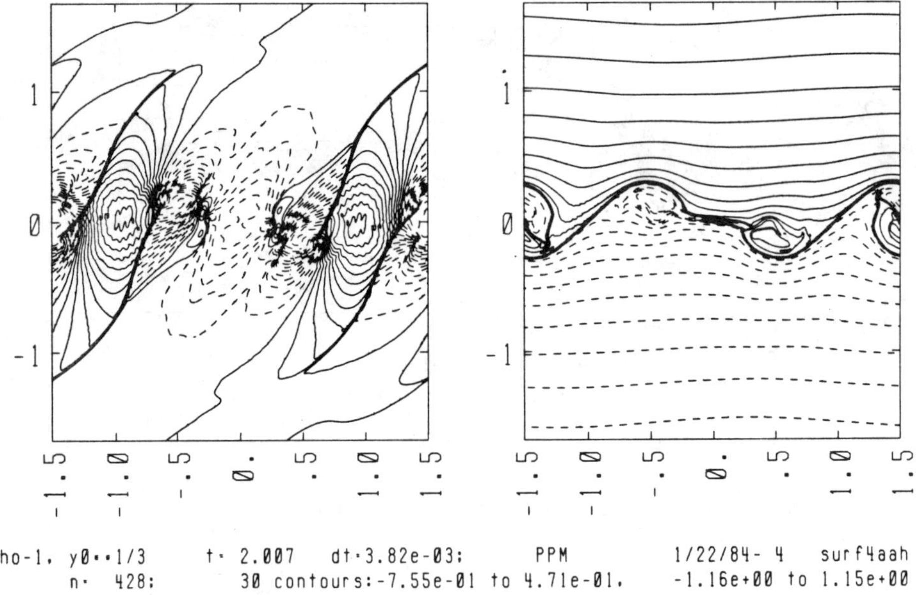

Fig. 6 The flow of Figs. 4 and 5 is here shown at time 2.01 in the same format as Fig. 5. The two wave systems have begun to interact through a collision of two vortices at x = ±1.5.

with pairs of transonic vortices located at the density minima in Fig. 5a. Embedded in each vortex is a shock which ends abruptly in a centered rarefaction at the slip surface. Before encountering these shocks the fluid is expanded into supersonic flow by rarefactions which also serve to turn the fluid into the slip surface so that some of it becomes entrained in the vortices. Note that both the slip surface and the shock waves are given very thin and well behaved numerical representations by PPM. Also note in this and figures to follow that PPM advection with its discontinuity steepener does a very good job of keeping track of y_o even though the flow becomes extremely convoluted.

As the kinks in the shear layer grow they eventually interact. The first such interaction, at time 2, is shown in Fig. 6, with the same format as Fig. 5. Two of the transonic vortices collide at x = ±1.5 to produce a complex vortex with two embedded strong shocks and two much weaker subsidiary shocks. By time 3.25, in Fig. 7, this composite vortex has split into two principal vortices and one smaller one just beginning to spin up at x = ±1.5. In the meantime the other

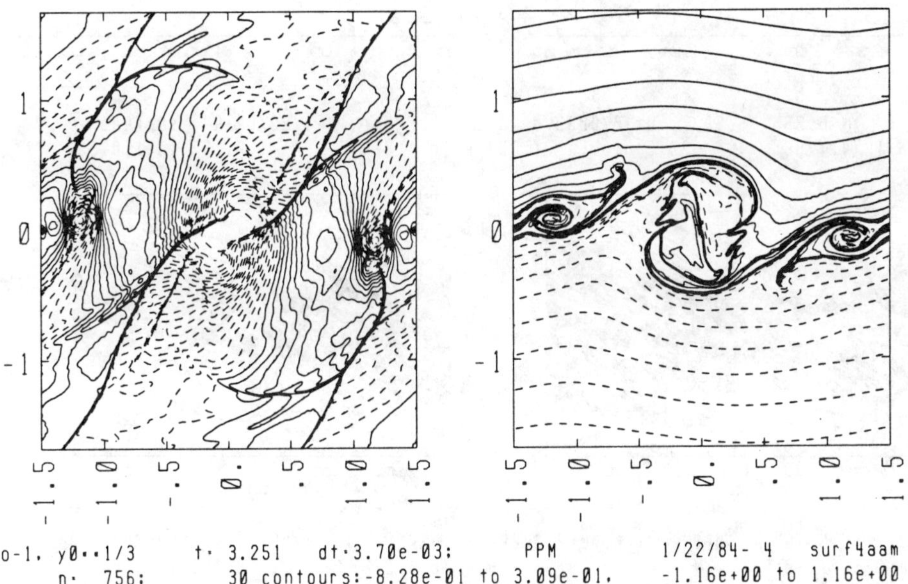

Fig. 7 The same flow is here shown at time 3.25. The vortex collision at x = ±1.5 has yielded 3 vortices, at x = ±1.5 (just beginning to spin up) and at x = ±1.2. Another vortex collision is proceeding at x = 0.

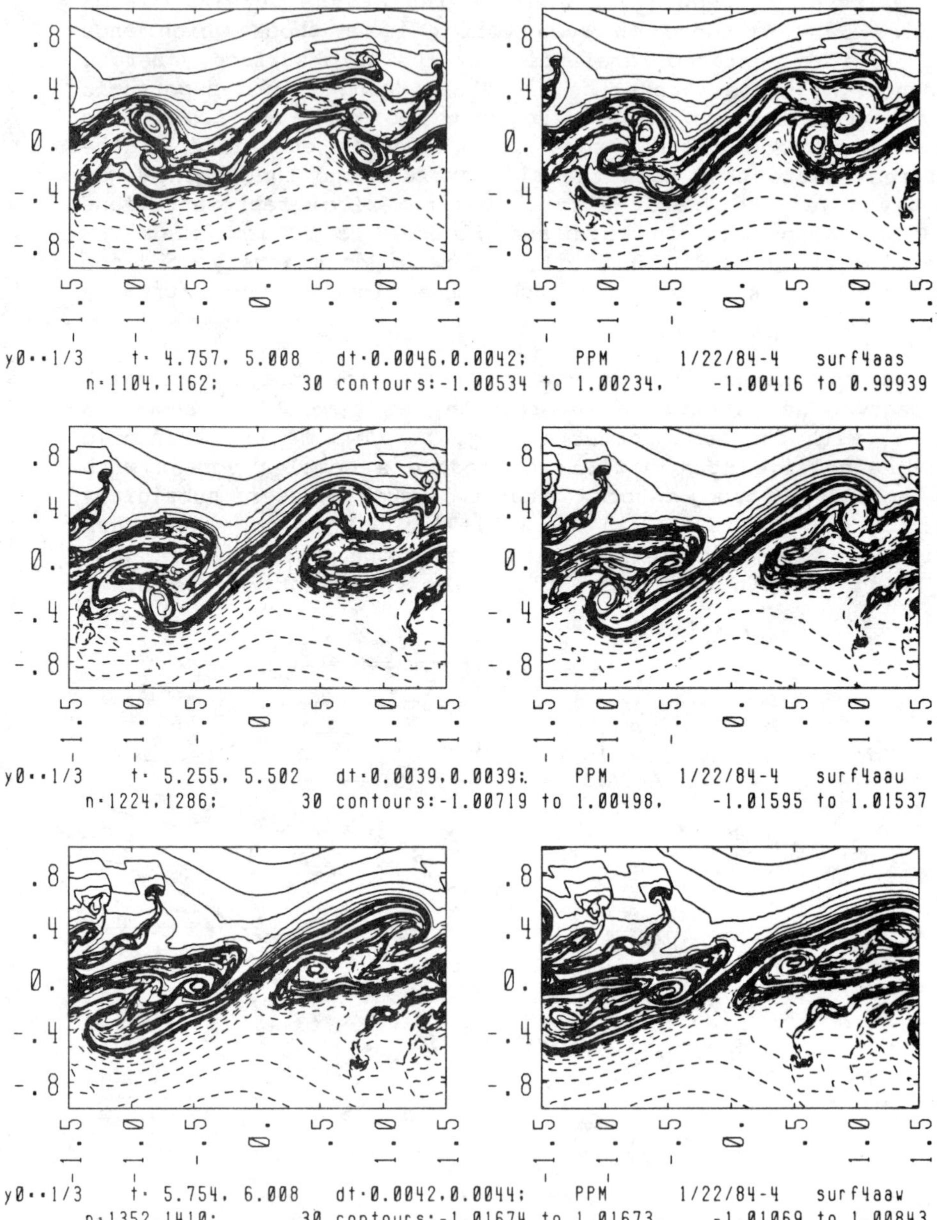

Fig. 8 The flow of Figs. 4-7 is shown here at times 4.76, 5.01, 5.26, 5.50, 5.75, and 6.01. The four principal vortices coming out of the vortex collision at x = 0 in Fig. 7 here collide with the vortices at x = ±1.2 in that figure. The result is just four principal vortices, rather than the original six, and of course many very small vortices. Note the vortex at x = ±1.5.

original pair of vortices have collided at x = 0. This collision produces four major vortices which ultimately merge into two vortices by interacting with the other vortex pair as shown in Fig. 8. The result is the very symmetrical system of four vortices shown in Fig. 9 at time 13.5. The initial transient evolution thus consists of vortex collisions which generate additional vortices, while the subsequent gradual evolution involves the merging of vortices to arrive again at an ordered flow which can now persist for a very long time. The formation of the original kinks in the slip surface at nearly symmetrical locations has caused a harmonic mode of oscillation of the shear layer to be excited. Nevertheless, there is a small asymmetry in the flow in Fig. 9, and it is reasonable to expect that eventually a more ordered flow with only two vortices per period will develop. Given the imposed periodicity of the flow, this configuration should be very long lasting. However, such an outcome has not occurred by time 25, when the computation was terminated, as is evident from the flow at time 24, shown in Fig. 10. The principal difference between this flow and that in Fig. 9 is the increased mixing of the fluid originally above and below the slip surface which has occurred near the large vortices.

A number of features of this compressible flow are familiar from the incompressible flow simulations appropriate to very subsonic relative slip between the two fluids. In particular, the roll up of the slip surface for finite amplitude perturbations at two points where the surface tilts in the same direction occurs also in that case (see [13]). Also, the merging of vortices to give large coherent structures in the flow is familiar from the incompressible case (see [14]). The most intriguing modification of the flow brought about by compressibility is the necessity of generating vortices in pairs associated with oppositely facing shocks on either side of the slip surface. The appearance of these nonlinear vortex pairs and their associated shocks out of the very smooth initial flow is a bit surprising. Even when a tanh(y)-type velocity profile is used to spread the slip surface out over a few zones these isolated nonlinear wave systems appear. Although these wave systems may conceivably be excited in this simulation by some large, local perturbation associated with the grid, after a short period of apparent self-similar growth they are well resolved and should be computed correctly.

A number of experiments were devised to see if the wave systems shown in Fig. 5 represent a characteristic response of the slip surface to large, local perturbations. The results indicate that this is the case. The symmetry of the upper and lower half planes in the calculations resulted in symmetrical perturbations which generally excited two wave systems, each

Fig. 9 The flow achieves a relatively ordered appearance by time 13.5, shown here.

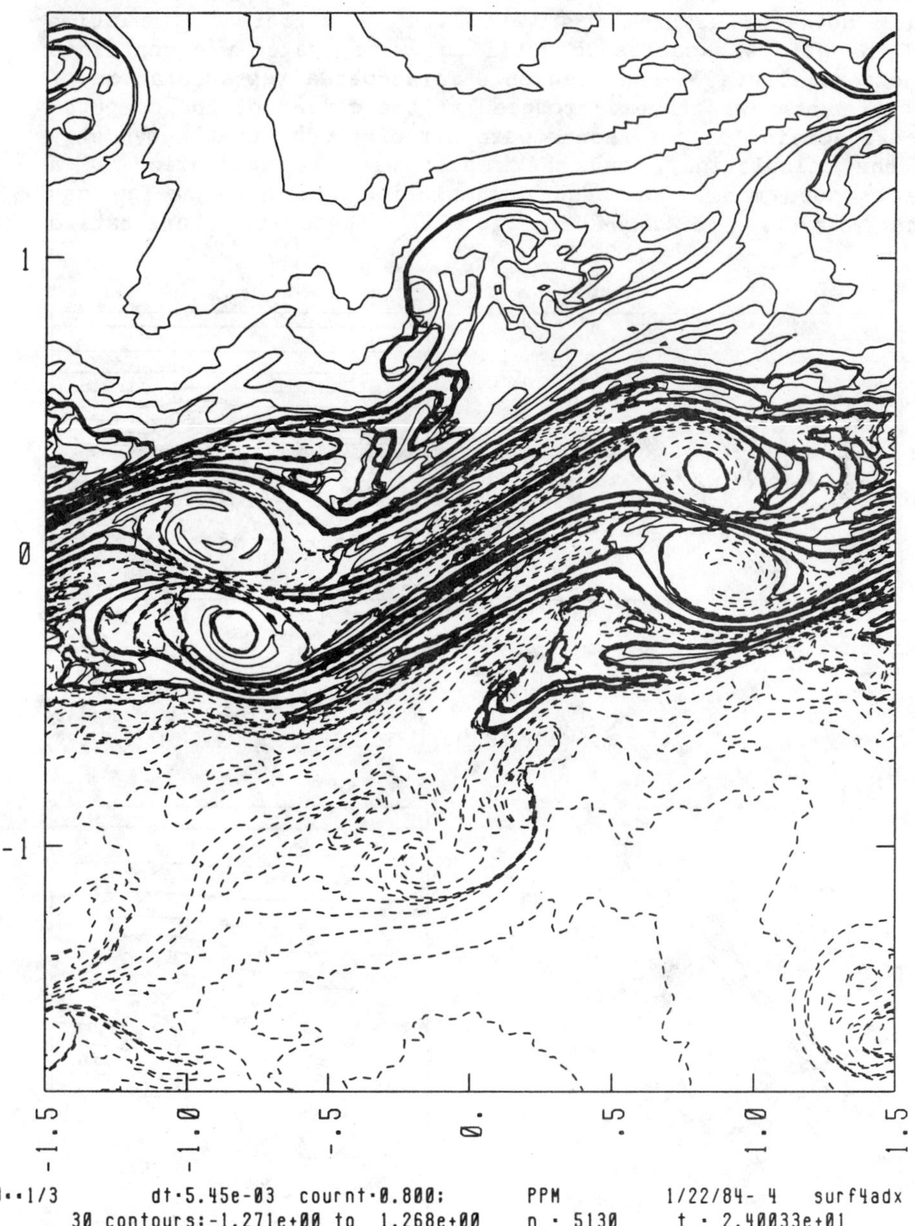

Fig. 10 By time 24, shown here, substantial mixing of the fluid has taken place within the 4 main vortices, but the flow is otherwise similar to that at time 13.5.

containing two vortices. To excite only a single system, a fully developed system was taken from a different simulation and the grid was coarsened until the wave system was contained in essentially a 38-zone region. This coarse representation of a wave system was then introduced at the center of the computational domain (only 19 zones were perturbed due to the symmetry of the calculation), and the result was the excitation of a nearly pure mode. One such calculation for a gamma-law gas with $\gamma = 1.4$ is shown in Fig. 11. Note the qualitative

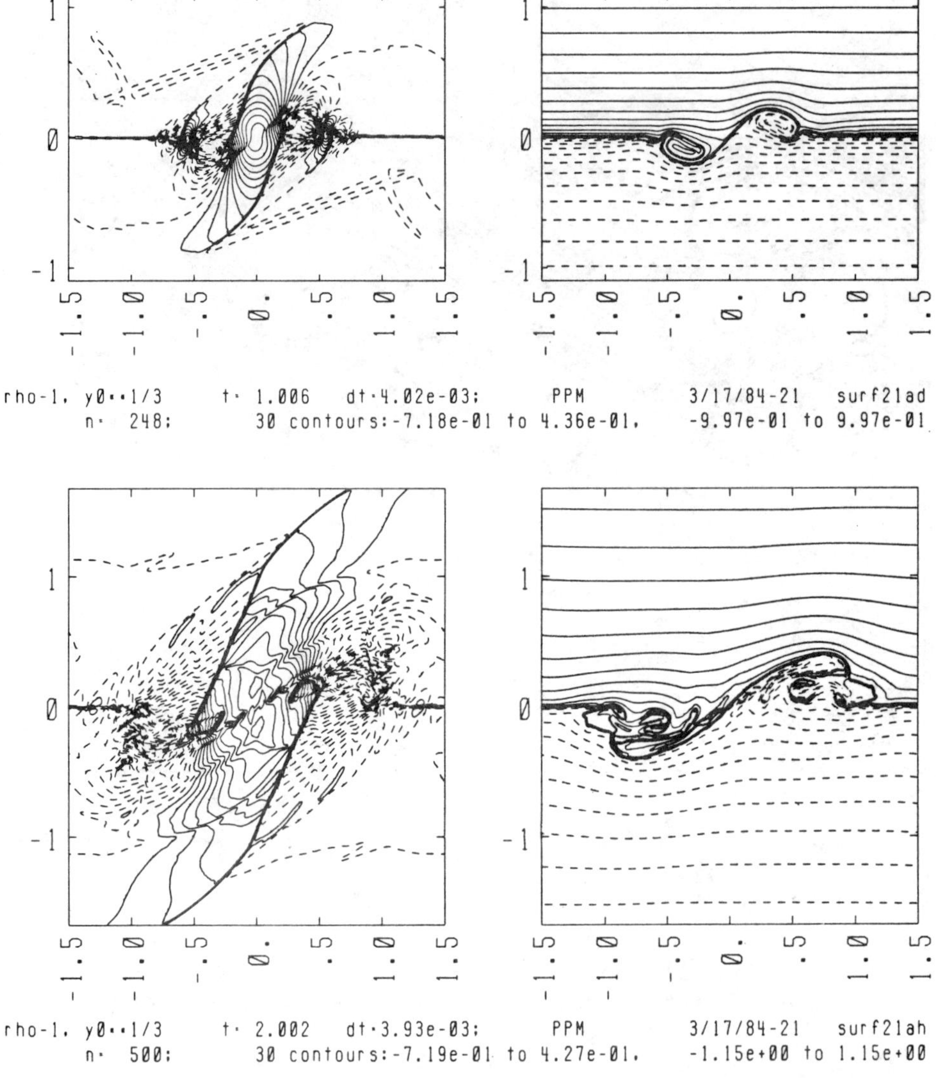

Fig. 11a,b.

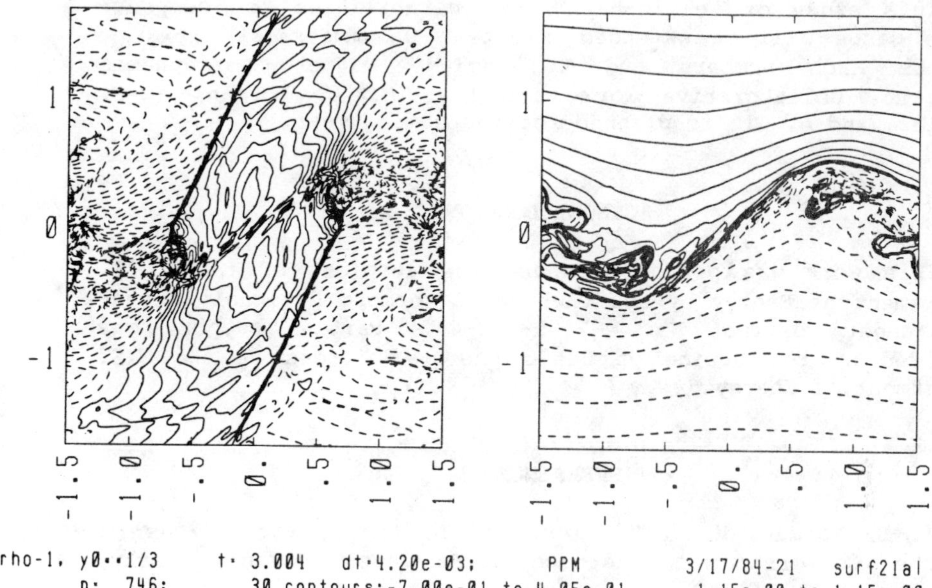

Fig. 11c. Here a single nonlinear wave system has been excited by a local strong perturbation of a Mach 2 slip surface near $x = 0$ in a gamma-law gas with $\gamma = 1.4$. The wave system grows nearly self-similarly until it eventually interacts with itself via the periodic boundary condition. The process ultimately should result in a mixing layer similar to half of that shown in Fig. 10 (i.e., a fundamental rather than a harmonic mode of oscillation should be excited). Note the similarity between this flow at time 1 and the isothermal flow in Fig. 5.

similarity between this and the isothermal case in Fig. 5. In fact, a run just like that of Fig. 11 was performed for the isothermal case, and it gave results strikingly similar in detail to those for the gamma-law gas. Apparently the detailed relation between pressure and density has a minor effect upon these flows at modest Mach numbers, despite the obvious importance of the compressibility of the gas. The nearly self-similar growth of the nonlinear wave system excited in the flow in Fig. 11 is marred by the excitation of similar, smaller wave systems. These generate small vortices which become entrained in the larger ones to produce a chaotic flow. It is not clear whether these smaller wave systems are excited by the larger one or instead by small signals generated as a result of the coarse representation of the initial perturbation. A search for nonlinear solutions to the fluid equations formulated in terms of the similarity variables x/t and y/t seems the most fruitful approach for settling this question.

This study of Kelvin-Helmholtz instabilities for compressible gases will be extended to treat unsymmetrical flows at various Mach numbers and to include magneto-hydrodynamic effects. Collaborative work with Drs. M. L. Norman, K.-H. Winkler, and A. Giz is planned for the coming year.

ACKNOWLEDGMENT

This work was performed under the auspices of the U. S. Department of Energy by the Lawrence Livermore National Laboratory under contract No. W-7405-ENG-48. Partial support was provided by the Office of Basic Energy Sciences of the U.S. Department of Energy.

REFERENCES

1. M. L. Norman, K.-H. Winkler, and L. Smarr, Proceedings of the Conference on Astrophysical Jets, Torino, Italy, October 1982.
2. P. R. Woodward, "Piecewise-Parabolic Methods for Astrophysical Fluid Dynamics," to appear in the Proceedings of the NATO Advanced Workshop in Astrophysical Radiation Hydrodynamics, Munich, West Germany, August 1982.
3. P. R. Woodward, "Kelvin-Helmholtz Instability of Supersonic Jets," in preparation.
4. P. R. Woodward and P. Colella, J. Comp. Phys. **54** (1984), in press.
5. P. Colella and P. R. Woodward, J. Comp. Phys. **54** (1984), in press.
6. B. van Leer, J. Comp. Phys. **23**, 276 (1977).
7. B. van Leer, J. Comp. Phys. **32**, 101 (1979).
8. B. van Leer and P. R. Woodward, Proc. TICOM Conference, Austin, Texas (1979).
9. P. Colella, SIAM J. Sci. Stat. Computing, in press.
10. S. K. Godunov, Mat. Sb. **47**, 271 (1959).
11. P. R. Woodward and P. Colella, Lecture Notes in Physics **141**, 434 (1981).
12. R. A. Gerwin, Rev. Mod. Phys. **40**, 652 (1968).
13. R. Krasny, "A Numerical Study of Kelvin-Helmholtz Instability by the Point Vortex Method," University of California-Berkeley Center for Pure and Applied Mathematics Report No. PAM-193, December 1983.
14. P. C. Patnaik, J. Fluid Mech. **73**, 215 (1976).